W9-BVH-736

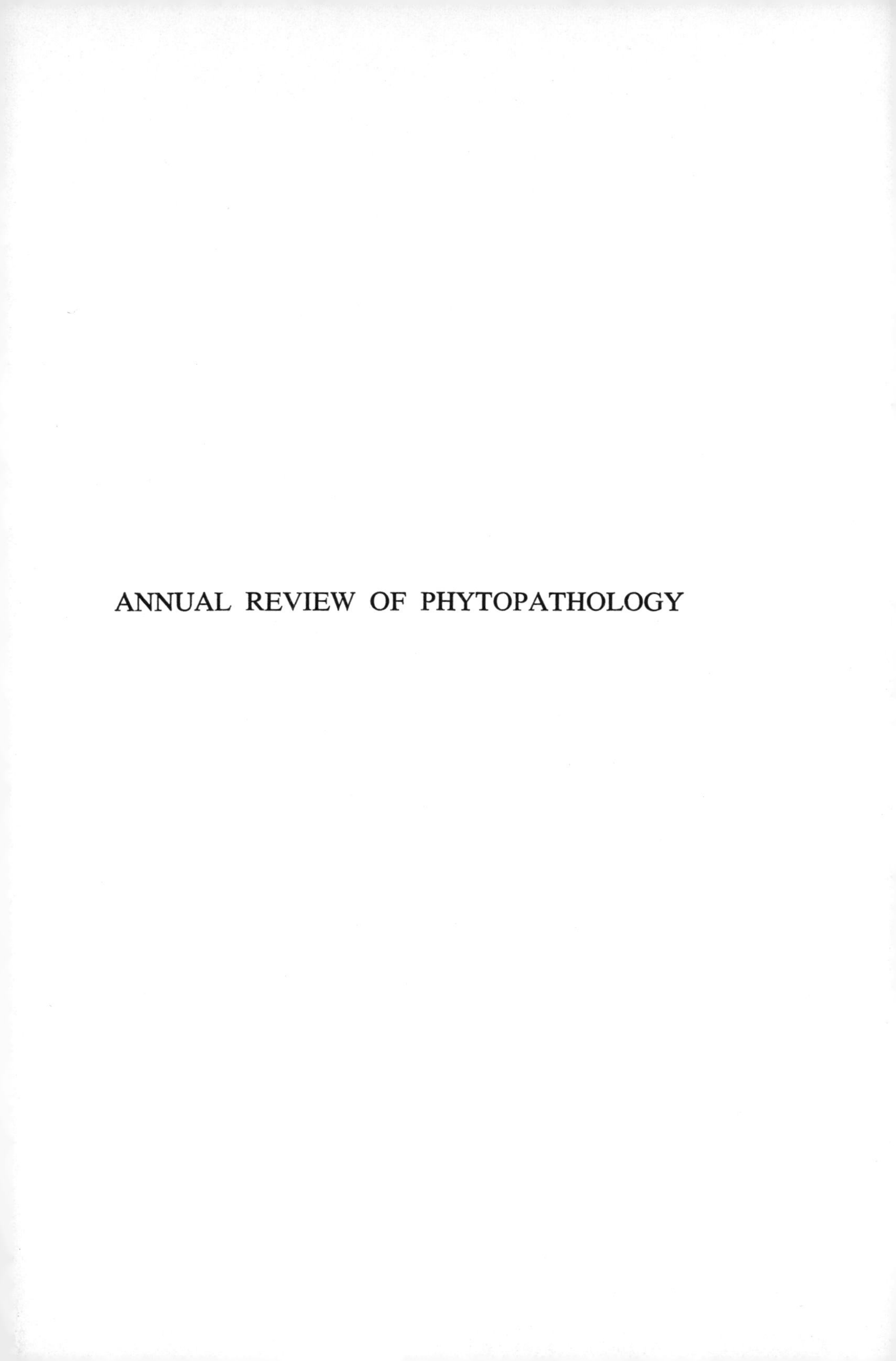

ANNUAL REVIEW OF PHYTOPATHOLOGY

ANNUAL REVIEW OF PHYTOPATHOLOGY

KENNETH F. BAKER, *Editor*
The University of California, Berkeley

GEORGE A. ZENTMYER, *Associate Editor*
The University of California, Riverside

ELLIS B. COWLING, *Associate Editor*
North Carolina State University

VOLUME 10

1972

ANNUAL REVIEWS INC.
4139 EL CAMINO WAY
PALO ALTO, CALIFORNIA, USA

ANNUAL REVIEWS INC.
PALO ALTO, CALIFORNIA, USA

Standard Book Number: 8243–1310–0
Library of Congress Catalog Card Number: 63–8847

FOREIGN AGENCY
Maruzen Company, Limited
6, Tori-Nichome, Nihonbashi
Tokyo

PRINTED AND BOUND IN THE UNITED STATES OF AMERICA BY
GEORGE BANTA COMPANY, INC.

PREFACE

Annual Review of Phytopathology was the twelfth among the eighteen progeny of Annual Reviews Inc. On this, its tenth birthday, it is appropriate to examine its record, to determine how well it is filling the needs of the profession of plant pathology.

There have been 188 reviews published, covering the wide range of subjects shown in Volume 5 pages 468–70, and Volume 10, pages 554–57. We are pleased that reader comments and reviews have indicated general approval of the extent, depth, and type of coverage provided.

Authors from 20 foreign countries contributed 41.5% of these papers; these countries also had about 44% of the world's 10,800 plant pathologists, according to L. Chiarappa (Volume 8, pages 421–22). Great Britain, Canada, Germany, Australia, Japan, The Netherlands, Italy, Hungary, Israel, and New Zealand contributed most frequently. We are gratified that the journal continues to be international in both authors and readers, and that its content is generally thought to be satisfactory.

There have been many changes in the Editorial Committee during the past year. J. G. Horsfall concluded ten years of distinguished and effective service as Editor, during which time he established this as one of the leading phytopathological journals. K. F. Baker became the new Editor. G. A. Zentmyer, who was appointed an Associate Editor last year, continues in that capacity, and E. B. Cowling replaces Baker as an Associate Editor. H. D. Thurston replaced Zentmyer on the Committee, and J. F. Fulkerson replaced Cowling. We welcome them to the group.

We thank W. M. Epps, of Clemson University, South Carolina, who served as guest committeeman in planning this volume.

Mrs. Virginia Hoyle, Assistant Editor of the Review, has, with unfailing good humor, contributed greatly to the smooth functioning of the editorial process. Our sincere appreciation is again expressed for the excellent subject index prepared by D. C. Hildebrand, of the University of California, Berkeley.

Finally, we acknowledge our deep indebtedness to the 257 authors of the first 10 volumes. They have determined the quality of the journal; the Editors have merely made their papers available to world pathologists.

THE EDITORIAL COMMITTEE

CONTENTS

REPRINTS

The conspicuous number (**3539** to **3561**) aligned in the margin with the title of each review in this volume is a key for use in the ordering of reprints.

Reprints of most articles published in the *Annual Reviews of Biochemistry* and *Psychology* from 1961, and the *Annual Reviews of Microbiology* and *Physiology* from 1968 are now maintained in inventory. Beginning with July 1970 this reprint policy was extended to all other *Annual Reviews* volumes.

Available reprints are priced at the uniform rate of $1 each postpaid. Payment must accompany orders less than $10. The following discounts will be given for large orders: $5–9, 10%; $10–24, 20%; $25 and over, 30%. All remittances are to be made payable to Annual Reviews Inc. in U.S. dollars. California orders are subject to sales tax. One-day service is given on items in stock.

For orders of 100 or more, any Annual Reviews article will be specially printed and shipped within 6 weeks. Reprints that are out of stock may also be purchased from the Institute for Scientific Information, 325 Chestnut Street, Philadelphia, Pa. 19106. Direct inquiries to the Annual Reviews Inc. reprint department.

The sale of reprints of articles published in the Reviews has been expanded in the belief that reprints as individual copies, as sets covering stated topics, and in quantity for classroom use will have a special appeal to students and teachers.

S. D. Garrett

ON LEARNING TO BECOME A PLANT PATHOLOGIST

3539

S. D. Garrett
Botany School, Cambridge, England

Scientists are said to be uncommunicative about the thought-processes and other circumstances leading to their discoveries. One historian of science that I have read went on to add, rather cruelly I thought, that those who were most communicative generally had least of value to say. I remembered this comment when I was invited to contribute this prefatory chapter; my pleasure at being asked to follow so distinguished a line of predecessors was mixed with doubt as to whether I could find anything new to say. I had already written more than enough about my particular field of research on pathogenic root-infecting fungi, and others have thought more deeply than I on teaching and research in general. So I finally decided to take the risk implied by my opening statement and give a frankly autobiographical account of my early difficulties in research, which were considerable enough to make something to write about.

My three years (1926–29) as an undergraduate at the University of Cambridge were unexpectedly happy ones, first because I had not foreseen that I should be treated, straight from school, as a mature individual responsible for making his own decisions about his course of study, and subjected to minimal restrictions of liberty; a small proportion of modern undergraduates seems to find present restrictions, still further reduced since my time, a challenge additional to that of achieving academic success. Second, I greatly enjoyed reading the subject of botany, part-time for two years and whole-time for the third; botany was taught in few boys' schools at that time, and so the subject carried for me all the excitement of a new field of study. Four lecturers in particular impressed me, the first being Sir Albert Seward, who as Professor of Botany gave the course of elementary lectures. Seward was a man of impressive personality and presence, which duly became him when he served a term as Vice-Chancellor of the University, and he had enough of the actor in him to make a first-class lecturer. Later on, after I had married, I discovered that Seward was a first cousin of my wife's grandmother, who used to refer to him as 'Cousin Bertie.' F. F. Blackman, on the other hand, needed no assistance from the art of presentation; his lectures impressed us by their extraordinary clarity and he was able to make the most difficult subject appear quite simple. Harry Godwin's lectures on plant ecology have subsequently influenced me the most, because the interest he aroused in his subject later led me to seek for parallels between vegetational ecology and the microbial ecology of the soil. Last but not least, it was F. T. Brooks's lectures on mycology and plant pathology that caused me to devote more attention to his subject than to any other branch of botany. Even now, I find this difficult to explain. Brooks had a good loud voice, but few of the other arts of oratory. Nor was his subject-matter particularly inspiring in the intellectual sense; he gave us a preview of the book he had just finished writing, entitled 'Plant Diseases' (3). This book is perhaps the most concise encyclopedia of plant diseases caused by fungi that has yet been published and I have found it, and its second edition of 1953, an invaluable reference work throughout my life. But in both book and lectures, the facts were sternly left to speak for themselves. So I can only conclude that it was Brooks's unbounded enthusiasm for mycology that infected me, distilling over from his somewhat arid recitals of vital statistics for a host of fungi. But he could be witty on occasion, as in his presidential address to the British Mycological Society in 1922 (2), when he remarked 'The fungi are a remarkable group of organisms of which we, as mycologists, are rightly proud.'

In those days, the course for third-year undergraduates specializing in botany was a very comprehensive one; so it is in the Cambridge Botany School today, though with the arising of new branches of botanical science, some of the older ones now get less time. Now, as then, overspecialization among third-year students is discouraged. For future plant pathologists, it is particularly important that they should thoroughly understand the development and working of the healthy plant, both singly and in communities; only if the norm is understood can departures from it in disease be properly elucidated. For this reason, I have never wanted to press for more teaching-time for our subject of mycology and plant pathology; a well trained general botanist will be able to pick up what he needs for teaching and research as he goes along, though post-graduate instruction can certainly make this easier for him, and so we provide it.

The second, and perhaps paradoxical, reason I so enjoyed reading botany at Cambridge was because at no time while I was an undergraduate did I expect to follow a botanical career, and so I was not unduly worried by anxiety over examinations. The study of botany was for me a pleasant but incidental means to the end of obtaining a sufficiently respectable honours degree to get a job. Right up to the finish, I had intended to get a job in business of one kind or another. But when I went to the University Appointments Board to see about this, the advisory officer, after a few minutes chat, said 'I don't think you would be much good in business; you had better go back to the Botany School and see if they can find something for you.' For many years, I credited this adviser with unusual acumen; more recently, with increasing administrative experience myself, I have suspected that he was merely trying to shift the load to a region of lower pressure. Be that as it may, I went back to the Botany School to see F. T. Brooks; after reproaching me for my lack of foresight, he said he would recommend me for the post of Assistant Plant

Pathologist at the Waite Agricultural Research Institute of the University of Adelaide. The appointing authorities duly accepted Brooks's recommendation, but during the next two years their faith in him as a referee must have been badly shaken.

I arrived at the Waite Institute in October 1929, with only the haziest notions as to what a life of research in plant pathology would be like. Undergraduate learning is indeed an essential preparation for research, but it is only a continuation of school learning; the most critical transition in the development of a research worker still awaited me, but of this I was blissfully unaware, and the adjustment to life in Australia was quite enough to occupy me for the time being. Nor could I then appreciate my particular good fortune in being an early member of the staff of the Waite Institute, which had started work under its Director, A. E. V. Richardson, in March 1925. I could indeed sense the remarkable enthusiasm of all members of the research staff, but at that time I did not realize that the initial momentum of a new research institute is not automatically maintained, and so I took this invigorating atmosphere quite for granted. Much of the credit for this undoubtedly belonged to the Director, Professor Richardson; I was neither old enough nor close enough to him to make a critical assessment of his achievement, but all of us were impressed by his extraordinary energy and by the resulting rapid growth in the size and scope of the Institute's activities. This furious activity was maintained outside as well as inside the Institute; catching him twice in the same day for speeding, a traffic policeman remarked to him 'So it's you again.' Another of Richardson's good qualities as a director was the trouble he took to keep in personal touch with the younger members of his research staff, and the direct interest he took in their work. Some of us younger men, unfettered by wife and family, used to go back to our laboratories most evenings of the week; more often than not, Richardson would be in his office and would look in on us before he went home. At that time, I did not realize how much additional effort this must have required of him, nor that it was unusual for a man burdened with great administrative responsibilities. But it was a great encouragement to us, and he set an example that I have failed to follow as closely as I should have liked in my own, smaller sphere of responsibility.

The first head of the plant pathology section at the Waite Institute was Geoffrey Samuel, who had been transferred from the staff of the University Botany Department under Professor T. G. B. Osborn, the plant ecologist. Although only seven years older than I was, Samuel seemed much older than that to me, as he was already an accomplished plant pathologist, with published work on fungal and viral diseases, as well as on the manganese deficiency disease of oats, to his credit. Undoubtedly he was one of the best all-round plant pathologists I have known, ready to turn his hand to any problem, as was necessary in a comparatively young country where agricultural practice was still rapidly evolving, and where important plant disease problems were more numerous than pathologists. His transfer from research to administration in England in 1937 was a loss that Australia could ill afford at that time. Soon after I had arrived, Samuel told me that my work was to be on the foot-rot diseases of cereals, of which the most important in South Australia was take-all, caused by *Ophiobolus graminis;* the second was the so-called 'no growth' disease, which he had described the year before (6), and attributed to a root-infecting species of *Rhizoctonia.*

About six months later, Samuel departed on a year's sabbatical leave, leaving me to fend for myself, though indeed I had a welcome companion in adversity, J. G. Bald, who also was struggling in the early stages of research. During the year that followed, I could hardly have pursued a more disastrous course had I deliberately set out to do so. I now think there were two reasons for this. In the first place, a university training for undergraduates does not fully prepare them for research work; it provides the necessary background of knowledge and understanding, but often little guidance as to how this knowledge and understanding was obtained by the original research workers. It was not merely that I knew nothing of the art of scientific investigation; worse than that, I failed to realize that there was an art that had to be learned. Normally, a research supervisor teaches his research students this art; nowadays we have an excellent book on this very subject by one of my Cambridge colleagues, Professor W. I. B. Beveridge (1). My position was rather that of many young bachelors of my generation, who supposed that a young woman with her first baby was endowed by nature with all the necessary instinctual behavior to look after it, and needed little or no instruction in the art of child care. My ideas about scientific research consisted of a vague feeling that the most important thing was to strive for originality; the easiest way for a young research worker to be original is to prove his supervisor wrong, just because his supervisor's ideas happen to be the nearest object of attack. This was my second obstacle, and one that I had placed in my own way, because a supervisor is much more likely than his research student to be right, as I later had to admit to myself. Later on, I have seen this tendency operating, perhaps quite unconsciously, in several of my own research students; naturally I understand and sympathize with it. So if they want to go against my advice, I say 'I don't think that will work. But if you don't believe me, try it and see; only in that way will you be able to convince yourself one way or the other.'

Before leaving, Samuel had left me two clear directives: first that take-all was the most important cereal foot-rot disease and that I should pay most attention to this, and second that his conclusion concerning a *Rhizoctonia* species as the cause of the no-growth disease still needed confirmation. I soon forgot about these directives, in the excitement of pursuing an idea of my own, which was original only in the degree of its wrongness; I thought I could infallibly ascertain the real causes of these cereal foot-rots if I made tissue-platings from diseased plants on a large enough scale. As I had expected, the isolation plates in their hundreds soon showed me which were the dominant fungi developing from the tissue-platings, and dominance sometimes exceeded 90% of colonies; in no plate did either *Ophiobolus graminis* or *Rhizoctonia solani* appear, and so I thought I was making a real breakthrough. And so I continued to think, until the results of inoculation trials came to hand, when it was quite obvious that my agar-plate dominants were either weak parasites or saprophytes. Nor did my efforts in advisory work do anything to restore my self-esteem at that time, but rather the reverse. Attempts to isolate and prove a fungal pathogen from the roots of some poorly growing cereal plots at the Roseworthy Agricultural College ceased abruptly when someone else showed that the pathogen was not a fungus, but a nematode. My humiliation was increased by the fact that this someone else was not a plant pathologist at all, but an agricultural chemist. Among my other failures, the one I remember best was an experiment with *Fusarium roseum* f.sp. *cerealis* (or *F. culmorum,* as we then called it). In an inoculum-dosage trial, using cultures of the fungus on an autoclaved mixture of oat and barley grain, the most severe incidence of wheat seedling blight was caused by the smallest inoculum dose, whereas the series receiving the largest dose appeared better than the uninoculated controls. Later it was possible to explain this, when we realized that this particular inoculum, even after apparently complete colonization by *F. roseum,* must have contained a substantial proportion of unconsumed nutrients; after mixing with soil, there was a profuse mycelial development of saprophytic fungi on the surface of the inoculum grains, and this must have led to antibiotic inactivation of the pathogen. But at that time, I could not explain these results; when asked to provide a demonstration for a party of visitors, I just switched round the inoculum-dosage labels and received, with a wry smile, congratulations on a most interesting demonstration.

As the time for Samuel's return drew near, I awaited it with considerable apprehension. My fears were groundless; Samuel quickly appreciated the situation, and treated me with great tact and kindness, never even asking to see the pile of useless records lying in my desk. Useless they were indeed at that time, but later I profited by these mistakes on so large a scale, as they provided a foundation for my subsequent interest in the succession of fungi colonizing different substrates. I had had enough of liberty for the time being, and was now glad to learn all I could from Samuel. In the next two years, we published papers on the no-growth disease of cereals (7), and on ascospore discharge from the perithecia of *Ophiobolus graminis* (8), for both of which Samuel provided the ideas and direction. I thus began to gain confidence, though I still wished as strongly as ever to find my own feet and produce some useful ideas of my own. About this time, I began to realize that there was such a thing as the art of scientific investigation and I sought for a way of teaching myself this art. By good fortune, the means were already to hand. Not long before, I had read with profound interest all the papers by William Brown that I could find; somewhat surprisingly, they had not come to my attention as an undergraduate at Cambridge. I realized that Brown's papers would be an admirable model for me, and so I read a number of them again, but in a different way. I said to myself 'If he can do that, why can't I?' So I dissected the papers to find out how Brown did it, rather as one might take a piece of machinery to pieces to find out how it worked. I could not have selected a better model; for many years now, I have regarded William Brown as one of the few mycologists approaching de Bary in stature, and time has merely strengthened my conviction. When I left the Waite Institute after four years' work there, I went to work for two years as one of Brown's research students at the Imperial College in London, and this further increased my already profound debt to him. In appreciation of his ideas that had continued so strongly to influence my own thinking, I later dedicated my second book (5) to him.

This realization was the turning point in my struggle to make myself into a research worker. I soon saw that the methods of experimentation we were using for work on root-infecting fungi were crude and inadequate, and so I conceived the idea of small containers filled with soil maintained in a uniform environment in the laboratory, under much more strictly controlled conditions than we had yet employed. For this purpose, I chose cylindrical-form glass tumblers, because containers produced for general use are cheaper than scientific glassware; when I later required rather larger containers, I employed glass jam-jars for the same reason. Along with this idea of reduction in dimensions of soil containers came the conception of a parallel reduction in the time-scale of experimentation; I thought it might be possible to determine by microscopical assessment the effect of soil conditions on the rate of spread of *Ophiobolus graminis* along wheat seedling roots in as little as two weeks, and so to achieve both greater accuracy and a substantial saving of time over that required for glasshouse experimentation. This expectation was fulfilled before I left the Waite Institute (4). Thus began a life-long search for ways of bringing root-disease problems of diverse kinds within the sphere of laboratory experimentation, with the object of increasing accuracy and saving time. Success in this venture has brought me more lasting satisfaction than has anything else, because the devising of a really new and useful technique is perhaps the most original contribution most research workers are likely to make to their science.

Devising a new technique is the most difficult part of experimentation; more often than not, it is the only difficult part. Indeed, most experimentalists are likely to be remembered longer for their contributions to technique than for anything else; most good techniques have a life longer than that of most hypotheses.

Samuel received these ideas with warm encouragement, the first that I had truly earned, and soon after that I designed the first experiment of my own that really satisfied me. The results of this particular experiment were of no great scientific importance, and indeed I did not deem the small investigation of which they were a part to be worth publishing. But for me these results were the most important I have ever obtained, and gave me the greatest pleasure. For much of my life, either my particular interests at the time, or my later responsibilities as a university teacher and administrator, have condemned me to a succession of rather long-term experiments for my personal research; results from these have unfolded so gradually that there have been few moments of drama. In later life, the moments of intense joy that are the reward for much thinking and much preparation in research have arrived, for the most part, not with the results of an experiment but earlier, when the solution to a problem suddenly occurred to me. At such times, experimental verification has certainly been necessary, but has carried with it no more than the usual satisfaction in the exercise of sound craftmanship.

LITERATURE CITED

1. Beveridge, W. I. B. 1950. *The Art of Scientific Investigation*. London: Heinemann, 172 pp.
2. Brooks, F. T. 1924. Presidential Address. Some present-day aspects of mycology. *Trans. Brit. Mycol. Soc.* 9:14–32
3. Brooks, F. T. 1928. *Plant Diseases*. Oxford Univ. Press, 386 pp.
4. Garrett, S. D. 1934. Factors affecting the severity of take-all. I. The importance of micro-organisms. *J. Agr. S. Australia* 37:664–74
5. Garrett, S. D. 1956. *Biology of Root-infecting Fungi*. Cambridge Univ. Press, 293 pp.
6. Samuel, G. 1928. Two 'stunting' diseases of wheat and oats. *J. Agr. S. Australia* 32:40–43
7. Samuel, G., Garrett, S. D. 1932. *Rhizoctonia solani* on cereals in South Australia. *Phytopathology* 22:827–36
8. Samuel, G., Garrett, S. D. 1933. Ascospore discharge in *Ophiobolus graminis*, and its probable relation to the development of whiteheads in wheat. *Phytopathology* 23: 721–28

MAIN TRENDS IN THE DEVELOPMENT OF PLANT PATHOLOGY IN HUNGARY

3540

ZOLTÁN KIRÁLY

Research Institute for Plant Protection, Budapest, Hungary

Development or history?—According to Berdyaev (4) "History is not an objective empirical datum; it is a myth. . . . A purely objective history would be incomprehensible."

This is the reason why I am outlining here the *development* of plant pathology in Hungary rather than its history. I would like to give objective data on plant pathology in order to avoid any aspect of myth. Most of the events of plant pathology are not sensational enough in this small country to be mythic, i.e. to be historical indeed. Also, development of basic as well as applied plant pathology and the organization of plant protection in Hungary are much more interesting than a historical picture. I hope that other small countries and developing nations will learn something from this myth-less development of plant pathology in Hungary. Naturally, the present-day situation will be treated more thoroughly than the early periods.

RESEARCH

Usual and unusual pioneering.—The father of basic as well as practical plant pathology in Hungary is Professor Gy. Linhart (1844–1926). As the pupil of Brefeld and Kühn he introduced scientific thinking with regard to the study of diseases of plants. The plant-disease situation in Hungary in those early days was very primitive, as was the case in Europe and America. For example: the journal *Magyar Gazda (Hungarian Farmer)* published a short note in 1847 on the smut disease of wheat. According to the author, who signed himself only "F.F.", the causal agent of the smut disease is a parasitic plant which is able to produce new parasitic agents. Therefore, cereal grain infested with smut will produce smutted plants. An editorial comment followed expressing public opinion which said: "Every point of your article, permit us Sir, is a complete paradox; nevertheless we shall publish it in accordance with your esteemed wishes." From this comment it can be seen that the basic concepts for understanding plant diseases had not yet been laid. To lay this foundation Linhart appeared and taught the basic principles of scientific plant pathology (mostly mycology); he introduced practical control measures for different diseases and founded a collection of plant parasitic fungi called *Fungi Hungarici.* Perhaps the most important event in his career was the organization of the Experiment Station for Plant Physiology and Plant Pathology (1897). This institution was one of the forerunners of the present Hungarian Research Institute for Plant Protection.

Another scientist unusual in his approach to plant pathology was Professor G. Doby (1877–1968), a biochemist at the Experiment Station for Plant Physiology and Plant Pathology. He worked mainly on the enzymology of diseased plants but particularly on the leaf roll disease of potatoes. At that time, the viral nature of the causal agent of the disease was not known. As early as 1912, Doby (10) claimed that leaf roll was a metabolic disease, since he found that the activities of peroxidase and polyphenoloxidase were augmented in the tuber tissues of diseased plants. His approach to research was unusual in that it was so completely modern and years before its time that it remained unappreciated and not understood. Consequently, he had little influence on plant pathology either in Hungary or in general. Thus, Doby remained an "historical" rather than an influential figure. His work on the increased activity of oxidative enzymes in 1912 remained only a "myth." His original ideas were re-investigated and partly re-discovered 40 years later (cf. 14, 15). In the year 1910, neither Hungary nor the world was ready for the concept of pathophysiology.

During the first decade of this century, the anatomical-morphological trend lost its leading role in human pathology and was replaced by the physiological approach using the disciplines of chemistry and physical chemistry. In plant pathology however, it took 45 years more before a similar change occurred. This physiological approach to plant pathology is perhaps now in the peak of its development.

The era of mycology.—Plant pathology in Hungary remained synonymous with applied mycology between the two World Wars. In 1924 the first chair of plant pathology at Budapest University was held by Professor K. Schilberszky (1863–1935). He discovered a new potato disease (black wart) and the causal agent of it, *Synchytrium (Chrysophlyctis) endobioticum* (57). Schilberszky trained a series of applied mycologists who were "crop oriented." Thus, plant pathology in Hungary during the first part of this century concentrated on the diseases of *plants* and not on *disease* per se.

A brilliant mycologist, Gy. Istvánffi (1860–1930), a pupil of Brefeld and later the director of the Institute of Viticulture in Budapest, was another giant among Hungarian plant pathologists. His work centered on diseases of grape and he published a series of papers on powdery mildew, downy mildew, the *Pseudopeziza* and *Phyllosticta* diseases of grape. The method developed by Istvánffi & Pálinkás (28) for determining the incubation period of *Plasmopara viticola* was of great practical importance all over the world.

Professor B. Husz (1892–1954) was an influential figure in Hungarian phytopathology in the 1930s and 1940s. He was an excellent and impressive teacher. The present generation of plant pathologists originated mostly from his school. One of the most important events in his life was a study-tour at the University of Minnesota, in the laboratory of Professor Stakman. Husz learned many modern ideas concerning rust and smut diseases at the University of Minnesota and put them into practice in Hungary. As a consequence of his activity, plant breeders began to deal with breeding for disease resistance and the determination of physiologic races of rusts. His book on "*The Diseased Plant and Its Therapy*" was the only source of modern information for students during and after the second World War (27).

Fungicide research was also fairly well developed in Hungary before the Second War. Bodnár & Terényi (5–7) published a series of fascinating papers on the mechanism of action of metal cations, like copper and mercury, on smut spores. According to their theory cationic metals act primarily at or on the surface of the spore, replacing such nontoxic cations as hydrogen, calcium, magnesium, or potassium by base exchange. If, however, the spores are subsequently leached with water or acid, hydrogen will in turn replace the metal cation (copper) of the fungicide, and the spores can be reactivated. They were also able to demonstrate a synergistic action between two fungicides as early as 1931 (65).

Speculations of a genius in plant pathology.—The Hungarian biochemist, the Nobel laureate A. Szent-Györgyi had some slight contact with plant pathology when he was working on the mechanism of respiration in animal tissues. He was engaged for a short while in a study of the oxidation-reduction reactions catalyzed by phenol-phenoloxidases. Since potato was a good source of this enzyme and its substrates, he became a "plant biochemist" at a certain moment of his life. It must be stressed, however, that he worked quite independently from plant pathologists, and was not aware of the results and ideas of Doby. In a short paper from the Szeged University in Hungary (63) he presented a speculation concerning the resistance of plants to infection. The experimental basis of his ideas was the finding that polyphenoloxidase, which does not act on its substrate in the plant because of compartmentization, suddenly oxidizes the phenols in potato tubers if the tissue is mechanically injured or infected. According to his idea: since phenol oxidation products (quinones) are strong antiseptic agents, phenol oxidation must have an important role in resistance to infections. In fact this was experimentally demonstrated 20 years later by other workers. Still, his "mythic," and, therefore, historic speculations on the relation of injury and phenol oxidation to disease resistance in plants were of primary interest at least for a small group of enthusiastic Hungarian pathophysiologists in the 1950s. Darwin was right when he said in a letter to A. R. Wallace (Dec. 22, 1857) "I am a firm believer that without speculation there is no good or original observation."

The pathophysiological trend.—During the past 20 years a new scientific school developed in Hungary as well as in many other countries, called physiological pathology or pathophysiology. The new approach stemmed from the teamwork of young plant pathologists (mycologists), bacteriologists, serologists, virologists, and plant biochemists who were enthusiastic enough also to be interested in related fields of science. In addition, they were on friendly terms with each other, and worked in the same laboratory in the Research

Institute for Plant Protection in Budapest. Thus both the conditions and the time became ripe for the application of methods of plant physiology and biochemistry to investigating plant diseases. So they were interested in *disease* per se, i.e. in the physiology of symptom expression and disease resistance. Somehow, circumstances became suddenly favorable for following the early speculations of Doby and Szent-Györgyi.

The first investigations were concerned with the augmented anabolic and catabolic changes in smut- and rust-infected wheat (13, 34, 35). Increased rate of respiration seemed to be a general phenomenon in infected plants, particularly when necrosis of tissues takes place around the infection site (14). This was the case not only with fungus diseases but also in the cases of viral or bacterial infections (16, 53). It turned out that the increased respiration was related to increased synthesis of nucleic acid, protein, phenol, and starch around the infection site (17, 36, 58–60), at least in the early stages of the disease.

The synthesis of nucleic acids and proteins is controlled hormonally, at least in the case of rust diseases (32). In leaves infected by rusts an increased cytokinin activity parallels an increase in nucleic acid and protein synthesis in the "green island" area around the infection site (39). The movement of nutrients to the infected area is also controlled by this hormone (56).

Synthetic activities utilizing ATP cause uptake of oxygen (respiration) indirectly, thereby permitting processes that lead to the formation of new ATP. In the late stages of infection, however, uncoupling of respiration from phosphorylation leads to destruction of ATP, and again, this causes uptake of oxygen indirectly just as in the former case (55). Thus, in the early stage of disease, anabolic processes induce high rates of respiration, whereas in the late stages, injury and decomposition of tissues lead to increased respiration. Canadian, American, and Japanese workers also presented very similar results.

Plant defense reaction, tissue necrosis, and phenol metabolism.—The results discussed above focused attention on the phenol metabolism of plants during hypersensitive reactions. It was known that one of the most common plant defense responses, the "hypersensitive" reaction, was connected to the necrosis of the infected tissues. It was possible to show unequivocally that tissue necrosis (local lesion formation) was related to the oxidation of phenols, at least in the case of viral infections (16, 61). Systemic infections, on the other hand, increased the activity of polyphenoloxidase only slightly. The importance of this system was stressed by the discovery that reducing agents, like ascorbic acid etc., are able to reduce phenol-oxidation products, and, at the same time, suppress local lesion development.

The classical case of plant hypersensitivity is with rust diseases. This conspicuous form of resistance was "hyper-susceptibility" as early as the first decade of this century but the reason for this designation was more or less an intuitive speculation. There was no indication that this form of plant disease resistance was really a hypersensitive response. The pathophysiology group in Budapest (and other research groups in Canada and Germany) in the 1950s and 1960s was also interested in the investigation of the mechanism of the hypersensitive response of wheat to rust infections. It turned out that changes in the respiration rate, in phenol oxidation (29) as well as accumulation of phenols (36) were more pronounced in the hypersensitive (resistant or incompatible) than in the compatible (susceptible) host-pathogen complexes. Actually, resistance and tissue necrosis were related to the stimulation of physiological responses and somehow to phenol metabolism. Thus it was experimentally demonstrated that the theory of the hypersensitive nature of rust resistance was based on sound speculations. Again, "without speculation there is no good or original observation" (Darwin).

The accumulation of phenolic compounds after infection, as in the case of the hypersensitive reaction (15) is also of interest from the point of view of the phytoalexin theory of plant resistance first proposed by the school of K. O. Müller, and later Cruickshank. More and more evidences were presented that most of the phytoalexins, which produced a necrosis in response to infection in resistant plants, were indeed phenol derivatives.

From these experiments it was concluded that both the hypersensitive type of plant resistance and tissue necrosis involve changes in phenol metabolism (15). This, however, only means that tissue necrosis is the final expression of a hypersensitive reaction on the tissue level; and altered phenol metabolism may have nothing to do with hypersensitivity per se. The accumulation and oxidation of phenolic compounds might be a consequence of incompatibility (hypersensitivity). Hence, tissue necrosis may be an end result, i.e. a visible expression, and not the cause of the real and unknown hypersensitive reaction. In the past we investigated incompatibility only on a tissue level (macroscopic observation) and not in cell:cell systems or at the molecular level.

This scepticism was further promoted by another finding of the Budapest group in connection with bacterial infections of plants (cf. 46). The hypersensitive reaction caused by bacteria does not resemble hypersensitivity as induced by viruses or fungi, except that a very rapid tissue necrosis occurs. It was supposed that cell death occurred so rapidly that it was extremely difficult to demonstrate enzymatic changes or other characteristic alterations in phenol metabolism (54). The secondary nature of at least some of the metabolic changes observed in diseased tissues is supported by comparative analyses of the physiology of detached leaves and diseased ones (12). So, the picture of hypersensitivity is not as clear today as it appeared to be at first glance. We are now of the opinion that the primary events of hypersensitive reaction remain unknown, because our past work was concerned with secondary phenomena i.e. necrosis of cells related to the changes in the oxidation-reduction systems and the production of phytoalexins. Hypersensitive response as a form of plant resistance is part of the fundamental biological phenomena of incompatibility (cf. 1, 8). It is my feeling that application of serological and electron microscopic procedures and other modern methods of molecular biology will be necessary to gain a deeper insight into the mechanism of incompatibility in plant-pathogen relations.

Plant defense reaction in relation to protein and nucleic acid metabolism.—To avoid disappointments from investigating secondary phenomena, pathophysiologists turned their attention to proteins and nucleic acids hoping to get closer to the primary events of resistance (hypersensitivity). The fascinating results of previous (11, 18) and recent (9) workers called attention to the role of "common antigens" in plant disease resistance. The conclusion was reached that the presence of some proteins in hosts and pathogens might be important in compatibility and their absence in incompatibility. Using stem rust of wheat and serological analysis with rabbit antisera it was indeed possible to show that loss of certain proteins parallels the loss of rust resistance in detached (senescent) leaves. However, the prevention of the loss of these proteins by cytokinins inhibits the breakdown of resistance to rust (cf. 20, 31). The character of these protein components is unknown as yet. It is my feeling that enzymes that are able to oxidize phenols are involved, but this remains to be shown.

The resistance of tobacco and bean plants to virus infections also seems to be related to the protein and nucleic acid metabolism of the hosts. This was shown first by changes in the relation of age of tobacco leaves to their resistance to infection by tobacco mosaic virus (TMV) (33, see also 25). Very young leaves with a high rate of protein and nucleic acid synthesis are more resistant to infection than the mature, slightly senescent leaves with low rates of syntheses. This was also demonstrated experimentally. Treatment of leaves resulting in juvenility, such as the removal of the terminal bud or application of cytokinins to the leaves, promotes ribonucleic acid and protein synthesis and increases resistance to TMV infection. On the other hand, various treatments that suppress ribonucleic acid and protein synthesis of the host, induce a slight senescence, and therefore decrease resistance to infection (33). All these results support the concept that intensive host syntheses (in the state of juvenility) are unfavorable for viral infections, thereby causing resistance. A decrease in host syntheses (a slight senescence) is advantageous for infection of viruses, and, therefore, increases susceptibility of the host (33). This concept might be of importance from the point of view of virus chemotherapy, which is still an unsolved problem of phytopathology.

A further step in this line was the finding that inhibitors of protein synthesis usually promote leaf senescence, and the formation of local lesions (2, 19).

A new line in the investigation of plant defense reactions to infection was initiated recently by Hungarian plant pathophysiologists interested in the role of living and heat-killed as well as saprophytic bacteria in inducing acquired resistance (47, 51). Later it turned out that one can get the same effect by infiltrating foreign proteins into tobacco or bean leaves (37). A similar role of nucleic acids in the induction of acquired resistance to viral infections was shown mainly by Israeli workers. However, the reason for the "state of immunity" induced by the first infection or infiltration of leaves with foreign proteins and nucleic acids is far from being understood. Nevertheless, the application of basic concepts of "plant immunity" to the induction of resistance in plants to pathogens may be of practical importance in the future.

The trend of phytobacteriology.—The hypersensitive reaction caused by bacteria was intensively investigated by members of the Plant Protection Institute in Budapest. Later a fruitful collaboration came to existence between this Institute and the Department of Plant Pathology in Columbia, Missouri. The first experimental evidence for the occurrence of the hypersensitive reaction in bacterial diseases of plants was given in the 1960s (44, 48, 49). This reaction develops in plants if (*a*) phytopathogenic bacteria are introduced into nonhost plants, (*b*) if a virulent bacterial strain infects a resistant host strain (i.e. cultivar), or (*c*) if a bacterial strain that lost its pathogenicity infects an originally susceptible host strain (45). The first case is important from a practical point of view also, because the principle that most of the phytopathogenic bacteria are able to induce a visible reaction in nonhosts, can be utilized for diagnostic purposes. By the injection-infiltration method (42) it is easy to introduce a mass of unidentified bacterial cells into tobacco leaves or other plants having succulent leaves. If the bacterium in question is a pathogenic one, it will induce atypical tissue necrosis (hypersensitive reaction) in the nonhost plant (or perhaps typical symptoms if the plant happens to be the host). Saprophytes do not cause any symptom of necrosis or other visible alterations in the tissues. Thus, bacteriologists are able to select out saprophytes from a collection of unknown bacteria obtained from the field.

Bacteriological experiments are extremely suitable for the studies of plant defense reactions. One can introduce a definite number of bacteria into attached leaves by the injection-infiltration method, and the population changes during the defense reaction in the plant can be followed by colony counts on agar-plates. In addition, by injecting antibiotics into the intercellular spaces, multiplication of the pathogen can be stopped at will. It was possible to distinguish three phases, the induction-phase, the latent period, and the collapse of tissues, in the development of the hypersensitive reaction as induced by bacteria, but the events of incompatibility leading to the hypersensitive reaction remained unknown (43, 62).

Plant virology research.—This trend of research on the level of identification, host range determination, and detection of infection in crops was initiated in the 1930s in Hungary, and was very extensive and successful in particular with potato diseases (21–24, 64).

Organization

Hungarian Research Institute for Plant Protection.—This Institute is the center of basic plant pathological research in Hungary. It was formed 40 years ago by integrating three Experiment Stations: Entomology, Plant Physiology and Pathology, and Plant Biochemistry. At present this Institute is concerned only with basic research and adaptation of new principles of plant protection to special European or Hungarian conditions. Extension work, the application of plant protection technologies, screening of pesticides, and quarantine administration belong to the Plant Protection Service with an extensive network of Plant Protection Stations, as well as to the Agricultural Experiment Stations.

Basic research conducted in the Research Institute for Plant Protection represents the first stage of a three-level scientific program of plant protection launched in 1971 by the Hungarian Ministry of Agriculture and Food. Results of basic research, including new methods for investigating plant pathogens and host plants for disease resistance, are applied in the second level of the program by the Agricultural Experiment Stations. This consists of working out plant protection technologies, e.g. spraying schedules and other control measures. The third level is made up of extension work, screening of pesticides, toxicology of pesticides as well as public health aspects, quarantine routine, etc. All these are conducted in the Plant Protection Stations.

The Research Institute for Plant Protection has six Departments: Entomology, Plant Pathology, Pathophysiology, Weed Research, Analytical Chemistry, and Organic Chemistry. A total of 45 scientists, 20 of them holding Ph.D. degrees in phytopathology, conduct research in the Plant Protection Institute. Graduate students (aspirants) usually conduct experiments in the laboratories of this Institute because university departments (chairs) of plant protection are engaged in training plant protection engineers according to the practical requirements of the large farms. Aspirants, under the guidance of scientific advisers, work for a three to four year period as a rule, in the Plant Protection Institute, before they defend their Theses and acquire the scientific degree of Candidate of Science (C.Sc.) which is equivalent to the Ph.D. Practically speaking, the center of the graduate school for plant pathology, agricultural entomology, and pesticide chemistry is the Plant Protection Institute. The next degree is the D.Sc. which, of course, can be acquired after a fairly long and successful scientific career.

During the past decade, plant pathologists of the Research Institute have published a series of handbooks, laboratory manuals, and symposium volumes both in Hungarian (3, 26, 31, 50, 52, 66–68) and in English: *Proceedings of the Conference on Scientific Problems of Plant Protection* (30), *Host-Parasite Relations in Plant Pathology* (41), *Biochemical and Ecological Aspects of Plant-Parasite Relations* (40), *Methods in Plant Pathology* (38), *The Biochemistry and Physiology of Infectious Plant Disease* (20).

Plant Protection Service.—This organization, with a network of 20 Plant Protection Stations, applies principles and new results of basic plant pathology to the daily practice of plant protection. Each Station has about 30 agriculturists, mostly plant protection engineers, who are engaged in extension work and quarantine routine. Laboratories of these Stations are equipped well enough to be able to detect residues of pesticides in crops, check the toxicological properties of pesticides, and give local forecasts for particular epidemics. Field trials for screening fungicides and other pesticides are also carried out in the Plant Protection Stations. Altogether 600 agriculturists, most of them plant protection engineers, comprise the staff in these stations. This number of experts seems to be nearly ideal for a small country of ten million population.

EDUCATION

Plant Protection Engineers or Plant Doctors?—In practical agriculture as well as in education, plant pathology is considered an integral part of plant protection. Education at the universities, therefore, is determined by the needs of daily practice. The separate Chairs (Departments) of Plant Pathology and Entomology were combined 20 years ago to create Chairs of Plant Protection. Consequently, the subject of plant pathology is subordinated to the requirements of plant protection at the agricultural universities. This aspect of plant pathology is something different from that of the academic subject. Plant pathology is "crop oriented"; professors and students are interested in diseases and insect pests of crops, in the application of pesticides, and in the complex (or integrated) protection of the most important cultivated plants. Plant pathology as a science is somewhere in the background providing the necessary basis for the protection of particular crops. Students who complete five semesters in plant pathology, entomology, pesticide chemistry, practical plant protection, and economics of control measures acquire the diploma of "plant protection engineer." This type of expert is usually very familiar with schedules of control measures, new pesticides, toxicological aspects of crop protection, economics of control, etc., but much less with basic concepts of plant pathology, the biology of the pathogens and pests, and the mode of action, as well as the mechanism of action of pesticides. They are practicing plant doctors, as is sometimes said. Indeed, some of them would like to be regarded as plant doctors, but in fact, they are not. The task of plant protection engineer has a technical aspect today rather than a biological one. Thus, plant protection engineer seems to be an appropriate term at present.

As is seen from the foregoing, education in the science of plant pathology is not connected with the university. Young scientists are trained in the Research Institute for Plant Protection in the framework of the graduate school, or aspirantura, as we call it. This dual system of education, in plant protection on the one hand and in basic plant pathology on the other, corresponds with the needs of a small country like Hungary.

LITERATURE CITED

1. Allen, P. J. 1959. Physiology and biochemistry of defense. In *Plant Pathology, An Advanced Treatise,* ed. J. G. Horsfall, A. E. Dimond, 1:435–67. New York: Academic
2. Balázs, E., Gáborjányi, R., Tóth, Á., Király, Z. 1969. Ethylene production in Xanthi tobacco after systemic and local virus infections. *Acta Phytopathol. Hung.* 4:335–58
3. Beczner, L., Bodor, J., Paizs, L. 1970. *A zöldségfélék növényvédelme (Control of Vegetable Diseases).* Budapest: Mezőg. Kiadó 313 pp.
4. Berdyaev, N. 1962. *The Meaning of History.* Cleveland & New York: Meridian. 191 pp.
5. Bodnár, J., Terényi, A. 1930. Biochemie der Brandkrankheiten der Getreidearten, II. Mitt. Biophysikalische und biochemische Untersuchungen über die Kupferadsorption der Weizensteinbrandsporen [*Tilletia tritici* (Bjerk) Winter]. *Z. Phsiol. Chem.* 186:157–82
6. Bodnár, J., Terényi, A. 1932. Biochemistry of the smut diseases of cereals, Note 4. The mechanism of the action of mercury salts on the spores of wheat bunt [*Tilletia tritici* (Bjerk) Winter]. *Z. Physiol. Chem.* 207:78–92
7. Bodnár, J., Villányi, I., Terényi, A. 1927. Biochemie der Brandkrankheiten der Getreidearten. I. Mitt. Die Kupferadsorption der Weizensteinbrandsporen [*Tilletia tritici* (Bjerk) Winter] aus Kupferverbindungen. *Z. Physiol. Chem.* 163:73–93
8. Burnet, F. M. 1971. "Self-recognition" in colonial marine forms and flowering plants in relation to the evolution of immunity. *Nature* 232:230–35
9. DeVay, J. E., Schnathorst, W. C., Foda, M. S. 1967. Common antigens and host-parasite interactions. In *The Dynamic Role of Molecular Constituents in Plant-Parasite Interaction,* ed. C. J. Mirocha, I. Uritani, 313–28, St. Paul: Am. Phytopathol. Soc. Inc.
10. Doby, G. 1912. Biochemische Untersuchungen über die Blattrollkrankheit der Kartoffel. II. Die Oxydasen der ruhenden und angetriebenen Knollen. *Z. Pflanzenkr.* 21:321–36
11. Doubly, J. A., Flor, H. H., Clagett, C. O. 1960. Relation of antigens of *Melampsora lini* and *Linum usitatissimum* to resistance and susceptibility. *Science* 131:229
12. Farkas, G. L., Dézsi, L., Horváth, M., Kisbán, K., Udvardy, J. 1963–64. Common pattern of enzymatic changes in detached leaves and tissues attacked by parasites. *Phytopathol. Z.* 49:343–54
13. Farkas, G. L., Király, Z. 1955. Studies on the respiration of wheat infected with stem rust and powdery mildew. *Physiol. Plant.* 8:877–87
14. Farkas, G. L., Király, Z. 1958. Enzymological aspects of plant diseases. I. Oxidative enzymes. *Phytopathol. Z.* 31:251–72
15. Farkas, G. L., Király, Z. 1962. Role of phenolic compounds in the physiology of plant diseases and disease resistance. *Phytopathol. Z.* 44:105–50
16. Farkas, G. L., Király, Z., Solymosy, F. 1960. Role of oxidative metabolism in the localization of plant viruses. *Virology* 12:408–21
17. Farkas. G. L., Solymosy, F. 1962. Hydrogen and electron transport systems in hypersensitive host-virus combinations. II. Electron transport systems. *Proc. Conf. Czechosl. Plant Virol.,* 5th, Prague. Pp. 45–47
18. Fedotova, T. I. 1940. Immunological character of various species and varieties of wheat in relation to wheat leaf rust (*Puccinia triticina*). *Vestnik Zashch. Rast.* 4:123–30 (In Russian)
19. Gáborjányi, R., Balázs, E., Király, Z. 1971. Ethylene production, tissue senescence and local virus infections. In *Biochemical and Ecological Aspects of Plant-Parasite Relations,* ed. Z. Király, L. Szalay-Marzsó, 51–55. Budapest: Akad. Kiadó.
20. Goodman, R. N., Király, Z., Zaitlin, M. 1967. *The Biochemistry and Physiology of Infectious*

Plant Disease. Princeton: Van Nostrand. 354 pp.
21. Horváth, J. 1962. Reliability of the Igel-Lange test for potato leafroll virus (*Corium solani* Holmes) infection. *Növénytermelés* 11:257–66 (In Hungarian)
22. Horváth, J. 1967. Separation and determination of viruses pathogenic to potatoes with special regard to potato virus Y. *Acta Phytopathol. Hung.* 2:319–60
23. Horváth, J. 1968. Susceptibility and hypersensitivity to tobacco mosaic virus in wild species of potatoes. *Acta Phytopathol. Hung.* 3:35–43
24. Horváth, J. 1968. Susceptibility, hypersensitivity and immunity to potato virus Y in wild species of potatoes. *Acta Phytopathol. Hung.* 3:199–206
25. Horváth, J. 1969. Die Anfälligkeit von *Chenopodium amaranticolor* Coste et Reyn. gegenüber dem Kartoffel-Y-Virus im Hinblick auf die Blattsequenz. *Acta Bot. Hung.* 15:71–77
26. Horváth, J. 1971. *Növényvirusok, vektorok, virusátvitel. (Plant Viruses, Vectors and Virus Transmission)* Budapest: Akad. Kiadó
27. Husz, B. 1941. *A beteg növény és gyógyitása. (The Diseased Plant and Its Therapy.)* Budapest: Term. Tud. Társ. 343 pp.
28. Istvánffi, Gy., Pálinkás, Gy. 1913. Études sur le mildiou de la vigne. *Ann. Inst. Centr. Ampel. Roy. Hongrois* 4:1–122
29. Király, Z. 1959. On the role of phenoloxidase activity in the hypersensitive reaction of wheat varieties infected with stem rust. *Phytopathol. Z.* 35:23–26
30. Király, Z. (ed.) 1961. *Proc. Conf. on Scientific Problems of Plant Protection.* Vol. 1: *Phytopathology*. Budapest: Növényvéd. Kut. Int. Kiadv. 212 pp.
31. Király, Z. 1968. *A növényi betegségellenállóság élettana. (Physiology of Plant Disease Resistance.)* Budapest: Akad. Kiadó. 138 pp.
32. Király, Z., El Hammady, M., Pozsár, B. 1967. Increased cytokinin activity of rust-infected bean and broad bean leaves. *Phytopathology* 57:93–94
33. Király, Z., El Hammady, M., Pozsár, B. 1968. Susceptibility to tobacco mosaic virus in relation to RNA and protein synthesis in tobacco and bean plants. *Phytopathol. Z.* 63:47–63
34. Király, Z., Farkas, G. L. 1953. Preliminary investigations on winter hardiness of winter barley infected by *Ustilago nuda* (Jens.) Rostr. *Növénytermelés* 2:130–33 (In Hungarian)
35. Király, Z., Farkas, G. L. 1957. On the role of ascorbic oxidase in the parasitically increased respiration of wheat. *Arch. Biochem. Biophys.* 66:474–85
36. Király, Z., Farkas, G. L. 1962. Relation between phenol metabolism and stem rust resistance. *Phytopathology* 52:657–64
37. Király, Z., Gáborjányi, R., Stamova, Luba. 1970. Defense reactions induced by foreign proteins infiltrated into leaves prior to inoculation with viruses. *Abstr. Conf. Biochem. Ecol. Aspects Plant-Parasite Relations,* Budapest, p. 23
38. Király, Z., Klement, Z., Solymosy, F., Vörös, J. 1970. *Methods in Plant Pathology with Special Reference to Breeding for Disease Resistance.* Budapest: Akad. Kiadó. 509 pp.
39. Király, Z., Pozsár, B., El Hammady, M. 1966. Cytokinin activity in rust-infected plants: Juvenility and senescence in diseased leaf tissues. *Acta Phytopathol. Hung.* 1:29–37
40. Király, Z., Szalay-Marzsó, L. (ed.) 1971. *Biochemical and Ecological Aspects of Plant-Parasite Relations. A Symposium.* Budapest: Akad. Kiadó. 433 pp.
41. Király, Z., Ubrizsy, G. (ed.) 1964. *Host-Parasite Relations in Plant Pathology. A Symposium.* Budapest: Növényvéd. Kut. Int. 257 pp.
42. Klement, Z. 1963. Rapid detection of the pathogenicity of phytopathogenic pseudomonads. *Nature* 199:299–300
43. Klement, Z. 1971. The hypersensitive reaction of plant to bacterial infections. In *Biochemical and Ecological Aspects of Plant-Parasite Relations,* ed. Z. Király, L. Szalay-Marzsó, 115–18
44. Klement, Z., Farkas. G. L.. Lovrekovich, L. 1964. Hypersensitive reaction induced by phytopatho-

genic bacteria in the tobacco leaf. *Phytopathology* 54:474–77
45. Klement, Z., Goodman, R. N. 1966. Hypersensitive reaction induced in apple shoots by an avirulent form of *Erwinia amylovora*. *Acta Phytopathol. Hung.* 1:177–84
46. Klement, Z., Goodman, R. N. 1967. The hypersensitive reaction to infection by bacterial plant pathogens. *Ann. Rev. Phytopathol.* 5:17–44
47. Klement, Z., Király, Z., Pozsár, B. 1966. Suppression of virus multiplication and local lesion production in tobacco following inoculation with a saprophytic bacterium. *Acta Phytopathol. Hung.* 1:11–18
48. Klement, Z., Lovrekovich, L. 1961. Defense reactions induced by phytopathogenic bacteria in bean pods. *Phytopathol. Z.* 41:217–27
49. Klement, Z., Lovrekovich, L. 1962. Studies on host-parasite relations in bean pods infected with bacteria. *Phytopathol. Z.* 45:81–88
50. Lehocky, J., Reichart, G. 1968. *A szölö védelme. (Control of grape diseases and pests.)* Budapest: Mezög. Kiadó. 264 pp.
51. Lovrekovich, L., Farkas, G. L. 1965. Induced protection against wildfire disease in tobacco leaves treated with heat-killed bacteria. *Nature* 205:823–24
52. Németh, Mária V. 1961. *A gyümölcsfák virusbetegségei. (Virus Diseases of Fruit Trees.)* Budapest: Mezög. Kiadó. 296 pp.
53. Németh, J., Klement, Z. 1967. Changes in respiration rate of tobacco leaves infected with bacteria in relation to the hypersensitive reaction. *Acta Phytopathol. Hung.* 2:303–08
54. Németh, J., Klement, Z., Farkas, G. L. 1969. An enzymological study of the hypersensitive reaction induced by *Pseudomonas syringae* in tobacco leaf tissues. *Phytopathol. Z.* 65:267–78
55. Pozsár, B., Király, Z. 1958. Effect of rust infection on oxidative phosphorylation of wheat leaves. *Nature* 182:1686–87
56. Pozsár, B., Király, Z. 1966. Phloem-transport in rust infected plants and the cytokinin-directed long-distance movement of nutrients. *Phytopathol. Z.* 56:297–309
57. Schilberszky, K. 1930. *Die Gesamtbiologie des Kartoffelkrebses.* München: Freising. 72 pp.
58. Solymosy, F., Farkas, G. L. 1962. Simultaneous activation of pentose phosphate shunt enzymes in a virus-infected local lesion host plant. *Nature* 195:835
59. Solymosy, F., Farkas, G. L. 1962. Hydrogen and electron transport system in hypersensitive host-virus combinations. I. Substrate level. *Proc. Conf. Czechosl. Plant Virol.*, 5th, Prague. 41–44
60. Solymosy, F., Farkas, G. L. 1963. Metabolic characteristics at the enzymatic level of tobacco tissues exhibiting localized acquired resistance to viral infection. *Virology* 21:210–21
61. Solymosy, F., Farkas, G. L., Király, Z. 1959. Biochemical mechanism of lesion formation in virus-infected plant tissues. *Nature* 184:706–07
62. Süle, S., Klement, Z. 1971. Effect of high temperature and the age of bacteria on the hypersensitive reaction of tobacco. In *Biochemical and Ecological Aspects of Plant-Parasite Relations,* ed. Z. Király, L. Szalay-Marzsó, 119–22
63. Szent-Györgyi, A., Vietorisz, K. 1931. Bemerkungen über die Funktion und Bedeutung der Polyphenoloxydase der Kartoffeln. *Biochem. Z.* 233:236–39
64. Szirmai, J. 1969. Further experiments on the glass tube method to control degeneration of potatoes. *Zentralbl. Bakt. Paras. Inf. Hyg.* 2. Abt. 123:336–39
65. Terényi, A. 1931. Die Wirkungserhöhung der zur Saatbeizung gebrauchten Kupfer-Salzlösungen. *Chem. Rundschau* 5:123–25
66. Ubrizsy, G. (ed.) 1965. *Növénykórtan. (Plant Pathology)* Vol. 1: 579 pp. Vol. 2:942 pp. Budapest: Akad. Kiadó
67. Ubrizsy, G. (ed.) 1968. *Növényvédelmi enciklopédia. (Encyclopedia of Plant Protection)* Vol. 1:486 pp. Vol. 2: 530 pp. Budapest: Mezög. Kiadó
68. Ubrizsy, G., Vörös, J. 1968. *Mezö gazdasági mikológia. (Agricultural mycology)* Budapest: Akad. Kiadó. 576 pp.

THE HISTORY OF PLANT PATHOLOGY IN INDIA 3541

S. P. Raychaudhuri, J. P. Verma, T. K. Nariani, and Bineeta Sen
Division of Mycology and Plant Pathology, Indian Agricultural Research Institute, New Delhi, India

Plant pathology in India began with the establishment in 1905 of the Indian (then Imperial) Agricultural Research Institute at Pusa, Bihar (now at New Delhi) and the appointment of E. J. Butler (later Sir Edwin) as the first Imperial Mycologist. The credit for laying the foundation of plant pathology goes to him, and he may aptly be called the "father of Indian plant pathology." Before he departed from India he published in 1918 a book on 'Fungi and Diseases in Plants' which remains a classic on the subject.

The first Indian universities that were established in 1857 at Calcutta, Madras, and Bombay, emphasized taxonomy of fungi. Plant pathology as a University Science became established at Lucknow, Allahabad, and Madras Universities (founded in 1921, 1887, and 1857 respectively), only in the 1930s. There are now 16 Agricultural Universities, each with a Plant Pathology Department. Work on plant pathology is now also carried out at various State Departments of Agriculture and many other institutions.

Three plant disease epidemics stimulated greatly the growth of plant pathology in India. These were the great Bengal famine of 1942 caused by a *Helminthosporium* blight of rice, the severe wheat shortage in Madhya Pradesh during 1946 and 1947 due to wheat rust, and the red-rot epidemic in 1938–42 on sugar cane in Uttar Pradesh and Bihar State. Losses from diseases generally are enormous. Annual losses due to wheat rust alone have been estimated to be about 0.3 million tons of wheat—worth about 32.2 million rupees. The total overall decrease in production of food grains due to fungal diseases has been considered to be about 5 million tons a year (130).

The increasing importance of plant pathology led to the foundation of the Indian Phytopathological Society in 1947 by B. B. Mundkur under the chairmanship of S. R. Bose. Initially there were 20 members; now the total is 1095 in India and abroad. In addition there are 340 subscribers to the Society journal.

The Indian (then Imperial) Council of Agricultural Research was established to promote, guide, and coordinate agricultural and animal husbandry research. Later, education and development work were also brought within the scope of the Council, which operates numerous institutes over the country on various crops. Recently, under the leadership of the Indian Council of Agricultural Research, multidisciplinary all-India Coordinated Research Projects have been drawn up on various crops like rice, maize, cotton, wheat, fruits, etc. with an emphasis on plant pathology.

Dasgupta (31) presented in 1958 the history of mycology and plant pathology in India, including Burma and Ceylon. Since it included a complete review of the literature it is extremely useful to research workers in plant pathology. Raychaudhuri (129) dealt in 1967 with the development of plant pathological research, education, and extension work in India. We will here highlight in historical perspective the work done on important developments in plant pathological research.

Diseases Caused by Fungi

E. J. Butler stayed at the Indian Agricultural Research Institute for 16 years (1905–1921) and established a strong school of mycology and plant pathology. His book, published in 1918, served as the major source of literature and inspiration to budding plant pathologists. With it as a base, Mundkur (81) wrote in 1949 a smaller edition of *Fungi and Plant Diseases,* which was revised by Chattopadhyay in 1967.

Cereal rusts received attention in India as early as 1907 when Milligan reported a heavy outbreak of rust on wheat in Punjab. Studies on the epidemiology of rusts and methods of their control comprise some of the most important contributions to plant pathology carried out in India. Mehta (68) initiated in 1929 a series of experiments on the annual recurrence of black stem rusts on wheat in India, and published his results in a monograph in 1948. He showed (68–72) that *Berberis* and other alternate hosts do not play any significant role in the perpetuation of wheat rusts in India, and that there is no local source of infection to account for the recurrence of rusts in the plains, where the intense summer heat destroys all the rust inoculum of the preceding season. He showed that the main source of inoculum responsible for the fresh infection in the plains is derived from rust infection on the hills, where the crop is grown early. As a result of extensive surveys made periodically all over the wheat-growing areas of India, it was found (56) that the black stem rust, which survives in the Himalayas, is not an important source of infection; the immense bulk of inoculum comes from the south (Nilgiri and Pulney hills).

Certain newer diseases of wheat have come into prominence in recent years because of new agricultural practices, such as extensive use of fertilizer, intensive irrigation, and use of the new high-yielding varieties (130). Leaf blight of wheat caused by *Alternaria triticina* is one such example. Mitra (74) recorded *Tilletia indica*, a new bunt (Karnal bunt) on wheat. The disease was thought, erroneously, to be soil borne (75–77). Later studies showed that it was air borne, and the infection was not systemic (79, 80).

The most important disease of rice in India is the *Helminthosporium* leaf spot caused by *H. oryzae*. This occurs endemically almost every year (155, 167), and was responsible for the severe famine conditions in Bengal in 1942. The nature and extent of damage caused by this disease were investigated, and trials laid for controlling the primary seed-borne infection (15, 97).

Green-ear disease is serious on bajra (*Pennisetum typhoides*). Chaudhuri in 1931 (18–20) first showed that the oospores of the pathogen, *Sclerospora graminicola,* in the soil were responsible for its propagation. Artificial infection of the host could be accomplished only under very humid conditions (171).

Red rot (*Colletotrichum falcatum*) causes enormous loss to the sugarcane crop in India. It was first studied by Butler & Khan (9) at Pusa in 1913. Infection was found to occur through seed, soil, and irrigation water (23–25). Simple practical schedules for its control were suggested (24), and the disease is now totally controlled by a hot-air treatment of the setts (156).

Butler (8) first described the smut disease of sugar cane caused by *Ustilago sacchari.* The mode of infection (40), life history, and method of perpetuation of the pathogen have been investigated (65).

A monograph on potato diseases was written by Butler (7) in 1903. Dastur (39) reported that potato blight due to *Phytophthora infestans,* though uncommon in the Indian plains is found on the hills, and caused severe damage to crops grown at an elevation of 6000 feet and above. Late blight was subsequently found to occur in the plains also (62, 178). Dastur (42) described two new diseases of potato: leaf rot caused by a species resembling *P. parasitica,* and tuber rot caused by a new species, *P. himalayensis.*

Malformation is a devastating and somewhat mysterious disease of mango (*Mangifera indica*). During the last six decades, different workers have attributed it to different causes: nutritional imbalance, virus, eriophyid mites, etc. Affected shoots from mango trees recently have yielded *Fusarium moniliforme* (=*Gibberella fujikuroi*) which reproduces the disease in healthy inoculated seedlings grown in the glasshouse and kept free from mites (166). The disease has recently been shown to be systemic in branches (174), and in preliminary trials good results have been obtained with benlate and aphidan for control of this malady. It is suggested that a judicious combination of pruning, and application of insecticides, fungicides, and growth regulators may prove effective in controlling the disease (175).

Strong schools for fundamental pathology, especially the biochemistry of host-parasite interaction, were established at Lucknow and Madras Universities under the leadership of S. N. Dasgupta and T. S. Sadasivan, respectively. Dasgupta studied the role of enzymes in pathogenicity; a general high metabolic rate and higher level of several enzymes were shown in virulent strains (33, 185, 186). Sadasivan's school developed the concept of vivotoxins and worked out the mechanism of cotton wilt caused by *Fusarium vasinfectum.* The production of fusaric acid by *F. vasinfectum* as a vivotoxin has been demonstrated (57). It was shown (147) that iron, although present in sufficient quantity in a living system, may be bound up with fusaric acid as a chelate and, therefore, unavailable to the host plant. A correlation existed between the amount of fusaric acid produced in vivo and the amount of heavy metals chelated. Further investigations revealed that chelation of zinc could also be one of the causes favoring wilt (58, 146).

Research in forest pathology was initiated by Bagchee at the Forest Research Institute, Dehra Dun in the Himalayas. He reviewed (3) the work done on the coniferous rusts, root- and stem-rotting fungi, canker pathogens, nursery diseases, timber diseases, and the ecology and habits of forest fungi. He also listed rusts and polypores attacking forest trees. The principal diseases of oak in India were also studied (4).

Some of the historical events in the use of fungicides should be mentioned. In 1885 Ozanne (96) first used a fungicide in India for control of a crop disease when he used copper sulphate against sorghum smut. Lawrence (96) used Bordeaux mixture for the first time in 1904 against Cercospora leaf spot of groundnut. Coleman (28) claimed control in 1915 of *Phytophthora omnivora* var. *arecae* on arecanut with Bordeaux mixture plus resin. Cotton anthracnose was controlled (41) by seed dressing with an organomercurial fungicide after delinting the seed with sulphuric acid (96). Narasimhan suggested in 1930 the use of linseed oil in Bordeaux mixture for improving coverage and tenacity (96).

Wide use of antibiotics for the control of fungal diseases of crop plants is fairly new in India. Lately, Hindustan Antibiotics has taken a leading role in the commercial manufacture of antibiotics, and one of their products that has been used extensively as a seed dressing (47), a spray on standing crops (113), and a post-harvest dip (48, 168) is aureofungin.

Diseases Caused by Viruses

Plant viruses did not receive much attention in India until about three decades ago, though one of the earliest records in the country is the spike disease of sandalwood, first reported by Coleman in 1917 (29) to be a graft-transmissible virus disease, but now known to be caused by a mycoplasma (53, 173).

One of the first diseases to be investigated in India in the late thirties was tobacco leaf curl caused by a virus transmitted by the whitefly, *Bemisia tabaci* (98, 115). The virus-vector relationship was studied, and the virus has been shown to have a very wide host range, including tomato, papaya, sannhemp, chilli, and a number of weeds and ornamental plants (73, 84, 183).

The next two decades were mostly devoted to investigations on dissemination and control of a number of virus diseases of economic crop plants such as sugar cane mosaic (116), stenosis or small leaf of cotton (170), yellow vein mosaic of bhindi (11), tomato leaf curl (183), *Katte* disease of small cardamom (177), papaya mosaic and leaf curl (12, 84), and virus diseases of temperate fruits (2, 6, 83, 133, 134). The investigations mostly concerned the vectors and sources of resistance. As a result, resistant varieties were reported for several viruses, and other sources of resistance in wild species such as *Carica cauliflora* for papaya mosaic (13), *Abelmoschus manihot* var. *pungens* for yellow vein mosaic of bhindi (93), *Lycopersicon peruvianum* for tomato leaf curl (94), varieties Ichinose and Kairyonezumegaishi of *Morus alba,* and Oshimasho and Kosen of *M. latifolia* for mulberry mosaic (137). Whitefly, *Bemisia tabaci* Gen., was found to be an effective vector of a number of virus diseases of crop plants and to carry several viruses simultaneously (176).

During the last decade a large number of virus diseases affecting cereals (26, 46, 82, 125, 136, 140, 151, 152, 164, 192), legumes (1, 22, 85, 86, 89, 148, 159, 193), and plantation crops such as cardamom (132), citrus (90, 91, 179), and coconut (165) have been studied. Fundamental problems such as purification, morphology of virus particles, serology, tissue culture, inhibition, and virus-vector relationships have received increasing attention.

A notable feature of these findings is the elucidation of the cause of some serious diseases of complex etiology such as citrus die-back, coconut root wilt, and sandal spike. The citrus die-back was shown to be caused by a complex in which greening disease (now shown to be due to a mycoplasma transmissible by psylla, *Diaphorina citri* Kuway (10), and fungi such as *Colletotrichum gloeosporioides, Diplodia natalensis, Curvularia tuberculata,* and *Fusarium* sp.) plays a major role (141). The coconut root wilt, which has seriously affected the economy of the coconut industry in South India, has been associated with a rod-shaped virus (165), and the sandalwood spike with a mycoplasma (173).

Antisera and particle morphology has been used in identification and detection of plant viruses in recent years, and fundamental studies on purification and serology have received attention with the introduction of the ultracentrifuge in most laboratories. A number of viruses, such as mosaic streak of wheat and tungro of rice (135), various cucurbit mosaics (153), cowpea mosaic (22), coconut root wilt (165), barley mosaic (46), and sannhemp mosaic (87, 128), have been purified, their particle morphology studied, and antisera prepared.

Tissue-culture techniques opened the way for investigations concerning host-parasite relationships and inhibition studies, as well as production of virus-free plants by meristem-tip culture. A number of viruses such as chilli mosaic, sannhemp mosaic, tobacco mosaic, and potato virus X, have been successfully cultivated in tissue culture (139) and attempts are in progress to get virus-free plants of citrus by tissue-culture technique.

During the last few years inactivation of plant viruses has been attempted with plant extracts, microbial growth products, antimetabolites, growth regulators, and ultraviolet and gamma radiations, to find some effective inhibitors of viruses for their control. Thiouracil and 8-azaguanine have proved effective in reducing virus concentrations and suppressing symptoms in several cases (138, 145). Inhibition of plant viruses has also been reported for extracts of chilli, acacia, Datura, deodar, and papaya latex (52, 99, 131, 144, 181), growth products of *Trichothecium roseum, Aspergillus niger,* and *Bacillus subtilis* (144), and ultraviolet and gamma radiations (88, 114, 143).

Recent studies have been focused on diseases caused by mycoplasma-like organisms that have been reported in sandalwood affected with spike disease, brinjal affected with little leaf (173), and greening disease of citrus. The mycoplasma of the sandalwood spike and the greening disease of citrus have been cultured in the laboratory (51, 95). The effect of tetracycline antibiotics has been studied on these diseases, and suppression of symptoms reported (53, 92).

Diseases Caused by Bacteria

Several bacterial diseases have long been known in India, and there are reasons to believe that some of them (citrus canker, leaf spot of mango, and black arm of cotton) may have originated there. Although bacterial diseases were important and widespread, very little work was done until 1950, when M. K. Patel established (103, 105) a school of plant bacteriology at the College of Agriculture, Poona. He reported the first new species, *Xanthomonas uppalii* on *Ipomoea muricata* in 1948 (100). Patel and his associates have established more than 40 new species of plant-pathogenic bacteria, mostly belonging to the genus *Xanthomonas.* Other active centers have since been developed at the Division of Mycology and Plant Pathology at the Indian Agricultural Research Institute, New Delhi, Poona, and Bangalore, and many bacterial diseases of cereals, sugarcane, fiber, oilseeds, vegetables, legumes, and fruit trees have been studied in detail.

The first disease suspected to be caused by bacteria was 'bangle blight' of potato in Bombay. Cappel reported it in 1892 to be prevalent in Poona and other places in Bombay, and considered the causal organism to be a fungus (78). Butler (7) suggested in 1903 that the disease was similar to bacterial wilt of potato caused by *Bacillus solanacearum* (syn: *Pseudomonas solanacearum*). However, Coleman (27) in 1909 was the first to establish the bacterial association with bangle blight. Mann & Nagpurkar isolated the organism, established its pathogenicity, and studied its biochemical and physiological properties (66). The pathogen has now been shown to cause brown rot or wilt of potato, which has been reported from Madras (8), Uttar Pradesh, Bihar, Bengal (66), and other parts of India (120). The same organism also affects banana and brinjal (17, 30) and causes wilt of tomato and tobacco (8, 54). Hutchinson in 1913 (54) developed his concept of toxins while working on this disease.

Hutchinson (55) first reported the yellow ear rot or 'tundu' disease of wheat (*Triticum vulgare*), which had occurred in Punjab since 1908. Chaudhuri (21) suggested in 1935 that *Corynebacterium tritici* causes the disease, but later studies by Vasudeva & Hingorani (180) showed that the presence of both a nematode (*Anguina tritici*) and the bacterium is necessary.

Red stripe of sugarcane, caused by *X. rubrilineans,* was first reported in 1933 from India (44). Gummosis of sugarcane caused by *X. vasculorum,* a very serious disease of the crop, was only reported from Madras in 1960 (119).

Patwardhan (112) in 1928 reported black rot of cabbage in India; *X.*

campestris caused black rot of other cruciferous plants on inoculation (101). Severe black-vein disease of cabbage was later reported in Bombay (101) and West Bengal (16).

Black arm of cotton was reported in Madras in 1918. Although Uppal (169) found the disease to be of minor importance, severe epidemics were reported from Madras and Bombay (107, 111, 118). The disease has become very serious since the recent replacement of resistant, diploid, rainfed, indigenous cottons (*Gossypium arboreum* and *G. herbaceum*) with high-yielding but susceptible, irrigated, tetraploid cottons (*G. hirsutum* and *G. barbadense*) (190, 191). Knight (59) considered that the disease originated in India because the Indian cottons, which have been subjected to selection pressure against bacterial blight for a longer period than the New World cottons, are more resistant. However, Indian work tends to indicate that the disease was introduced into this country through exotics in the middle of the 19th century (108). Recently certain varieties of *G. hirsutum* (101–102B, Reba. B-50, BJA-592, HG-9, and P.14T.128) have been demonstrated to be resistant to all the races present in India (157, 188). The disease is generally believed to be parenchymatous, but vascular infection has been reported (5, 187). It has been demonstrated for the first time that certain systemic fungicides (plantvax and vitavax) can check the secondary spread of the disease and to some extent aid in curing the plants (158, 189).

The canker disease of citrus caused by *X. citri* may have originated in India, since the disease occurs on Kew specimens collected at Dehra Dun in the Himalayas during 1827–1831 (50). The severity of the disease in India was first emphasized by Luthra & Sattar (64) at Punjab, and it is now well established in citrus orchards all over the country.

Although leaf spot of mango was first observed in Poona and Dharwar in 1947 (110), it probably occurred much earlier. Similar spots were observed on leaves in the herbarium of the Forest Botanist, Dehra Dun, collected in Bihar in 1881 and at Dehra Dun in 1908.

The bacterial blight of rice, thought to be confined to Japan, was reported in 1959 in Bombay (160) and has in a short time become very serious and widely distributed, attacking the commonly cultivated varieties (161, 162). It is seed borne. Variation in isolates of the pathogen, *X. oryzae,* has been observed (14, 45, 154). *Indica* varieties IRRI 69/469, 70/470, and M. Sung Song, and Japonica type Tainan-3 and Chianung-242 have been reported to be resistant (127, 163).

Several new and important diseases have been described in recent years on bajra and ragi (117, 122, 123), castor bean, (43, 102), sesamum (49, 126), and legumes (60, 104, 106, 121, 172). A complete list of bacterial diseases is given by Rangaswami (120) and certain reviews (67, 109).

Diseases Caused by Phanerograms

Very little information is available on phanerogamic plant parasites in India. Luthra (63) described *Striga* as a root parasite of sugar cane. Since then three species (*S. densiflora, S. euphrasioides,* and *S. lutea*) have been reported on sugar cane, rice, and sorghum (61). Among other genera common in India are *Dendrophthoe* spp. (especially on mango) and *Orobanche* (on brinjal, tomato, cauliflower, turnip, etc.).

Nonparasitic Diseases

A school for the study of deficiency diseases was established in 1940 by S. N. Dasgupta at Lucknow University. The main work done was on tip necrosis of mango. Although this disease was first recorded in 1909 by Woodhouse it became prevalent in the 1930s (37, 149).

It was observed quite early that orchards situated nearest operating brick kilns sustained greatest damage. Both Sen and Dasgupta reported the production of limited necrosis when healthy mangoes or trees were exposed to coal fumes. The earlier observation that SO_2 could produce mango necrosis could not be confirmed (38, 150, 184).

There was evidence that ethylene, another constituent gas of the brick kiln fumes, and a mixture of ethylene and SO_2 in certain proportions, when administered in large doses continuously for a number of days produced typical black tip (124). It was found that ethylene in small doses (1:10,000) induced ripening, while heavier concentrations (10:10,000) produced dark brown lesions around lenticels all over the skin of the fruit. Finally the spots coaelesced, forming dark brown patches.

Other known deleterious constituent gases from brick kiln fumes, e.g. carbon monoxide and fluorine, also failed to produce typical symptoms (35). It was, however, observed that an extract of the necrotic tissue showed successful infectivity. This leads to the possibility that certain infective principles were present in the brick kiln fumes. Petroleum ether and ether-soluble fractions of brick kiln fumes were isolated and formed the crude residue that could be resolved into crystalline and liquid fractions. These produced typical symptoms on injection into healthy tissue and the highest necrotic activity was shown by the crystalline fraction (34). It was, however, demonstrated that a spray with boron in the form of borax could successfully prevent necrosis (36). Mango necrosis is now designated as probably a boron-deficiency disease (32).

Vasudeva & Raychaudhuri (182) reported a serious disease of guava from Rajasthan, characterized by severe reduction in leaf size, interveinal chlorosis, suppression of growth, dieback of shoots, and cracking of fruits. Foliar sprays, applications of zinc sulphate in soil, and shoot or trunk injections cured the disease (142). The disease was concluded to be due to zinc deficiency.

Concluding Remarks

It is clear from the foregoing account that although the studies on fungal diseases were started in India quite early, plant virology and plant bacteriology received proper attention only in the last two decades, while research on nonparasitic diseases and phanerogamic parasites has yet to develop. It may, therefore, be concluded that in the Formative Period of plant pathology in India (1905–1950) work was mainly confined to mycology with special emphasis in the early period on symtomatology and description of the pathogen. Later, stress was laid on the physiology of disease, development of resistant varieties, and chemical control. The next period may be called the Modern Period (1950–1970) where the new techniques evolved were applied in the fields of mycology, virology, and bacteriology. Recent years have witnessed the identification of well-defined races of a few important bacterial plant pathogens like *Xanthomonas oryzae* and *X. malvacearum*, and the demonstration of the usefulness of systemic fungicides like plantvax and vitavax in bacteriology. Several viruses have been purified, antisera prepared, and techniques developed for their quick identification. Virus inhibitors have also been described. The demonstration of mycoplasmal etiology of several diseases is another landmark. The culture of Mycoplasma shows the advanced stage reached in Indian plant pathology.

During the next decade, epidemiology and forecasting of plant diseases will receive greater attention. Studies on air pollution and deficiency diseases may be established on a firmer basis. Research on vegetable pathology, transit and storage diseases, and integrated disease control will be intensified. Detection of plant diseases by remote sensing has been initiated. To prove Koch's postulates in case of Mycoplasma diseases will be the goal of several plant pathologists. These years will be critical to Indian agriculture, and plant pathology is going to play an important role in establishing agriculture in India on a sound footing.

LITERATURE CITED

1. Azad, R. N., Nagaich, B. B., Sehgal, O. P. 1961. Broad bean mosaic virus in India. *Indian Phytopathol.* 14:169–73
2. Azad, R. N., Sehgal, O. P. 1958. Some virus diseases of temperate fruits. *Indian Phytopathol.* 11:159–64
3. Bagchee, K. D. 1939. Indian forest mycology with special reference to forest pathology. Presidential Address (Sect. Bot.) *Proc. 26th Indian Sci. Congr.* Part II:141–67
4. Bakshi, B. K., Bagchee, K. D. 1950. Principal diseases and decays of oaks in India. *Indian Phytopathol.* 4:162–69
5. Bhagwat, N. Y., Bhide, V. P. 1962. Vascular infection of some cottons by *Xanthomonas malvacearum*. *Indian Cotton Grow. Rev.* 16:80–82
6. Bhargava, K. S., Bhist, N. S. 1957. Three virus diseases of hill fruits in Kumaon. *Curr. Sci.* 26: 324–25
7. Butler, E. J. 1903. Potato diseases of India. *Agr. Ledger* 4:112–19
8. Butler, E. J. 1918. *Fungi and disease in plants.* Thacker, Spink & Co., Calcutta. 547 pp.
9. Butler, E. J., Khan, A. H. 1913. Red rot of sugarcane. *Dep. Agr. India, Bot. Mem.* 6:151–78
10. Capoor, S. P., Rao, D. G., Viswanath, S. M. 1967. *Diaphorina citri* Kuway a vector of the greening disease of citrus in India. *Indian J. Agr. Sci.* 37:572–76
11. Capoor, S. P., Varma, P. M. 1950. Yellow vein mosaic of *Hibiscus esculentus* L. *Indian J. Agr. Sci.* 20:217–30
12. Capoor, S. P., Varma, P. M. 1958. A mosaic disease of papaya in Bombay. *Indian J. Agr. Sci.* 28: 225–33
13. Capoor, S. P., Varma, P. M. 1961.

Immunity to papaya mosaic virus in the genus *Carica*. *Indian Phytopathol.* 14:96–97
14. Chakravarti, B. P., Rangarajan, M. 1966. Streptocycline an effective antibiotic against bacterial plant pathogens. *Hindustan Antibiotic Bull.* 8:209–11
15. Chattopadhyay, S. B. 1952. Trial of different seed treating fungicides for control of primary seed-borne infection of *Helminthosporium oryzae* Breda de Haan. *Proc. 39th Indian Sci. Congr.* Part III:36
16. Chattopadhyay, S. B., Mukherjee, K. 1955. Black vein disease of cabbage in West Bengal. *Sci. Cult.* 21:107–08
17. Chattopadhyay, S. B., Mukhopadhyay, N. 1968. Outbreaks and new records. *F.A.O. Plant Prot. Bull.* 16:53–58
18. Chaudhuri, H. 1928. Green ear disease of bajra (*Pennisetum typhoideum* Rich.). *Proc. 15th Indian Sci. Congr.* 223
19. Chaudhuri, H. 1932. Further studies on "green ear" of bajra. *Proc. 19th Indian Sci. Congr.* 299
20. Chaudhuri, H. 1932. *Sclerospora graminicola* on bajra (*Pennisetum typhoideum*). *Phytopathology* 22:241–46
21. Chaudhuri, H. 1935. A bacterial disease of wheat in Punjab. *Proc. Ind. Acad. Sci.* 1B:579–85
22. Chenulu, V. V., Sachchidananda, J., Mehta, S. C. 1968. Studies on mosaic disease of cowpea from India. *Phytopathol. Z.* 63:381–87
23. Chona, B. L. 1943. Red rot of sugarcane and its control. *Indian Farming* 4:27–32
24. Chona, B. L. 1947. Simple practical schedules for protection against red rot, smut and mosaic of sugarcane. *Indian Farming* 8:630–32
25. Chona, B. L. 1950. Studies in the diseases of sugarcane in India III. Sources and modes of red rot infection. *Indian J. Agr. Sci.* 20:363–85
26. Chona, B. L., Seth, M. L. 1960. A mosaic disease of maize (*Zea mays*) in India. *Indian J. Agri. Sci.* 30:25–32
27. Coleman, L. C. 1909. The ring disease of potatoes. *Mysore Dep. Agr., Mycol. Ser. Bull.* 15
28. Coleman, L. C. 1915. The control of koleroga of the areca palm, a disease caused by *Phytophthora omnivora* var. *arecae*. *Agr. J. India* 10:129–36
29. Coleman, L. C. 1917. Spike disease of sandal. *Mysore Dep. Agr. Bull. Mycol. Ser.* 3
30. Das, C. R., Chattopadhyay, S. B. 1955. Bacterial wilt of egg plant. *Indian Phytopathol.* 8:130–35
31. Dasgupta, S. N. 1958. History of botanical researches in India, Burma and Ceylon. Part 1, Mycology and plant pathology. Spec. Publ. *Ind. Bot. Soc.* 118 pp.
32. Dasgupta, S. N. 1959. Mango-necrosis—A boron deficiency disease. Presidential address, *Nat. Acad. Sci. India,* 34–38
33. Dasgupta, S. N. 1969. Changing concept of phytopathogenicity. *Indian Phytopathol.* 22:28–42
34. Dasgupta, S. N., Iyer, S. N., Verma, G. S. 1956. Studies in the diseases of *Mangifera indica* Linn. IX. Isolation of brick kiln fume constituents causing mango necrosis. *Ind. J. Agr. Sci.* 26:259–66
35. Dasgupta, S. N., Sen, C. 1960. Studies in the diseases of *Mangifera indica* Linn. XII. Further studies in the effect of boron on mango necrosis. *Proc. Nat. Inst. Sci. India* 26B:80–87
36. Dasgupta, S. N., Sen, C. 1960. Studies in the diseases of *Mangifera indica* Linn. XI. The effect of boron on mango necrosis. *Phytopathology* 50:431–33
37. Dasgupta, S. N., Verma, G. S. 1939. Studies in the diseases of *Mangifera indica* L. I. Preliminary observations on the necrosis of the mango fruit with special reference to the external symptoms of the disease. *Proc. Ind. Acad. Sci.* 9B:13–28
38. Dasgupta, S. N., Verma, G. S., Sinha, S. 1941. Studies in the diseases of *Mangifera indica* Linn. III. Investigations into the effect of sulphur dioxide gas on the mango fruit. *Proc. Ind. Acad. Sci.* 13B:71–82

39. Dastur, J. F. 1915. The potato blight in India. *Mem. Dep. Agr. India Bot.* 7:163–76
40. Dastur, J. F. 1920. The mode of infection by smut in sugarcane. *Ann. Bot. Lond.* 34:391–97
41. Dastur, J. F. 1934. Cotton anthracnose in the Central Provinces. *Ind. J. Agr. Sci.* 4:100–20
42. Dastur, J. F. 1948. *Phytophthora* spp. of potatoes (*Solanum tuberosum* L.) in the Simla hills. *Indian Phytopathol.* 1:19–26
43. Desai, M. V., Shah, H. M. 1963. Bacterial leaf blight of castor beans. *Curr. Sci.* 32:474–75
44. Desai, S. V. 1933. Occurrence of the red stripe disease of sugar cane in India. *Proc. 20th Ind. Sci. Congr.*, 61–62
45. Devadath, S., Padmanabhan, S. Y. 1969. A preliminary study on the variability of *Xanthomonas oryzae* on some rice varieties. *Plant Dis. Reptr.* 53:145–48
46. Dhanraj, K. S., Raychaudhuri, S. P. 1969. A note on barley mosaic in India. *Plant Dis. Reptr.* 53:967–68
47. Dharam Vir, D., Raychaudhuri, S. P. 1969. Antibiotics in plant disease control. I. Relative efficacy of antibiotics for the control of covered smut of oats (*Ustilago hordei* Pers.) Lagerh. *Hindustan Antibiot. Bull.* 11:166–68
48. Dharam Vir, D., Raychaudhuri, S. P., Thirumalachar, M. J. 1968. Studies on *Diplodia* rot of mango and *Alternaria* rot of tomato fruits and their control. *Hindustan Antibiot. Bull.* 10:322–26
49. Durgapal, J. C., Rao, Y. P. 1967. Bacterial leaf spot of Sesamum (*Sesamum orientale* L.) in India. *Indian Phytopathol.* 20:178–80
50. Fawcett, H. S., Jenkins, A. E. 1933. Records of citrus canker from herbarium specimens of the genus Citrus in England and the United States. *Phytopathology* 23:820–24
51. Ghosh, S. K., Raychaudhuri, S. P., Varma, A., Nariani, T. K. 1971. Isolation and culture of mycoplasma associated with citrus greening disease. *Curr. Sci.* 40:299–30
52. Gupta, V. K., Raychaudhuri, S. P. 1971. Nature of virus inhibitors in *Acacia arabica* Wild. *Ann. Phytopathol. Soc. Japan* 37:124–27
53. Hull, R., Horne, R. W., Nayar, R. M. 1969. Mycoplasma-like bodies associated with sandal spike disease. *Nature, Lond.* 224:1121–22
54. Hutchinson, C. M. 1913. Rangpur tobacco wilt. *Dep. Agr. India, Bact. Ser., Mem.* 1:67–83
55. Hutchinson, C. M. 1917. A bacterial disease of wheat in the Punjab. *Dept. Agr. India, Bact. Ser., Mem.* 1:169–75
56. Joshi, L. M., Saari, E. E., Gera, S. D. 1971. Epidemiology of wheat rusts in India. *Symp. Epidemiol., Forecasting Control Plant Dis.* Lucknow, 43–45 (Abstr.)
57. Kalyanasundaram, R. 1955. Antibiotic production by *Fusarium vasinfectum* Atk. in soil. *Curr. Sci.* 24:310–11
58. Kalyanasundaram, R., Saraswathi Devi, L. 1955. Zinc in the metabolism of *Fusarium vasinfectum* Atk. *Nature, Lond.* 175:945
59. Knight, R. L. 1948. The genetics of blackarm resistance. VII. *Gossypium arboreum* L. *J. Genet.* 49:109–16
60. Kulkarni, Y. S., Patel, M. K., Abhyankar, S. G., 1950. A new bacterial leaf spot and stem canker of pigeon pea. *Curr. Sci.* 19:384
61. Kumar, L. S. S. 1940. Flowering plants which attack economic crops. I. *Striga. Indian Farming* 1:593
62. Lal, T. B., 1948. Occurrence of late blight of potato in the plains of India. *Indian Phytopathol.* 1:164
63. Luthra, J. C. 1921. Striga as a root parasite of sugarcane. *Agr. J. India* 16:519–23
64. Luthra, J. C., Sattar, A. 1940. Further studies on the blight disease of gram [*Ascochyta rabiei* (Pass.) Lab.] in the Punjab—control by resistant types. *Proc. 27th Ind. Sci. Cong.* Part III: 228–29 *Abstr.*
65. Luthra, J. C., Sattar, A., Sandhu, S. S. 1938. Life-history and modes of perpetuation of smut of sugarcane. (*Ustilago scitam-*

inea Syd.). *Indian J. Agr. Sci.* 9:849–61

66. Mann, H. H., Nagpurkar, S. D. 1921. The ring disease of potato. *Dep. Agr. Bombay Bull.* 102:38–57
67. Mathur, R. S., Swarup, J., Sinha, S. K. 1965. Bacterial plant pathogens of India. *Labdev. J. Sci. Technol.* 3:1–13
68. Mehta, K. C. 1929. The annual recurrence of rusts on wheat in India. Presidential Address (Sect. Bot.) *Proc. 16th Indian Sci. Congr.* 199–223
69. Mehta, K. C. 1931. Annual outbreaks of rusts on wheat and barley in the plains of India. *Indian J. Agr. Sci.* 1:297–01
70. Mehta, K. C. 1931. The cereal rust problem in India. *Indian J. Agr. Sci.* 1:302–05
71. Mehta, K. C. 1937. Dissemination of wheat rust. *Proc. 24th Indian Sci. Congr.* 259
72. Mehta, K. C. 1941. The wheat rust problem of India. *Curr. Sci.* 10:357–61
73. Mishra, M. D., Raychaudhuri, S. P., Jha, A. 1963. Virus causing leaf curl of chilli (*Capsicum annuum* L.) *Indian J. Microbiol.* 3:73–76
74. Mitra, M. 1931. A new bunt on wheat in India. *Ann. Appl. Biol.* 18:178–79
75. Mitra, M. 1937. Effect of bunt (*Tilletia indica*) on wheat. *Proc. 24th Indian Sci. Congr.* 374
76. Mitra, M. 1937. Studies on the stinking smut or bunt of wheat in India. *Indian J. Agr. Sci.* 7: 459–78
77. Mitra, M. 1937. Soil infection as a factor in the transmission of wheat bunt. *Proc. 24th Indian Sci. Congr.* 374
78. Mollison, J. 1901. *A text book of Indian Agriculture.* Advocate of India Steam Press, Bombay, 283 pp.
79. Mundkur, B. B. 1943. Karnal bunt, an air-borne disease. *Curr. Sci.* 12:230–31
80. Mundkur, B. B. 1943. Studies in Indian cereal smuts. V. Mode of transmission of the Karnal bunt of wheat. *Indian J. Agr. Sci.* 13:54–58
81. Mundkur, B. B. 1949. *Fungi and Plant Diseases.* Macmillan & Co. Ltd., London, 246 pp.
82. Nagaich, B. B., Vashisth, K. S. 1963. Barley yellow dwarf: a new viral disease in India. *Indian Phytopathol.* 16:318–19
83. Nagaich, B. B., Vashisth, K. S. 1965. Additional viral diseases of temperate fruits in Simla Hills. *Indian Phytopathol.* 18: 288–90
84. Nariani, T. K. 1956. Leaf curl of papaya. *Indian Phytopathol.* 9: 151–57
85. Nariani, T. K. 1960. Yellow mosaic of Mung (*Phaseolus aureus* L.) *Indian Phytopathol.* 13:24–29
86. Nariani, T. K., Kandaswamy, R. K. 1961. Studies on a mosaic disease of cowpea (*Vigna sinensis* Savi). *Indian Phytopathol.* 14:77–82
87. Nariani, T. K., Kartha, K. K., Prakash, Nam. 1970. Purification of southern sannhemp mosaic virus using butanol and differential centrifugation. *Curr. Sci.* 39:539–41
88. Nariani, T. K., Paliwal, Y. C. 1963. Inhibition of sannhemp mosaic virus by ultraviolet and gamma radiations. *Indian Phytopathol.* 16:282–84
89. Nariani, T. K., Pingaley, K. V. 1960. A mosaic disease of soybean (*Glycine max* (L.) Merr.). *Indian Phytopathol.* 13: 130–36
90. Nariani, T. K., Raychaudhuri, S. P., Sharma, B. C. 1968. Exocortis in citrus in India. *Plant Dis. Reptr.* 52:834
91. Nariani, T. K., Raychaudhuri, S. P., Bhalla, R. B. 1967. Greening virus of citrus in India. *Indian Phytopathol.* 23:146–50
92. Nariani, T. K., Raychaudhuri, S. P., Viswanath, S. M. 1971. Response of greening pathogen to certain tetracycline antibiotics. *Curr. Sci.* 40:552
93. Nariani, T. K., Seth, M. L. 1958. Reaction of *Abelmoschus* and *Hibiscus* species to 'yellow vein' mosaic virus. *Indian Phytopathol.* 11:136–43
94. Nariani, T. K., Vasudeva, R. S. 1963. Reaction of *Lycopersicon* spp. to tomato leaf curl virus. *Indian Phytopathol.* 16:238–39
95. Nayar, R., Ananthapadmanabha, H. S. 1970. Isolation, and pathogenicity trials with myco-

plasma-like bodies associated with sandal spike disease. *Indian Acad. Wood Sci.* 1:59–61

96. Nene, Y. L. 1971. *Fungicides in plant disease control.* Oxford & IBH Publishing Co., New Delhi, Bombay Calcutta, 385 pp.
97. Padmanabhan, S. Y., Roychoudhuri, K. R., Ganguly, D. D. 1948. *Helminthosporium* disease of rice. I. Nature and extent of damage caused by the disease. *Indian Phytopathol.* 1: 34–47
98. Pal, B. P., Tandon, R. K. 1937. Types of leaf-curl in Northern India. *Indian J. Agr. Sci.* 7: 363–93
99. Paliwal, Y. C., Nariani, T. K. 1965. Effect of plant extracts on the infectivity of sannhemp (*Crotalaria juncea* L.) mosaic virus. *Acta Virol.* 9:261–67
100. Patel, M. K. 1948. *Xanthomonas uppalii* sp. nov. pathogenic on *Ipomoea muricata. Indian Phytopathol.* 1:67–69
101. Patel, M. K., Abhyankar, S. G., Kulkarni, Y. S. 1949. Black rot of cabbage. *Indian Phytopathol.* 2:58–61
102. Patel, M. K., Bhatt, V. V., Kulkarni, Y. S. 1951. Three new bacterial diseases of plants from Bombay. *Curr. Sci.* 20:326–27
103. Patel, M. K., Bhatt, V. V., Kulkarni, Y. S. 1951. Three new bacterial diseases of plants from Bombay. *Indian Phytopathol.* 4: 144–51
104. Patel, M. K., Dewan, N. D. 1949. Bacterial blight of cowpea. *Indian Phytopathol.* 2:75–80
105. Patel, M. K., Dhande, G. W., Kulkarni, Y. S. 1951. Studies in some species of *Xanthomonas. Indian Phytopathol.* 4:123–40
106. Patel, M. K., Dhande, G. W., Kulkarni, Y. S. 1952. Bacterial leaf spot of *Cyamopsis tetragonoloba* (L.) Taub. *Curr. Sci.* 21: 183
107. Patel, M. K., Kulkarni, Y. S. 1948. *Xanthomonas malvacearum* (E. F. Smith) Dowson on exotic cottons in India. *Curr. Sci.* 17:243–44
108. Patel, M. K., Kulkarni, Y. S. 1950. Bacterial leaf-spot of cotton. *Indian Phytopathol.* 3:51–63
109. Patel, M. K., Kulkarni, Y. S. 1958. A review of bacterial plant disease investigation in India. *Indian Phytopathol.* 6: 131–40
110. Patel, M. K., Kulkarni, Y. S., Moniz, L. 1948. *Pseudomonas mangiferae-indicae,* pathogenic on mango. *Indian Phytopathol.* 1:147–52
111. Patel, P. N., Trivedi, B. M. 1969. *Half-yearly scientific report for the period ending 31.6.1969. Div. Mycol & Plant Pathol., Indian Agr. Res. Inst.,* 1969
112. Patwardhan, G. B. 1928. Field, garden and orchard crops of Bombay presidency. *Dep. Agr. Bombay Bull.,* 30
113. Pavgi, M. S., Mandokhot, A. M. 1969. Aureofungin spray in the control of leaf rust of wheat. *Hindustan Antibiot. Bull.* 11: 180–85
114. Prasad, R. N., Raychaudhuri, S. P., 1961. Mosaic disease of *Zinnia elegans* Jacq. *Indian Phytopathol.* 14:123–26
115. Pruthi, H. S., Samuel, C. K. 1937. Entomological investigations on the leaf curl disease of tobacco in North Bihar. I. Transmission experiments with some suspected insect vectors. II. An alternate host of the virus and the insect transmitter. *Indian J. Agr. Sci.* 7:659–70
116. Rafay, S. A. 1935. Physical properties of sugar cane mosaic virus. *Indian J. Agr. Sci.* 5:663–70
117. Rajagopalan, C. K. S., Rangaswami, G. 1958. Bacterial leaf spot of *Pennisetum typhoides. Curr. Sci.* 27:30–31
118. Ramakrishnan, T. S., Ramakrishnan, K. 1950. Observations on the blackarm of cotton in Madras State. *Indian Phytopathol.* 3:64–74
119. Rangaswami, G. 1960. Studies on two bacterial diseases of sugar cane. *Curr. Sci.* 29:318–19
120. Rangaswami, G. 1962. *Bacterial plant diseases in India.* Asia Publishing House, Bombay, 163 pp.
121. Rangaswami, G., Gowda, S. S. 1963. On some bacterial diseases of ornamentals and vegetables in Madras State. *Indian Phytopathol.* 16:74–85

122. Rangaswami, G., Prasad, N. N., Eswaran, K. S. S. 1961. A bacterial leaf blotch disease of cumbee (*Pennisetum typhoides* Stapt. & Hubbard). *Madras Agr. J.* 48:180–81
123. Rangaswami, G., Prasad, N. N., Eswaran, K. S. S. 1961. Bacterial leaf spot diseases of *Eleusine coracana and Setaria italica* in Madras State. *Indian Phytopathol.* 14:105–07
124. Ranjan, S., Jha, V. R. 1940. The effect of ethylene and sulphur dioxide on the fruits of *Mangifera indica. Proc. Ind. Acad. Sci.* 6B:267–88
125. Rao, D. G., Varma, P. M., Capoor, S. P. 1965. Studies on mosaic disease of *Eleusine* in the Deccan. *Indian Phytopathol.* 18:139–50
126. Rao, Y. P. 1962. Bacterial blight of sesamum (*Sesamum orientale* L.). *Indian Phytopathol.* 15:297–98
127. Rao, Y. P., Srivastava, D. N. 1970. Resistance in rice to bacterial blight by *Xanthomonas oryzae* (Uyeda & Ishiyama) Dowson. *Indian Phytopathol.* 23:154
128. Raychaudhuri, S. P. 1947. A note on the mosaic virus in sannhemp (*Crotalaria juncea* Linn.) and its crystalization. *Curr. Sci.* 16:26–28
129. Raychaudhuri, S. P. 1967. Development of mycological and plant pathological research, education, and extension work in India. *Rev. Appl. Mycol.* 46: 577–83
130. Raychaudhuri, S. P. 1968. Role of plant pathology in crop production in India. *Indian Phytopathol.* 21:1–13
131. Raychaudhuri, S. P., Chadha, K. C. 1965. Deodar fruit extract —an inhibitor of chilli mosaic virus. *Indian Phytopathol.* 18: 96A
132. Raychaudhuri, S. P., Chatterjee, S. N. 1961. Chirke—a new virus threat to cardamom. *Indian Farming* 11:11–12
133. Raychaudhuri, S. P., Chatterjee, S. N., Dhar, H. K. 1961. Preliminary note on the occurrence of the yellow-net vein disease of mulberry. *Indian Phytopathol.* 14:94–95
134. Raychaudhuri, S. P., Chatterjee, S. N., Dhar, H. K. 1962. A mosaic disease of mulberry. *Indian Phytopathol.* 15:187–88
135. Raychaudhuri, S. P., Chenulu, V. V., Ganguly, B. 1969. Electron microscopic studies on rice and wheat viruses in India. *Proc. Int. Symp. Electron microscopy Life Sci.* 1–5 Dec., Calcutta, 1969
136. Raychaudhuri, S. P., Ganguly, B. 1968. A mosaic streak of wheat. *Phytopathol. Z.* 62:61–65
137. Raychaudhuri, S. P., Ganguly, B., Basu, A. N. 1965. Further studies on mosaic of mulberry. *Plant Dis. Reptr.* 49:981
138. Raychaudhuri, S. P., Mishra, M. D. 1965. Effect of some metabolites and their analogues on the infectivity of tobacco tissue culture containing chilli mosaic virus. *Virology* 25:483–84
139. Raychaudhuri, S. P., Mishra, M. D. 1965. Cultivation of some plant viruses in callus tissue cultures. *Indian Phytopathol.* 28: 50–53
140. Raychaudhuri, S. P., Mishra, M. D., Ghosh, A. 1967. Preliminary note on the occurrence and transmission of rice yellow dwarf virus in India. *Plant Dis. Reptr.* 51:1040
141. Raychaudhuri, S. P., Nariani, T. K., Lele, V. C. 1969, Citrus dieback problem in India. *Proc. First Int. Citrus Symp. Riverside,* California. 3:1433–37
142. Raychaudhuri, S. P., Nariani, T. K., Joshi, H. C. 1961. Deficiency disease of guava in Rajasthan and its control. *Indian Phytopathol.* 14:134–38
143. Raychaudhuri, S. P., Prasad, H. C. 1960. Inhibition of radish mosaic virus in *Brassica* spp. by ultraviolet irradiation. *Indian Phytopathol.* 13:175–78
144. Raychaudhuri, S. P., Prasad, H. C. 1965. Effect of plant extracts and microbial growth products on the infectivity of radish mosaic virus. *Indian J. Microbiol.* 5: 13–16
145. Raychaudhuri, S. P., Sharma, D. C. 1962. Thiouracil an inhibitor of ring spot strain of potato virus X. *Indian J. Microbiol.* 2: 169–72

146. Sadasivan, T. S. 1958. Moulds, metabolites and tissues. Presidential address, *45th Indian Sci. Congr.*, 12 pp.
147. Sadasivan, T. S., Subramanian, C. V. 1954. Recent advances in the study of soil-borne Fusaria. *J. Indian Bot. Soc.* 33:162–76
148. Sahare, K. C., Raychaudhuri, S. P. 1963. Mosaic disease of urid (*Phaseolus mungo* L.). *Indian Phytopathol.* 16:316–18
149. Sen, P. K. 1941. Black-tip of mango. *Sci. Cult.* 7:56
150. Sen, P. K. 1943. Black-tip disease of mango. *Ind. J. Agr. Sci.* 13: 300–33
151. Seth, M. L., Raychaudhuri, S. P., Singh, D. V. 1971. A streak disease of bajra (*Pennisetum typhoides*) (Burun. Stap. and Hubb.) in India. *Curr. Sci.* 40: 272
152. Seth, M. L., Raychaudhuri, S. P., Singh, D. V. 1971. Studies on bajra mosaic disease. *2nd Int. Symp. Plant Pathol.*, New Delhi, 113 (Abstr.)
153. Shanker, G., Nariani, T. K., Prakash, N. 1969. Detection and identification of cucurbit viruses by electron microscopy and serology. *Proc. Int. Symp. Electron Microscopy in Life Sci.* Calcutta, 1–5 Dec., 1969, 27
154. Shekhawat, G. S., Srivastava, D. N. 1968. Variability in Indian isolates of *Xanthomonas oryzae* (Uyeda & Ishiyama) Dowson, the incitant of bacterial leaf blight of rice. *Ann. Phytopathol. Soc. Japan* 34:289–97
155. Singh, J. 1934. Leaf spot disease of rice. *Proc. 21st Indian Sci. Congr.* 297
156. Singh, K. 1968. Treatment of sugarcane setts against a few important diseases. *Proc. 1st Summer Inst. Plant Dis. Control,* 149
157. Singh, R. P., Verma, J. P. 1971. Reaction of genetic stocks to black arm of cotton (*Xanthomonas malvacearum*) (E. F. Smith) Dowson. *Indian Phytopathol.* 24:193
158. Singh, R. P., Verma, J. P. 1971. Control of bacterial blight of cotton. *Pesticides.* In press
159. Sreenivasan, T. N., Nariani, T. K. 1966. Studies on a mosaic disease of pea (*Pisum sativum* L.). *Indian Phytopathol.* 19:189–93
160. Srinivasan, M. C., Thirumalachar, M. J., Patel, M. K. 1959. Bacterial blight disease of rice. *Curr. Sci.* 28:469–70
161. Srivastava, D. N., Rao, Y. P. 1963. Bacterial blight of guar. *Indian Phytopathol.* 16:69–73
162. Srivastava, D. N., Rao, Y. P. 1968. Epidemiology and prevention of bacterial blight of rice in India. *Int. Rice Comm. Newslet.*, March 28–33
163. Srivastava, D. N., Rao, Y. P., Durgapal, J. C., Jindal, J. K. Singh, W. 1967. Screening rice varieties for resistance to bacterial blight. *Indian Farming* 17: 25–28
164. Subbayya, J., Raychaudhuri, S. P. 1970. A note on a mosaic disease of ragi (*Eleusine coracana*) from Mysore, India. *Indian Phytopathol.* 23:144–48
165. Summanwar, A. S., Raychaudhuri, S. P., Jagadish Chandra, K., Prakash, N., Lal, S. B. 1969. Virus associated with coconut root (wilt) disease. *Curr. Sci.* 38:208–10
166. Summanwar, A. S., Raychaudhuri, S. P., Phatak, H. C. 1966. Association of the fungus *Fusarium moniliforme* Sheld. with the malformation in mango (*Mangifera indica* L.). *Indian Phytopathol.* 29:227–28
167. Sundararaman, S. 1922. Helminthosporium disease of rice. *Bull. Agr. Res. Inst. Pusa* 128:7
168. Swarup, V., Raghava, S. P. S. 1970. Control of gladiolus diseases by aureofungin. *Hindustan Antibiot. Bull.* 12:63–65
169. Uppal, B. N. 1948. *Diseases of cotton in India.* Indian Central Cotton Comm., Bombay, 32 pp.
170. Uppal, B. N., Capoor, S. P., Raychaudhuri, S. P. 1944. 'Small leaf' disease of cotton. *Curr. Sci.* 13:284–85
171. Uppal, B. N., Kamat, M. N. 1928. Artificial infection of *Pennisetum typhoideum* by *Sclerospora graminicola*. *Agr. J. India* 23: 309–10
172. Uppal, B. N., Patel, M. K., Nigam, B. G. 1946. Bacterial blight of French beans in Bombay Province. *Proc. Nat. Inst. Sci.* (India) 12:351–59
173. Varma, A., Chenulu, V. V., Ray-

chaudhuri, S. P., Prakash, N., Rao, P. S. 1969. Mycoplasma like bodies in tissue infected with sandal spike and brinjal little leaf. *Indian Phytopathol.* 22: 289–91

174. Varma, A., Raychaudhuri, S. P., Lele, V. C., Ram, A. 1969. Towards the understanding of the problem of mango malformation. *Int. Symp. Mango & Mango Culture,* Indian Council Agric. Res. 31 (Abstr.)
175. Varma, A., Raychaudhuri, S. P., Lele, V. C., Ram, A. 1971. Preliminary investigation on epidemiology and control of mango malformation. *Proc. Ind. Nat. Sci. Acad.* In press
176. Varma, P. M. 1955. Ability of the whitefly to carry more than one virus simultaneously. *Curr. Sci.* 24:317–18
177. Varma, P. M., Capoor, S. P. 1958. Mosaic disease of cardamom and its transmission by the banana aphid (*Pentalonia nigronervosa* Coq.). *Indian J. Agr. Sci.* 28:97–108
178. Vasudeva, R. S., Azad, R. N. 1949. Late blight of potatoes in the plains. *Indian Farming* 10: 345
179. Vasudeva, R. S., Capoor, S. P. 1958. Outbreaks and new records. India. Citrus decline in Bombay State. *FAO Plant Prot. Bull.* 6:91
180. Vasudeva, R. S., Hingorani, M. K. 1952. Bacterial disease of wheat caused by *Corynebacterium tritici* (Hutchinson) Bergey et al. *Phytopathology* 42:291–92
181. Vasudeva, R. S., Nariani, T. K. 1952. Host range of bottle gourd mosaic virus and its inactivation by plant extracts. *Phytopathology* 42:149–52
182. Vasudeva, R. S., Raychaudhuri, S. P. 1954. Guava disease in Pushkar valley and its control. *Indian Phytopathol.* 7:78
183. Vasudeva, R. S., Samraj, J. 1948. A leaf curl disease of tomato. *Phytopathology* 38:364–69
184. Verma, G. S. 1952. The formation of lesions by gases on mango fruits. *J. Ind. Bot. Soc.* 31:316–41
185. Verma, J. P. 1964. Role of enzymes in pathogenicity. *Bull. Bot. Soc.* Bengal, 18:149–56
186. Verma, J. P. 1971. Studies in the enzyme make-up of *Alternaria* IX. Nucleophosphatase activity. *Ind. Phytopathol.* 24:43–49
187. Verma, J. P., Singh, R. P. 1970. Quantitative estimation of the degree of susceptibility of certain popular varieties of cotton to black arm disease. *Sci. Cult.* 36:565–67
188. Verma, J. P., Singh, R. P. 1970. Two new races of *Xanthomonas malvacearum,* the cause of blackarm of cotton. *Cotton Grow. Rev.* 49:203–05
189. Verma, J. P., Singh, R. P. 1971. Busan: an effective chemical to control bacterial blight of cotton. *Cotton Grow. Rev.* 48:60–62
190. Verma, J. P., Singh, R. P. 1971. Epidemiology and control of bacterial blight of cotton. *Proc. Ind. Nat. Sci. Acad.* In press
191. Verma, J. P., Singh, R. P. 1971. Problems in controlling bacterial blight of cotton. *2nd Int. Symp. Plant Pathol.,* New Delhi, 109–10
192. Yaraguntaiah, R. C., Keshavamurthy, K. U. 1969. Transmission of the virus component of the ragi disease complex in Mysore. *Plant Dis. Reptr.* 53:361–63
193. Yaraguntaiah, R. C., Nariani, T. K. 1963. Bean mosaic virus in India. *Indian J. Microbiol.* 3: 147–50

THE IMPACTS OF THE SOUTHERN CORN LEAF BLIGHT EPIDEMICS OF 1970–1971[1] **3542**

A. J. ULLSTRUP

Department of Botany and Plant Pathology, Purdue University
Lafayette, Indiana

INTRODUCTION

Corn (*Zea mays* L.) ranks third in value among the cultivated crops of the world; it is exceeded only by wheat and rice. In the United States both the land area planted to corn and its total yield of grain are greater than for any other cultivated crop. About 80% of this crop is fed to livestock; the remainder is used in industry and for human food.

Corn generally has been considered a relatively healthy crop. On the assumption that the amount of pathological research done on a crop can be taken as a measure of its general health, I counted the number of abstracts concerned with diseases of various crops in the *Review of Applied Mycology*. The total number published in the years 1935, 1940, 1950, 1960, and 1969 was 804 abstracts on diseases of corn compared to 1967 on diseases of wheat, 1554 on diseases of tobacco, and 1365 on diseases of tomato. Using a similar index (pages published about diseases of various crops divided by the annual value of the crop in the U.S.), Stevens (65) calculated the following values for various grain crops: rice 4.9, barley 3.5, wheat 3.4, sorghum 2.3, oats 1.8, rye 1.5, and corn 0.8. He attributed the small index values for rye and corn to the fact that they are mainly cross-pollinated whereas the other five crops listed are mainly self-pollinated. The implication from this correlation is that diversity in a crop is a good general defense against disease. As shall be seen shortly, the lesson implicit in this correlation, which was published in 1939, had to be learned all over again in 1970.

Corn is susceptible to many fungus pathogens, to a lesser number of pathogenic bacteria, nematodes, and viruses, to at least one mycoplasma-like agent, and to one parasitic higher plant. Past epidemics of several corn diseases have stimulated searches for sources of genetically-controlled resistance. These sources of resistance were then incorporated into hybrid corn varieties of good agronomic quality. Examples of such epidemics include those of the northern corn leaf blight caused by *Helminthosporium turcicum* in 1939–43, maize dwarf mosaic virus in the 1960s, corn stunt—a disease caused by a mycoplasma-like entity—in the 1960s, and Stewart's wilt—a bacterial disease of sweet corn. Fortunately, none of these diseases became suffi-

[1] Journal Paper No. 4620, Purdue Agricultural Experiment Station.

ciently severe or widespread to have a significant effect on the corn markets of the United States. All of these diseases have been controlled or potentially can be controlled through genetic resistance. In 1970, corn was considered so healthy a crop that the United States Department of Agriculture was supporting no full time corn pathologists in the Corn Belt.[2]

Since both the relative healthiness of corn prior to 1970 and its hypersusceptibility to a disease in 1970 were determined largely by the techniques of corn breeding, certain aspects of this subject will be reviewed briefly.

Production of Hybrid Seed Corn

When hybrid seed corn was first produced on a commercial scale 30–35 years ago, the fields in which such seed was grown were usually planted so that 6 rows of the seed-parent (female) alternated with 2 rows of the pollen-parent (male). Tassels on plants in the seed rows were removed before shedding pollen. Such detasseled plants would then be pollinated by pollen only from the pollen-parent plants. This insured the desired cross and eliminated self-fertilization of seed-parent plants.

Occasionally a rare corn plant is found that produces no viable pollen. This condition, called male sterility, has obvious advantages in corn breeding —it simplifies the production of hybrid seed by eliminating the necessity of detasseling (29, 30). Several years ago a male-sterile plant was found in Texas (57); its sterility was found to be governed by as yet unknown factors contained in the cytoplasm rather than by genes contained in the nuclei of the plant. This source of male sterility, designated as Texas cytoplasm male sterility (Tcms), was transferred into many inbred lines of corn by backcrossing, usually 6–8 times, with pollen of the inbred lines. Its progeny are all male-sterile. If such seed were planted by a farmer there would be no pollen in his field to effect fertilization and hence no seed would be produced on the ears. To avoid this the farmer may buy a blend of Tcms-seed and normal-cytoplasm seed that produces plants bearing fertile pollen. These pollen-fertile plants provide sufficient pollen in the farmer's field to fertilize all male-sterile plants and thus produce a seed set.

Another way of producing hybrid seed is to allow the Tcms plants in the seed-parent rows to be fertilized with pollen from plants in the pollen-parent rows that contain a *genetic* factor (a *restorer gene,* designated as Rf_1) that is capable of restoring fertility to the progeny of the male-sterile seed parent (31). In a cross of this kind—Tcms seed-parent X Rf_1-containing pollen-parent—all of the progeny are male-fertile, and shed functional pollen. Thus there is no need for the farmer to buy a blend to insure pollination and hence a seed set.

The restoration of fertility by the gene Rf_1 has no apparent effect on susceptibility of Tcms-corn to Race T of *H. maydis*. This suggests that cyto-

[2] The Corn Belt of the United States includes the states of Iowa, Illinois, Indiana, Ohio, Minnesota, Nebraska, Missouri, South Dakota, and Wisconsin.

plasmically-controlled sterility and susceptibility to this pathogen may be determined by different mechanisms.

By 1970, the use of inbred lines and hybrids containing Tcms was adopted so widely that about 85% of the hybrid seed corn produced in the United States was of this type.

When Tcms-corn was first introduced into production of hybrid seed, it obviously could not be foreseen that this type of cytoplasm carried with it hyper-susceptibility to a then unknown physiologic race of *Helminthosporium maydis,* the causal agent of southern corn leaf blight (SCLB). The first evidence of this hyper-susceptibility was reported in the Philippines in 1961 (48). This was confirmed and reported twice in 1964 and again in 1965 (1, 71, 72). It should be pointed out that in one of these reports (71) corn rust (the specific pathogen not indicated) was equally pathogenic on Tcms corn and normal corn. These observations were made in the "dry season" of 1961–62 and 1962–63, and SCLB was not severe enough to differentiate between the 2 types of cytoplasm. In 1969, it was also discovered that Tcms-corn was hyper-susceptible to another fungus disease—yellow leaf blight caused by *Phyllosticta maydis* (3, 58, 59).

As is recognized now, the widespread use of Tcms corn was potentially hazardous. The necessary combination of the physiologically specialized race of the pathogen and favorable weather developed in 1970. The result was not a merely serious epidemic of SCLB but one of the most damaging and widely dispersed epidemics in the history of plant pathology. The amount of food energy lost to disease was many times larger than that lost during the historic famine-producing epidemic of potato blight in the 1840s.

Origin of the Epidemic in Retrospect

In 1969 a disease of corn leaves and ears was observed in a few localized areas in southern Iowa, Illinois, and Indiana. In October, of that year, I examined infected ears from a large seed field in southeastern Iowa. The ears showed a grayish-black rot. Morphology of the conidia indicated that *H. maydis* was involved. The Tcms-plants in this field apparently had become badly infected sometime between late August and late September. Ears and stalks were rotted and much lodging resulted. Plants of normal cytoplasm in this field showed no ear infection and only very few leaf lesions.

The same disease was reported in fields of dent corn in southern Iowa.[3] Here again, the disease appeared only in fields of Tcms corn. The disease was reported in southern Illinois (60), and I observed it in seed harvested in three widely separated fields in southern Indiana.

Pure cultures of the pathogen were isolated and inoculated onto seedlings of Tcms and normal-cytoplasm versions of several inbred lines. These tests demonstrated that the pathogen was a distinctive physiologic race of *H. maydis.* It vigorously attacked seedlings containing Tcms but caused only mild

[3] Verbal communication from C. A. Martinson, Iowa State University.

infection of plants with normal-cytoplasm. Parallel inoculations with an isolate of the common physiologic race showed no evidence of differential parasitism—both versions were equally susceptible (38). The two races were described and designated "Race T" for the new race that was highly virulent on Tcms corn and "Race O" for the old race of *H. maydis* that had been recognized for years in many regions of the world (61).

Descriptions and illustrations of symptoms caused by Race T of *H. maydis* on Tcms corn and on the resistant normal-cytoplasm corn have been published (38, 68). This race also produces a toxin that is host-specific, i.e., it affects Tcms corn but not the normal-cytoplasm, resistant corn (50). Race O, in addition to its lack of differential parasitism on Tcms and normal-cytoplasm versions of near-isogenic inbred lines, attacks ears only rarely (4, 56). It is primarily a leaf-infecting pathogen.

The origin of Race T is still not fully understood. Recent isolations from stored corn grown in Iowa in 1968 indicate that Race T was present in that state as early as 1968 (36). Based on studies of mating types, it has been postulated that Race T was either introduced into the Corn Belt, or arose there by mutation, within a few years prior to 1968 and that it was transported from the Corn Belt into the southern United States on infected seed corn (47). This race may have been inadvertently introduced into this country in seed by direct or indirect routes from parts of the world where it already existed. Race T has now been observed or identified in Asia, Africa (25), Europe, North, Central, and South America.[4]

The Epidemic of 1970

In January of 1970, corn pathologists began to receive reports and specimens from Florida indicating that a leaf disease was causing severe damage on Tcms corn. Leaf lesions were numerous, and decay of the ears, infection of the kernels, and rotting of the stalks also were observed. Very soon, the same unusual symptoms also were observed in southern Alabama and Mississippi (51). Here, too, the hyper-susceptibility of Tcms corn was evident and the ears and stalks were attacked.

The spring of 1970 was unusually wet in much of the southern United States. In the Corn Belt the weather during the summer was also favorable for rapid development of SCLB. As a result, the disease progressed northward from the South into the Corn Belt states (6, 51). By mid-July it was well established in southern Illinois and Indiana, but still not in severe proportions.

Although the disease continued to spread and intensify, the potential destruction it was to create in the Corn Belt was not apparent, or expected, even as late as the end of July when much of the corn in the central Corn Belt was in full silk or even a week beyond this stage. In Indiana, it was not until the first week of August that the seriousness of the epidemic finally be-

[4] Personal observations and identifications.

came evident. The disease became very severe in Illinois, eastern Iowa, Indiana, southern Ohio, and parts of Missouri (5). It was present, although less severe, in Minnesota (27), Wisconsin, Michigan, southern Ontario Canada, and eastward to southern New England. In 1970 SCLB was reported in nearly every state east, and several states west of the Mississippi River.

In the South many fields of corn were a total loss; some were plowed under and planted to other crops when it became apparent that the corn crop would fail (62). In some regions of the Corn Belt, yields per acre were reduced by 50%. The standard test weight for marketing of grain (56 pounds per bushel) was reduced to 45–50 pounds. Not only were yield and weight of grain reduced, but the ears and stalks of corn continued to rot until harvest. Much additional grain was also lost because lodging made harvesting difficult. Fungicidal sprays were used in some areas of the Corn Belt; in most instances, however, sprays were applied too late to be effective.

Impacts.—The seriousness of SCLB epidemic was reflected in marked increases in the price of corn during August, 1970. The price of corn "futures" on the Chicago grain market, the largest market dealing in this commodity in the United States, soared from about $1.35 per bushel in late July to $1.68 per bushel by September 20 (8, 9, 13, 14, 75–82). This is the first time a disease had seriously affected the price of corn. These dramatic increases in price also had their effects abroad—British farmers shifted from corn to barley as livestock feed; as they increased demand for barley, its price also increased.[5]

Many farmers became concerned over the possible hazard of feeding infected grain or corn forage to livestock. As soon as the disease became severe in the South, tests were undertaken at the State Experiment Stations in Florida, Georgia, Alabama, and Mississippi. Fortunately, no toxicity symptoms were observed in a variety of livestock. Similar tests at Purdue University confirmed these results (37). Had the diseased corn been toxic to farm animals SCLB could have had an even more serious effect on livestock farmers.

The epidemic of SCLB stimulated an immense amount of publicity (10, 11, 12, 14, 74, 83). No plant disease in the United States has ever received so much attention from the press and other communications media. During the period from August to November 1970, the Chicago Tribune published 37 articles on the disease. The New York Times, the Wall Street Journal, and many of the other influential daily newspapers and weekly news magazines kept the public aware of the gravity of the disease (67, 70). Local papers serving small communities in the Corn Belt devoted much space to the disease and its development (32, 34). Radio and television programs frequently gave descriptions of the SCLB situation. Many bulletins and circulars on the disease were issued by Agricultural Experiment Stations (55, 64, 66).

[5] Verbal communication from Sir Frederick C. Bawden, Rothamstead Experiment Station.

A closed-circuit television broadcast was made from Purdue University on August 25, 1970. With technical supervision from the Department of Electrical Engineering, faculty members from the departments of Botany and Plant Pathology, Agronomy, Agricultural Engineering, Agricultural Economics, and Animal Sciences explained the nature of the disease, the factors responsible for its prevalence and severity, and how to cope with it. The program was beamed to receiving stations throughout Indiana. At the end of the program, answers to questions received by telephone were televised back to the audiences (63).

Numerous conferences and meetings were held throughout the epidemic area. In October 1970, a symposium was held by the American Phytopathological Society during its meeting in Hot Springs, Arkansas. Here, Agricultural Experiment Station personnel and representatives from the seed-corn industry met to review the nature and effects of the epidemic and to project potential seed supplies for 1971 (49). At this meeting a voluntary system for labeling of corn seed was announced by the vice president of the American Seed Trade Association (73). Each bag of seed would be labeled conspicuously with the letter "T" for Tcms-cytoplasm, "B" for blends of normal- and Tcms-cytoplasm, and "N" for seed containing normal cytoplasm.

In December 1970, a national "Leaf Blight Informational Conference" was called by Secretary of Agriculture, Clifford Hardin. A panel of speakers informed the Secretary, and about 150 large corn and livestock farmers from the South and the Midwest, about the prospects of a renewed epidemic in 1971, the projected seed supply, the possible effects on the price of corn, and the expected performances of different types of seed. President Nixon's presence at the meeting was indicative of the impact of the disease on the nation's agriculture.

Losses due to the epidemic were officially estimated at nearly one billion dollars in the nation as a whole. In the South, reductions in yield were generally much greater than in the Corn Belt. In parts of southern Illinois and Indiana, losses of 50–100% were sustained. The average loss in yield in these States was 20–30%.

Storage and milling problems with infected corn were anticipated but did not prove to be serious. When properly dried, infected corn stored as well as healthy corn. Milling difficulties were overcome by blending infected with noninfected corn.

During the harvest period, there were several reports of farm workers who suffered or died, as a result of inhaling the spores that rose in clouds around picking machines working in fields of heavily infected corn (33, 35, 42). To my knowledge none of the reports of death caused by the fungus were confirmed. Respiratory difficulties, including symptoms of asthma and hay fever as well as some skin irritations were attributed to the fungus. Inhalation of spores can be prevented by wearing a respirator and thorough washing usually will allay skin irritation. There is at least one report in the literature of lung infection by an undetermined species of *Helminthosporium* (28), but this appears not to be *H. maydis* or a plant-pathogenic species.

During the winter, many dire predictions of failure were made for the 1971 corn crop. A report was circulated that normal-cytoplasm corn died soon after planting in soil taken from a field where the blight was severe in 1970. Although the cause of death was not reported, this story received a great deal of publicity in northwestern Indiana. Reports were also circulated about the susceptibility of wheat and oats to Race T of *H. maydis* (15). This also caused a flurry of concern among farmers. To date there has been no substantiated record that Race T of *H. maydis* can cause disease in these crops under field conditions.

During the late summer of 1970, the National Aeronautics and Space Administration conducted a study to determine the feasibility of detecting SCLB from aircraft at various altitudes. Although these experiments were begun well after onset of the disease, different levels of disease severity could be distinguished from the air using appropriate cameras and other devices for measurement of spectral reflectance.

In November 1970, SCLB was observed in Brazil but the disease did not become as severe as it had in the United States. The Brazilian crop included only about 20% Tcms corn. Also, weather conditions were less favorable than in the United States. At considerable financial sacrifice, seed producers in Brazil withheld their Tcms-corn from the market in 1971 and began at once to multiply normal-cytoplasm seed stocks in the northern part of their country.

Seed supplies for the 1971 crop.—During the late winter and early spring of 1971, seed supplies in the United States were estimated to consist of about 25% normal-cytoplasm hybrids, 25% Tcms hybrids and 40% blends. The remainder consisted of open-pollinated varieties and normal cytoplasm F_2 seed of single- and double-cross hybrids harvested out of production fields in 1970.

The demand for normal-cytoplasm seed was far beyond the supply. To offset this deficit partially, many seed producers increased their normal-cytoplasm hybrids and inbred lines during the winter months of 1970–71 in Hawaii (7), Florida, Central and South America, and in some of the Caribbean Islands. Some of these plantings were successful; others were not. Seed was also imported from Argentina, Hungary, and Yugoslavia (43). Some of these hybrids were made up, in part, of inbred lines that were originally developed in the Corn Belt of the United States. These generally performed satisfactorily, but some were not well adapted and yields of seed were poor.

Prices for 50 pounds of seed corn in 1971 ranged from $20 to $30 for seed with Tcms-cytoplasm, $28 to $32 for blends, and $30 to $39 for normal-cytoplasm seed. Prices varied among seed producers and also among hybrids, depending on their performance record. The extreme shortages of normal-cytoplasm seed stimulated some stealing of seed supplies (16) and the unscrupulous purchase of empty "N" bags which were then refilled with Tcms seed or low performance F_2 seed (44–46). Such abuses were few, in part because they were given very wide publicity. Regulations regarding labeling of hybrid seed were established in the States of North Carolina, Mississippi, and Illinois. The Alabama Department of Agriculture prohibited the planting of Tcms corn in 1971.

The Epidemic of 1971

Race T of *H. maydis* survived the winter of 1970–71 in most of the South and as far north as South Dakota, Minnesota, Wisconsin, and Michigan. Tests conducted in 10 mid-western States indicated the pathogen remained viable in corn debris on the soil surface (69). With the exception of a test in South Dakota, the fungus failed to survive in *buried* refuse in the Corn Belt.

Successful overwintering of the fungus portended an early onset of the disease in 1971. Since weather conditions are important in determining the occurrence of SCLB a great deal of attention was paid to the weather during the early part of the growing season. Weather conditions in the early spring and summer were dry in much of the South. In the Corn Belt, localized infections appeared in May and June (19, 20) on volunteer corn that had sprouted in fields where the disease was severe in 1970 and debris remained on the surface. From such infected debris and diseased volunteer corn, the pathogen spread to young planted corn. Localized epidemics developed when cribs of infected corn were emptied and shelling operations began, or where shelled corn was moved from bins. The spores were released from the infected corn and carried by air currents to nearby fields where young plants became infected.

By June, the disease had become established in much of Illinois, Indiana, eastern Iowa, parts of Missouri, Kansas, and Ohio. As in 1970, it was less severe in Minnesota, Nebraska, Wisconsin, Michigan, and Pennsylvania.

The National Aeronautics and Space Administration launched an extensive program of overflights in seven Corn Belt states. Flights were made in a definitely patterned grid. Photographic and spectral reflectance measurements made from the air were compared with ground observations collected by crews who recorded the condition of the corn, i.e., size and stage of growth, disease development, nutrient deficiencies, and other factors that were thought to affect spectral reflectance. Intensive overflights at an altitude of 60,000 feet were made in a strip along the western side of Indiana (40).

Epidemiological studies were made in several states. These data and general observations on the progress of the disease were transmitted to a data-collecting center at Ames, Iowa. Here they were assembled and dispersed to research and extension personnel over the area where the disease was present. A "Blight Information Center" in the U. S. Department of Agriculture kept in constant communication with observers in the Corn Belt, and was thus able to disseminate current information on the status of the disease. A "Dixie Early Warning" (DEW) system was established to relay current information on the disease in the South. A computerized epidemiological study called "Epimay" was undertaken to evaluate the influence of various meteorological factors on development of the disease. It was anticipated that short-term forecasts of the disease could be made from this study.

Fortunately, the overall severity of the disease in 1971 did not equal that in 1970. Some fields in the southern and south-central reaches of the Corn Belt were damaged severely, but the weather in July and August was generally cool. Many record-low temperatures (42–52°F [5.5–11.0°C] night temperature) were reported in the central Corn Belt. These conditions were not favorable for rapid reproduction of the pathogen. Also much more of the resistant, normal-cytoplasm seed was planted than had been anticipated from pre-season estimates. Finally, it is possible that so little Tcms-corn was planted in the South that this eliminated the possibility of massive amounts of inoculum being carried into the Corn Belt.

Impacts in 1971.—When it was reported that the pathogen could overwinter in the Corn Belt, and again when the disease first appeared in June, many predictions of another severe epidemic were made (17, 18, 41, 53). The corn "futures" market began a sharp rise from about $1.30 per bushel in late May to about $1.63 per bushel in late June (21, 22); thereafter, it declined steadily. The corn market during the summer of 1971 was very sensitive (24). Announcement of the approval of a new fungicide to control corn blight—"Citcop 4E"—caused a 3- to 4-cent drop in the price per bushel.

In spite of some losses in yield due to the disease, the cool weather in July and August, and some deficiencies in rainfall, the 1971 corn crop in the United States was a near-record (23, 54, 85, 86). The generally good growth of the crop coupled with an estimated 9% increase in acreage planted, made possible total production of about 5.4 billion bushels compared to about 4.1 billion bushels in 1970. The average yield in Indiana was about 97 bushels per acre in 1971 compared to 74 in 1970, and 81 in 1969 (26). Not only was the production of commercial grain excellent, but the supply of resistant normal-cytoplasm seed corn was increased sufficiently to meet the demand for 1972 planting. Some early published price lists for 50 pounds of corn seed ranged from about $13 to $31, most of the variation being determined by the particular hybrid and the seed producer.

Future Control of SCLB

In the years immediately ahead it is expected that SCLB will be controlled by a complete change-over from Tcms-corn to normal-cytoplasm corn. Probably no catastrophic plant disease of an important crop has ever been brought under such relatively certain control so rapidly over so wide an area. This was accomplished quickly because susceptibility was inherent in the Tcms-corn, and the resistant, normal-cytoplasm corn was available and could be increased enough to meet demands for seed within a period of one year. During 1971, the rows of "seed parent" plants in seed-production fields were detasseled, as had been done before Tcms-corn came into use. This provided summer employment for many thousands of teenage boys and girls who pulled tassels in the fields of seed corn. Tassels were also removed with various types of cutting equipment.

The critical year for supplies of normal-cytoplasm seed stocks was 1971. Had susceptibility to this disease been determined by nuclear instead of cytoplasmic factors, the change-over to resistant seed would have taken much longer.

Controlled and replicated field experiments with fungicidal sprays were conducted by Purdue University during 1971. They showed significant reductions in severity of the disease and a corresponding increase in yield with applications of Dithane M-45, Manzate 200, and Citcop 4E. The first two fungicides had been approved earlier for use on corn by the Food and Drug Administration. Especially in Indiana, thousands of acres of corn were sprayed with fixed-wing aircraft and helicopters (39).

In the future, disease resistant sources of male-sterility may be employed. Such sources are at hand, but their usefulness and potential hazards must be examined very carefully. Chemical gametocides also may become useful in hybrid seed production, but these, too must be tested thoroughly. So far no chemical has been found that will destroy or sterilize pollen with the thoroughness needed for large scale seed production.

Conclusions

The effect of SCLB on the agriculture of the United States in 1970 is now history. But history should provide guidance for future decisions that will enable plant pathologists and plant breeders to *prevent* future epidemics of plant disease. It is good that we understand much about why and how this epidemic developed.

In terms of human suffering, it is particularly fortunate that this epidemic occurred in a developed nation with a highly diversified agriculture. This disease cut deeply into the agricultural economics of the South and the Corn Belt of the United States. But it did not cause the widespread malnutrition, starvation, and great emigration of people that followed the potato blight epidemics in Ireland in 1845 and 1846. Emerging and developing nations, especially those whose agriculture is dependent on only a few crops, will be wise to learn from the example of the SCLB epidemics in the United States how important it is to diversify their agriculture and maintain adequate genetic (and cytoplasmic) diversity in their major crops.

This is the first and most important lesson to be learned. Never again should a major cultivated species be molded into such uniformity that it is so universally vulnerable to attack by a pathogen, an insect, or environmental stress. *Diversity must be maintained in both the genetic and cytoplasmic constitution of all important crop species.* Sorghum, a major crop closely related to corn, is now in almost exactly the same vulnerable condition corn was in the winter of 1969–70. Very limited sources of male sterility have been incorporated into sorghum as well.

The National Academy of Science has appointed a special committee to study the genetic vulnerability of crops and to make specific recommendations to minimize this vulnerability (84). This report (52) should be studied thoroughly by every plant pathologist and plant breeder. Another report emphasizing the need for genetic and cytoplasmic diversity in our crops as a safe-guard against diseases and other hazards (2) has recently been published.

The second major lesson to be learned has to do with cooperation among scientists within and between disciplines, and with the way that agricultural research is organized, its results communicated, and incentives provided for scientific achievement. The epidemics of SCLB in 1970 and 1971 stimulated an extremely rapid mobilization of industrial, Federal, State, and University research, and extension personnel. The efforts of the seed-corn industry in mobilizing to increase resistant seed stocks are especially commendable. Scientists and educators in various disciplines within all of these agencies worked together, as never before, to do what could be done to control the disease and disseminate information about it. It was gratifying to see and to participate in this kind of teamwork among scientists whose traditional loyalties and incentives for achievement usually reside within their particular discipline, company, or research agency. Having once achieved this degree of coordination in both the public and private sectors, and particularly between plant pathologists and plant breeders, every effort should be made to insure that similar cooperation will continue and occur more easily in the future. Coordinated vigilance for the safety of our crops is essential if future epidemics of this kind are to be prevented.

The recent epidemics of SCLB in the United States point up the importance of the science of plant pathology in the economy of this nation and the world. They also emphasize the need for constant vigilance over the health of crops throughout the world. This, in turn, will require even closer cooperation among nations.

LITERATURE CITED

1. Aala, F. T. 1964. The corn leaf blight disease: A problem in the production of hybrid corn seed involving male sterility. *Philippine J. Plant Ind.* 29:115–22
2. Adams, M. W., Ellingboe, A. H., Rossman, E. C. 1971. Biological uniformity and disease epidemics. *Bioscience* 21:1067–70
3. Ayers, J. E., Nelson, R. R., Coons, C., Scheifele, G. L. 1970. Reaction of various inbred lines and single crosses in normal and male-sterile cytoplasm to the yellow leaf-blight organism (Phyllosticta sp.). *Plant Dis. Reptr.* 54:277–80
4. Burton, C. L. 1968. Southern corn leaf blight on sweet corn ears in transit. *Plant Dis. Reptr.* 52: 847–51
5. Chicago Tribune, August 6, 1970. New fungus periling midwest corn crop
6. Chicago Tribune, August 14, 1970. Corn fungus peril still grows

7. Chicago Tribune, August 20, 1970. Corn flown to Hawaii in face of blight
8. Chicago Tribune, August 20, 1970. Price of September corn up limit of 8 cents
9. Chicago Tribune, August 26, 1970. Indiana blight boosts corn
10. Chicago Tribune, September 15, 1970. Illinois farm chief hits U.S. corn study
11. Chicago Tribune, September 28, 1970. Rains add woes to corn crop
12. Chicago Tribune, September 30, 1970. Corn blight farmers' main topic of talk
13. Chicago Tribune, October 7, 1970. Ten counties' corn is cut 21%
14. Chicago Tribune, October 9, 1970. Corn yield goes down, down
15. Chicago Tribune, November 8, 1970. Report corn leaf blight may damage oats, wheat
16. Chicago Tribune, February 21, 1971. Bizarre corn thefts kindled by fear of seed shortage
17. Chicago Tribune, April 19, 1971. Prepare to battle corn blight
18. Chicago Tribune, April 19, 1971. Corn blight expected again this year
19. Chicago Tribune, May 25, 1971. Find blight in 3 Illinois counties
20. Chicago Tribune, May 30, 1971. Southern Illinois farm first hit by blight in '71
21. Chicago Tribune, June 11, 1971. Corn, wheat futures soar on blight news
22. Chicago Tribune, July 2, 1971. Puts corn blight, not steel pact, as inflation key
23. Chicago Tribune, August 1, 1971. 1971 could see bumper corn crop —despite blight
24. Chicago Tribune, August 22, 1971. Blight or not, corn off 5 cents
25. Craig, J., 1971. Occurrence of Helminthosporium maydis Race T in West Africa. *Plant Dis. Reptr.* 55:672–73
26. Crop Production, 1971. U.S.D.A. Statistical Reporting Service, Crop Reporting Board, January, 1972
27. Crozier, W. F., Braverman, S. W. 1971. Helminthosporium maydis in seeds of Minnesota-grown field corn. *Phytopathology* 61: 427–28
28. Dolan, C. T., Weed, L. A., Dines, D. E. 1970. Bronchopulmonary helminthosporiosis. *Am. J. Clin. Pathol.* 53:235–42
29. Duvick, D. N. 1959. The use of cytoplasmic male-sterility in hybrid seed production. *Econ. Bot.* 13: 167–95
30. Duvick, D. N. 1965. Cytoplasmic pollen sterility in corn. *Advan. Genet.* 13:1–56
31. Duvick, D. N., Snyder, R. J., Anderson, E. G. 1961. The chromosomal location of Rf_1, a restorer gene for cytoplasmic pollen sterile maize. *Genetics* 46:1245–52
32. Evansville (Ind.) Courier, August 14, 1970. Corn leaf blight is critical
33. Evansville (Ind.) Courier, September 15, 1970. Experts say blight, illness not linked
34. Evansville (Ind.) Courier, September 24, 1970. Corn price star of farm outlook
35. Evansville (Ind.) Courier, September 21, 1970. Corn blight suspected in farmer's illness
36. Foley, D. C., Knaphus, G., 1971. Helminthosporium maydis Race T in Iowa in 1968. *Plant Dis. Reptr.* 55:855–57
37. Foster, J. R. 1971. Feeding blighted corn to hogs. AS-394, Cooperative Extension Service, Purdue University
38. Hooker, A. L., Smith, D. R., Lim, S. M., Beckett, J. B. 1970. Reaction of corn seedlings with male-sterile cytoplasm to Helminthosporium maydis. *Plant Dis. Reptr.* 54:708–12
39. Indianapolis News, June 24, 1971. Corn farmers wonder, "to spray or not?"
40. Indianapolis News, June 24, 1971. Hoosier cornfields to be on camera
41. Indianapolis Star, June 13, 1971. Unemployment, blight danger grow. Increase in food production brings about new problems
42. Kokomo Tribune, September 18, 1970. Two deaths, illness linked to harvesting blighted corn
43. Lafayette (Ind.) Journal & Courier, March 2, 1971. Imported seed corn headed for Kentland
44. Lafayette (Ind.) Journal & Courier, March 4, 1971. Seed corn bags are "bootlegged"
45. Lafayette (Ind.) Journal & Courier, March 5, 1971. False-label

seed peddlers reported seeking profit on corn blight fears
46. Lafayette (Ind.) Journal & Courier, March 12, 1971. Seed corn bag-tag warning out
47. Leonard, K. J. 1971. Association of virulence and mating type among Helminthosporium maydis isolates collected in 1970. *Plant Dis. Reptr.* 55:759–60
48. Mercado, A. C., Lantican, R. M. 1961. The susceptibility of cytoplasmic male-sterile lines of corn to Helminthosporium maydis Nisikado & Miy. *Philippine Agr.* 45:235–43
49. Miller, P. R. (Ed.) 1970. Southern corn leaf blight—Special issue. *Plant Dis. Reptr.* 54:1099–1136
50. Miller, R. J., Koeppe, D. E. 1971. Southern corn leaf blight: Susceptible and resistant mitochondria. *Science* 173:67–69
51. Moore, W. F., 1970. Origin and spread of southern corn leaf blight in 1970. *Plant Dis. Reptr.* 54:1104–08
52. National Academy of Science, National Research Council. 1972. Genetic Vulnerability of Major Crops. Washington, D. C. 250 pp.
53. New York Times, April 18, 1971. A triumph of genetics threatens disaster
54. New York Times, May 30, 1971. Midwestern corn farmers expect record corn crop, barring blight
55. Robbins, P. R. February 20, 1971. The southern corn leaf blight—some economic considerations. *Economic and Marketing Information for Indiana Farmers.* Purdue University, Lafayette, Ind.
56. Robert, A. L. 1956. Helminthosporium maydis on sweet corn ears in Florida. *Plant Dis. Reptr.* 40: 991–95
57. Rogers, J. S., Edwardson, J. R. 1952. The utilization of cytoplasmic male-sterile inbreds in the production of corn hybrids. *Agron. J.* 44:8–13
58. Scheifele, G. L., Nelson, R. R. 1969. The occurrence of Phyllosticta leaf spot of corn in Pennsylvania. *Plant Dis. Reptr.* 53:186–89
59. Scheifele, G. L., Nelson, R. R., Coons, C. 1969. Male-sterility cytoplasm conditioning susceptibility of resistant inbred lines of maize to yellow leaf blight caused by Phyllosticta zeae. *Plant Dis. Reptr.* 53:656–59
60. Scheifele, G. L., Whitehead, W., Rowe, C. 1970. Increased susceptibility to southern leaf spot (Helminthosporium maydis) in inbred lines and hybrids of maize with Texas male-sterile cytoplasm. *Plant Dis. Reptr.* 54:501–03
61. Smith, D. R., Hooker, A. L., Lim, S. M. 1970. Physiologic races of Helminthosporium maydis. *Plant Dis. Reptr.* 54:819–22
62. Southeast Farm Weekly (Knoxville, Tenn.) August 27, 1970. Blight blasts corn crop. Feed and seed shortage likely
63. Southern corn leaf blight in Indiana. A backgrounder. 1970. ID 76, Cooperative Extension Service, Purdue Univ.
64. Southern corn leaf blight. 1970. Univ. Nebraska.
65. Stevens, N. E. 1939. Disease, damage and pollination types in "grains". *Science* 89:339–40
66. The southern corn leaf blight puzzle. A 1971 management guide to minimize risks. 1970. Cooperative Extension Service, Univ. Ill.
67. Time, March 8, 1971. The farm plague
68. Ullstrup, A. J. 1971. Southern corn leaf blight and other corn leaf diseases. BP-5–19, 7 pp. Cooperative Extension Service, Purdue Univ.
69. Ullstrup, A. J. (compiler) 1971. Overwintering of Race T of Helminthosporium maydis in midwestern United States. *Plant Dis. Reptr.* 55:563–65
70. U. S. News & World Report. May 17, 1971. A threat to U. S. food supply
71. Villareal, R. L., Lantican, R. M. 1964. The effect of "T" cytoplasm on yield and other agronomic characters in corn. *Philippine Agr.* 48:144–47
72. Villareal, R. L., Lantican, R. M. 1965. The cytoplasmic inheritance of susceptibility to Helminthosporium leaf spot in corn. *Philippine Agr.* 49:294–300
73. Walker, D. D. 1970. Labeling of 1971 seed corn. *Plant Dis. Reptr.* 54:1114–17

74. Wall Street Journal, August 17, 1970. Much of nation's crop of corn called periled by southern leaf blight
75. Wall Street Journal, August 18, 1970. Grains, soybeans rise daily limit on corn blight
76. Wall Street Journal, August 19, 1970. Blight kills off 25% of corn in Illinois, official says
77. Wall Street Journal, August 19, 1970. Grain, soybeans rise on Monday reversed by profit takers
78. Wall Street Journal, August 20, 1970. Corn-product prices are raised as blight imperils crop and pushes grain quotes up
79. Wall Street Journal, August 20, 1970. Corn, related futures resume their advance after Tuesday declines
80. Wall Street Journal, August 24, 1970. Corn futures declines spark lower prices in other contracts
81. Wall Street Journal, August 25, 1970. Blight in corn fields may bring sharp rises in prices of meat, eggs
82. Wall Street Journal, August 26, 1970. Concern about blight raises corn futures as much as 6½ cents
83. Wall Street Journal, September 28, 1970. About 60% of seed corn set for '71 crop to be vulnerable to southern leaf blight
84. Wall Street Journal, March 15, 1971. Crop-failure danger due to new grains to get urgent study
85. Wall Street Journal, June 21, 1971. Corn blight or bumper crop in 1971? Many farmers expecting record harvest
86. Wall Street Journal, July 6, 1971. Growing conditions take turn for better spurring forecasts of larger crop yields

DOTHISTROMA BLIGHT OF PINUS RADIATA

3543

I. A. S. Gibson

Scientific Liaison Officer, Commonwealth Mycological Institute, Ferry Lane, Kew, Surrey, England

The exploding world population produces a constantly rising demand on plant resources, which include wood.

Pinus radiata D. Don has played a dramatic role in meeting this need. Its annual increment can be as high as 24 m/ha (hectare) and it is very easy and cheap to establish, so that it is an excellent plantation species. In its natural state *P. radiata* is confined to a small area of about 4000 hectares in the coastal region of California and the offshore island of Guadeloupe, where it is not outstanding in size or form. When introduced into a wide range of countries towards the end of the 19th century it generally grew much better than at home, and it has been adopted for large-scale commercial planting in Chile, New Zealand, Australia and, during later years, in East Africa.

Presently the most extensive estates of *P. radiata* are in Chile and New Zealand, where over a quarter of a million hectares in each country support lumber and paper industries. Over 100,000 ha of the species are found in Australia, largely distributed between New South Wales, Victoria, South Australia, and Tasmania; there are about 50,000 ha in Africa, divided mostly between the Cape region of South Africa and the highland areas of Kenya, Uganda, and Tanzania, and in the northern hemisphere *P. radiata* is planted over large areas in northern Spain.

The wisdom of planting extensive even-aged crops of a single forest species has often been questioned by the conservative forester. He would argue, rightly, that their uniform nature, with long rotational periods, carries greater risks of serious outbreaks of pests and diseases than are incurred by the farmer with his annual turnover. These risks are enhanced in most forest crops—and in *P. radiata* in particular—by the lack of information on their pests and diseases, so that there is little chance to predict or evaluate them in advance.

Despite these circumstances, no disease problems of more than local importance occurred in the rapidly expanding forest plantations of the southern hemisphere from the time that they were started until about 15 years ago. Then the needle disease caused by *Dothistroma pini* (dothistromal blight) appeared and was soon recognized as potentially important to *P. radiata* at an international level. Since then an increasing volume of research has been directed mainly to an economic solution of the problem, with useful support from more basic investigations.

It is the first object of this paper to review the results of this work and to define the extent of our knowledge—or ignorance—of the disease. In addition, we shall try to identify mistakes that have been made and lessons learned, so as to be better prepared for similar emergencies in the future.

As happens frequently with pests and pathogens that achieve prominence quickly, mycologists still disagree on the correct nomenclature for the causal fungus. This will be discussed in a later section of the paper and, for convenience, the fungus will be referred to in the general text by its most widely-used name—*Dothistroma pini* Hulbary.

Further information on the botany and economics of *P. radiata* may be obtained from Scott's monograph (83) and the authors listed in his bibliography.

The History of the Disease

Recognition of *Dothistroma* blight as a critically important forest disease dates from 1957, when a necrotic condition of the foliage appeared in young *P. radiata* plantations of the Lushoto region of the Usumbara Mountains in Tanganyika (now Tanzania). However, the cause was not recognized until nearly 3 years later, partly because saprophytic fungi—in particular *Pestalotiopsis* spp—made the true pathogen hard to find on diseased foliage, and partly because insufficient attention was given to the problem. Indeed, for a brief but regrettable period the condition was known locally as '*Pestalotiopsis* Disease' and '*Pestalotiopsis* Needle Blight' (34), despite the fact that pathogenicity tests of several of these species were negative.

In 1960 blight spread from this very limited locality to the *P. radiata* plantations of the southeastern Aberdares in Kenya, where it was particularly favored by temperatures generally below 20°C and evenly distributed rainfall in the region of 2000 mm per annum. From this center it moved to other exotic plantation areas in East Africa, so that by 1964 all major plantings of *P. radiata* were infected (37).

The pathogen was finally recognized during this period of spread, first as *Actinothyrium marginatum* Sacc. (35) and later—more correctly—as *Dothistroma pini* Hulbary (36).

Blight also appeared and spread rapidly in the highland plantations of *P. radiata* in Nyasaland (now Malawi) and Southern Rhodesia (now Rhodesia) between 1960 and 1962, causing moderate to severe damage (6).

At this point it is necessary to recount the history of *P. radiata* as a major plantation species in Central Africa, in order to appreciate our present views on the history of blight on the Continent.

Before the second World War *P. radiata* was regarded as the most promising exotic softwood for plantation development in Nyasaland and Southern Rhodesia, and steps were taken to promote the planting of the species. However, about 1940 these young plantations were severely damaged by foliage disease that was believed to be a condition well known in South Africa, caused by the fungus *Diplodia pinea* (Desm.) Kickx, in association with hail damage and marginal site conditions. In view of South African experience any further planting of *P. radiata* was suspended in Central Africa and was not resumed until about 1955, following its success in the East African highlands.

The prompt and widespread appearance of blight in these new *P. radiata* plantations in Central Africa suggested strongly that the pathogen was already established in the region. This hypothesis received some support when the reports of the previous outbreak of disease in *P. radiata* of 1940 were reexamined and found to correspond in many ways with the field symptoms of dothistromal blight. In addition, herbarium material of diseased foliage which had been collected at that time and diagnosed as infected by a *Septoria* sp, was later found to be identical to *D. pini* in morphology and host symptoms (37). These observations strongly suggest that blight has been present in Central Africa since the early 1940s and that it spread to Tanzania in the following 15 years.

The outbreaks of blight in East and Central Africa resulted in further reduction in planting *P. radiata* with its substitution by other species not susceptible to the disease.

Christensen recorded *D. pini* in South Africa in 1965, on necrotic foliage of *P. canariensis* in a small plantation near the Cape (40) but he failed to find the fungus when he revisited the same locality in 1966 and 1967. This record stimulated a search through pine foliage collections at the Plant Protection Institute, Pretoria, by Miss B. A. Louwrens, for possible cases of blight that might have been collected but overlooked or misidentified. None were found and, to date, Christensen's record appears to be the only one from South Africa, despite a careful watch kept for the disease by G. C. A. van der Westhuizen[1] and C. S. Moses.[2]

Dothistroma blight has been confirmed from Swaziland in nursery stock of *P. elliottii* (7), and from diseased foliage of *P. radiata* collected by T. Middleton from some trees near Addis Ababa, Ethiopia. While dothistromal blight was spreading through the pine crops of Central and East Africa, the disease suddenly appeared in Chile and New Zealand.

P. radiata was first introduced into Chile about 1890 but was not adopted for large-scale forest development until about 1935, when exotic forestry was greatly extended in that country. This estate is now by far the most extensive of its kind in South America, although neighboring nations have started similar schemes.

The majority of Chilean pine plantations extend from a point about 100 miles north of Concepcion to the town of Valdivia in the south. In 1964–65 Dubin & Staley reported that Chilean plantations were suffering from a serious needle cast disease, identified as dothistromal blight (20). Since then it has spread and now occurs in most crops of young *P. radiata* from Concepcion to the wet regions of Puerto Montt, south of Valdivia. However, it has

[1] Head, Mycology Section, Plant Protection Institute, Pretoria, South Africa.
[2] Forest Pathologist, Department of Forestry, Pretoria, South Africa.

not been found further north of Concepcion, where *P. radiata* grows under warmer and drier conditions (21).

After dothistromal blight was discovered in the Chilean plantations of *P. radiata,* evidence was obtained that it had probably been in the country for several years. Dubin (21) mentions that Oehrens (67) collected material of the fungus from Chilean *P. radiata* in 1962 but had identified it as *Septoria acicola,* and that Yudelevich had observed very similar symptoms on *P. radiata* in 1957, at a time when the crop was thought to be suffering from dieback caused by *Diplodia pinea.* From these observations it would seem that these very early histories in Chile and Africa have much in common. Blight has also been found in Brazil, Argentina, and Uruguay (26, 27, 75) but none of these countries has appreciable crops of susceptible pines and the disease has no great potential for economic loss.

The history of pine plantation development in New Zealand and Australia has followed much the same pattern as that of Chile, except that a greater diversity of species was planted in the former countries. There was little loss from true disease in these plantations in their early years, although serious damage was caused in New Zealand by the wasp *Sirex noctilio* Fabricio and its fungal associate *Amylostereum areolatum* Boidin (44). However, in 1964 blight was identified in a *P. radiata* hybrid plantation near Tokorua in the North Island of New Zealand and by 1966 the disease had spread from this center, first to pine plantations in the southeast and southwest of the island and then to Nelson and other localities in South Island.

The disease also moved north to the Auckland district over the same period, so that by 1967 some 8000 ha of pines were severely damaged out of a total of nearly 33,000 ha of infected plantations. In his review of the history of blight in New Zealand, Gilmour (45, 46) expresses the view that the disease was probably present in the country for several years before it became a serious problem, but was masked by the presence of other common foliage pathogens, such as *Lophodermium pinastri* (Fr.) Chev. and *Naemacyclus niveus* (Fr.) Sacc. until circumstances favored its build-up and spread. At the time of writing, dothistromal blight does not appear to have reached Australia.

No outbreaks have been recorded in plantations in the northern hemisphere on the same scale as those in the south, even in the considerable areas of *P. radiata* plantations in Spain. However, there are records of blight in *P. radiata* in British Columbia (69), California (90), Oregon (90), and the Palni Hills of India (5).

The Name of the Pathogen

The nomenclature of the causal agent of blight has had a confused history, especially during the early years before it became notorious as a crop pathogen. Even now taxonomists disagree.

We are indebted to Shishkina & Tsanava (80, 81) for indicating that the earliest record of the fungus was probably made by Doroguin in 1911 and 1912

(18), when he described *Cytosporina septospora.* However, Doroguin apparently changed his views a number of years later, when he included it under *Brunchorstia pinea* (Karst.) v. Hohn (19) in 1926, and it is interesting to note that this identity for the blight fungus is still accepted in certain parts of Europe, although the two fungi are quite distinct (48).

The next record of the fungus was made in 1920 by Saccardo, who described it as *Actinothyrium marginatum* from collections of diseased pine foliage from Idaho (79). However, valid doubts were cast on this description by Sydow & Petrak in 1926 (86), who regarded it as a *nomen confusum,* as Saccardo's material consisted of two discordant elements, one of which was the fruiting body of *Leptostroma decipiens* Petr., while the other comprised spores identified at the time as belonging to *Lecanosticta acicola* (Thüm.) Syd. Later, Dearness (15) and Hedgcock (49) agreed with this conclusion but referred the sporal component to *Cryptosporium acicolum* Thüm. and *Septoria acicola* (Thüm.). Sacc., respectively. In 1944 Siggers (85) reexamined a range of material that had been disposed as *L. acicola, C. aeicolum, S. acicola,* and *A. marginatum,* and concluded that all corresponded most closely to *Dothistroma pini,* a fungus described by Hulbary in 1943 from collections from *P. nigra* Arn. v. *austriaca* Aschers and Graebn. in Illinois (52). A similar conclusion was reached by Murray & Batko (65), from a comparison between collections of *A. marginatum* and *D. pini* in the U.S.A. and their own material from England.

At about this point *Dothistroma* blight was recognized as a serious threat to the *P. radiata* crops of East Africa and the causal agent was identified first as *A. marginatum* (35) and later as *D. pini* (36). The latter epithet has been generally accepted as the correct name for the fungus until recently, when Shiskina & Tsanava (82) proposed that it should be changed to *Dothistroma acicola* (Thüm.). Then Morelet suggested that *Dothistroma septospora* (Doroguin) Morelet, was a more correct name for the imperfect fungus (64).

Three varieties of *D. pini* have been described. The first two were proposed by Thyr & Shaw in 1964 (88) as v. *pini,* based on Hulbary's type material and collections from Illinois, Kansas, and Kentucky, and as v. *linearis,* based on the type material of *A. marginatum* and collections from Idaho and Montana. This was followed by a study of a somewhat wider range of material by Ivory in 1967 (53), who proposed a third variety—*keniensis*—intermediate between v. *pini* and v. *linearis,* to include a wider range of forms.

There is still some doubt on the validity of these varieties, and some workers, such as Gadgil (29), feel that the characters by which they are distinguished are insufficiently reliable, despite the careful controls imposed by the authors in the course of their observations.

In 1966 Funk & Parker (28) described *Scirrhia pini* from diseased foliage of a range of pines, including *P. radiata,* collected on Vancouver Island. These authors carefully distinguished their fungus from *Scirrhia acicola* (Dearn.) Siggers [syn. *Systremma acicola* (Dearn.) Wolf & Barbour], and showed it to be the perfect state of a conidial form of *D. pini* closely resembling *D. pini* v. *linearis*. Since then *S. pini* has been collected from diseased foliage of a range of pine species and localities, notably a single collection from *P. radiata* in New Zealand (2), and collections from *P. nigra* v. *austriaca* in France (63), where it was associated with *D. pini* v. *pini* and from *P. clausa* in Tanzania[3] and *P. radiata* in Rhodesia,[3] in association with *D. pini* v. *keniensis*. However, there is no proof, beyond the indication of the association, that *S. pini* can be regarded as the perfect state of the varieties of *D. pini,* other than *linearis.*

Shishkina & Tsanava have also reported the discovery of a perfect state for the blight pathogen in Georgia, USSR. They have identified this fungus with *S. acicola,* and on this basis they proposed to change the name of the imperfect state from *D. pini* to *D. acicola* (82).

By 1967 the nomenclature of the causal agent of blight had become so complicated that Murray was prompted to make an urgent plea for some agreement on the correct names for the perfect and imperfect states of the fungus backed, if necessary, by a thorough examination of the whole problem (66). There was no response to this until 1969, when Dr. E. Punithalingam, of the Commonwealth Mycological Institute, started an exhaustive study of a wide range of collections of *D. pini,* which included its type material and those of other species associated with it in previous taxonomic studies. At the time of writing this work is not complete.

The Nature of the Pathogen

D. pini is a primary pathogen that invades and kills pine foliage. It sporulates shortly after the death of these tissues, by the formation of groups of minute black stromata, which emerge through the dead epidermis. These contain conidial masses and are often associated with a red staining of the substrate, which has led to the common names for the disease of 'red band' or 'red spot,' adopted in various parts of the world. The perfect state of the pathogen is formed in a similar way and consists of linear black ascostromata, which burst through the epidermis of the dead leaves. Endospores, similar to those described by Crosby (14) from *Scirrhia acicola* (Dearn.) Siggers, have been observed in cultures of *S. pini* by Ivory (57) and Gadgil (29). They have been seen to germinate but their role in the life cycle of the fungus is not known.

Water appears to be essential for the dispersal of conidia, and may be necessary for dispersal of ascospores as well. Stromata on infected needles swell when placed in water, so that they rupture and liberate masses of conidia. Under field conditions these would then be carried from the foliage by drainage and dispersed by a splash take-off mechanism, as various investigators have shown (37, 78). This conforms with the results of most field studies in Chile, East Africa, New Zealand, and the U.S.A. (21, 37, 45, 73), where maximum conidial dispersal is found to take place under light rain or

[3] Herbarium records, Commonwealth Mycological Institute.

heavy mist conditions and few, if any, airborne conidia occur in infected *P. radiata* plantations during dry periods.

An exception is found in the work of Cobb et al (11), who studied the dispersal of conidia and ascospores in infected *P. radiata* plantations in California. These workers report that few ascospores were trapped, relative to conidia, and that the latter were trapped at all times irrespective of rain or wind conditions; they found high relative humidity and temperature extremes were associated with low concentrations of airborne spores. As far as is known, no explanation has been offered for these observations.

However, all this evidence makes it clear that conidia are the most important spore form for dispersal of the fungus and account for the movement of the pathogen over short distances. We are less well-informed on how it has covered distances of hundreds of miles, which must have occurred on several occasions during its history.

While there is no doubt that human agencies—and aircraft in particular—could have played an important part in the long-distance spread of inoculum of *D. pini,* it is possible also that this might have been achieved by airborne conidia alone. Gibson et al (37) in their work in Kenya, showed that conidia were taken up into clouds in the course of formation from heavy mist on hilltops, on which heavily-infected pine plantations were growing. Under these conditions the spores were liberated from the wet foliage and, once incorporated in the cloud, would encounter temperatures low enough to ensure their survival for extended periods without germination (57). It is also possible that conidia of *D. pini* are sufficiently resistant to extremes of temperature, humidity, and irradiation to survive appreciable periods of airborne exposure under drier and warmer conditions. Gadgil (30) has shown that conidia dried on glass slides can survive as long as 16 weeks in the dark, at a temperature of 18°C and 35% RH., but rather less in other conditions of light, temperature, and humidity.

The survival of conidia of *D. pini* in infected pine foliage is dependent on the environmental conditions. Early studies of the disease showed that conidia of *D. pini* remain viable in dry needles for more than eleven months at room temperature (18°C) and for five months at 30°C (37). Later, Ivory showed that similar material would survive several days of dry heat at 35°C and desiccation at 30°C for 9 weeks (37). Free conidia will also survive prolonged exposure to low temperatures—0°–5°C—under wet or dry conditions (and are thus well-suited for cloud transport) but will not survive freezing (37, 54).

In contrast, the life of inoculum of *D. pini* in litter on the damp plantation floor is relatively limited. In a careful study in New Zealand Gadgil (31) has shown that the pathogen often disappears from the substrate after 2 months under these conditions and never survives longer than 4 months, but can be recovered from infected needles that have been suspended above the plantation floor for 6 months or more. We conclude from this investigation, and from some earlier observations in East Africa (37), that the disappearance of *D. pini* from infected needles on the plantation floor is probably due to microbial competition.

Gibson et al (37) and Ivory (54) found that water with nutrient traces is necessary for germination of conidia of *D. pini* v. *keniensis* in East Africa, both in vitro and on host tissues, and that it takes about 24 hours at 18°C for the 50% level to be reached, with some variations between varieties of *D. pini.* However, liquid water may not be essential for conidial germination under all circumstances, as Sheridan & Yen (84) working with an isolate of *D. pini* from New Zealand which resembled v. *keniensis* in its spore dimensions, obtained good germination at about 95% RH, and some germination at humidities as low as 76%.

Conidia frequently produce more than one germ tube and Gadgil (29) has reported that secondary conidia were frequently produced on the germ tubes formed by his isolates shortly after they started to elongate, but this was only observed very rarely by Ivory (57) and is not mentioned by Peterson (72, 74).

D. pini has no special nutrient requirements and can grow on agar medium containing simple carbohydrates, nitrogen salts, phosphates, and major mineral elements (54). Growth is extremely slow, so that colonies only become visible on agar plates 7 or 8 days after they have been sown with conidia and incubated at 18°C, and Ivory notes that the same slow linear growth rate also occurs when *D. pini* is colonizing infected host tissues (54). The colonies of *D. pini* in agar culture appear at first like bacterial growth but later develop white aerial mycelium; the fungus produces a dark red pigment in the medium and conidia occur most abundantly in the first few weeks of growth.

The inability of *D. pini* to survive appreciable periods on natural substrates in competition with other saprophytes, combined with its slow growth rate and its lack of any special nutrient requirements, make it an excellent example of the type of pathogen defined by Garrett (33) as an ecologically obligate parasite, where the host provides the shelter of a noncompetitive habitat rather than some special nutrient factor.

Studies of the effects of the environment on *D. pini* have been very largely confined to conidial germination and mycelial growth and little work has been done on the factors controlling production of conidia.

Ivory found that conidia of *D. pini* v. *keniensis* germinate between 8°C and 25°C, with an optimum of about 18°C (54). This is reasonably close to the results of Gadgil (29), working with *D. pini* v. *pini,* and Sheridan & Yen (84) and Cobb et al (10), who studied isolates similar to *D. pini* v. *linearis* in New Zealand and California respectively. Peterson found somewhat higher temperature limits with *D. pini* v. *pini* (73) in Nebraska, which may be due either to experimental conditions or a considerable latitude of physiologic as well as morphologic variation within the species.

The only detailed work on the effect of temperature and other environmental factors on the mycelial growth of *D. pini* appears to be that of Ivory

(54) working in East Africa with *D. pini* v. *keniensis* in liquid culture. He showed that the mycelium could grow in a wider range of temperature than those limiting conidial germination, with a lower optimum of 13°C. He also showed that the conidia would germinate over a pH range of 2.2–5.5, with an optimum of 3.5, and that subsequent mycelial growth occurred between pH 2 and 7 with the same optimum. No consistent effects of light or spore density on conidial germination were found (37, 54, 57).

Our knowledge of the infection process is also derived from the work of Gadgil, Ivory, and Peterson, whose results do not agree in all respects. All three authors observe that infection occurs exclusively by the penetration of leaves by conidial germ tubes by way of stomatal pores, a process taking a minimum of about 3 days but often longer when temperature and humidity conditions are suboptimal (29, 57, 71, 74). However, Gadgil (29) working in New Zealand with *D. pini* v. *pini* and Ivory (57), using *D. pini* v. *keniensis,* both observed that there was no influence of the stoma on the direction of growth of the germ tube in experiments using *P. radiata,* while Peterson (74) in his work with *P. ponderosa* v. *nigra* and *P. nigra* v. *austriaca,* found that the stoma exerted a marked effect on the direction of growth of the germ tube. Ivory (57) and Peterson (74) have both suggested that this discrepancy may be related to the methods employed. Thus, Peterson's studies were made in the field, where moisture gradients from the stoma might influence the direction of germ tube growth, while Ivory and Gadgil used detached needles in a saturated atmosphere where such gradients would be destroyed.

Alternatively, the difference between the results may be related to the host species used, as other differences are known to exist between *P. radiata* on the one hand, and *P. ponderosa* and *P. nigra* v. *austriaca* on the other, in their reaction to blight infection. Gadgil also reported that vesicle-like appressoria were regularly formed by the germ tube over the stomatal pore before the foliage was penetrated, but Ivory only rarely observed structures of this kind, and Peterson makes no reference to them at all (29, 57, 74).

There seems to be little information on the history of the pathogenic mycelium after it has penetrated the host and before the first appearance of lesions. However, after penetration the mycelium is confined almost exclusively to the mesophyll, where it grows intercellularly at a rate of 0.1 mm per diem in the dead tissues. This has been interpreted as evidence of the production of a phytotoxin by the fungus, which Ivory (57) believes to be nonspecific to the host.

The period between infection, the first appearance of symptoms, and the time taken after this for the pathogen to sporulate in *P. radiata* are variable and depend on temperature and light, as well as host provenance and possibly other factors.

In Kenya and Tanzania the period between infection and the appearance of the first symptoms varies between 5 weeks to 2½ months (37, 51) and seems to be related to temperature, as the shorter period is associated with warmer climates. Under these conditions the pathogen may sporulate 1–2 weeks after appearance of first symptoms, completing the infection cycle. Dubin has also reported an incubation period for blight of 2 months for *P. radiata* in Chile (21), but a later report (77) gives 4 months for this; no information is available on the period between symptom expression and sporulation on the host in this country. Under the seasonal climate of New Zealand the incubation period varies with time of year, being shortest at 5 weeks in the summer and extending to 16 weeks in autumn and early winter. Sporulation appears to be delayed in similar manner, as it occurs about 2 weeks after symptom expression in summer, but when lesions appear in the autumn sporulation may only follow in the next spring.

Light modifies the rate and extent of lesion expression of blight in *P. radiata.* This was observed first in East Africa, where trees of heavily-infested young plantations were found to show very little sign of blight when they were growing in sheltered or shaded positions. Gibson et al (41) showed that this was an effect of sunlight on host reaction and the sporulation of the pathogen on the host, and not in any way an effect of shelter on dispersal of the fungus or on its infection process. These studies were followed by more detailed work by Ivory (57), who found that light had no effect on the penetration of the foliage by the pathogen, but showed that symptom expression depended on the degree of irradiation of the foliage.

The great majority of *Dothistroma* blight of pines has occurred either in temperate latitudes between 25° and 40° north or south of the equator or in tropical highlands where the climate is relatively cool and wet. This agrees fairly well with the conditions found by most investigators to favor germination of conidia of *D. pini* in the laboratory, but it would be unwise to assume on these grounds that we know enough about the effects of environment on the processes of infection, symptom expression, and subsequent sporulation of the pathogen, to define confidently those conditions likely to lead to high incidence of the disease.

It is worth noting that divergent results have even been obtained where investigators have attempted to define conditions favoring outbreaks of blight on the basis of field studies. Thus, Gilmour (45, 46) agrees with Peterson (73) that several days of rain or overcast, humid weather are needed to establish infections of pine after a period of conidial dispersal, and suggests that moderate blight hazard conditions for *P. radiata* in North Island, New Zealand, are provided by a well-distributed rainfall between 1270 and 1520 mm per annum, with temperatures between 10°C and 18°C for half the year. This contrasts with the field results obtained by Hocking & Etheridge in Tanzania (51), from which they conclude that temperatures of over 18°C and rainfall as low as 13 mm per week are sufficient to permit multiple blight infection of *P. radiata.*

The Disease in the Field, its Impact and Control

Dothistromal blight is largely a disease of pines, although *Pseudotsuga menziesii* (Mirb.) Franco, and *Larix decidua* Mill. are also mildly susceptible to attack by *D. pini* v. *pini* (3, 22).

Gilmour, Ivory, and Peterson (47, 55, 72) have published recent lists of pine hosts for *D. pini* on the basis of field observations, and the former two authors have attempted classifications according to susceptibility. In addition, Cobb et al (10), have tested the susceptibility of a range of pine seedlings by exposing potted seedlings to natural infection in the field. All these authors provide evidence that *P. radiata* is by far the most economically important susceptible pine species, while *P. patula,* rated by Ivory as immune and by Gilmour as mildly susceptible, is probably the most important resistant species.

However, the results obtained by Gilmour, Ivory, and Cobb et al show considerable discrepancies otherwise. While these might be partially explained by differences in the methods and circumstances of the three studies, it is more likely that they are due to variation in the host range and reaction between different strains of the pathogen, as Cobb et al have proposed, or to the range of host reactions to blight that may be found within provenances of a single pine species. The latter has been observed by the author to occur in *P. caribaea* and *P. elliottii* and is demonstrated well in Peterson's study of the blight reactions of a range of *P. nigra* of different geographical origins (76).

There appear to be two types of reaction of pine hosts to attack by *D. pini.* Some species, such as *P. radiata,* become increasingly more resistant as they grow older, while others, like *P. ponderosa,* remain equally susceptible irrespective of age and size. The age at which *P. radiata* shows a significant increase in blight resistance varies with the degree of infection hazard. Where temperature and moisture are moderately favorable to infection the first signs of mature plant resistance may be found at about 8 years, but this age may increase to as much as 15 when conditions allow heavy and persistent re-infection by *D. pini.* Similarly, recovery of *P. radiata* from blight as it grows older is much slower where early and severe attack has become established than in crops where this is lighter.

Despite careful studies, such as those of Ivory (57), we still know very little about the mechanism of blight resistance in pines or of the nature of the development of mature plant resistance in *P. radiata,* beyond the fact that it is a function of the whole tree and is transferred by grafting, as Garcia & Kummerow (32) have shown. Nevertheless, mature plant resistance has proved to be a most valuable feature of *P. radiata* and has led to a means of blight control which will be discussed later.

Observations on the initial spread of blight in young, healthy *P. radiata* have been made in East Africa, New Zealand, and elsewhere. These agree on all general points and describe how the first signs of disease appear on the oldest and most sheltered foliage, from where it spreads progressively into the younger needles, until a stable, chronic state of defoliation is established or, exceptionally, the tree dies. Disease development within the tree and the crop follows a typical sigmoid curve, with an exponential stage where very high rates of spread can occur. This was observed when blight first spread through East Africa, and is described well by Gilmour as 'explosive' (45). It is likely that Cobb et al (12) were also observing the explosive exponential stage of blight in young *P. radiata* in California when they found values greater than 4 for van der Plank's "r" (89).

The first serious outbreaks of blight in *P. radiata* took place under conditions highly favorable to the disease, where a significant proportion of young trees died from defoliation. This gave *D. pini* a disproportionate reputation as a killer, which has persisted in some quarters until today. However, it is now abundantly clear that the most serious results of blight are through retardation of growth from defoliation and not death to trees.

The first quantitative studies of the effect of blight on the growth of *P. radiata* were made by Gibson et al in Kenya, and showed that a general reduction in height increment was related to disease severity in plants a little over a year old (37). Later, in Tanzania, Etheridge showed that a close linear relation existed between height growth and degree of blight attack in plants between 1 and 2 years old (23).

Two studies of 3–4 year old *P. radiata* in Kenya (9) and Tanzania (51), gave a sigmoid relation between severity of blight and reductions in the rate of increment of height and diameter. The early stage of attack, which was concentrated in the oldest parts and did not involve more than about 20% of the foliage of the tree, had little effect on diameter and height growth, but its later spread into younger foliage was accompanied by marked reductions in growth rate. Generally, diameter increment fell by one half where severe needle cast had affected half the foliage, and a further increase in disease to more than the 85% level resulted in almost complete cessation of diameter growth. Reduction in rate of height growth was not so marked, but this reached nearly 50% when blight exceeded the 75% level.

A similar relation has been found between blight attack and diameter growth of *P. radiata* in Chile (77), where 80% needle attack was related to reduction in diameter increment of nearly 75% in 7-year-old trees, and a 60% attack in 9-year-old trees led to a growth check of nearly one-third.

Whyte (91), in a much more extensive study in New Zealand, also finds that increment loss is slight during early stages of disease but increases when the disease attacks more than 25% of the foliage and younger needles are involved.

All these results indicate that the effects of blight on the growth of *P. radiata* are the direct result of destruction of photosynthetic tissues. Thus we find a linear relationship for young plants where all the foliage is equally efficient, but a curved relation in older trees, where the first infections occur on senile foliage and no significant growth-checks occur until the disease has spread to the younger needles. Similar results have been obtained from measurements of the effects of artificial defoliation on the growth of pines (13, 60, 61, 68).

Any statement of the effects of dothistromal blight on *P. radiata* should include some account of the secondary effects of the disease. Extensive defoliation would be expected to encourage weed competition and to predispose the trees to attack by weak facultative pathogens. No work appears to have been done to separate the weed competition component from the over-all effects of blight on the growth of young *P. radiata,* but Etheridge (24) studied the incidence of secondary pathogens in diseased *P. radiata* in New Zealand and found that these effects are small relative to those of the primary pathogen. It is not yet possible to determine the longterm effects of the disease on the crop, but it is likely that these will be low in Chile, where the young crop severely infected in 1964 had largely recovered by 1968 (8, 77).

The control of blight.—The first attempts to evolve practical control measures for dothistromal blight of *P. radiata* were made in East Africa about 3 years after the disease had been recognized as a serious threat to the species. The delay was inevitable as no serious work could be planned without some background knowledge on the nature of the problem. During this interval, also, the planting of *P. radiata* was suspended as a precautionary measure in the East African highlands and its place was taken by extended planting of *Cupressus lusitanica* Mill. and *P. patula.* Thus, it was not until the latter half of 1963 that experiments were started to explore the use of shade, fungicides, and the development of resistant cultivars of *P. radiata* for control of blight. A first account of this work appeared in 1965 (38), and later papers have dealt with different aspects of the program in greater detail.

The discovery that full symptom expression of blight of *P. radiata* depended on the degree of irradiation received by the foliage led to attempts to control the disease in the young crop by interplanting nursery stock of this pine either in cut lines in bamboo thicket or mixed with 4-year-old *C. lusitanica* to provide shade. This promising approach had to be abandoned, however, as young *P. radiata* proved to be very light-sensitive and the degree of shade needed to control the disease produced severe growth checks (41).

During the earlier investigations of blight in East Africa, trials of fungicides for field control of the disease were confined to one or two materials with alleged systemic activity (which gave negative results), and in other experiments fungicides were employed only as a research tool to maintain healthy stock for growth studies in infective areas (37). It was thought at that time that fungicides were most unlikely to provide a practical solution to dothistromal blight, so the extensive screening of conventional surface active fungicides was not attempted until 1963, when Gibson et al (39), Hocking (50), and Etheridge (23) set up field trials in East Africa. These showed that fungicides based on copper compounds were particularly effective against blight and, with their relative cheapness, offered considerable promise as an economically acceptable control measure in the field. A program of field trials followed in Kenya in 1964, based on four copper fungicides applied from the air in two operations per year, to give a total dosage of a little over 4 kg of copper equivalent per hectare per annum. The first treatment was timed to coincide with the end of the dry season, when inoculum would be at its lowest in the crop and to anticipate the advent of the main rains when dispersal would take place; the second treatment followed after 2 or 3 months and served as a booster for the first. Treatments were started on 2–3 year-old *P. radiata* crops with a low disease rating, and were continued until the 8th or 9th year, when the limit of acceptable expenditure had been reached and it was hoped that the crop would be sufficiently mature and healthy to continue normal growth thereafter. It was realized that at this time aircraft were only suitable for the treatment of forest crops that provided good ground cover and that alternative means would be required for *P. radiata* plantations less than 3 years old, where this cover did not exist. The program was extended therefore, to include trials of copper fungicides applied to 1- and 2-year-old *P. radiata* by teams of men with knapsack sprayers. This method was chosen because wheeled or tracked vehicles were unlikely to prove economical or suitable for the difficult terrain of many plantation areas, and men on foot could provide placement of fungicide on individual trees, which would not be possible by other means.

The results of all these trials showed that copper fungicides applied at the prescribed rates could provide sufficient blight control to maintain normal growth (43), but it was clear that considerable improvements in operational techniques would be needed before this method could be used in general practice in the East African highlands. However, these results have found application in New Zealand, where a large industrial investment in forestry, a well-developed agricultural aviation industry, and a forest estate presenting few topographical difficulties, combined to make the use of fungicides for blight control both economically and technically acceptable.

It was under these circumstances that a comprehensive series of field trials of copper fungicides for the control of dothistromal blight of *P. radiata* was started in New Zealand in late 1965; these gave such satisfactory results that within a year routine spray operations had been adopted by government and private forest concerns for the protection of their young plantations. After intensive development in following years, the timing of sprays and other operational aspects of the work have been so improved that adequate protection can be obtained now by as few as 3 or 4 operations, each applying copper fungicides at 2.24 kg per acre Cu equivalent, spread over the first 15 years of growth (4), after which the crop is thought to be sufficiently resistant. Effective schedules for treatment of nursery stock by copper fungicides have also been worked out (59).

There has been relatively little work so far to explain this outstanding efficiency of copper fungicides in the control of dothistroma blight. However, a pilot investigation in Kenya has given results that suggest that this may be due largely to the high sensitivity of the germinating conidia of *D. pini* to copper ion and the extended redistribution of spray deposit dissolved in leaf moisture at levels of copper toxic to the fungus. There is some evidence, also, that copper may reduce the sporulation of the pathogen in established lesions (42).

It is interesting to note that Peterson's studies of blight of *P. ponderosa* and *P. nigra* v. *austriaca* in the USA, included tests of various fungicides in which copper-based materials were outstandingly successful (73). No trials of fungicides for blight control have been made in Chile and it seems that such measures would prove economically unacceptable.

To date, the development of cultivars of *P. radiata* with resistance to dothistromal blight has been largely concentrated in East Africa. Some work on these lines has been carried out in New Zealand, but this has been limited by the small area of plantations that had been left unsprayed and were suitable for searches for potentially resistant material. The development of blight-resistant cultivars does not appear to have been attempted on a large scale in Chile or elsewhere.

The earliest efforts to obtain blight-resistant selections of *P. radiata* were made in Lushoto plantations shortly after the disease had been correctly diagnosed in East Africa. In this work, and its later extension in 1963 by Ivory and Paterson in other parts of Africa, severely diseased plantations between the ages of 4 and 10 years were searched for trees showing outstanding health, combined with acceptable size and form. Such trees were regarded as likely to carry some inherent resistance to blight, as their position in the midst of a severely infected crop made it most unlikely that they were chance escapes. Ivory & Paterson's project was directed mainly to the evaluation and propagation of these selections, but had the secondary objectives of determining the nature of resistance and attempting the formulation of techniques for early detection and measurement.

An interim report on this work has been published recently, which shows that useful progress has been made (56). This describes how scion material from select trees was grafted in test orchards to compare its disease reaction and growth rate with control grafts from unselected trees. All selections were found to have a variable but useful partial resistance, transferable by grafting, but liable to influence by the nature of the stock. Blight-resistant trees were found to be generally more vigorous than average, but trees of superior vigor were not necessarily resistant to the disease (57). Completely resistant trees were never found and it appeared that the blight resistance corresponded with the 'horizontal' type defined by van der Plank (89).

Ivory's (57) experimental demonstration that there was a stock/scion interaction establishes the systemic nature of this resistance which, in other respects, seems to be definable as a precocious development of mature plant resistance. Thus, where unselect *P. radiata* does not show evidence of increasing blight resistance until after its 8th or 9th year, when grown under conditions typical of most of the East African highlands, the select trees show this quality to a significant degree as early as their 4th year. Partial blight resistance has been found to be transmitted by open-pollinated seed also, giving increased chances for the raising of plant stocks with this character on a large scale in the relatively near future. However, this improved stock will still be blight susceptible in the nursery and during its first 3–4 plantation years, when it will need fungicidal protection. This leaves an urgent need for improvement in methods of application of copper fungicides to these young crops—a research field full of difficulties, that has scarcely been broached to date. Of course, this need could disappear altogether if a means was found for the large-scale propagation of cuttings from trees old enough to have developed resistance, which would preserve the character of the parent. The propagation of *P. radiata* from rooted cuttings has been shown to be practicable in Australia, New Zealand, and East Africa (25, 87, 58), and recent progress has been made by Barnes (1) in Rhodesia, who has established a successful but small mixed plantation of *P. radiata* and *P. patula,* in which the former species has been grown from cuttings from old trees and has retained its blight-resistant character completely. With further improvement in the techniques for clonal propagation of cuttings this approach may provide a complete answer to the blight problem and reinstate *P. radiata* as a major plantation species.

Discussion

In the past, foresters have been largely concerned with the management of tree crops on sites that were either unsuitable or marginal for agriculture. This has allowed little scope for site improvement and, in matters of crop health, site factors have generally been considered before pathogens. This has led to ideals based on the natural forest and to the view that any departure from this type of environment—such as plantation systems—will inevitably lead to serious outbreaks of disease and economic disaster. These views prevailed until the present, despite attempts to rationalize them (70), so that there has been no lack of warning on the risks of disease that are inherent in the establishment of extensive estates of exotic forest species. The appearance of dothistromal blight of *P. radiata* in the southern hemisphere, therefore, provided ample vindication of these prophecies and it is instructive to examine the precautions taken in New Zealand, East Africa, and Chile against such an emergency, and evaluate their effectiveness.

Specialists in forest pests and diseases were recruited in New Zealand early in the development of exotic conifer plantations, and in 1955 de Gryse (16) reviewed the risks involved and the best means to combat them. He recommended the establishment of a continuous survey for dangerous forest insects and fungi (on the lines of a similar system in Canada) with an extension of fundamental research on the control and other aspects of known pests and diseases. This advice was taken and backed by legislation to reduce the risks of importation of dangerous forest pathogens and, later, to restrict movement of infected plants within the country (4).

Much simpler action was taken in East Africa, with the appointment of four entomologists and a single pathologist to various forest departments and institutes between 1950 and 1960. These officers maintained a forest insect and fungus survey, with some research on known pests and pathogens.

Chile seems to have been least well-equipped for the early detection and investigation of new forest diseases, which were largely the responsibility of the local agricultural plant protection services. In addition, suitable import controls were gradually brought into force in Chile and East Africa to lessen the risk of introducing exotic tree pests and pathogens.

While this legislation was necessary and correct, its effectiveness in all these regions was limited by the means available for its enforcement, and the lack of information on the important diseases of species such as *P. radiata*. It was small wonder, therefore, that dothistromal blight passed unnoticed in its early days and that when it was finally noted, its cause was confused with other well-known pathogens and saprophytes.

At best, plant import control is an expedient to gain time for improved methods of control before the arrival of new, serious plant diseases. It is regrettable, therefore, that in the days before the emergence of dothistromal blight so little attention was paid to the development of methods of disease control in forest crops. This was largely due to the general assumption that any supplementary control measures for new, serious tree diseases would be economically prohibitive, so that the affected crop species would have to be replaced by some alternative until disease-resistant varieties were developed. It is now clear from recent work in East Africa and New Zealand that chemical control of forest tree diseases can be acceptable and that where the crop is part of extensive industrial development, as in New Zealand, expensive control measures are readily justified. It is equally possible that fungicides might have provided economic control of blight in East Africa if the problems of application to highland forest crops had been foreseen and tackled earlier. Instead, *P. radiata* was replaced by less productive alternative species very soon after blight appeared in East Africa and it is still an open question whether the losses to be expected from disease would have exceeded those accepted in this change of planting policy.

Hopefully, those countries that have extensive forest plantations but have yet to encounter serious disease problems, may learn from this experience and start research now to anticipate possible technical problems in disease control operations. Indeed, this may have happened already in Victoria, Australia, where forest airspray feasibility trials have been made in the absence of a major foliage disease problem (17).

Finally, an answer must be attempted to the two questions of 'whither'? and 'whence'? as they apply to dothistromal blight.

Despite the attention that *D. pini* has attracted in recent years we still know very little about its distribution before 1940. It appears to have been a very mild pine needle pathogen, distributed widely in the temperate zones of North America and Europe. At the end of World War II, and probably aided by the vast increase in air traffic at that time, it found its way almost simultaneously into extensive and susceptible *P. radiata* plantations in South America, Africa, and New Zealand. The work of Ivory (53) shows that *D. pini* occurs in the southern hemisphere in two distinct groups, one in Africa (characterized by the variety *keniensis*) and the other in South America and New Zealand dominated by the variety *pini*, indicating two possible courses of invasion.

This course of events, and the resulting serious diseases of *P. radiata*, can be well accounted for by environmental factors and does not need to be explained by any inherent change in the fungus. However, the emergence of *D. pini* as a serious pine pathogen in North America and its spread to parts of Canada, California, and Alaska over the same period, does suggest that a change in the nature of the fungus towards greater virulence may have taken place at the same time. Some support for this view is obtained from the high values for 'r' obtained by Cobb et al in their California work on blight (12), which Merrill (62) believes are characteristic of a recently introduced pathogen.

So far there are three regions—Australia, South Africa, and Spain—where extensive estates of *P. radiata* have been established and where blight has not yet been recorded. Although it is rash to predict future blight outbreak centers from our inadequate information on the epidemiology of blight, it seems likely that the rainfall regimes of the Cape in South Africa may not favor the disease in that region. On the other hand, parts of Australia—particularly Victoria and Tasmania—seem to be particularly well-suited for the establishment and spread of *D. pini* in *P. radiata*, having moderate temperatures with ample and well-distributed rainfall. Similar risks exist in northwest Spain, where *P. radiata* is grown in an annual rainfall of about 1500 mm and a temperature between 8°C and 18°C. In this region there is the possibility of spread of the pathogen from France, where it is already well-established (64).

LITERATURE CITED

1. Barnes, B. D. 1970. The Prospects of Re-establishing *Pinus radiata* as a Commercially Important Species in Rhodesia. *S. Afr. Forest. J.* 72:17–19
2. Bassett, C. 1967. Forest Pathology. *Dothistroma pini* Project: Distribution. *Rept. Forest Res. Inst. (N.Z.)* p. 48
3. Bassett, C. 1969. *Larix decidua*. A New Host for *Dothistroma pini*. *Plant Dis. Reptr.* 53:706
4. Bassett, C., Zondag, R. 1968. Protection Against Fungal and Insect Attack in New Zealand Forests. *Brit. Commonw. Forest. Conf. IXth* (India)
5. Bakshi, B. K., Singh, S. 1968. Dothistroma Blight, a Potential Threat to *Pinus radiata* Plantations in India. *Indian Forester* 94:824–25
6. Bates, G. R. 1962. Botany, Plant Pathology and Seed Services. *Rept. Minist. Agr. Rhodesia* 46–55
7. Browne, F. G. 1968. *Pests and Diseases of Forest Plantation Trees.* Oxford Univ. Press
8. Carvalho, N. Personal communication
9. Christensen, P. S., Gibson, I. A. S. 1964. Further Observations in Kenya on a Foliage Disease of Pines caused by *Dothistroma pini* Hulbary. I. The Effect of Disease on Height and Diameter Increment in Three- and Four-year old *Pinus radiata. Commonw. Forest Rev.* 43:326–31
10. Cobb, F. W. Jr., Uhrenholdt, B., Krohn, R. F. 1970. Spore Dimensions, Spore Germination and Virulence of *Scirrhia pini* on Pines in California. *Phytopathology.* In press
11. Cobb, F. W. Jr., Uhrenholdt, B., Murray, J. A. 1968. Aerial Dispersal of *Dothistroma pini* Spores. *Phytopathology* 58:1047 Abstr.
12. Cobb, F. W. Jr., Uhrenholdt, B., Krohn, R. F. 1969. Epidemiology of *Dothistroma pini* Needle Blight on *Pinus radiata. Phytopathology* 59:1021–22 Abstr.
13. Craighead, F. C. 1940. Some Effects of Artificial Defoliation on Pine and Larch. *J. Forest.* 38: 885–88
14. Crosby, E. S. 1966. Endospores in *Scirrhia acicola. Phytopathology* 56:720
15. Dearness, J. 1928. New and Noteworthy Fungi: V. *Mycologia* 20: 235–46
16. de Gryse, J. J. 1955. Forest Pathology in New Zealand. *Bull. N. Z. Forest Serv.* (Wellington) No. 11
17. Dexter, B. D. 1970. Aerial Application of Fungicide. *Research Activity 69 Forests Commission.* (Victoria, Australia) 34–36
18. Doroguin, G. 1912. Fungus Disease of Mountain Pine. *Forest J.* (Russia)
19. Doroguin, G. 1926. Note on *Cytosporina septospora.* Plant Diseases No. 1 *Bolez Rast.* 15:48–50
20. Dubin, H. J., Staley, J. 1966. *Dothistroma pini* on *Pinus radiata* in Chile. *Plant Dis. Reptr.* 50:280
21. Dubin, H. J. 1967. Preliminary Information about Dothistroma Blight in Chile. *Congr. Int. Un. Forest. Res. Org. XIV* 5:209–20
22. Dubin, H. J., Walper, S. 1967. *Dothistroma pini* on *Pseudotsuga menziesii. Plant Dis. Reptr.* 51: 454
23. Etheridge, D. E. 1965. Report to the Government of Tanzania on Forest Tree Diseases. *F. A. O. Rept. No. 2056*
24. Etheridge, D. E. 1967. The Role of Secondary Organisms in Dothistroma-Infected *Pinus radiata. Forest. Res. Inst. N. Z. Forest. Serv. Forest Pathol. Rept. No. 24*
25. Fielding, J. E. 1963. The Possibility of using Cuttings for the Establishment of Commercial Plantations of Monterey Pine. *World Consult. Forest Genet. Tree Improvement* (Stockholm)
26. Figueiredo, M. B., Namekata, T. 1969. *Dothistroma pini* Hulbary Agente Causal da Quiema a Aciculas em *Pinus* spp. Fungo Recentamente Observado no Brasil. *Biologico* 35:179-81
27. Fresa, R. 1968. "Banda Roja" de los Pinos (*Dothistroma pini* Hulbary var *linearis* Thyr & Shaw). *Rev. Inventres Agrop. Serv.* 5:1–7
28. Funk, A., Parker, A. K. 1966. *Scirrhia pini,* new sp.; the Perfect

State of *Dothistroma pini* Hulbary. *Can. J. Bot.* 44:1171–76

29. Gadgil, P. D., 1967. Infection of *Pinus radiata* Needles by *Dothistroma pini. N. Z. J. Bot.* 5:499–503
30. Gadgil, P. D. 1971. Survival of Spores of *Dothistroma pini. Rept. Forest. Res. Inst.* (N.Z.) 48
31. Gadgil, P. D. 1970. Survival of *Dothistroma pini* on Fallen Needles of *Pinus radiata. N. Z. J. Bot.* 8:303–9
32. Garcia, J., Kummerow, J. 1970. Infection of Monterey Pine Graftings with *Dothistroma pini. Plant Dis. Reptr.* 54:403–4
33. Garrett, S. D. 1950. Ecology of Root-Inhabiting Fungi. *Biol. Rev.* 25:220–54
34. Gibson, I. A. S. 1962. A Notebook on Pathology in Kenya Forest Plantations. Edition II. *Govt. Printer* (Nairobi, Kenya)
35. Gibson, I. A. S. 1962. Actinothyrium Needle Blight of Pines. *Commonw. Phytopath. News* 8: 47–48
36. Gibson, I. A. S. 1963. A Further Note on Dothistroma (Actinothyrium) Blight of Pines in Kenya. *Commonw. Phytopath. News* 9:47–48
37. Gibson, I. A. S., Christensen, P. S., Munga, F. N. 1964. First Observations in Kenya on a Foliage Disease of Pines, caused by *Dothistroma pini* Hulbary. *Commonw. Forest. Rev.* 43:31–48
38. Gibson, I. A. S. 1965. Recent Research into Dothistroma Blight of Pines in Kenya. *Agr. Vet. Chem.* 6:39–42
39. Gibson, I. A. S., Kennedy, P., Dedan, J. K. 1966. Further Observations in Kenya on a Foliage Disease of Pines, caused by *Dothistroma pini* Hulbary. II. Investigations into Fungicidal Control of the Disease. *Commonw. Forest Rev.* 45:67–76
40. Gibson, I. A. S. 1967. The Distribution, Impact and Control of a Foliage Disease of Pines in Africa, caused by *Dothistroma pini* Hulbary. *Congr. Int. Un. Forest. Res. Org. XIV* 5:209–220
41. Gibson, I. A. S., Christensen, P. S., Dedan, J. K. 1967. Further Observations in Kenya on a Foliage Disease of Pines caused by *Dothistroma pini* Hulbary. III. The Effect of Shade on the Incidence of Disease in *Pinus radiata. Commonw. Forest. Rev.* 46: 239–47
42. Gibson, I. A. S., Howland, A. K., Munga, F. M. 1970. The Action of Copper Fungicides in the Control of Dothistroma Blight of Pines: a Pilot Study. *E. Afr. Agr. Forest. J.* 36:139–53
43. Gibson, I. A. S. 1971. Field Control of Dothistroma Blight of *Pinus radiata* using Copper Fungicide Sprays. *E. Afr. Agr. Forest. J.* 36:247–74
44. Gilmour, J. W. 1966. The Pathology of Forest Trees in New Zealand. The Fungal, Bacterial and Algal Pathogens. *Tech. Pap. Forest. Res. Inst.* (N. Z.) 48
45. Gilmour, J. W. 1967. Distribution, Impact and Control of *Dothistroma pini* in New Zealand. *Congr. Int. Un. Forest. Res. Org. XIV* 5:221–48
46. Gilmour, J. W. 1967. Distribution and Significance of the Needle Blight of Pines, caused by *Dothistroma pini* in New Zealand. *Plant Dis. Reptr.* 51:727–30
47. Gilmour, J. W. 1967. Host List for *Dothistroma pini* in New Zealand. *N.Z. Forest. Res. Serv.* Leaflet 16
48. Gremmen, J. 1968. Présence de *Scirrhia pini* Funk et Parker en Roumanie. (Stade Conidien: *Dothistroma pini* Hulbary). *Bull. Trimest. Soc. Mycol.* (France) 84:480–92
49. Hedgcock, G. C. 1929. *Septoria acicola* and the Brown-spot Disease of Pine Needles. *Phytopathology* 19:993–99
50. Hocking, D. 1967. Dothistroma Needle Blight of Pines. II. Chemical Control. *Ann. Appl. Biol.* 59:363–73
51. Hocking, D., Etheridge, D. E. 1967. Dothistroma Needle Blight of Pines. I. Effect and Etiology. *Ann. Appl. Biol.* 59:133–41
52. Hulbary, R. L. 1941. A Needle Blight of Austrian Pine. *Illinois Nat. Hist. Survey Bull.* 21:231–36
53. Ivory, M. H. 1967. A New Variety of *Dothistroma pini* in Kenya. *Trans. Brit. Mycol. Soc.* 50:289–97

54. Ivory, M. H. 1967. Spore Germination and Growth in Culture of *Dothistroma pini* var *keniensis*. *Trans. Brit. Mycol. Soc.* 50:563–72
55. Ivory, M. H. 1968. Reaction of Pines in Kenya to Attack by *Dothistroma pini* var *keniensis*. *E. Afr. Agr. Forest. J.* 33:236–44
56. Ivory, M. H., Paterson, D. N. 1969. Progress in Breeding *Pinus radiata* Resistant to Dothistroma Needle Blight in East Africa. *Silvae Genet.* 19:38–42
57. Ivory, M. H. 1970. Dothistroma *Needle Blight of* Pinus radiata *in Kenya. A Study of Infection and Blight Resistance.* Ph.D. Thesis. Univ. London
58. Ivory, M. H. 1971. A Technique for Rooting Cuttings of *Pinus radiata* in Kenya. *E. Afr. Agr. Forest. J.* 36:356–60
59. Jancarik, V. 1969. Control of *Dothistroma pini* in Forest Nurseries. *Res. Leafl. Forest. Res. Inst.* N.Z. Forest Serv. No. 24
60. Lanier, L. G. 1967. Influence des Aiguilles de Different Ages sur la Croissance du *Pin sylvestre* (*Pinus sylvestris* L.). *Congr. Int. Un-Forest. Res. Org. XIV* 5:209–220
61. Linzon, S. N. The Effect of Artificial Defoliation of Various Ages of Leaves upon Pine Growth. 1958. *Forest Chron.* 34:50–56
62. Merrill, W. 1967. Analyses of some Epidemics of Forest Tree Diseases. *Phytopathology* 57:822
63. Morelet, M. 1967. Une Maladie des Pins Nouvelles pour la France, due à *Scirrhia pini* Funk et Parker, et Son Stade Conidien: *Dothistroma pini* Hulbary. *Bull. Mens. Soc. Linn.* (Lyon) 36:361–67
64. Morelet, M. 1969. *Scirrhia pini:* Note Complémentaire. *Bull. Mens. Soc. Linn.* (Lyon) 38: 268–70
65. Murray, J. S., Batko, S. 1962. *Dothistroma pini* Hulbary: A New Disease of Pine in Great Britain. *Forestry* 35:57–65
66. Murray, J. S. 1967. *Dothistroma pini* Hulbary: Its Occurrence in Europe. *Congr. Int. Un. Forest. Res. Org. XIV* 5:265–68
67. Oehrens, B. E. 1962. Fitopathologia Fungosa Valdiviana. 3ra Contribucion *Revista Univ. La Universidad Catolica de Chile* (Santiago)
68. O'Neil, L. C. 1962. Some Effects of Artificial Defoliation on the Growth of Jack Pine. (*Pinus banksiana* Lamb.) *Can. J. Bot.* 40:273–80
69. Parker, A. K., Collis, D. G. 1966. Dothistroma Needle Blight of Pines in British Columbia. *Forest. Chron.* 30:160–61
70. Peace, T. R. 1957. Approach and Perspective in Forest Pathology. *Forestry* 40:47–56
71. Peterson, G. W. 1966. Penetration and Infection of Austrian and Ponderosa Pine by *Dothistroma pini. Phytopathology* 56:894–95
72. Peterson, G. W. 1967. Dothistroma Needle Blight of Pines in North America. *Congr. Int. Un. Forest. Res. Org. XIV* 5:269–78
73. Peterson, G. W. 1967. Dothistroma Needle Blight of Austrian and Ponderosa Pines: Epidemiology and Control. *Phytopathology* 57: 437–41
74. Peterson, G. W. 1969. Growth of Germ Tubes of *Dothistroma pini* Conidia Positively Directed Towards Stomata of Austrian and Ponderosa Pine Needles. *Phytopathology* 59:1044 Abstr.
75. Peterson, G. W. 1969. Outbreaks and New Records. *Plant Prot. Bull. F.A.O.* 17:43
76. Peterson, G. W. 1971. Resistance to *Dothistroma pini* within Geographic Sources of *Pinus nigra. Phytopathology* 61:149–50
77. Rack, K. 1971. Personal communication
78. Rogerson, C. T. 1953. Kansas Mycological Notes, 1951. *Trans. Kans. Acad. Sci.* 56:53–57
79. Saccardo, P. F. 1920. Mycetes Boreali-Americane. *Nuovo G. Bot.* (Italy) 27:75–88
80. Shishkina, A. K., Tsanava, N. I. 1966. On the Study of Red Spots on Pine Needles in Georgia. (In Russian). *Inst. Plant Prot.* (Georgia) Vol. 18
81. Shishkina, A. K., Tsanava, N. I. 1966. *Dothistroma pini* Hulbary ad Pinum in Gruzia. *Acad. Sci. USSR Nov. Sist. Niz. Rast.* 205–9
82. Shishkina, A. K., Tsanava, N. I. 1967. *Systremma acicola* (Dearn.) Wolf et Barbour. Sovenshen-

naya Stadiya *Dothistroma acicola* (Thüm.) A. Schischke et N. Tsan. *Nov. Sist. Niz. Rast.* 276–77

83. Scott, C. W. 1960. *Pinus radiata. F.A.O. Forestry and Forest Prod. Stud. No. 14* (Roma)
84. Sheridan, J. J., Yen, C. C., 1970. A Note on the Effect of Temperature and Humidity on the Germination of Conidia of a New Zealand Isolate of *Dothistroma pini* Hulbary. *N. Z. J. Bot.* 8:658–60
85. Siggers, P. V. 1944. The Brown-spot Needle Blight of Pine Seedlings. *Tech. Bull. U. S. Dept. Agr.* No. 870
86. Sydow, H., Petrak, F. 1924. Zweiter Beitrag zur Kenntnis der Pilzflora Nordamerikas, unsbesondere der Nord-westlichen Staaten. *Ann. Mycol.* 22:387–409
87. Thulin, I. J., Faulds, T. 1968. The Use of Cuttings in the Breeding and Afforestation of *Pinus radiata. N. Z. J. Forest.* 13:66–67
88. Thyr, B. D., Shaw, C. G. 1964. Identity of the Fungus causing Red Band Disease on Pines. *Mycologia* 56:103–9
89. van der Plank, J. E. 1969. Pathogenic Races, Host Resistance and an Analysis of Pathogenicity. *Neth. J. Plant Pathol.* 75:45–52
90. Wagener, W. W. 1967. Red Band Needle Blight of Pines: a Tentative Appraisal for California. *U. S. Forest. Serv. Res. Note* PSW 153
91. Whyte, A. G. D. 1968. Tree Growth in the Presence of *Dothistroma pini. Rept. Forest. Res. Inst. N. Z. Forest Serv.* 51–52

SOME PROPERTIES AND TAXONOMIC SUB-DIVISIONS OF THE GENUS *PSEUDOMONAS*

3544

N. J. PALLERONI AND M. DOUDOROFF[1]
Department of Bacteriology and Immunology,
University of California, Berkeley

We shall not attempt in this review to cover the vast literature dealing with the genus *Pseudomonas* or to present a taxonomic treatise on this uncomfortably large and rather heterogeneous group of bacteria. We shall, instead, concentrate on the few species that we have studied, with the hope that our observations and ideas may be useful to other workers, of whom many will be needed to unravel the nomenclature and phylogeny of this group.

PHENOTYPIC CIRCUMSCRIPTION OF THE GENUS AND ITS RELATION TO OTHER GENERA

The definition of the genus *Pseudomonas* to be included in the forthcoming edition of Bergey's Manual of Determinative Bacteriology is based on the description of "aerobic pseudomonads" as proposed by Stanier, Palleroni & Doudoroff (75). The adopted definition will be: "Cells single, straight or curved rods, but not helical. Dimensions, generally 0.5–1.0 micron by 1.5–4 microns. Motile by means of polar flagella; monotrichous or multitrichous. Do not produce sheaths or prosthecae. No resting stages known. Gram negative. Chemoorganotrophs: metabolism respiratory, never fermentative. Catalase positive. Some are facultative chemolithotrophs, able to use H_2 or CO as energy source. Molecular oxygen is the universal electron acceptor; some can denitrify, using nitrate as an alternate acceptor. Strict aerobes, except for those species which can use denitrification as a means of anaerobic respiration. The G + C content of the DNA of those species examined ranges from 58 to 70 moles percent."

This description of the genus is clearly artificial and is applicable to several other genera currently recognized (e.g., *Xanthomonas, Acetomonas, Comamonas,* etc.). In the 8th edition of Bergey's Manual, some of these genera (*Xanthomonas, Gluconobacter*) will be retained, and we shall, therefore, not include them in the present discussion. We shall, however, consider the polarly flagellated members of the genus *Hydrogenomonas* as *Pseudomonas*

[1] Most of the published and unpublished studies reported in the present review were carried out with the partial support of grants AI-1808 from the National Institutes of Health and GB 17517 from the National Science Foundation.

species, since it has been proposed that the genus *Hydrogenomonas* should be abandoned (13, 61). Aside from the exclusion of some specialized groups, the artificial nature of the generic description proposed, and the problems of species and strain allocation to the genus are illustrated by the inclusion of some bacteria that are permanently immotile or have variable flagellar insertion. Although polar flagellation is a key generic character, the immotile species *P. mallei* is included in the genus because of its clear phenotypic and phylogenetic relationship to flagellated species (62, 65). Immotile strains of other species are occasionally isolated from nature. In some cases, such strains can be readily recognized as members of flagellated species (e.g., *P. aeruginosa*); in other cases their assignment is extremely difficult.

Another problem of delimiting the genus *Pseudomonas* pertains to the DNA composition of the component species. The range of G + C content (ca. 58–70 moles percent) is already quite broad as compared with that of some bacterial genera, but is comparable to that of others (45). Furthermore, this range may need to be expanded to accommodate other species closely related to those presently included in the genus. A splitting of the genus as presently constituted into smaller genera on the basis of DNA composition alone does not seem to be warranted, because of the wide and overlapping G + C spectra of certain DNA homology groups within the genus.

PHENOTYPIC PROPERTIES USEFUL FOR SPECIES CHARACTERIZATION

Morphology and Pigmentation

Cell morphology.—The cells of most strains of *Pseudomonas* are straight or slightly curved rods with the dimensions indicated in the generic description. Members of some species (notably *P. syringae,* some strains of *P. putida,* etc.) have cells considerably longer than 4μ. Although the shape and dimensions of the cells vary with conditions of cultivation and age of cultures, these properties have never been analyzed systematically, and the morphology is usually reported for cells observed during exponential growth in complex media. The curvature of the cells in one plane cannot always be taken too seriously; in many cases it is possible to find vibrioid cells in populations where the majority of the cells are straight and vice versa.

Flagella.—The cells of *Pseudomonas* species are typically motile by means of polar flagella, which have a wave length relatively constant for each species. The cells may have one or several polar flagella, and the shape and number of flagella is a taxonomically useful character. In some cases, it is necessary to quantitate the number of flagella on a statistical basis (40). In addition, the point of insertion of the flagella is not invariably polar for some species of the genus, in some instances being sub-polar. The problem of determining type of flagellation is illustrated by an analysis of the number of

TABLE 1. Number and arrangement of flagella in some "hydrogen bacteria"

Species	Number of flagella per cell				Origin of each flagellum			Type of flagellation
	Average	Per cent of cells counted possessing			Per cent of all flagella counted			
		1	2	3 or more	polar	sub-polar	lateral	
Alcaligenes eutrophus	2.7	20	29	51	30	45	25	peritrichous
Alcaligenes paradoxus	1.54	62	25	13	32	48	20	degenerately peritrichous
Pseudomonas flava	1.06	94	6	0	69	24	7	sub-polar
Pseudomonas facilis	1.02	98	2	0	98	1	1	polar

flagella and their point of insertion for four species of hydrogen bacteria shown in Table 1 (13). It should be emphasized that percentages of sub-polar flagella reported in the table probably represent minimal figures because those cells that are oriented on the slide in such a way that the origin of the flagellum is above or below the cell would appear to be polarly flagellated. Other problems connected with the correct interpretation of two-dimensional images like those of flagella stain preparations were discussed by Hodgkiss (26).

Under some growth conditions, certain strains produce lateral flagella often easily distinguished from the "normal" or polar type not only by the point of insertion but also by their shorter wave length. Such lateral flagella have been observed in the species *P. stutzeri* and *P. mendocina,* particularly after growth for a short time on solid complex media. These flagella are easily shed, and in older cultures most cells retain only a single polar flagellum. It is not known whether lateral flagella are functional in the motility of the cells. The property of producing "mixed flagellation" is found not only in the genus *Psuedomonas;* it has also been reported in the genera *Aeromonas, Chromobacterium* (64, 72, 73), and *Beneckea* (6). In *Chromobacterium* Sneath (72, 73) has demonstrated that polar and lateral flagella are antigenically different, a fact that suggests the presence in the cells of two sets of genetic determinants.

Carbon reserve materials.—Only two specialized carbon reserve materials are known to occur in cells of *Pseudomonas:* poly-β-hydroxybutyrate (PHB) and a glucose polysaccharide which is probably identical with glycogen. Some species may accumulate both reserve materials, and the nature of the principal carbon reserve may vary depending on the exogenous carbon source used as nutrient (2). PHB accumulation is a very important diagnostic character for identifying species. Its accumulation as intracellular globules is enhanced by growth under conditions of nitrogen starvation and usually can be detected by microscopic observation, either after staining a smear with Sudan Black (9) or simply by phase contrast microscopy. The best available exogenous carbon source for accumulation of PHB is DL-β-hydroxybutyrate, used by all species capable of accumulating the polymer. Some species (e.g., *P. pseudoalcaligenes*) accumulate PHB in small amounts, and it may be necessary to complement the microscopic observation with the isolation of the granules and their chemical characterization (71, 86).

Pigmentation.—Because of its striking character, the production of pigments by microbial cultures has received a great deal of attention from taxonomists. Many strains of *Pseudomonas* produce pigments, some of which diffuse into the medium while others remain associated with the cells.

One group of diffusible pigments of toxonomic importance are the fluorescent pigments produced by the so-called fluorescent pseudomonads. Some *Pseudomonas* species (e.g. *P. cepacia, P. marginata, P. caryophylli*) produce diffusible yellow-green pigments that are sometimes mistaken for the fluorescent pigments; these can be readily distinguished from the fluorescent pigments by examination of cultures on solid media under a source of ultraviolet light of short wave length (ca. 254 nm), under which no fluorescence is observed. In order to elicit the production of fluorescent pigments, an appropriate culture medium is indispensable. The best media are those of very low iron content, e.g., the medium B of King, Ward & Raney (35) and the medium proposed by Garibaldi & Bayne (20), containing egg white. Unfortunately, in a survey carried out by Lelliot et al (42) on the reproducibility of taxonomic tests in different laboratories, there was a particularly poor agreement among different observers on the detection of fluorescent pigment production.

Also frequently found among *Pseudomonas* species are the phenazine pigments, to which belong pyocyanin of *P. aeruginosa* (87), the green pigment (chlororaphin) of *P. chlororaphis* (*P. fluorescens* biotype D) (38) and the orange pigment (phenazine-1-carboxylic acid) of *P. aureofaciens* (*P. fluorescens* biotype E) (22, 36). Some *P. aeruginosa* strains have also been reported to produce phenazine-1-carboxylic acid (11, 82), dihydroxyphenazine-1-carboxylic acid (39), chlororaphin (82), oxychlororaphin (11, 69), and the red pigments, aeruginosin A (27) and aeruginosin B (25). Strains of *P. cepacia* (= *P. multivorans*) also produce phenazine pigments (49, 75) with a bewildering variety of colors (yellow, green, red, purple) which diffuse into the medium, are confined to the colonies, or both, depending on the carbon source used for cultivation. For phenazine pigment production the medium of preference is medium A of King, Ward & Raney (35). Demonstration of the green pigment of *P. chlororaphis* is difficult and on occasion we have observed, instead, production of an unidentified diffusible yellow pigment somewhat resembling that of *P. aureofaciens*.

Various other pigments of unknown chemical nature produced by *Pseudomonas* species are the dark diffusible pigments of some strains of *P. marginata* (= *P. alliicola*), *P. solanacearum*, and other plant pathogenic species. These pigments may be melanins, as is probably the black pigment produced in tyrosine media by *P. lemoignei* (16); melanin production has been reported for some strains of *P. aeruginosa* (43, 47).

Carotenoid and carotenoid-like pigments are produced by some species of the genus, notably *P. vesicularis* (2), *P. mendocina* (58), *P. flava,* and *P. palleronii* (14). Carotenoid-like pigments are also characteristic of pseudomonads of the genus *Xanthomonas* (80). These pigments are still listed among the carotenoids of unknown structure (81). The carotenoid pigments, always associated with the cells, can be extracted with organic solvents and characterized by their visible absorption spectra. Very careful handling of the cell extracts is necessary to prevent chemical deterioration of pigments. Among other water-insoluble pigments that are associated with the cells and do not diffuse into the medium, the blue pigment of *P. lemonnieri* (*P. fluorescens* biotype F) should be mentioned (78). Its chemical nature has been elucidated (79). This pigment was believed in 1966 to be a phenazine, but it was discovered that medium A of King, Ward & Raney (35) which is generally used to elicit the production of phenazine pigments, was unsatisfactory, and a different medium, consisting of 0.5% Bacto-peptone, 1% glucose and 2% Bacto-agar in tap water, was recommended (75). The yellow pigment of *P. maltophilia.* although water soluble, is not released into the medium, and has not been characterized.

Physiological Properties

Nutritional properties.—Most *Pseudomonas* species grow on purely mineral media supplemented with an appropriate organic carbon compound as the sole source of carbon and energy. As shown by den Dooren de Jong (17), some species can use a great variety of carbon sources; others can use only a limited number of compounds. Three species require known growth factors in addition to the principal source of carbon; these are *P. maltophilia, P. diminuta,* and *P. vesicularis* (2, 75).

Because the spectrum of substrates that can be used as sole carbon sources is, to a large extent, characteristic of a given species and because relatively few other characters are available for species differentiation, we have relied heavily on nutritional analysis in our circumscription of taxa below the generic level. A simplified procedure of testing for the utilization of different compounds, based on the "replica plating" technique of Lederberg & Lederberg (41), makes it possible to accumulate a great deal of information with relatively little effort (75). Our considerable experience with the method, however, as applied not only to *Pseudomonas* strains, but also to other aerobic bacteria, has revealed certain pitfalls and imperfections that should be recognized by anyone using the technique.

To begin with, the solid mineral base originally proposed by us (75), and subsequently used in our own and other laboratories, does not appear to be optimal under all circumstances, probably because of the high concentration of the strong chelating agents used in its preparation. For some species, it is unsuitable for autotrophic growth with hydrogen, and in many cases it allows only very slow heterotrophic growth with certain aromatic compounds, terpenes, and sugars. In general, the following mineral base gives much more satisfactory results: Na-K phosphate buffer, pH 6.8, M/30; NH_4Cl, 0.1%; $MgSO_4$, $7H_2O$, 0.05%; ferric ammonium citrate, 0.005%; $CaCl_2$, 0.0005%; Ionagar, 1%.

Many carbon sources can be autoclaved in the mineral medium. Those requiring special treatment are listed in Table 2. In most cases, a final concentration of 0.1% of the carbon source is used. This concentration can generally support very good growth; however, in some cases (e.g., when the substrates are aromatic compounds, terpenes, fatty acids, dicarboxylic acids, etc.) it may be too high because of the toxicity of the substrate itself or of some metabolic intermediate that is accumulated by the cells (e.g., catechol

TABLE 2. Carbon compounds requiring special treatments for their use as substrates in nutritional screenings

Compound[a]	Method of sterilization[b]	Final concentration (% w/v)	Comments
Monosaccharides, disaccharides, sugar alcohols	F	0.1–0.2	
2-Ketogluconate, hydroxymethylglutarate, α-ketoglutarate, pyruvate, laevulinate, citraconate, itaconate, mesaconate, benzoylformate, *m*-hydroxybenzoate, quinate, L-threonine, δ-aminovalerate, L-histidine, L- and D-tryptophan, L-kynurenine, kynurenate, *m*-aminobenzoate, betaine, sarcosine, creatine, pantothenate, trigonelline	F	0.1	
Starch (soluble)	aut	0.2	
Poly-β-hydroxybutyrate	aut		In thin overlayer of mineral base agar at 0.25% concentration (16)
Alcohols	none	0.1–0.2	Added aseptically to sterile medium

[a] Acid or basic compounds, hydrochlorides, sulfates, etc. should be neutralized before sterilization.

[b] F = filtration through Millipore membranes, except for amines, as indicated; aut = sterilization in autoclave.

from various aromatic compounds; anthranilate from tryptophan). The optimal concentration for phenol, is ca. 0.025%. Some volatile materials (e.g., terpenes, naphthalene, camphor, and geraniol), even when provided in the atmosphere within the Petri dish (75), are sometimes toxic. Partial saturation of the medium can be obtained by leaving the compound on the Petri dish lid for a few hours, then removing the lid and replacing it with an empty one. When a toxic carbon source is used in the solid medium, growth is often more rapid if a large number of patches are printed on the plate, because the growing bacteria reduce the concentration of the substrate. Toxicity problems should not, however, be solved by overcrowding the plate with patches of different organisms. The toxic effect of some substrates is more noticeable in liquid than in solid media. In such cases, growth in a liquid medium can frequently be initiated at a very low substrate concentration, and the compound

TABLE 2.—(*Continued*)

Compound[a]	Method of sterilization[b]	Final concentration (% w/v)	Comments
Geraniol, terpenes, naphthalene	none		Placed in lid of Petri dish; see toxicity problems in text
Phenol	none	0.025	Added aseptically to sterile medium
Testosterone	none		Suspension in water is sonicated for 30 min.; used immediately in thin overlayer of mineral medium at 0.2% concentration
L-tyrosine	aut		In thin overlayer of mineral base agar at 0.2% concentration
Anthranilate	F	0.05–0.1	
Amines	F	0.1	Filtered through sterilizing sintered glass (Pyrex, UF grade)
Hydrocarbons	F		2–3 drops per tube of liquid mineral medium; incubated with shaking

can be added periodically in small increments; the total amount of substrate can thus be built up to 0.1% or more without much difficulty.

Placing too many patches on a plate is generally undesirable because (*a*) some organisms may be inhibited by toxic substances excreted by their neighbors, (b) slow-growing bacteria may starve and fail to grow if surrounded by rapidly growing organisms and (*c*) some patches may grow at the expense of products excreted by neighboring patches. Such "cross-feeding" is frequently seen on media containing carbon sources that are hydrolyzed to products readily usable by most species (disaccharides, polysaccharides, acetamide, etc.) and may be recognized from the heavier growth of a given patch at its edges proximal to other patches.

The results obtained in nutritional screening can be expressed in many ways, but we have adopted only three notations: $-$ for no growth, $+$ for growth and $\pm$ for limited growth, after 4–7 days of incubation at 30°C. There is little difficulty in identifying the two extreme situations when the plates are compared with a similarly incubated plate containing no carbon source. Such a control plate may show some marginal growth barely visible to the eye. Results scored as $\pm$ are more difficult to assess. Poor growth may be due to very slow substrate use, in which case the patch eventually becomes very dense, to the presence of a metabolizable impurity in the carbon source, or to the accumulation of a toxic product. In these last cases, the patch will reach a limit of density at some stage of its growth. Doubtful results should be checked individually by second transfers to homologous solid or liquid media, preferably containing different substrate concentrations. It should be kept in mind that a few generations of cells growing very slowly on solid medium will give a visible patch, while at the same rate of growth with a small inoculum a liquid culture might be scored as negative. Some species that require growth factors grow very slowly when provided with their minimal needs, as shown by Ballard et al (2).

Occasionally, in patches replicated on certain media on which growth is very scanty, mutant colonies appear which are composed of cells capable of rapid growth on the particular carbon substrate. It should be noted that such mutants, which we have scored as $+^{M}$, will appear only if a large enough inoculum is used, or a sufficient amount of "marginal growth" occurs on the plate.

Acid production from carbohydrates and sugar derivatives.—The production of acid in media containing different carbohydrates has been used since the last century for identifying bacterial genera and species. The tests devised by Hugh & Leifson (29), which generally give clear and reproducible results, have been used for characterization of *Pseudomonas* species in most descriptions and in the determinative keys in the 7th edition of Bergey's Manual (8). Unfortunately, the nature of the acids produced during growth is often not known, and in many cases production of acid is not correlated with other nutritional data. For example, some bacteria produce acids from substrates that are not used for growth, while others produce no acid from usable carbohydrates. Furthermore, there is a problem of redundancy, in that a single enzyme may effect the oxidation of several different sugars that may or may not serve as carbon sources (5, 57). For these reasons, we have not relied on acid production in characterization of our strains, but believe that the technique of Hugh & Leifson should be exploited, where possible, for species descriptions.

Nitrate reduction.—Denitrification is a character that is taxonomically very useful, although it tends to be lost by many strains after repeated subcultures in ordinary laboratory media. Before recording the test as negative, it should be performed after several transfers of the organism in the denitrification medium under semi-aerobic conditions, because it has been observed that in many instances the property can be regained by a culture, possibly by selection of denitrifying clones. Occasional cultures are able to grow well under anaerobic conditions in the presence of nitrate without producing visible gas. Some of these can be induced to produce gas after repeated subcultures in the same medium, but not always. It may be that such cultures produce nitrous oxide, which is very soluble in water, and cannot further reduce this compound to N_2. Although the glycerol denitrification medium proposed by Stanier, Palleroni & Doudoroff (75) is excellent in many cases, it should be remembered that other carbon sources must be substituted for glycerol when this compound cannot be used as a substrate by the organisms under study. The production of nitrite is a classical character used for species descriptions, but we have not generally relied on this character because in some cases nitrite is accumulated transiently and only in certain media.

Pathways of degradation of aromatic compounds.—The biochemical pathways for degradation of various aromatic compounds have been studied in considerable detail in members of the genus *Pseudomonas,* particularly those belonging to the fluorescent group (21, 53). The analysis has been very extensive in several cases, covering not only study of pathways, but also the control mechanisms governing synthesis of the enzymes involved (24, 51, 52, 59), the immunological properties of the enzymes (76), and genetic aspects (34, 66, 85). From the practical taxonomic point of view, cleavage mechanisms of hydroxylated aromatic intermediates have provided useful data which correlate very well with a number of other unrelated phenotypic characters. The methods for detection of *ortho* and *meta* cleavages of catechol and protocatechuic acids are simple and have been described (75). Tests for the gentisate pathway are far more complicated (84). The cells are grown in all cases in a mineral medium containing an appropriate aromatic compound as carbon source, but some strains, unable to use any aromatic compound for growth, may still be tested after growth with a suitable usable precursor of the hydroxylated intermediate (e.g., quinate). In some instances, inability to grow with benzoate is apparently due to the extreme toxicity of this compound for organisms that can nevertheless use mandelate and apparently metabolize it via benzoate (unpublished observations). It should also be noted that the same organism may use different pathways for metabolizing different carbon sources which are metabolized through common intermediates. Thus, *P. putida* uses the *ortho* cleavage of catechol when grown on benzoate, and the *meta* cleavage when grown on phenol (19). Hence, comparisons between different organisms are valid only when these are grown with the same substrate.

Arginine dihydrolase reaction.—We have, in all cases, tested for this reaction by measuring the disappearance of arginine, but we have found the method of Thornley (83) much less laborious, quite satisfactory, and yielding comparable results. Although some species that give an arginine dihydrolase negative reaction can grow with arginine as sole carbon source, we have never seen a strain that gives a positive reaction but cannot grow with arginine.

Animal and plant pathogenicity.—Although roughly one half of the strains studied by us were of clinical origin or known to be pathogenic for plants or animals, we have not tested the pathogenicity or host specificities of more than a few. While *P. mallei* is clearly recognized as a highly specialized animal parasite, there is relatively little information available on the relationships among the saprophytic, parasitic, and potentially pathogenic members of the genus. The host of origin and type of lesion produced have a high taxonomic value to plant pathologists, but in most cases, the host ranges have not been thoroughly investigated. Similar problems are encountered not only with strains isolated from soil, but also with those isolated from diseased plants, because of the practical limitations on the number of hosts that can be tested for susceptibility to a given strain. Although we hope that, in future, host specificity will be better understood and become a rational basis of taxonomy, we can, at present, only agree with Skerman (70) that "one can either test pathogenicity over the entire plant world or, more rationally, attempt identification by other means."

Other properties.—We have excluded from this review other methods of phenotypic characterization, based on oxidase reaction, production of extracellular enzymes, levan formation, and temperature relationships, because the procedures have been adequately described in the paper by Stanier, Palleroni & Doudoroff (75).

THE INTERNAL SUBDIVISION OF THE GENUS *PSEUDOMONAS*

Ideally, taxonomy should yield a natural system of classification that should take into account an evaluation of the relatedness of organisms that have become differentiated from one another through divergent evolution, or have acquired similar properties as a result of convergent evolution. Very little taxonomic work performed to date with bacteria can claim to approach the ideal of a natural system based on phylogeny. Several approaches to analysis of relatedness among procaryotes are listed and discussed by Stanier (74), and include analysis at the genetic level in vivo and in vitro, as well as what he calls the "epigenetic" level, based essentially on structural and functional similarities of homologous proteins and regulatory mechanisms operative in their synthesis. A comprehensive review of molecular biological approaches to bacterial taxonomy has recently been presented by Mandel (46).

Although genetic transfer has been observed in some species of *Pseudomonas* (28, 85) and may occur in others, it is not a suitable tool for the study of the phylogeny of such a large and diverse genus. Perhaps the most valuable method for assessing phylogenetic relationships is determination of nucleic acid homologies among strains by in vitro hybridization. We have used the approach of DNA-DNA hybridization in a survey of a large number of phenotypically well characterized strains belonging to different *Pseudomonas* species. Our choice has been the so-called competition method using reference DNA immobilized on nitrocellulose membranes (33). The DNA annealing has in all cases been conducted at a temperature 25°C below the T_M of the immobilized DNA (ca. 70–72°C) and in many cases at the more restrictive temperature of 80°C. The DNA homology values obtained at 80°C remain at about the same level as those obtained at the lower temperature when one is dealing with closely related strains, but decrease, often to zero, for more distantly related ones. The values obtained at 80°C are probably less reliable because of greater and less controllable loss of immobilized DNA from the filters during hybridization, but generally give a clearer discrimination among different homology clusters. The competition technique, in rare cases, does not give reproducible results or good agreement of values obtained in reciprocal annealing experiments. In such cases, the direct binding technique (50) is found to be preferable (57).

Most *Pseudomonas* species studied fall into several distinct DNA homology groups. All members of a given group are related directly or indirectly to other members of the same group at various levels of DNA homology, but show no detectable homology with members of other groups.

The *P. fluorescens* DNA Homology Complex

The largest DNA homology group that we have studied, and which we have called the "*P. fluorescens* complex" because *P. fluorescens* appears to occupy a central position, contains both fluorescent and nonfluorescent pseudomonads, and includes the type species of the genus, *P. aeruginosa* (75). Aside from DNA homology, phylogenetic relatedness of some species of the complex is evidenced by identity of the pathways used for degradation of aromatic compounds, immunological relatedness of isofunctional enzymes involved in these pathways (76), and by the pattern of feedback inhibition of the bio-synthetic enzyme, DAHP synthetase (31). These and other "epigenetic" characters of the group are discussed by Stanier (74).

The *P. fluorescens* complex contains species that are phenotypically diverse and fall into several subgroups, e.g., the fluorescent and alcaligenes groups (75), and the stutzeri group (58). For the entire complex the range of GC content of the DNA is between ca. 58 and 67 moles percent (45). With the exception of some strains of *P. pseudoalcaligenes,* none of the component species accumulate poly-β-hydroxybutyrate, which is the characteristic reserve material in almost all species outside this group that we have studied. Some nomenspecies (e.g., *P. aeruginosa, P. mendocina*) are very homogeneous with respect to phenotypic properties as well as to DNA composition and homology; others are heterogeneous and probably include several biologically distinct species that are now lumped together for pragmatic reasons of classification and identification. This has been graphically illustrated for *P. aeruginosa* and *P. stutzeri* by plotting phenotypic matching coefficients against DNA competition values for a selected reference strain of each species, as well as by cluster analysis based on similarity coefficients of several species presented in dendrogram form (55). The homogeneity of *P. aeruginosa* and the heterogeneity of *P. stutzeri* (as now constituted) are also evident from the frequency distribution of 29 strains of each species that can use different numbers of carbon sources for growth, as shown in Figure 1. The data for this figure have been taken from our previous publications (58, 75).

P. aeruginosa.—This is a distinct species that can usually be recognized by a constellation of several phenotypic characters, namely pigment production, monotrichous flagellation, growth at 41°C, denitrification, and the utilization of several substrates such as acetamide and geraniol. Most cultures also have a characteristic odor. The most comprehensive characterization of this species, based on the study of 354 strains, is that of Jessen (32), who recognized it as a homogeneous and easily identifiable group. Phages for *P. aeruginosa* are generally specific for all strains of the species, but not for other members of the fluorescent complex, although the phages isolated against other fluorescent organisms occasionally attack *P. aeruginosa* (28).

The P. fluorescens group.—*P. fluorescens,* as circumscribed by Stanier, Palleroni & Doudoroff (75) can be distinguished phenotypically from *P. aeruginosa* mainly on the basis of its multitrichous flagellation, lower maximum growth temperature, and absence of pyocyanine, and from *P. putida* on the basis of its capacity to liquefy gelatin. The internal subdivision of the species into a number of biotypes (A to G) was based mainly on levan formation, denitrification, pigment production, and a number of nutritional characters. The distinguishing features of the different biotypes are listed in Table 15 of Stanier, Palleroni & Doudoroff (75). DNA homology studies support the view that *P. fluorescens* biotypes, taken together, represent a subgroup of the *P. fluorescens* complex. This sub-group is internally heterogen-

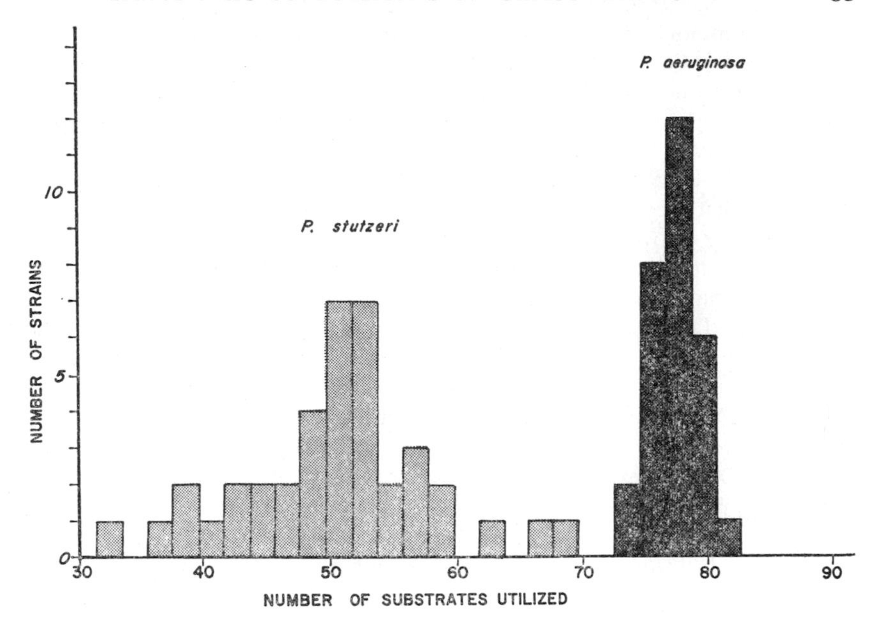

FIGURE 1. Frequency distribution of the numbers of carbon compounds that can be used for growth by 29 strains of *P. stutzeri* and 29 strains of *P. aeruginosa.* A total of 149 compounds were tested.

eous, but contains at least one very homogeneous cluster consisting of biotypes D and E (55).

Biotype A includes the typical *P. fluorescens* (63). Of the 24 strains assigned to this group (75) only three have been analyzed for DNA homology. These show an affinity to all other biotypes and the closest relationship to some strains of biotype B. The phenotypic homogeneity of this biotype is evident not only from our own studies (75), but also from the fact that all 18 strains from Jessen's collection that he had placed in one of his 82 biotypes of fluorescent pseudomonads (biotype 63, group V) (32) were found to fall into this group.

Biotype B differs phenotypically from biotype A mainly in its ability to denitrify and to grow with ethanol, propanol, and propyleneglycol. This group includes strains assigned to the phytopathogenic nomenspecies, *P. marginalis.* Sixteen of the nineteen strains assigned to this biotype were all classified by Jessen in his group V, biotypes 61 and 62.

The strains assigned to biotype C are denitrifiers, do not produce levan from sucrose, and do not grow on L-arabinose and D-xylose, while most strains of biotypes A and B are able to use these two sugars. Nine of the fifteen strains of C studied (75) grow on higher dicarboxylic acids, a property not commonly found in other members of the *P. fluorescens* complex with the exception of *P. aeruginosa.* Of a large number of Jessen's (32) strains that we have examined (unpublished observations) the great majority belong to his group IV (biotypes 49, 50, 51, and 52), but several strains of his other biotypes show a great resemblance to biotype C. Although only 3 strains of this biotype have been examined in DNA competition experiments, it appears that they represent a distinct homology cluster within the species. While, to our knowledge, no specific names other than *P. fluorescens* have been applied to biotype C, it seems likely that on further study, at least one new specific name may be warranted for its members.

Biotypes D and E differ from all other biotypes of *P. fluorescens* and from *P. putida* in the higher and identical G+C content of their DNA (63.5 moles percent) and in their capacity to produce closely related phenazine pigments. These two biotypes, to which the specific names *P. chlororaphis* and *P. aureofaciens* have been applied, form a single isolated cluster with respect to DNA homology and to a lesser extent with respect to numerical analysis of phenotypic characters. This is shown in Figure 2, based on data previously reported (55, 75) for 3 strains each of the two biotypes. It would appear that if *P. fluorescens* is to be dismembered, biotypes D and E might either be retained as separate nomenspecies or, more reasonably, combined in one species for which the name *P. chlororaphis* would be applicable on the basis of priority. It is worth mentioning that re-examination of one (denitrifi-

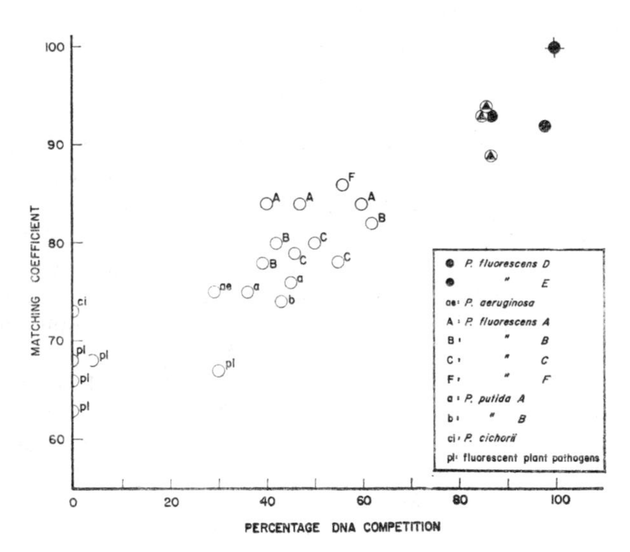

FIGURE 2. Matching coefficients versus DNA competition values (at $T_M - 25°C$), with *P. fluorescens* biotype D strain 31 as reference (55). Note that some members of the *P. fluorescens* DNA homology complex show no DNA homology with this particular reference strain.

cation) of the five phenotypic characters used (75) to differentiate between these biotypes, has shown that three of six strains of biotype E (strains 36, 38, and 41), which had been recorded as negative for denitrification, acquired the ability to produce gas after prolonged periods of "adaptation" in nitrate media. One other strain (strain 86) grew profusely under anaerobic conditions without visible gas production (unpublished observations).

Biotype F (nomenspecies *P. lemonnieri*) is usually identified by the production of a blue insoluble pigment (78, 79), only produced on certain media (see section on pigmentation). The single strain that has been analyzed by DNA competition and numerical taxonomy is clearly related to other biotypes.

Biotype G, as created by Stanier, Palleroni & Doudoroff (75) was recognized as a frankly provisional and heterogeneous group. A large number of strains that we have identified as members of this group (unpublished observations) were assigned by Jessen (32) to no less than 25 biotypes belonging to his groups II, III, and IV. Considerable work on this group will be needed to understand its internal subdivision and external relationships with other biotypes of *P. fluorescens.*

P. putida.—This species was redescribed by Stanier, Palleroni & Doudoroff (75), who recognized two biotypes, A and B, of the species. On the basis of DNA homology studies, however, it appears that these two sub-groups are not closely related, and that biotype B may have closer affinity to *P. fluorescens* (55). This was already noted in the original description of this biotype, and was later supported by immunological analysis of isofunctional enzyme (76; H. Kita, personal communication).

Biotype A, which should be regarded as typical of *P. putida,* is in fact fairly heterogeneous with respect to DNA homology, but relatively homogeneous with respect to phenotypic properties. This is illustrated in Figure 3, which also shows the phenotypic closeness of biotypes A and B and the DNA homology gap between them. Further evidence of phenotypic homogeneity of biotype A is apparent from the fact that all the strains received from Jessen were assigned by him to his group II (especially his biotypes 11, 12, and 13) (32; unpublished observations).

The fluorescent phytopathogens.—The fluorescent phytopathogenic species of *Pseudomonas,* characterized by the absence of the arginine dihydrolase system, represent a separate evolutionary branch of the *P. fluorescens* complex, as shown by DNA homology studies (55). With the exception of *P. cichorii,* members of this group are oxidase negative, which distinguishes them from all other members of the complex, as well as from most other *Pseudomonas* species. It is interesting to note that *P. cichorii* shows lower DNA homology with all oxidase negative strains that have been tested than these do with each other.

A fantastic number of specific names has been created by plant patholo-

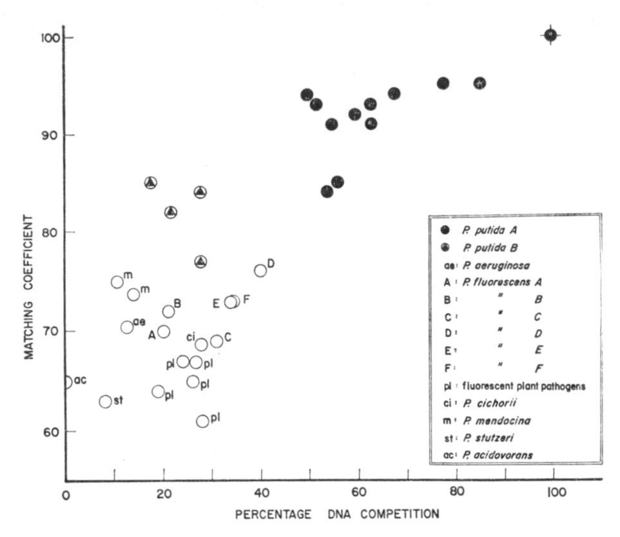

FIGURE 3. Matching coefficients versus DNA competition values (at T_M — 25°C), with *P. putida* biotype A strain 90 as reference (55).

gists for different members of this group, mainly on the basis of the plant host of origin, on the assumption that each organism is highly specific for its host, and in the type of lesion it produces. Of the more than 60 nomenspecies listed in the Seventh Edition of Bergey's Manual (8), most are unrecognizable because their phenotypic properties have not been thoroughly investigated, their host ranges have not been explored, and the type strains have been lost. Recent attempts have been made to characterize some members of the group on the basis of physiological tests, including those used by us in the study of saprophytic species (42, 48, 67). Although the numerical analysis of a large number of characters of many strains shows some clustering which corresponds to their nomenspecies assignments, there are also many inconsistencies in the nomenclature (67). On the basis of this numerical analysis, Sands, Schroth & Hildebrand (67) support the view expressed by Starr (77) and by Lelliott, Billing & Hayward (42) that the oxidase negative members of the group should, for the present, be lumped together in a single nomenspecies, *P. syringae,* which would contain a variety of pathotypes. Homology studies with a limited number of strains received under the designations of *P. syringae, P. phaseolicola, P. glycinea, P. mori, P. tomato,* and *P. savastanoi* also support this view, but indicate that there may be some homogeneous DNA homology clusters within such a species (55). Our feeling is that extensive physiological, DNA homology, and host-specificity studies will justify retention of some specific names (e.g., *P. phaseolicola*) regarded as synonyms according to the above proposal. We believe, however, that, with our present lack of knowledge and the confused state of the nomenclature, the advantages of lumping poorly characterized nomenspecies outweigh the advantages of retaining their individuality. Because of their special ecological niches, the fluorescent plant pathogens may offer some of the most interesting material for the study of bacterial speciation.

Nonfluorescent species of the P. fluorescens complex.—The nonfluorescent members of the *P. fluorescens* DNA homology complex include the species *P. alcaligenes, P. pseudoalcaligenes, P. mendocina,* and *P. stutzeri.* The first two species were originally both identified as *P. alcaligenes* by Hugh (1), and were subsequently separated on the basis of several phenotypic characters, including the G+C content of the DNA, but were kept in a single phenotypic group (45, 75). The last two species were assigned to the *stutzeri* group (58). Subsequent studies on DNA homology have indicated that *P. mendocina* is more closely related to *P. pseudoalcaligenes* than it is to *P. stutzeri* (55, 60).

P. stutzeri and P. mendocina.—As mentioned earlier in the discussion of the entire *P. fluorescens* complex, *P. stutzeri* is a very heterogeneous species with respect to phenotypic properties and DNA homology. It is the only *Pseudomonas* nomenspecies we have examined that has a broad span in the G+C content of the DNA (ca. 61–66 moles percent). The name *P. stanieri* was originally proposed for a cluster of low G+C content (45). Although the heterogeneity of *P. stutzeri* indicates that it may contain two or perhaps several species, we were unable to detect among many strains any clusters of different G+C contents that showed sufficiently clear-cut phenotypic differences to justify subdivision of the species at present (55, 58). Aside from the characteristic wrinkled appearance of the colonies of freshly isolated strains, some characters useful for identifying most strains of the species are the ability of hydrolyze starch and to grow at the expense of ethylene glycol.

Unlike *P. stutzeri, P. mendocina* is a homogeneous species that resembles *P. stutzeri* in its mixed type of flagellation on solid media (see section on morphology and pigmentation), the ability to denitrify, and to use ethylene glycol. The colonial morphology is, however, different, and colonies become pale yellow because of the presence of a carotenoid pigment in the cells. This species also differs from *P. stutzeri* in its ability to use geraniol, putrescine, spermine, betaine, and sarcosine, and in its inability to hydrolyze starch.

P. alcaligenes and P. pseudoacaligenes.—Of the many strains received by us as *P. alcaligenes*, only three could be identified as belonging to this species after an extensive phenotypic analysis, and even these three did not show great phenotypic similarity. The G+C content of the DNA of these strains is in the range ca. 66–68 moles percent (60). The type strain is described by Stanier, Palleroni & Doudoroff (75)

P. pseudoalcaligenes, described in 1966 (75) differs from *P. alcaligenes* in a number of phenotypic characters and in the lower G+C content of the

DNA (ca. 62–64 moles percent). Although phenotypically quite heterogeneous, this species is relatively homogeneous in DNA homology. As has been mentioned, DNA hybridization experiments show that *P. pseudoalcaligenes* is much more closely related to *P. mendocina* than to *P. alcaligenes*. A number of strains have been added to the collection originally consisting of only six strains; their phenotypic properties have been studied (60) and will be published in the near future. Aside from being the only nomenspecies of the *P. fluorescens* DNA homology complex that contains some strains that accumulate poly-β-hydroxybutyrate, *P. pseudoalcaligenes* is also the only species that is variable for the arginine dihydrolase reaction, another important diagnostic character in identification of *Pseudomonas* species. These two properties do not show any correlation with each other in different strains of the species, nor with DNA homologies between them.

The *pseudomallei-cepacia* DNA Homology Complex

The second largest group of species that has emerged as a separate DNA homology complex includes the animal pathogens *P. pseudomallei, P. mallei,* and the phytopathogens *P. cepacia* (= *P. multivorans*), *P. marginata* (= *P. alliicola*), and *P. caryophylli* (3, 65). The range of G+C content of the DNA for the entire group is ca. 65–70 moles percent (3, 45). None produce fluorescent pigments, although some strains produce greenish-yellow nonfluorescent diffusible pigments (see section on pigmentation). All species are multitrichous, and accumulate poly-β-hydroxybutyrate. All species of the group can use cellobiose, D-arabinose, and D-fucose as substrates for growth; this combination is rarely found in other *Pseudomonas* species. The distinctive characters of the species are listed by Stanier, Palleroni & Doudoroff (75) and by Ballard et al (3).

P. cepacia.—*P. cepacia* is the most omnivorous pseudomonad described to date. Of 136 organic compounds tested, more than a hundred could be used as sole carbon sources by a majority of the strains studied. This nutritional versatility led to the proposed name, *P. multivorans* (75), later rejected as a synonym of *P. cepacia* (3). This synonymy illustrates one of the major problems of bacterial taxonomy. *P. cepacia* was poorly described as a causative agent of onion rot by Burkholder in 1950 (10). Strains designated as *P. multivorans* were isolated from soil, water, and clinical specimens; not all of these cause rot of onion slices. Strains of this species produce phenazine pigments.

P. marginata.—*P. marginata* and *P. alliicola* were originally described as pathogens for *Gladiolus* and onions, respectively. Evidence of their identity is presented by Ballard et al (3), who regard the latter name as a synonym of the former. Nonfluorescent greenish-yellow diffusible pigments are produced in some media by many strains, and most strains received as *P. alliicola* produce a darkening of the medium, the bacterial mass on old slant cultures, or both.

P. caryophylli.—Only three strains of *P. caryophylli* have been thoroughly characterized. Except for some strains of *P. mallei,* this species is the least nutritionally versatile of the entire DNA homology complex. Although *P. caryophylli* is generally recognized as the agent of carnation wilt, it can cause rot of artificially inoculated onion slices, like *P. cepacia* and *P. marginata* (3).

P. pseudomallei and P. mallei.—The motile species, *P. pseudomallei,* which is an occasional animal pathogen causing melioidosis, and the nonmotile species, *P. mallei,* a specialized parasite that causes glanders in horses, are closely related, as evidenced by certain phenotypic characters (62), G+C content of the DNA (45), and DNA homology (64). While *P. pseudomallei* is versatile in its nutrition, *P. mallei* includes strains that have the most restricted nutritional spectrum in the entire pseudomallei-cepacia DNA homology complex. The taxonomy and ecology of the two species are discussed by Redfearn, Palleroni & Stanier (62).

The *acidovorans* DNA Homology Group

Two species, *P. acidovorans* and *P. testosteroni,* were recognized as belonging to a separate phenotypic group by Stanier, Palleroni & Doudoroff (75), who also characterized a large number of strains. In spite of the disparity in the DNA composition (67% and 62% G+C, respectively), the two species show DNA homology with each other and none with any other *Pseudomonas* tested (61). Other evidence of phylogenetic relationship between the two species, and of their isolation from other *Pseudomonas* species, can be adduced from studies of some enzymes involved in biosynthetic pathways. The pattern of feedback inhibition of the 3-deoxy-D-arabino-heptulosonate-7-phosphate synthetase, an enzyme of aromatic acid biosynthesis, is similar in *P. acidovorans* and *P. testosteroni,* and different from that found in *P. cepacia* and members of the *P. fluorescens* complex (31). The same is true for the feedback inhibition of the aspartokinase activity and the specificity for NADH of homoserine dehydrogenase activity (12). Among gross characters common to the group are accumulation of poly-β-hydroxybutyrate, use of higher dicarboxylic acids, *meta*-hydroxybenzoate and L-norleucine, and inability to use any aldose sugars, amines, or arginine. Unlike members of the *P. fluorescens* and pseudomallei-cepacia groups, both species metabolize *para*-hydroxybenzoate via the *meta* cleavage of protocatechuate.

P. acidovorans was named by den Dooren de Jong and thoroughly characterized by him in 1927 (17), using methods that we have largely adopted in our recent studies of the genus. Although this homogeneous and common species was one of the best described bacteria, its existence was ignored in subsequent taxonomic work until 1966 (75). *P. acidovorans* is nutritionally more versatile than *P. testosteroni,* and can use fructose, malonate, acetamide, and a number of other compounds not used by the latter species.

P. testosteroni is also a homogeneous species and, as the name implies, can grow with testosterone as sole carbon source. This nutritional property distinguishes it from *P. acidovorans* and from most other strains of *Pseudomonas* (75). Although both *P. acidovorans* and *P. testosteroni* metabolize *para*-hydroxybenzoate by a pathway involving the *meta* cleavage of protocatechuic acid, metabolism of *meta*-hydrozybenzoate is different in the two species (84).

The *P. facilis-delafieldii* DNA Homology Group

P. facilis was originally described as a species of the genus *Hydrogenomonos* (68) because of its ability to grow autotrophically in an atmosphere of H_2, O_2 and CO_2. *P. delafieldii,* which is not a "hydrogen bacterium," was isolated 13 years later from enrichment cultures with poly-β-hydroxybutyrate as the sole carbon source (16). This species was subsequently named by Davis et al (14), who recognized its phenotypic similarity to *P. facilis*, and characterized both species. Subsequent DNA homology studies have shown that these two species are, indeed, very closely related to each other, but not to any other hydrogen bacteria or *Pseudomonas* species that have been tested (61). The G+C content of the DNA ranges from 62–64 moles percent for *P. facilis* and 65–66 moles percent for *P. delafieldii.*

The *P. facilis-delafieldii* DNA Homology Group

This is a group consisting of two named species, *P. diminuta* and *P. vesicularis,* and probably at least one other unnamed species closely related to *P. diminuta.* The group has unique morphological and nutritional properties which, together with its isolation from other *Pseudomonas* species in DNA homology, may well justify its allocation to a separate genus. All strains of the group possess a single polar flagellum of very short wave length, and require growth factors, including pantothenate, biotin, and cyanocobalamin. They can produce both poly-β-hydroxybutyrate and glycogen as carbon reserves. They grow very slowly in synthetic media, have relatively long generation times even in rich media, and can use a limited number of carbon sources for growth. Nitrate cannot be used as a nitrogen source. With the exception of one aberrant strain (strain 235) which has a G+C content of 62 moles percent, all other strains of the group have a content of ca. 66–67 moles percent in their DNA (2). We have shown in unpublished studies that strain 235 shows DNA competition of 45% at 72°C with *P. diminuta,* while *P. vesicularis* gives a value of 27%. In a number of other DNA homology experiments, which, however, have not been exhaustive, we found no relationship between strains of either species to any other *Pseudomonas* species tested.

P. vesicularis differs from *P. diminuta* and from strain 235 in possessing a carotenoid pigment and in its ability to use sugars (glucose, galactose, maltose, and cellobiose) as carbon sources for growth. This species is also unusual in its ability to grow in defined media with 5% ethanol. A detailed phenotypic study of species of this group has been presented by Ballard et al (2).

Miscellaneous Species of Uncertain Phylogenetic Affinities

P. solanacearum.—This important plant pathogen, which has a remarkable host range, has been the subject of many studies (7, 23, 57). Hayward (23) proposed the recognition of four biotypes, which he distinguished on the basis of a few physiological characters. With the exception of his biotype 2, there was no correlation between the assignment of strains to these biotypes and the host of origin. A more extensive phenotypic analysis coupled with DNA homology studies has shown that the species represents an isolated taxon that is internally somewhat heterogeneous and contains two subgroups, one consisting of biotypes 1 and 2 of Hayward and the other, comprising his biotypes 3 and 4 (57). Support for this view is provided by studies on electrophoretic mobilities of some enzymes in this species (4). Although results of DNA competition experiments were unequivocal in separating *P. solanacearum* from most other *Pseudomonas* species (3, 55, 57, 61), inexplicable difficulties with some strains of *P. cepacia* were encountered, and the exclusion of the species from the pseudomallei-cepacia complex was based on direct DNA binding experiments (57). A new and as yet undescribed species of *Pseudomonas* isolated from hospital specimens and characterized by E. Ralston shows an appreciable DNA homology with *P. solanacearum* and a somewhat lower homology with *P. cepacia* and *P. mallei* (60). Further and more detailed studies may show that the new species may either belong to a "*P. solanacearum* DNA homology group" or that it may link *P. solanacearum* to the pseudomallei-cepacia complex.

P. saccharophila.—This distinctive species, described as a hydrogen bacterium in 1940 (18), has been the subject of many studies dealing with pathways and enzymes of carbohydrate metabolism. Until recently the species was represented by the single originally described strain which, unfortunately, was replaced in some culture collections by a strain of *P. stutzeri* (56). An ususual property for pseudomonads that *P. saccharophila* shares with *P. stutzeri* is the ability to hydrolyze starch. A second strain of *P. saccharophila* was isolated about a year ago from an enrichment for hydrogen bacteria. *Pseudomonos saccharophila* shows many phenotypic resemblances to the *P. facilis-P. delafieldii* DNA homology group, and is susceptible to some bacteriophages that attack these species, but no DNA homology could be detected between *P. saccharophila* and *P. delafieldii* (61). Only a few DNA hybridization experiments with other species of *Pseudomonas* have been performed to date; these have not shown any significant homology of *P. saccharophila* with any other group.

Other hydrogen bacteria.—Beside *P. saccharophila,* three other facultatively autotrophic hydrogen bacteria have been assigned to the genus *Pseudomonas* (14). These are *P. ruhlandii* (54), *P. flava* (37), and *P. palleronii.* Except for a few DNA hybridization experiments with *P. palleronii,* which showed no relationship of this species to any other tested (61), no DNA homology studies have been made with these species. Unlike *P. ruhlandii, P. flava* and *P. palleronii* have carotenoid-like pigments and the pigments of the latter species resemble those of the degenerately peritrichous hydrogen bacterium that has been named *Alcaligenes paradoxus* (13, 14).

P. lemoignei.—This species is remarkable in its very restricted nutritional spectrum; of 146 organic compounds tested, only 9 were found to be good carbon sources for growth, and four others were utilized poorly or not at all (16, 75). Although strains of this species were repeatedly isolated from enrichment cultures with poly-β-hydroxybutyrate as sole organic substrate, they were never isolated from any source of inoculum other than soil obtained from a particular garden in Berkeley (F. Delafield, personal communication). Aside from its peculiar nutritional properties, this organism is of interest mainly because of its enzymes involved in the metabolism of poly-β-hydroxybutyrate (15, 44). *P. lemoignei* has a G + C content of 58% in its DNA, which is among the lowest recorded for the genus. Its DNA homology has been tested with only two other species, *P. acidovorans* and *P. diminuta,* and found to be negligible (unpublished observations).

P. maltophilia.—This species is phenotypically homogeneous and easily recognizable (30). It is characterized by a requirement for methionine as a growth factor, its negative oxidase reaction, inability to use nitrate as a nitrogen source, and its relatively restricted spectrum of carbon sources suitable for growth. Among the carbon sources that can be used, however, are several carbohydrates, including lactose, which are seldom used by other species (75). Like almost all strains of the *P. fluorescens* DNA homology complex, *P. maltophilia* differs from other *Pseudomonas* species in its inability to accumulate poly-β-hydroxybutyrate. DNA-hybridization experiments have been performed with only one member of this complex, *P. pseudoalcaligenes* and two other *Pseudomonas* species, *P. diminuta* and *P. acidovorans,* and no significant homology was detected (unpublished observations).

Some Problems of Taxonomy and Species Identification

The problems of phenotypic characterization of bacterial taxa, and of setting up determinative keys for identification of strains are recognized by all bacterial taxonomists, and are well illustrated by studies on *Pseudomonas.* There are very few unambiguous characters (e.g., absence of prosthecae and a photosynthetic apparatus, and of the ability to grow anaerobically with energy derived from fermentation of organic compounds) that distinguish the genus as now defined from a few other bacterial genera. Separation of the genus from other genera such as *Xanthomonas, Zoogloea, Gluconobacter, Agrobacterium,* and *Alcaligenes* is based on very few characters perhaps more trivial and arbitrary than those used for the internal subdivision of *Pseudomonas*. While it can be argued that the genus *Pseudomonas* itself should be fragmented into several genera, especially in view of the DNA homology studies, this does not presently seem desirable, because no clear-cut characters or even combination of characters have been found to differentiate phylogenetically distinct species groups. The one exception may be the DNA homology group consisting of *P. diminuta* and *P. vesicularis,* which share some distinctive phenotypic properties and which, on further study, might be assigned to a separate genus.

Although we have mentioned here a few diagnostic traits of species that have been well characterized, we have not attempted to present any keys for species identification. We had reluctantly agreed to prepare a key system for the 8th edition of Bergey's Manual of Determinative Bacteriology, now in press, and are already somewhat unhappy with our efforts. Simple dichotomous keys based on single differential characters, such as those used in the 7th edition of the Manual (8), are unsatisfactory because such characters may be absent in any given strain isolated from nature or lost by mutation and selection in culture media. For this reason constellations of characters must be selected to circumscribe each species or biotype, and these can be conveniently presented only in table form, as we have done in our various taxonomic papers. Unfortunately, the addition of a species to any list of species in a table increases the number of characters that have to be included for its differentiation from all others. A key in table form prepared for many species becomes too complicated for practical purposes of identification, and its use requires a very large number of differential tests. Furthermore, as more strains belonging to any given species are studied, some differential characters may be found to be less useful, requiring a revision of the entire table to prevent confusion. As a compromise, we have devised an artificial system of keys based on a few characters for the primary differentiation of groups of species, and secondary keys in table form for the separation of the species within each group. That such a compromise is not very satisfactory became evident to us when a DNA homology study was made of the species *P. pseudoalcaligenes,* and a number of new strains were added to the collection. This species has emerged as a fairly homogeneous homology cluster, but heterogeneous with respect to poly-β-hydroxybutyrate (PHB) accumulation and the arginine dihydrolase reaction. PHB accumulation was used by us as a primary criterion for allocating species to a section of our system of keys, and the arginine dihydrolase reaction as one of the characters differentiating *P. pseudoalcaligenes* from other species in this section. Thus, although the keys are suitable for identifying a majority of available strains, including the type strain, they fail in identification of other strains.

The studies summarized in this review have barely scratched the surface of the problems of speciation in the genus *Pseudomonas*. Preliminary unpublished experiments on DNA-ribosomal RNA homologies have indicated that some DNA homology groups within the genus may be as distantly related to one another as they are to *Escherichia coli.* Future experiments of this type may throw new light on the phylogeny and taxonomy of the genus. Another approach to elucidating phylogenetic relationships has been made by Richard Ambler, who is determining the amino acid sequences of the cytochromes and azurins of a number of well characterized strains of *Pseudomonas.* Such studies, coupled with immunological investigations of homologous proteins and of nucleic acid homologies, will undoubtedly help to create a more natural system of classification than we have at present.

LITERATURE CITED

1. American Type Culture Collection. 1970. *Catalogue of Strains.* Ninth Edition, p. 54. Rockville, Md., U.S.A.
2. Ballard, R. W., Doudoroff, M., Stanier, R. Y., Mandel, M. 1968. Taxonomy of the aerobic pseudomonads: *Pseudomonas diminuta* and *P. vesiculare. J. Gen. Microbiol.* 53:349–61
3. Ballard, R. W., Palleroni, N. J., Doudoroff, M., Stanier, R. Y., Mandel, M. 1970. Taxonomy of the aerobic pseudomonads: *Pseudomonas cepacia, P. marginata, P. alliicola,* and *P. caryophylli. J. Gen. Microbiol.* 60: 199–214
4. Baptist, J. N., Shaw, C. R., Mandel, M. 1971. Comparative zone electrophoresis of enzymes of *Pseudomonas solanacearum* and *Pseudomonas cepacia. J. Bacteriol.* 108:799–803
5. Baumann, P., Doudoroff, M., Stanier, R. Y. 1968. A study of the *Moraxella* group. II. Oxidase-negative species (genus *Acinetobacter*). *J. Bacteriol.* 95:1520–41
6. Baumann, P., Baumann, L., Mandel, M. 1971. Taxonomy of marine bacteria: the genus *Beneckea. J. Bacteriol.* 107:268–94
7. Buddenhagen, I., Kelman, A. 1964. Biological and physiological aspects of bacterial wilt caused by *Pseudomonas solanacearum. Ann. Rev. Phytopathol.* 2:203–30
8. Breed, R. S., Murray, E. G. D., Smith, N. R. 1957. *Bergey's Manual of Determinative Bacteriology.* 7th Edition. Baltimore: Williams & Wilkins, 1094 pp.
9. Burdon, K. L. 1946. Fatty material in bacteria and fungi revealed by staining dried, fixed slide preparations. *J. Bacteriol.* 52:665–78
10. Burkholder, W. H. 1950. Sour skin, a bacterial rot of onion bulbs. *Phytopathology,* 40:115–17
11. Chang, P. C., Blackwood, A. C. 1969. Simultaneous production of three phenazine pigments by *Pseudomonas aeruginosa* Mac 436. *Can. J. Microbiol.,* 15:439–44
12. Cohen, G. N., Stanier, R. Y., Le Bras, G. 1969. Regulation of the biosynthesis of amino acids of the aspartate family in coliform bacteria and pseudomonads. *J. Bacteriol.,* 99:791–801
13. Davis, D. H., Doudoroff, M., Stanier, R. Y., Mandel, M. 1969. Proposal to reject the genus *Hydrogenomonas:* taxonomic implications. *Int. J. Syst. Bacteriol.,* 19:375–90
14. Davis, D. H., Stanier, R. Y., Doudoroff, M., Mandel, M. 1970. Taxonomic studies on some gram negative polarly flagellated "hydrogen bacteria" and related species. *Arch. Microbiol.,* 70:1–13
15. Delafield, F. P., Cooksey, K. E., Doudoroff, M. 1965. β-Hydroxybutyric dehydrogenase and dimer hydrolase of *Pseudomonas lemoignei. J. Biol. Chem.,* 240: 4023–28
16. Delafield, F. P., Doudoroff, M., Palleroni, N. J., Lusty, C. J., Contopoulou, R. 1965. Decomposition of poly-β-hydroxybutyrate by pseudomonads. *J. Bacteriol.,* 90:1455–66
17. den Dooren de Jong, L. E., 1927. Bijdrage tot de kennis van het mineralisatieproces. Rotterdam: Nijgh & Van Ditmar. 200 pp.
18. Doudoroff, M. 1940. The oxidative assimilation of sugars and related substances by *Pseudomonas saccharophila* with a contribution to the problem of the direct respiration of di- and polysaccharides. *Enzymology,* 9:59–72
19. Feist, C. F., Hegeman, G. D. 1969. Phenol and benzoate metabolism by *Pseudomonas putida:* regulation of tangential pathways. *J. Bacteriol.,* 100:869–77
20. Garibaldi, J. A., Bayne, H. G. 1962. The effect of iron on the *Pseudomonas* spoilage of farm washed eggs. *Poultry Sci.* 41: 850–53
21. Gunsalus, C. F., Stanier, R. Y., Gunsalus, I. C. 1953. The enzymatic conversion of mandelic acid to benzoic acid. III. Fractionation and properties of the soluble enzymes. *J. Bacteriol.,* 66:548–53
22. Haynes, W. C., Stodola, F. H., Locke, J. M., Pridham, T. G., Conway, H. F., Sohns, V. E., Jackson, R. W. 1956. *Pseudomo-*

nas aureofaciens Kluyver and phenazine-α-carboxylic acid, its characteristic pigment. *J. Bacteriol.*, 72:412–17
23. Hayward, A. C. 1964. Characteristics of *Pseudomonas solanacearum. J. Appl. Bacteriol.*, 27: 265–77
24. Hegeman, G. D. 1966. Synthesis of the enzymes of the mandelate pathway by *Pseudomonas putida. J. Bacteriol.*, 91:1140–54
25. Herbert, R. B., Holliman, F. G. 1964. Aeruginosin B. A naturally occurring phenazinesulfonic acid. *Proc. Chem. Soc.*, 1964, p. 19.
26. Hodgkiss, W. 1964. The flagella of *Pseudomonas solanacearum. J. Appl. Bacteriol.*, 27:278–80
27. Holliman, F. G. 1957. Pigments from a red strain of *Pseudomonas aeruginosa. Chem. Ind.* 28: 1668
28. Holloway, B. W. 1969. Genetics of *Pseudomonas. Bacteriol. Rev.*, 33:419–43
29. Hugh, R., Leifson, E. 1953. The taxonomic significance of fermentative versus oxidative metabolism of carbohydrates by various Gram-negative bacteria. *J. Bacteriol.*, 66:24–26
30. Hugh, R., Ryschenkow, E. 1961. *Pseudomonas maltophilia,* an alcaligenes-like species. *J. Gen. Microbiol.*, 26:123–32
31. Jensen, R. A., Nasser, D. S., Nester, E. W. 1967. Comparative control of a branch-point enzyme in microorganisms. *J. Bacteriol.*, 94:1582–93
32. Jessen, O. 1965. *Pseudomonas aeruginosa* and other green fluorescent pseudomonads. A taxonomic study. Copenhagen: Munksgaard, 244 pp.
33. Johnson, J. L., Ordal, E. J. 1968. Deoxyribonucleic acid homology in bacterial taxonomy: effect of incubation temperature on reaction specificity. *J. Bacteriol.*, 95: 893–900
34. Kemp, M. B., Hegeman, G. D. 1968. Genetic control of the β-ketoadipate pathway in *Pseudomonas aeruginosa. J. Bacteriol.*, 96:1488–99
35. King, E. O., Ward, W. K., Raney, D. E. 1954. Two simple media for the demonstration of pyocyanin and fluorescein. *J. Lab. Clin. Med.*, 44:301–07
36. Kluyver, A. J. 1956. *Pseudomonas aureofaciens* (nov. sp.) and its pigments. *J. Bacteriol.*, 72:406–11
37. Kluyver, A. J., Manten, A. 1942. Some observations on the metabolism of bacteria oxidizing molecular hydrogen. *J. Microbiol. Serol.*, 8:71–85
38. Kögl, F., Postowsky, U. J. 1930. Über das grüne Stoffwechselprodukt des *Bacillus chlororaphis. Ann. Chem. Liebigs,* 480:280–97
39. Korth, H. 1962. On the selective formation of α-phenazine carboxylic acid in *Pseudomonas aeruginosa* in acid medium and its property as redox catalyst. *Zentralbl. Bakteriol. Parasitenk.* I, 185:511–15
40. Lautrop, H., Jessen, O. 1964. On the distinction between polar monotrichous and lophotrichous flagellation in green fluorescent pseudomonads. *Acta Pathol. Microbiol. Scand.*, 60:588–98
41. Lederberg, J., Lederberg, E. M. 1952. Replica plating and indirect selection of bacterial mutants. *J. Bacteriol.*, 63:399–406
42. Lelliott, R. A., Billing, E., Hayward, A. C., 1966. A determinative scheme for the fluorescent plant pathogenic pseudomonads. *J. Appl. Bacteriol.*, 29:470–89
43. Liu, P. V. 1962. Non-motile varieties of *P. aeruginosa* producing a melaninlike pigment. *J. Bacteriol.*, 84:378
44. Lusty, C. J., Doudoroff, M. 1966. Poly-β-hydroxybutyrate depolymerases of *Pseudomonas lemoignei. Proc. Nat. Acad. Sci.*, 56: 960–65
45. Mandel, M. 1966. Deoxyribonucleic acid base composition in the genus *Pseudomonas. J. Gen. Microbiol.*, 43:273–92
46. Mandel, M. 1969. New approaches to bacterial taxonomy: perspective and prospects. *Ann. Rev. Microbiol.*, 23:239–74
47. Mann, S. 1969. Über melaninbildende Stamme von *Pseudomonas aeruginosa. Arch. Microbiol.*, 65: 359–79
48. Misaghi, I., Grogan, R. G. 1969. Nutritional and biochemical comparisons of plant-pathogenic and saprophytic fluorescent pseudomonads. *Phytopathology,* 59:1436–50

49. Morris, M. B., Roberts, J. B. 1959. A group of pseudomonads able to synthesize poly-β-hydroxybutyric acid. *Nature,* 183:1538–39
50. Okanishi, M., Gregory, K. F. 1970. Methods for the determination of deoxyribonucleic acid homologies in *Streptomyces. J. Bacteriol.,* 104:1086–94
51. Ornston, L. N. 1966. The conversion of catechol and protocatechuate to β-ketoadipate by *Pseudomonas putida.* IV. Regulation. *J. Biol. Chem.,* 241:3800–10
52. Ornston, L. N. 1971. Regulation of catabolic pathways in *Pseudomonas. Bacteriol. Rev.,* 35:87–116
53. Ornston, L. N., Stanier, R. Y. 1964. Mechanism of β-ketoadipate formation by bacteria. *Nature,* 204: 1279–83
54. Packer, L., Vishniac, W. 1955. Chemosynthetic fixation of carbon dioxide and characteristics of hydrogenase in resting cell suspensions of *Hydrogenomonas ruhlandii* nov. sp. *J. Bacteriol.,* 70:216–33
55. Palleroni, N. J., Ballard, R. W., Ralston, E., Doudoroff, M. 1972. Deoxyribonucleic acid homologies among some *Pseudomonas* species. *J. Bacteriol.,* 110:1–11
56. Palleroni, N. J., Doudoroff, M. 1965. Identity of *Pseudomonas saccharophila. J. Bacteriol.,* 89: 264
57. Palleroni, N. J., Doudoroff, M. 1971. Phenotypic characterization and deoxyribonucleic acid homologies of *Pseudomonas solanacearum. J. Bacteriol.,* 107: 690–96
58. Palleroni, N. J., Doudoroff, M., Stanier, R. Y., Solánes, R. E., Mandel, M. 1970. Taxonomy of the aerobic pseudomonads: the properties of the *Pseudomonas stutzeri* group. *J. Gen. Microbiol.,* 60:215–31
59. Palleroni, N. J., Stanier, R. Y. 1964. Regulatory mechanisms governing synthesis of the enzymes of tryptophan oxidation by *Pseudomonas fluorescens. J. Gen. Microbiol.,* 35:319–34
60. Ralston, E. 1972. *Some contributions to the taxonomy of the genus Pseudomonas.* Ph.D. Thesis, University of California, Berkeley, 114 pp.
61. Ralston, E., Palleroni, N. J., Doudoroff, M. 1972. DNA homologies of some so-called *Hydrogenomonas* species. *J. Bacteriol.,* 109:465–66
62. Redfearn, M. S., Palleroni, N. J., Stanier, R. Y. 1966. A comparative study of *Pseudomonas pseudomallei* and *Bacillus mallei. J. Gen. Microbiol.,* 43:293–313
63. Rhodes, M. E. 1959. The characterization of *Pseudomonas fluorescens. J. Gen. Microbiol.,* 21: 221–63
64. Rhodes, M. E. 1965. Flagellation for the classification of bacteria. *Bacteriol. Rev.,* 29:442–65
65. Rogul, M., Brendle, J. J., Haapala, D. K., Alexander, A. D. 1970. Nucleic acid similarities among *Pseudomonas pseudomallei, Pseudomonas multivorans* and *Actinobacillus mallei. J. Bacteriol.,* 101: 827–35
66. Rosenberg, S. L., Hegeman, G. D. 1969. Clustering of functionally related genes in *Pseudomonas aeruginosa. J. Bacteriol.,* 99: 353–55
67. Sands, D. C., Schroth, M. N., Hildebrand, D. C. 1970. Taxonomy of phytopathogenic pseudomonads. *J. Bacteriol.,* 101:9–23
68. Schatz, A., Bovell, C. R. 1952. Growth and hydrogenase activity of a new bacterium *Hydrogenomonas facilis. J. Bacteriol.,* 63: 87–98
69. Sierra, F., Veringa, H. A. 1958. Effect of oxychlororaphine on the growth in vitro of *Streptomyces* species and some pathogenic fungi. *Nature,* 182:265
70. Skerman, V. B. D. 1967. *A guide to the identification of the genera of bacteria.* 2nd Edition. Baltimore: Williams & Wilkins 303 pp.
71. Slepecky, R. A., Law, J. H. 1960. A rapid spectrophotometric assay of alpha, beta unsaturated acids and beta-hydroxy acids. *Anal. Chem.,* 32:1697–99
72. Sneath, P. H. A. 1956. The change from polar to peritrichous flagellation in *Chromobacterium* species. *J. Gen. Microbiol.,* 15:99–105
73. Sneath, P. H. A. 1960. A study of the bacterial genus *Chromobacterium. Iowa State J. Sci.,* 34: 243–500
74. Stanier, R. Y. 1971. Toward an

evolutionary taxonomy of the bacteria. *Recent Advan. Microbiol., 10th Int. Congr. Microbiol.* pp. 595–604

75. Stanier, R. Y., Palleroni, N. J., Doudoroff, M. 1966. The aerobic pseudomonads: a taxonomic study. *J. Gen. Microbiol.,* 43: 159–271
76. Stanier, R. Y., Wachter, D., Gasser, C., Wilson, A. C. 1970. Comparative immunological studies of two *Pseudomonas* enzymes. *J. Bacteriol.,* 102:351–62
77. Starr, M. P. 1959. Bacteria as plant pathogens. *Ann. Rev. Microbiol.,* 13:211–38
78. Starr, M. P., Blau, W., Cosens, G. 1960. The blue pigment of *Pseudomonas lemonnieri. Biochem. Z.,* 333:328–34
79. Starr, M. P., Knackmuss, H. J., Cosens, G. 1967. The intracellular blue pigment of *Pseudomonas lemonnieri. Arch. Mikrobiol.,* 59:278–94
80. Starr, M. P., Stephens, W. L. 1964. Pigmentation and taxonomy of the genus *Xanthomonas. J. Bacteriol.,* 87:293–302
81. Straub, O. 1971. Lists of natural carotenoids. In *Carotenoids* ed. O. Isler, Birkhäuser Verlag, Basel and Stuttgart), 771–850
82. Takeda, R. 1958. *Pseudomonas* pigments. II. Two pigments, 1-phenazine-carboxylic acid and hydroxychlororaphine, produced by *Pseudomonas aeruginosa* T 359. *Hakko Kogaku Zasshi,* 36:286–90
83. Thornley, M. J. 1960. The differentiation of *Pseudomonas* from other Gram-negative bacteria on the basis of arginine metabolism. *J. Appl. Bacteriol.,* 23:37–52
84. Wheelis, M. L., Palleroni, N. J., Stanier, R. Y. 1967. The metabolism of aromatic acids by *Pseudomonas testosteroni* and *P. acidovorans. Arch. Mikrobiol.,* 59: 302–14
85. Wheelis, M. L., Stanier, R. Y. 1970. The genetic control of dissimilatory pathways in *Pseudomonas putida. Genetics,* 66:245–66
86. Williamson, D. H., Wilkinson, J. F. 1958. The isolation and estimation of poly-β-hydroxybutyrate inclusions of *Bacillus* species. *J. Gen. Microbiol.,* 19:198–209
87. Wrede, F., Strack, E. 1929. Über das Pyocyanin, den blauen Farbstoff des *Bacillus Pyocyaneus.* IV. Die Konstitution und Synthese des Pyocyanin. *Z. Physiol. Chem.* 181:58–76

THE ROLE OF MIXED INFECTIONS IN THE TRANSMISSION OF PLANT VIRUSES BY APHIDS[1]

3545

W. F. Rochow

Plant Science Research Division, Agriculture Research Service, U.S. Department of Agriculture, and Department of Plant Pathology, Cornell University, Ithaca, New York

Introduction

> I suspect that many plant diseases are influenced by associated organisms to a much more profound degree than we have yet realized, not only as to inhibition, but as to acceleration, of the processes. It may be that a number of diseases may require an association of organisms for their occurrence and cannot be produced by infection of one organism alone. These considerations appear to indicate an inviting field for much more extended research. (22)

Since 1930, when H. S. Fawcett made these comments, studies of mixed virus infections have played important roles in the development of plant virology. Identification of diseases caused by simultaneous infection by more than one virus, interactions among related viruses in cross-protection tests, and physiological studies of interactions among unrelated viruses all have contributed to current concepts in plant virology. The range and importance of these interactions have been discussed in other reviews (9, 17, 40, 76, 77). This review, which is intended to be illustrative rather than complete, will cover just one small aspect of plant virus interactions in mixed infections, that related to virus transmission by aphids. This "narrow" topic seems worthy of study now for at least three reasons. First, research on the role of mixed infections is a useful approach to understanding some basic aspects of the relationships between viruses and aphid vectors. Second, certain mixed infections may represent a place where current interest in heterogeneity in the assembly of virus protein and nucleic acid—phenomena such as transcapsidation and phenotypic mixing—can focus on a practical or functional role in the plant virus transmission process. Recent work on the existence of multiple genetic components in apparently homogeneous virus populations (8, 48, 79) underscores the importance of heterogeneity of plant virus populations even in singly-infected plants. Third, the mixed infections may serve

[1] Cooperative Investigation, Plant Science Research Division, U.S. Department of Agriculture, and Cornell University Agricultural Experiment Station. Supported in part by NSF grant GB-21013.

key roles in nature where the great economic importance of aphid-transmitted plant viruses emphasizes the need for a better understanding of the ecology of the viruses, a topic also recently reviewed by others (21, 63, 89).

It is usually difficult, if not impossible, to sort out all the variables that influence aphid transmission of viruses from mixed infections. A multitude of factors can influence any one of the three biological systems that interact when an aphid transmits a virus from one plant to another (67, 88, 102). Progress has been made in recent years in elucidating mechanisms of virus interactions in mixed infections, but the viruses most studied were not those transmitted by aphids. Any consideration of aphid transmission must begin with an understanding of the range in type of transmission mechanisms, recently discussed by Watson & Plumb (100). Most of the viruses spread by aphids are transmitted in the stylet-borne (nonpersistent) manner characterized by the ability of aphids to acquire and transmit virus within a few minutes or seconds during brief probes in epidermal tissue and by loss of the ability to transmit virus within a short time. Other viruses are transmitted in the circulative (persistent) manner characterized by aphid feeding in phloem tissue and by retention of transmissibility following a molt and often for the life of the insect. Between the two extremes many workers recognize a semipersistent virus-vector relationship characterized by retention of the ability to transmit virus for longer periods than in the typical stylet-borne manner, but by inability of aphids to transmit virus following a molt.

Mixed plant virus infections are common in nature. Emphasis in this review will be on singly- and doubly-infected plants because most experimental work with multiple infections has involved two viruses. Mixed infections in nature are not necessarily so limited. Bennett (9) pointed out that sugar beets are often found to be simultaneously infected with curly top, mosaic, yellownet, and dodder latent mosaic viruses. Examples of interactions involving three viruses are known from a variety of crops (5, 6, 74). The common occurrence of mixed infections in the field provides both the source of most double infection systems studied and a focus for experimental work.

Virus transmissions after exposure of aphids to doubly-infected plants usually follow predictable patterns. The presence of one virus usually seems to have no measurable effect on the transmission of another virus, whether the infections involve related strains of one virus or mixtures of unrelated viruses. In recent years, however, a growing list of exceptions to these expected events has emerged from observations on a range of crop plants and on a number of aphid species. These exceptions, to be discussed in subsequent sections, are associated with the observation that an aphid can often transmit one virus only if the virus occurs in the presence of a second virus. Examples of such dependent aphid transmissions have been mentioned in recent reviews (26, 40, 55, 85, 100), but the systems have frequently been discussed more as interesting curiosities than as fruitful approaches to the study of several aspects of interactions among viruses and vectors. The exceptions to the predictable events in aphid transmission of viruses are examples of the potential importance of associated pathogens described so clearly more than 40 years ago (22).

INDEPENDENT TRANSMISSION FROM MIXED INFECTIONS

Selective transmission of one virus.—Some of the earliest work on aphid transmission of viruses illustrated the selectivity of the process, and demonstrated how such selectivity could aid in identifying diseases caused by more than one virus. Hoggan (32) showed that colonies of *Myzus persicae* (Sulzer) transmitted cucumber mosaic virus, but not tobacco mosaic virus, from tobacco plants infected by both viruses. This selective feature of aphid transmission of viruses was used by Smith (82) and other early workers (44, 92) to demonstrate the composite nature of potato rugose mosaic; from the diseased potatoes aphids transmitted potato virus Y but not potato virus X. Other examples are known for such selective transmission from plants infected by two viruses, only one of which is normally aphid transmissible. Thus, aphids transmit tobacco etch virus but not tobacco mosaic virus from plants infected by both (36), despite the fact that the two viruses can occur in the same cell (27, 56). The selectivity is not confined to stylet-borne viruses. Kassanis (38) showed that aphids regularly transmitted only potato leaf roll virus from plants also infected by potato virus X.

The independent transmission of one virus from a mixture also occurs when the mixture involves strains of one virus. Castillo & Orlob (14) showed that most individual *M. persicae* transmitted either one or the other of two strains of cucumber mosaic virus from mixed infections. Watson (98) found that one strain of cucumber mosaic virus was transmitted no more readily by aphids from plants that also contained other strains of the virus than from plants containing only the one strain. *M. persicae* usually transmitted a strain of alfalfa mosaic virus independently from mixed infections containing two strains of the virus (14). Allen (3) used *Rhopalosiphum prunifoliae* (Fitch) to separate distinct isolates of barley yellow dwarf virus from doubly-infected plants. When severe and mild strains of beet yellows virus were mixed, *M. persicae* usually, but not always, recovered one or the other of the original strains from the mixture (10).

Another basis for independent transmission of viruses from mixed infections is specificity in aphid-virus interactions. Doncaster & Kassanis (20) showed that *Myzus ascalonicus* Doncaster transmitted only cucumber mosaic virus from plants doubly infected with potato virus Y, and that the same aphid species transmitted only henbane mosaic virus from mixed infections with tobacco etch virus. Since *M. ascalonicus* transmits neither potato virus Y nor tobacco etch virus from single infections, the selective transmission from mixed infections is not surprising. Similarly, from lettuce doubly infected with dandelion yellow mosaic virus and lettuce mosaic virus, *M. persicae* transmitted only lettuce mosaic virus; *M. ornatus* Laing transmitted only dandelion yellow mosaic virus (37). Studied by Kvicala (46, 47) of mixed infections by cabbage black ringspot virus and cauliflower mosaic virus emphasized the importance of the aphid species. *M. persicae* and *Brevicoryne brassicae* (L.) transmitted both viruses from doubly-infected plants, but *M. ornatus* transmitted cauliflower mosaic virus selectively from the mixture. For circulative viruses, vector specificity among isolates of barley yellow dwarf virus often enables identification of mixed infections in field-collected plants. For example, although both *Rhopalosiphum maidis* (Fitch) and *R. padi* (L.) may recover virus from a field-collected plant, subsequent tests often show that *R. maidis* recovered a virus distinct from the one transmitted from the original plant by *R. padi* (74, 75). Vector specificity of several isolates of barley yellow dwarf virus was the same in single and double infections in recent tests made in Manitoba (28).

Another kind of selective transmission from mixed infections results from differences in the mechanism of transmission between viruses. It is usually fairly easy to separate a stylet-borne virus from a circulative virus. For example, Mellor & Fitzpatrick (57) separated at least two component viruses, one that was stylet-borne and one that was circulative, by making progressive transfers of *Capitophorus fragaefolii* (Ckll.) fed on yellows-infected strawberries. Use of combinations of feeding periods helped Prentice (66) identify five viruses in strawberry complexes. Similar separations have been achieved even when differences were less extreme than for the typical stylet-borne and circulative mechanisms. Watson (93) used a combination of acquisition feedings from 5 minutes to 1 day and inoculation test feedings from 10 minutes to successive transfers of 1 day to separate beet yellows and beet mosaic viruses from mixed infections in sugar beets. Such separations, which occurred in tests with both *Aphis fabae* Scopoli and *M. circumflexus* (Theobald), were possible because beet mosaic virus has a typical stylet-borne relationship with the vector, but beet yellows virus is semipersistent in the aphid. Maassen (50) was able to separate two viruses from a strawberry complex by means of aphids because vein necrosis virus had a semipersistent relationship with the vector and epinasty virus was persistent. In general, the virus-vector relationship of a given plant virus is the same, regardless of whether the virus is acquired from singly- or doubly-infected plants.

Nonselective transmission of two viruses.—Although aphids often transmit only one virus from doubly-infected plants, these insects can also transmit both viruses together. Schroeder et al (80) found that individual *Acyrthosiphon pisum* (Harris) could transmit a combination of viruses (red clover vein mosaic virus and bean yellow mosaic virus) to produce the typical streak disease of peas. Another example for stylet-borne viruses is the simultaneous transmission of rubus yellow-net virus and black raspberry necrosis virus by *Amphorophora rubi* Kalt reported by Stace-Smith (86). As would be expected in such tests, groups of aphids are more likely to transmit both viruses than are individual aphids. For example, when five aphids were fed on each test plant, rubus yellow-net virus was usually transmitted together with black raspberry necrosis virus; but when individual aphids were used, the possibility of both viruses being transmitted was reduced (86).

Aphids also are capable of transmitting more than one strain of a single virus. Castillo & Orlob (14) found that individual *M. persicae* occasionally transmitted both strains of alfalfa mosaic virus from double infections, although usually only one strain was transmitted. Simultaneous transmissions of viruses that have a circulative relationship with the aphid also occur. MacKinnon (51) reported that 64% of individual *M. persicae* transmitted both potato leaf roll virus and turnip latent virus from doubly-infected plants in one set of experiments.

Quantitative effects.—The concentration of a virus may be quite different in doubly-infected plants from that in comparable singly-infected ones (76, 77). Quantitative aspects of virus content in source plants probably often play a role in virus transmission by aphids. Zitter (105) found that the ability of aphids to transmit cucumber mosaic virus or western celery mosaic virus from celery paralleled virus concentration in source plants. One effect of mixed infections with the two viruses was a change in the interval between inoculation and the time needed for cucumber mosaic virus to reach its highest concentration in the plant. For singly-infected plants the maximum concentration of cucumber mosaic virus occurred two weeks after inoculation, but for plants also infected by western celery mosaic the interval was three weeks. Simons (81) suggested that reduced aphid transmissions from plants doubly-infected by potato virus Y and cucumber mosaic virus resulted from corresponding decreases in virus titer of doubly-infected plants, but differences in age of singly- and doubly-infected plants clouded his comparison. The demonstration by Kassanis (36) of selective transmission of tobacco etch virus from plants also infected by henbane mosaic virus is an extreme example of a quantitative effect because tobacco etch virus suppressed henbane mosaic virus in the mixed infections.

The interaction of pea streak and alfalfa mosaic viruses provides a good example of a quantitative effect. Hampton & Sylvester (30) found that the frequency of transmission of pea streak virus by *A. pisum* dropped from 72% when the virus was alone to 46% in the presence of alfalfa mosaic virus. On the other hand, the transmission frequency of alfalfa mosaic virus was 8% from single infections but 22% from mixed infections with pea streak virus. Tobacco wilt virus may be transmitted less readily from plants also infected by potato virus Y than from singly-infected plants (5).

There is at least one example of a possible quantitative effect of prior acquisition of one virus on transmission of a second virus. MacKinnon (52) found that *M. persicae* reared on *Physalis floridana* Rybd. infected with potato leaf roll virus were better vectors of turnip latent virus than were aphids reared on healthy leaves. That the effect was not solely a result of the presence of potato leaf roll virus was indicated by the fact that aphids reared on potato leaf roll virus-infected *Datura stramonium* L. were no better vectors of turnip latent virus than those reared on healthy leaves.

Transmissions Following Alternate Acquisition Feedings

Aphids are often exposed to two viruses, not only by feeding on a doubly-infected plant, but also by feeding first on one, and then on another, singly-infected plant. Usually such alternate acquisition feedings appear to have as little influence on virus transmission as do acquisitions from doubly-infected plants. Sylvester (90) found that previous acquisition by *M. persicae* of either beet yellows virus or beet yellow-net virus did not affect acquisition and transmission of the other virus. Similarly, MacKinnon (51) found that previous acquisition of either potato leaf roll virus or turnip latent virus did not affect acquisition and transmission of the other virus by *M. persicae.* When Harrison (31) tested individual *M. persicae,* he found that aphids infective for a mild strain of potato leaf roll virus could still acquire and transmit a virulent strain. No significant difference was noted in transmission of either of two strains of cucumber mosaic virus or of two strains of alfalfa mosaic virus, regardless of the sequence of acquisition by *M. persicae* (14).

An important exception to the general lack of evidence for virus interactions following alternate acquisition feeding by aphids is provided by recent work of Kassanis & Govier (43). Although *M. persicae* was unable to transmit virus from plants infected by potato aucuba mosaic virus, aphids were capable of transmitting this virus if they first fed on tobacco infected by potato virus Y. Since the interaction of potato aucuba mosaic virus and potato virus Y is an example of dependent transmission, the observations will be discussed in the following section.

Dependent Transmission from Mixed Infections

Terminology.—The major exceptions to predictable transmissions from mixed virus infections are systems in which one virus is transmitted by aphids from plants only in the presence of a second virus. Various terms have been used to identify and describe the two interacting viruses. The virus regularly transmissible only in the presence of a second one has been referred to as the assisted virus, carried virus, aided virus, helped virus, and dependent virus. The second virus, which makes the first transmissible by aphids, has been called the assisting virus, carrier virus, assistor virus, and helper virus. The whole system of interaction of the two viruses with their aphid vectors has been referred to as virus complex, composite disease, the helper relationship, and dependent transmission. For consistency in this discussion, the systems will be referred to as dependent transmission (26), the first virus (the one that depends on a second for transmission) will be referred to as the dependent virus, and the second virus will be called the helper virus (85).

Examples.—Probably the first illustration that aphid transmission of virus from doubly-infected plants was not always predictable resulted from a chance observation. In 1936 Clinch et al (15) tested the possibility of virus transmission by *M. persicae* from potatoes with interveinal mosaic, a complex disease caused by potato virus X and potato aucuba mosaic virus. Since no virus transmission occurred from the diseased potatoes, they initially thought "that no insect-borne element was present" (15). In one test, however, the source plants also contained potato virus A. Aphids from this source transmitted potato virus A to 17 of 28 test plants, but four of the plants also became infected by potato aucuba mosaic virus. In another test, when aphids acquired virus from plants doubly infected by potato virus A and potato aucuba mosaic virus, 14 of 20 test plants became infected by both viruses. These results suggested that aphids could transmit potato aucuba mosaic virus, but only if the virus source plants were also infected by potato virus A. More extensive tests by Kassanis (39) established the dependence of potato aucuba mosaic virus transmission on potato virus A, showed that the phenomenon varied with the strain of potato aucuba mosaic virus, and demonstrated that potato virus Y could also function as a helper virus. Watson (95) described a second example for stylet-borne viruses that also involved potato virus Y as a helper virus. Potato virus C was not transmitted by aphids from singly-infected plants, but when aphids fed on plants infected by both potato viruses Y and C, they transmitted both viruses (Table 1).

In 1945 Smith (83) provided the first experimental evidence for depen-

TABLE 1. Examples of Dependent Transmission by Aphids in which Transmission of One Virus Depends on Presence of a Second (Helper) Virus[a]

Disease	Dependent Virus	Helper Virus	Vector	References
Tobacco Rosette	Tobacco Mottle (C) (M)	Tobacco Vein-Distorting (C)	*Myzus persicae*	83, 84
Carrot Motley Dwarf	Carrot Mottle (C) (M)	Carrot Red-Leaf (C)	*Cavariella aegopodii*	59, 101
Parsnip Yellow Fleck	Parsnip Yellow Fleck (SP) (M)	Anthriscus Yellows (C)	*Cavariella aegopodii*	58
Groundnut Rosette	Groundnut Rosette (C) (M)	Groundnut Rosette Assistor (C)	*Aphis craccivora*	34, 62
Tobacco Yellow Vein	Tobacco Yellow Vein (C) (M)	Tobacco Yellow Vein Assistor (C)	*Myzus persicae*	2
Barley Yellow Dwarf	MAV (C)	RPV (C)	*Rhopalosiphum padi*	68, 69, 72
"Systemic Necrosis"	Potato C (StB) (M)	Potato Y (StB) (M)	*Myzus persicae*	43, 95
Potato Aucuba Mosaic	Potato Aucuba Mosaic (StB) (M)	Potato A or Y (StB) (M)	*Myzus persicae*	15, 39, 43

[a] Designations under the virus name indicate that its aphid-virus relationship is circulative (C), semipersistent (SP), or stylet-borne (StB). (M) indicates that the virus is also known to be mechanically transmissible, at least to some hosts.

dent transmissions, and established the phenomenon for circulative viruses. He found that tobacco rosette resulted from mixed infection by tobacco mottle virus and tobacco vein-distorting virus (83, 84). Although tobacco mottle virus was sap-transmissible, it was transmitted by *M. persicae* or *M. convolvuli* (=*pseudosolani* Theob.) only if the source plant was also infected by tobacco vein-distorting virus. Other examples of dependent transmission have been recorded more recently from a variety of crop plants and for several additional aphid species (Table 1). The major features of carrot motley dwarf, groundnut rosette, and tobacco yellow vein are similar to those for tobacco rosette. In all four instances the virus-vector relationship is circulative, the dependent virus is mechanically transmissible, and the helper virus has not been mechanically transmitted. Parsnip yellow fleck is similar, but parsnip yellow fleck virus is considered semi-persistent, although few data on the aphid-virus relationship are available (58). The barley yellow dwarf system differs from the others because the dependent virus (MAV) has not been mechanically transmitted to plants, but it can be transmitted from singly- or doubly-infected plants by an aphid species (*Macrosiphum avenae* Fabricius) different from the one (*R. padi*) that transmits MAV only from mixed infections (Table 1). The two barley yellow dwarf viruses are differentiated on the basis of their relative vector specificity and their distinct serological properies (1, 71). MAV is transmitted specifically by *M. avenae;* RPV is transmitted specifically by *R. padi.*

In addition to the known cases of dependent transmission listed in Table 1, others may illustrate the phenomenon, but evidence for them is incomplete. For example, Watson (97) suggested that yellow-net mild yellows virus might be a helper virus without which yellow-net virus could not be transmitted from sugar beets. Other tests of yellow-net (65) suggested that each virus could be transmitted separately if young test plants were used. Another system that might involve dependent transmission is the turnip latent complex studied by MacKinnon (53) and MacKinnon & Lawson (54). At least three viruses are involved. The possible role of the vein blotch virus or the mild chlorosis virus in the aphid transmission of turnip latent virus has yet to be determined (J. P. MacKinnon, personal communication). Another possibility of dependent transmission is the interaction of tobacco mosaic virus and cucumber mosaic virus. Recent studies by G. B. Orlob and John Lojek revealed that aphids transmitted tobacco mosaic virus from doubly-infected tobacco plants to *Nicotiana glutinosa* L. (G. B. Orlob, personal communication). Thus, cucumber mosaic virus might serve as a helper virus for aphid transmission of tobacco mosaic virus, a system of special interest because of many past attempts to learn why tobacco mosaic virus is generally not aphid transmissible. A type of dependent transmission might explain the intermediate isolates obtained in tests by Björling (10) of mixed infections with severe and mild strains of beet yellows virus.

Recent studies at Cornell University have uncovered another unpublished example of dependent aphid transmission for barley yellow dwarf. In several experiments, *R. maidis* transmitted the MAV isolate of barley yellow dwarf virus from mixed infections despite its inability to transmit MAV alone (71). The helper virus in this case was RMV, an isolate of barley yellow dwarf virus transmitted specifically by *R. maidis*. In one test, for example, comparative aphid transmissions by means of *R. maidis* and *M. avenae* were made from 45 MAV-infected oats and from 45 oats initially inoculated with both MAV and RMV. From the MAV-infected plants only *M. avenae* transmitted virus (to 134 of 134 plants); *R. maidis* did not transmit virus to any of 135 test plants, each having been infested with about 10 aphids for a 5-day test feeding. From almost all of the doubly-inoculated oats, both aphids transmitted virus. Virus recovered from the doubly-infected plants by *M. avenae* was subsequently transmissible by *M. avenae* but not by *R. maidis*. Thus, *M. avenae* appeared to recover MAV specifically from the mixed infection, a result identical to results from similar tests of mixtures of RPV and MAV (68). Subsequent comparative aphid transmission tests from plants infected by means of *R. maidis* that fed on the original doubly-infected oats, however, showed that *R. maidis* transmitted MAV in addition to RMV from at least 25 of the 45 plants. This conclusion was based on virus transmission by *M. avenae* to 111 of 111 test plants in subsequent tests of the 25 cases. None of 66 control plants in the tests became infected. Although results of this and other transmission experiments are clear, identification of the MAV transmitted from the mixed infections by *R. maidis* has not yet been confirmed by serological tests.

Comparative features.—The importance of the helper virus is a common feature in all examples of dependent transmission by aphids. But beyond this observation generalizations among the systems are hard to find. As expected, many features of the diseases involving stylet-borne viruses are quite different from those where the viruses have a circulative relationship with their vectors. Within each of these two major groups, similarities among systems are often obscured by variations in the kinds of data available and by the distinctive features of each. A discussion of some aspects of dependent transmission may illustrate the important factors, the kinds of variation, and the current level of understanding of the phenomenon.

Even the basic interaction of the dependent virus with the helper virus probably varies from cases of absolute dependence to those of relative dependence. Smith (84) made many attempts to obtain aphid transmission of the tobacco mottle virus alone. No virus transmissions occurred, even when mass infestations of aphids were used. The dependence of tobacco mottle virus on tobacco vein-distorting virus appears to be absolute. In contrast, dependence of MAV on RPV in barley yellow dwarf is relative. Occasional transmissions of MAV by *R. padi* from singly-infected plants have occurred from time to time during the 15 years in which the system has been studied, but they have been rare and inconsistent compared with the regular transmission of MAV from most doubly-infected plants. Similarly, occasional transmissions of potato virus C alone have been obtained from some virus source plants (94). Dependent transmission systems probably form a spectrum from absolute dependence at one end, through relative dependence, to cases where virus interactions have only minor quantitative effects on virus transmission by aphids.

Most of the systems represent virus mixtures discovered by study of diseases in the field. Many authors have stressed the potential practical importance of dependent transmission in natural spread of viruses, but little direct information is available on the role of the phenomenon in the field. The mixed infections can be very common. For example, in one study of carrots, about half of the plants with symptoms of virus infection contained a mixture of carrot mottle virus and carrot red-leaf virus (99). Groundnut rosette has been a widespread disease in several areas of the world for many years (11, 12, 45, 62). Dependent transmission generally should increase spread of viruses in a field because the phenomenon provides a vector for viruses not normally transmitted by aphids. But dependent transmission might even account for a reduction in spread of a virus. For example, one reason for relatively little spread of potato aucuba mosaic virus in Britain could be the freedom of most commercial potatoes from the helper viruses needed for its transmission by aphids (39).

The dependent virus and the helper virus do not always have the same host range; this factor could be important in understanding the role of dependent transmission in the field. Murant & Goold (58) found that anthriscus yellows virus did not infect parsnips; thus, aphid transmission of parsnip yellow fleck virus from parsnip to parsnip may not occur in the field. The economic importance of parsnip yellow fleck might be a function of the aphids that enter parsnip fields following feeding on source plants, such as cow parsley [*Anthriscus sylvestris* (L.) Bernh.], that are susceptible to both the dependent and helper viruses (58). That susceptibility of plants to the helper virus is not a necessary feature of dependent transmission is also illustrated by the observation that groundnut and soybean are susceptible to the dependent virus but not to the helper virus of tobacco yellow vein (2).

For many of the virus disease complexes known to involve dependent transmission, symptoms of doubly-infected plants are more severe than those of comparable singly-infected ones. This relative increase in symptom severity of doubly-infected plants may be great enough to be a synergistic reaction (1, 68). Such increases in symptom severity are not a necessary feature of the dependent systems; the helper viruses in groundnut rosette and tobacco yellow vein induce no obvious symptoms. The dependent systems appear to represent stable, consistent interactions, although some changes, such as the attenuation of carrot motley dwarf after subculturing for some months, have been observed (101). Smith (85) noted the stability of tobacco rosette "in experiments performed at intervals over a period of years." The interaction of MAV and RPV in Coast Black oats has been a consistent, reproducible system in our greenhouse and growth chambers for more than 10 years.

Relationships between viruses that interact in dependent transmission extend from viruses closely related to those that appear to be completely unrelated. Potato virus Y can serve as a helper virus for serologically related potato virus C as well as for the serologically unrelated potato aucuba mosaic virus. Most of the viruses transmitted in the circulative systems seem to be unrelated, but little is known about the relatedness of many. For barley yellow dwarf the unrelatedness of MAV and RPV has been shown both by serological methods and by in vivo interactions (1). In a recent comparison among three separate systems, no cross-protection occurred between the components of tobacco rosette and tobacco yellow vein virus, nor between tobacco yellow vein virus and groundnut rosette virus (2). Several of the viruses in the circulative group are polyhedral particles about 25–30 nm in diameter. Carrot mottle virus is about 50 nm in diameter and contains lipid (59). Rod-shaped particles may be involved in groundnut rosette (11, 12). The morphology of many viruses in the circulative group is unkown. Viruses in the stylet-borne systems are flexuous, rod-shaped particles.

Smith (84) focused attention on virus specificity in dependent transmission in early studies of tobacco rosette. None of four other viruses functioned as a helper virus to make tobacco mottle virus transmissible by aphids, nor did tobacco vein-distorting virus serve as a helper virus for any of three others with which it was combined. The current picture is somewhat different because tobacco yellow-net virus, which is similar to tobacco vein-distorting virus, can also serve as a helper for tobacco mottle virus (96); tobacco bushy-top virus, which may be a strain of tobacco mottle virus, is transmitted by aphids only in the presence of tobacco vein-distorting virus (16); groundnut rosette assistor virus can be a helper virus for tobacco yellow vein virus (2); and either RPV or RMV can be a helper virus for MAV, although a different aphid species is involved in each of the barley yellow dwarf cases. Thus, present information suggests a lack of specificity in the interaction between helper and dependent viruses, but relationships among the different helper viruses are unknown, and in relatively few systems have tests for specificity been made.

A partial specificity exists among viruses in the stylet-borne dependent transmission systems. Potato aucuba mosaic virus became aphid transmissible, not only in mixed infections with potato virus A or potato virus Y (39) but also in the presence of bean yellow mosaic, turnip mosaic, and henbane mosaic viruses (100). There is a level of virus specificity even in this instance because potato aucuba mosaic virus was not transmitted from doubly-infected plants in tests with five other potential helper viruses (39). Potato virus C, which is "helped" by potato virus Y, was not transmitted from mixed infections with henbane mosaic virus, which is similar to potato virus Y (95). Specificities in the relationships even vary with virus strains. Kassanis (39) found that two strains of potato aucuba mosaic virus were "helped" by potato virus Y but not by potato virus A, although other strains were transmitted in the presence of either helper virus.

Another feature of dependent transmission important in any attempt to understand the virus interactions concerns maintenance of the mixed infection through subcultures. Since the viruses that have a circulative relationship with their vectors all have helper viruses that have not been transmitted mechanically, the mixtures cannot be maintained by mechanical inoculations. The mixed infections generally are retained in subsequent transfers made by means of aphids because the vectors usually transmit the helper virus together with the dependent virus to recreate the mixture each time a transfer is made. There may be important differences, however, among the various systems with respect to the probability that both viruses will be transmitted. For example, *R. padi* continues to transmit both RPV and MAV through many subcultures (68), but *R. maidis* seems to transmit both MAV and RMV only through one or two subsequent serial transfers. Such differences could greatly influence detection and function of dependent aphid transmission in nature. Since viruses in the stylet-borne systems are also mechanically transmissible, they are usually maintained in this manner. Watson (95) found that mixtures of potato viruses Y and C were still aphid transmissible after two years of subculturing by mechanical transmissions to *N. glutinosa.* A dependent virus is not changed following aphid transmission from the mixture. For example, Kassanis (39) showed that potato aucuba mosaic virus transmitted from mixed infections with potato virus Y did not remain aphid transmissible after being separated from potato virus Y.

The most important question in any understanding of dependent aphid transmission is whether the virus interaction occurs in the plant, in the aphid, or in both. Current information suggests that the answer for circulative dependent systems is different from that for stylet-borne systems. Dependent transmission of circulative viruses apparently is based on interaction of the viruses in the plant and not in the aphid. In the early experimental work on dependent transmission, Smith (84) found that tobacco vein-distorting virus served as a helper virus for tobacco mottle virus only if the two viruses occurred together in the source plant. Alternate acquisition feeding on plants infected by each of the viruses alone did not result in aphid transmission of tobacco mottle virus. Carrot mottle virus has not been transmitted by aphids exposed alternately to the two components of carrot motley dwarf (100).

Tests of the RPV and MAV interaction in barley yellow dwarf also support the observation that presence of both viruses in the source plants is a basic feature of circulative dependent transmission. In many experiments I have tried to obtain transmission of MAV by *R. padi* by permitting interaction of MAV and RPV in the vector. In one series of tests, for example, in only 2 of 265 experiments was MAV transmitted by *R. padi* when the aphids fed first on plants infected by one of the viruses and then were exposed to the other virus by feeding on infected leaves or by injection with concentrated inocula (69). The interaction also failed to occur when *R. padi* fed through membranes on, or were injected with, concentrated inocula made by mixing partially purified preparations of each of the separate viruses. The rare transmission of MAV by *R. padi* in tests of possible interactions between RPV and MAV in the vector probably simply reflects the relative nature of the specificity. In another discussion (70) I pointed out the possible significance of the fact that these rare transmissions generally occurred in experiments when *R. padi* was exposed to MAV before RPV, and I then developed a theoretical model to explain these observations. The model was based on possible interaction between the protein capsid of the two viruses and the salivary membranes of the aphid. Although the model may apply to some cases of dependent transmission, it is based only on rare events that are in sharp contrast to the common events of dependent transmission systems. It will not be discussed here.

Early studies (40) on potato aucuba mosaic virus also suggested that the dependent virus could be transmitted by aphids only from doubly-infected plants, but recent work on the two dependent systems that involve stylet-borne viruses has shown that aphids do not have to acquire the helper and dependent viruses from the same plant. Kassanis & Govier (43) found that groups of aphids (*M. persicae*) could transmit potato aucuba mosaic virus if first allowed to probe for 1–2 minutes into a leaf infected with potato virus Y before probing into leaves infected by potato aucuba mosaic virus. Potato virus Y served as a helper virus about as readily in such sequential acquisition feedings as it did when aphids probed into plants infected by both viruses. Similarly, potato virus C was transmitted by aphids, not only following probes into plants doubly infected by potato viruses C and Y, but also when the aphids were allowed to probe first into plants containing potato virus Y before they were exposed to potato virus C. The order of alternate acquisition feeding was critical. Neither potato aucuba mosaic virus nor potato virus C was transmitted if aphids probed first into plants infected by the dependent virus before they probed into plants infected by the helper virus. The finding that dependent transmission is not confined to doubly-infected source plants opens a whole new range of possible roles for the phenomenon.

Mechanisms.—Much interest in dependent transmission by aphids has been directed toward an understanding of the processes that control the phenomenon. Many mechanisms have been suggested. The suggestions range from complex possibilities, such as the helper virus functioning to aid multiplication of the dependent virus in the aphid vector (101); to relatively simple ones, such as increased mutation rates for the dependent virus in mixed infections (95) or a change in the location of the dependent virus in doubly-infected cells to make virus available to aphids (39). Four mechanisms represent the more feasible possibilities; the two currently supported by experimental evidence suggest that one explanation applies to systems involving the stylet-borne relationship, and another applies to systems where the aphid-virus relationships are circulative. This generalization probably will prove to be an over-simplification, but at least it provides a starting point for discussion and future research. The various suggestions are not mutually exclusive, however, and it may be that more than one is correct even for transmission of a single dependent virus.

The first suggested mechanism emphasized the quantity of virus in the doubly-infected plant (83). Since virus concentration is sometimes increased in mixed infections, it is possible that doubly-infected plants merely contain more of the dependent virus than singly-infected ones. A critical increase in virus concentration may enable aphids to acquire and transmit the virus. Tests for the quantitative explanation have been carried out on at least three of the dependent systems; none of the tests has supported the explanation. Smith (84) made dilution experiments and found no difference between the concentration of tobacco mottle virus from plants infected only by this virus and that from plants doubly infected by it and the helper virus. For barley yellow dwarf, tests of the quantitative explanation have been based on comparing the amount of MAV extracted from MAV-infected oats with the amount extracted from oats infected by both MAV and RPV. No increases in MAV content of doubly-infected plants over that in singly-infected plants have been detected either in tests based on infectivity assays made by feeding *M. avenae* through membranes on diluted extracts or in tests based on analytical sucrose gradient centrifugation of the two kinds of preparations (Rochow, unpublished data). The concentration of potato aucuba mosaic virus often increased in mixed infections with either potato virus A or potato virus Y, but virus concentration and aphid transmissibility in individual experiments were not correlated (39). In some experiments the content of potato aucuba mosaic virus in doubly-infected plants was four times that in singly-infected ones, but aphid transmission did not occur. In other experiments the concentration of potato aucuba mosaic virus in doubly-infected plants was lower than that in singly-infected ones, but aphids transmitted potato aucuba mosaic virus only from the doubly-infected plants.

The present lack of evidence does not necessarily eliminate the quantitative explanation. It is possible that increases in the concentration of the helper virus in mixed infections occur only in specific tissues, in certain cells, or even at critical sites within certain cells. In at least one case, mixed virus infections did alter location of virus within a cell; only in mixed infections with cucumber mosaic virus was tobacco mosaic virus frequently observed within the nucleus (33). Recent studies have emphasized the potential importance of different sites within cells or tissues for acquisition of virus by aphids (13, 60, 104). Even comparisons of virus concentration from epidermal tissues of singly- and doubly-infected plants (40) may be too crude to reflect the importance of the site of virus acquisition by aphids.

A second explanation for dependent transmission, which resulted from studies of potato virus C, was based on genetic recombination of potato viruses Y and C in the doubly-infected plant. Watson (95) originally favored this explanation partly because it accounted for perpetuation of aphid transmissibility of the mixed infection. Since recent work has shown that the helper virus and the dependent virus do not need to be acquired from the same plant (43), genetic recombination cannot be a necessary factor in the dependent transmission of potato virus C. Although genetic recombination may play a role, more simple explanations are equally plausible.

A third possible mechanism for dependent transmission is based on the suggestion that some kind of complex or aggregate of the dependent and helper viruses forms and that the dependent virus is merely carried along as a result of aphid transmission of the helper virus. Kassanis & Govier (43) favored this idea as the most likely explanation for the dependent transmission of potato aucuba mosaic virus and potato virus C.

> Differences in surface structure of the stylets might account for differential absorption and elution, i.e. potato aucuba mosaic virus and potato virus C by themselves are not absorbed to the stylets but potato virus Y is, and when it is, it modifies the surface structure or charge of the stylets, thus allowing potato aucuba mosaic virus or potato virus C to be adsorbed either on to the modified stylet surface or on to the potato virus Y particles (43).

Support for this mechanism resulted from finding that the two stylet-borne, dependent viruses could be transmitted by aphids following sequential probes first into plants infected by the helper virus and next into plants infected by the dependent virus, as well as following probing into plants infected by both viruses. Because potato viruses C and Y are very closely related, and probably irregularly distributed in leaf tissue, Kassanis & Govier (43) believe that transmission of potato virus C from doubly-infected plants also probably happens after two successive probes, the first in a cell containing potato virus Y, and second in a cell containing potato virus C. Although this is an attractive possibility, it does not necessarily follow that the mechanism for transmission following alternate acquisitions is the same as that for transmission of the dependent virus from doubly-infected tissue.

The fourth mechanism, supported by results from studies of barley yellow dwarf (72), involves possible interaction of the two viruses during the assembly stage of virus synthesis in doubly-infected plants. The notion is that, during simultaneous synthesis of the two viruses, some nucleic acid of the dependent virus becomes incorporated into virus particles containing the protein capsid of the helper virus. Such "mixed" virus particles can then be transmitted by the appropriate vector species because the protein capsid is similar to that of the helper virus regularly transmitted by the aphid. Evidence for this explanation is based on use of antiserum specific for the MAV and RPV isolates of barley yellow dwarf virus (72, 73). When virus preparations made from doubly-infected plants were treated with MAV antiserum, and aphids were allowed to feed through membranes on the treated preparations, *M. avenae* did not transmit virus; MAV in the preparation was neutralized by the specific antiserum. From the same preparation, however, *R. padi* transmitted both RPV and MAV to about one-third of the test plants. Perhaps the "mixed" virus particles function in the aphid like RPV (because of the RPV protein capsid) and in the plant like MAV (because of the MAV nucleic acid). The incorporation of nucleic acid of one virus into particles coated with the protein of another virus represents a well-known type of interaction for bacterial and animal viruses. Studies of similar interactions among plant viruses are still in the early stages. Since the interaction is of such potential importance in plant virology, the phenomenon will be discussed briefly in the next section.

Heterologous Encapsidation

Terminology.—In 1946 Delbruck & Bailey (18) found that anomalous particles of the bacterial virus T2 were liberated from mixed infections of T2 and T4. When Novick & Szilard (61) studied the interaction of T2 and T4 in mixed infections in bacteria, they found that some virus particles had the phenotype (protein) of T4, but the genotype (nucleic acid) of T2. Further studies of the interaction of these two bacterial viruses led to the concept of phenotypic mixing (87). Some of the virus particles produced in the mixed infection contained the nucelic acid of T2, but the protein of both T2 and T4. Such phenotypically mixed particles were doubly neutralizable, i.e. they could be neutralized by antiserum specific either for T2 or T4. The phenomenon was soon observed for animal viruses, where phenotypic mixing was found to be common, especially among certain groups (23, 24). The term phenotypic mixing originally was used in a general sense "to describe the association of phenotypes with nonhomologous genotypes" (23). As studies of different virus systems progressed, however, the complexity of virus interactions in doubly-infected cells became clear. For example, most phenotypically mixed influenza virus particles also had mixed genomes (23). In some cases the "mixing" between two viruses involved the nucleic acid of one virus and the entire protein capsid of another virus, rather than a "mosaic" of two capsids. The terms transcapsidation (24) and genomic masking (49, 103) have been used to designate the enclosure of the genome of one virus within the complete capsid of another virus. Thus, it seems best now to reserve the term phenotypic mixing to describe virus particles that contain the nucleic acid of one virus enclosed in a capsid containing protein synthesized under the direction of two viruses. Since a general term is needed, I will use heterologous encapsidation for the general phenomenon of nucleic acid of one virus becoming enclosed within a protein capsid derived wholly or in part from synthesis of a second virus. Phenotypic mixing and transcapsidation will describe two types of heterologous encapsidation (Figure 1).

Isolated nucleic acid and protein of certain viruses have been combined in vitro to form a spectrum of combinations similar to those depicted in Figure 1 (7, 25). The fact that the interactions can occur under certain conditions in vitro complements and lends confidence to studies on heterologous encapsidation in vivo, but the factors responsible for virus reconstitution and self-assembly in vitro are not necessarily those controlling the in vivo maturation of virus. As Bancroft (7) has pointed out:

> Implicit in the evaluation of results gained from the study of in vitro assembly processes is the possibility that the events in such processes may turn out to resemble those that occur in vivo.

Role in plant virology.—Although heterologous encapsidation of bacte-

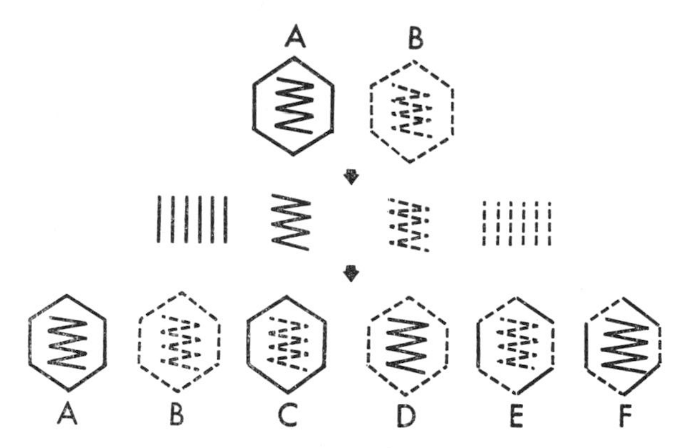

FIGURE 1. Diagrammatic representation of some combinations of nucleic acid and protein capsid for two viruses (A, B) synthesized in a mixed infection. Progeny of the mixed infection could include particles similar to the originals (A, B) which would represent homologous encapsidation. Four additional types of virus particles illustrate heterologous encapsidation. Two of them (C, D) represent transcapsidation (genomic masking); two others (E, F) represent phenotypic mixing.

rial viruses has been studied for some time and is an important phenomenon for animal viruses (23, 24), many unsuccessful searches have been made for examples of the phenomenon for plant viruses. In addition to evidence for transcapsidation between two serologically distinct barley yellow dwarf viruses (72), several other studies have provided data for a kind of heterologous encapsidation between temperature-sensitive and normal strains of tobacco mosaic virus. Results of work in three different laboratories showed that unusual strains of tobacco mosaic virus, which normally remain uncoated in single infections, become coated by protein of a second strain during simultaneous synthesis in mixed infections (4, 41, 42, 78). A unique system of mixed infections in barley with tobacco mosaic virus and barley stripe mosaic virus (29) also provided evidence for transcapsidation (genomic) masking). Dodds & Hamilton (19) showed that mixed infections of the two viruses in barley resulted in transcapsidation of some nucleic acid of tobacco mosaic virus with the protein of barley stripe mosaic virus. In another study, transcapsidation of nucleic acid of barley stripe mosaic virus with the protein coat of brome mosaic virus may have occurred (64).

The scarcity of evidence for heterologous encapsidation among plant viruses may simply be a reflection of where researchers have looked. Most studies have been directed toward mechanically transmissible viruses characterized by high yields in virus purification. Evidence for transcapsidation in the barley yellow dwarf system supports the possibility that among plant viruses the likely place for heterologous encapsidation to be an important, functional aspect of the transmission or infection process is in the relationship between plant viruses and their vectors. A concept of heterologous encapsidation that might prove especially useful for plant viruses is a spectrum of categories for the phenomenon (91). In some situations the mixed virus particles may be required for viral systems to function, and in others the phenomenon may merely be an inconsequential result of a mixed infection.

The various systems of dependent transmission by aphids evoked many predictions for heterologous encapsidation (7, 34, 39, 40, 70) even before some direct evidence (72) became available. The basis for importance of heterologous encapsidation in dependent transmission is the critical role of the protein capsid in aphid-virus specificity. This function of the protein capsid is currently supported only indirectly by experimental results, but expanded in vitro studies of viruses transmitted by aphids should soon help provide more direct data. For example, study of self-assembly of cucumber mosaic virus has been mentioned as one such approach for stylet-borne viruses (7). An observation possibly related to the role of virus capsid protein is that in all dependent transmission systems, except possibly parsnip yellow fleck, the dependent virus acquires a virus-vector relationship similar to that of the helper virus. Since aphids are animals, additional support for the importance of capsid protein comes from analogies with other virus-animal systems in which interactions between virus protein and cells are known to be critical (23, 24, 35). Perhaps the best general support for the potential importance of heterologous encapsidation in dependent transmission is that the process is basically a simple one and there seems to be no reason why it should not occur!

Another step in any speculation on the role of heterologous encapsidation in plant virology concerns the tremendous potential of the phenomenon in the ecology of plant viruses. The simple possibilities illustrated in Figure 1 suggest a fascinating range of qualitative and quantitative effects. For example, heterologous encapsidation could be a mechanism for maintenance of a plant virus in the field in the absence of its regular aphid vector. Although the MAV isolate of barley yellow dwarf virus is regularly transmitted only by *M. avenae,* mixed infections with either RPV or RMV essentially triple the number of effective vector species because *R. padi* or *R. maidis* can transmit MAV from the corresponding mixed infections. It could be significant that most of the mixed infections of barley yellow dwarf viruses found in our tests of field-collected winter grains during the past six years have involved RMV as one component of the mixture (74, 75).

Another possible role of heterologous encapsidation is alteration in host range of a virus and creation of a "new" disease. A phenotypically mixed virus particle (such as Figure 1, E) might become transmissible by an aphid species unrelated to the species that normally transmits the virus (Figure 1, B). If this "new" vector happens to feed on susceptible plants different from those colonized by the common vectors, the virus genome could be introduced into a new host from which still other aphid species might transmit virus from plant to plant. Thus, a barley yellow dwarf virus, usually restricted to small grains and grasses possibly because of the host range of the aphid vectors, might be transmitted to a plant such as sugar beet and appear to be a distinct virus. One intriguing aspect of such speculation is that, at least theoretically, the phenomenon could occur as a result of interaction among only one infected plant, only one aphid, and only one phenotypically mixed virus particle!

These are just a few examples of the potential importance of heterologous encapsidation of plant viruses. The phenomenon is only one of several explanations for dependent transmission of viruses by aphids. Aphids are only one example on a long list of vectors of plant viruses. The total plant pathological picture of possible interactions among vectors, plants, and mixed virus infections can stagger the imagination. As Fawcett (22) pointed out,

> Work with mixtures, however, will not make the already complex problem of plant pathology as a whole any easier or less complex, but it may throw much light on certain relationships, relationships which will probably never be discovered by the use of pure cultures of single organisms. We cannot, therefore, in my judgement, avoid entering actively into this largely unexplored field.

Acknowledgements

I am grateful to R. Hull, J. P. MacKinnon, and G. B. Orlob for providing unpublished material or comments. The helpful suggestions of Robert M. Goodman, R. I. Hamilton, Irmgard Muller, and D. A. Roberts are also gratefully acknowledged.

LITERATURE CITED

1. Aapola, A. I. E., Rochow, W. F. 1971. Relationships among three isolates of barley yellow dwarf virus. *Virology* 46:127–41
2. Adams, A. N., Hull, R. 1972. Tobacco yellow vein—another disease complex with an assistor virus. *Ann. Appl. Biol.*, in press
3. Allen, T. C., Jr. 1957. Strains of the barley yellow-dwarf virus. *Phytopathology* 47:481–90
4. Atabekov, J. G., Schaskolskaya, N. D., Atabekova, T. I., Sacharovskaya, G. A. 1970. Reproduction of temperature-sensitive strains of TMV under restrictive conditions in the presence of temperature-resistant helper strain. *Virology* 41:397–407
5. Badami, R. S., Kassanis, B. 1959. Some properties of three viruses isolated from a diseased plant of *Solanum jasminoides* Paxt. from India. *Ann. Appl. Biol.* 47:90–97
6. Bagnall, R. H. 1956. Three viruses isolated from Irish Cobbler potatoes affected with the interveinal mosaic disease. *Diss. Abstr.* 16:1761
7. Bancroft, J. B. 1970. The self-assembly of spherical plant viruses. *Advan. Virus Res.* 16: 99–134
8. Bancroft, J. B. 1971. The significance of the multicomponent nature of cowpea chlorotic mottle virus RNA. *Virology* 45: 830–34
9. Bennett, C. W. 1953. Interactions between viruses and virus strains. *Advan. Virus Res.* 1:39–67
10. Björling, K. 1970. Observations on the viral progeny of mixed infections with beet yellows virus. *Handlingar* II 24(1):1–11
11. Bock, K. R., Ngugi, E., Ambetsa, T., Mwathi, G. K. 1970. Groundnut viruses. *Record of Research, Annual Report, 1969,* East African Agr. & Forestry Res. Organ., 85
12. Bock, K. R., Perry, J., Waindi, E. N., Ambetsa, T., Mwathi, G. K. 1969. Plant pathology. Viruses of legumes. *Rec. Res., Ann. Rept., 1968,* East African Agr. & Forestry Res. Organ. 90–91
13. Bradley, R. H. E., 1968. Tobacco leaf veins as sources of a cucumber mosaic virus for aphids. *Virology* 34:172–73
14. Castillo, M. B., Orlob, G. B. 1966. Transmission of two strains of cucumber and alfalfa mosaic viruses by single aphids of Myzus persicae. *Phytopathology* 56:1028–30
15. Clinch, P. E. M., Loughnane, J. B., Murphy, P. A. 1936. A study of the aucuba or yellow mosaics of the potato. *Roy. Dublin Soc. Sci. Proc.* 21:431–48
16. Cole, J. S. 1962. Isolation of tobacco vein-distorting virus from tobacco plants infected with aphid-transmissible bushy-top. *Phytopathology* 52:1312
17. Costa, A. S. 1969. Conditioning of the plant by one virus necessary for systemic invasion of another. *Phytopathol. Z.* 65:219–30
18. Delbruck, M., Bailey, W. T. 1946. Induced mutations in bacterial viruses. *Cold Spring Harbor Symp. Quant. Biol.* 11:33–37
19. Dodds, J. A., Hamilton, R. I. 1971. Evidence for possible genomic masking between two unrelated plant viruses. *Phytopathology* 61:889–90 (Abstr.)
20. Doncaster, J. P., Kassanis, B. 1946. The shallot aphis, *Myzus ascalonicus* Doncaster, and its behaviour as a vector of plant viruses. *Ann. Appl. Biol.* 33: 66–68
21. Duffus, J. E. 1971. Role of weeds in the incidence of virus diseases. *Ann. Rev. Phytopathol.* 9:319–40
22. Fawcett, H. S. 1931. The importance of investigations on the effects of known mixtures of microorganisms. *Phytopathology* 21:545–50
23. Fenner, F. 1968. *The Biology of Animal Viruses. Molecular and Cellular Biology.* New York: Academic Press, 1:474 + 27 pp.
24. Fenner, F. 1970. The genetics of

animal viruses. *Ann. Rev. Microbiol.* 24:297–334
25. Fraenkel-Conrat, H. 1970. Reconstitution of viruses. *Ann. Rev. Microbiol.* 24:463–78
26. Freitag, J. H. 1969. Interactions of plant viruses and virus strains in their insect vectors. In *Viruses, Vectors, and Vegetation,* 303–25. Ed. K. Maramorosch. New York: Interscience
27. Fujisawa, I., Hayashi, T., Matsui, C. 1967. Electron microscopy of mixed infections with two plant viruses. I. Intracellular interactions between tobacco mosaic virus and tobacco etch virus. *Virology* 33:70–76
28. Halstead, B. E., Gill, C. C. 1971. Effect of inoculation of oats with paired combinations of barley yellow dwarf virus isolates. *Can. J. Bot.* 49:577–81
29. Hamilton, R. I., Dodds, J. A. 1970. Infection of barley by tobacco mosaic virus in single and mixed infection. *Virology* 42: 266–68
30. Hampton, R. O., Sylvester, E. S. 1969. Simultaneous transmission of two pea viruses by Acyrthosiphon pisum quantified on sweetpea as diagnostic local lesions. *Phytopathology* 59: 1663–67
31. Harrison, B. D. 1958. Ability of single aphids to transmit both avirulent and virulent strains of potato leaf roll virus. *Virology* 6:278 86
32. Hoggan, I. A. 1929. The peach aphid (Myzus persicae Sulz.) as an agent in virus transmission. *Phytopathology* 19:109–23
33. Honda, Y., Matsui, C. 1969. Occurrence of tobacco mosaic virus within the nucleus. *Virology* 39:593–94
34. Hull, R., Adams, A. N. 1968. Groundnut rosette and its assistor virus. *Ann. Appl. Biol.* 62: 139–45
35. Joklik, W. K. 1968. Virus-host interactions. In *The Biological Basis of Medicine, chap. 11,* 2: 341–72. Ed. E. E. Bittar. New York: Academic Press
36. Kassanis, B. 1941. Transmission of tobacco etch viruses by aphides. *Ann. Appl. Biol.* 28: 238–43
37. Kassanis, B. 1947. Studies on dandelion yellow mosaic and other virus diseases of lettuce. *Ann. Appl. Biol.* 34:412:21
38. Kassanis, B. 1952. Some factors affecting the transmission of leafroll virus by aphids. *Ann. Appl. Biol.* 39:157–67
39. Kassanis, B. 1961. The transmission of potato aucuba mosaic virus by aphids from plants also infected by potato viruses A or Y. *Virology* 13:93–97
40. Kassanis, B. 1963. Interaction of viruses in plants. *Advan. Virus Res.* 10:219–55
41. Kassanis, B., Bastow, C. 1971. *In vivo* phenotypic mixing between two strains of tobacco mosaic virus. *J. Gen. Virol.* 10:95–98
42. Kassanis, B., Bastow, C. 1971. Phenotypic mixing between strains of tobacco mosaic virus. *J. Gen. Virol.* 11:171–76
43. Kassanis, B. Govier, D. A. 1971. New evidence on the mechanism of aphid transmission of potato C and potato aucuba mosaic viruses. *J. Gen. Virol.* 10:99–101
44. Koch, K. 1931. The potato rugose mosaic complex. *Science* 73:615
45. Kousalya, G., Ayyavoo, R., Bhaskaran, S., Krishnamurthy, C. S. 1970. Rosette disease of groundnut—transmission studies. *Madras Agr. J.* 57:172–78
46. Kvicala, B. 1945. Selective power in virus transmission exhibited by an aphis. *Nature* 155:174–75
47. Kvicala, B. A. 1948. Studies on the composite nature of cauliflower mosaic with a special regard to the selective transmission of both viruses in this complex disease by certain aphids. *Brünn Vysoka Skola Zemed. Sbornik. Rada C.* 40:1–87
48. Lane, L. C., Kaesberg, P. 1971. Multiple genetic components in bromegrass mosaic virus. *Nature New Biol.* 232:40–43
49. Ling, C. M., Hung, P. P., Overby, L. R. 1970. Independent assembly of QB and MS2 phages in doubly infected *Escherichia coli. Virology* 40: 920–29
50. Maassen, H. 1968. Untersuchungen zur analyse eines erdbeervi-

ruskomplexes und nachweis eines neuen persistenten erdbeervirus. *Phytopathol. Z.* 62: 343–50
51. MacKinnon, J. P. 1960. Combined transmission by single aphids of two viruses that persist in the vector. *Virology* 11:425–33
52. MacKinnon, J. P. 1961. Transmission of two viruses by aphids reared on different hosts. *Virology* 13:372–73
53. MacKinnon, J. P. 1965. A mild chlorosis virus of Physalis floridana found in a turnip latent virus complex. *Can. J. Bot.* 43: 509–17
54. MacKinnon, J. P., Lawson, F. L. 1966. Transmission of vein blotch virus of Physalis floridana by aphids. *Can. J. Bot.* 44:1219–21
55. Matthews, R. E. F. 1970. *Plant Virology.* New York: Academic Press, 778 pp.
56. McWhorter, F. P., Price, W. C. 1949. Evidence that two different plant viruses can multiply simultaneously in the same cell. *Science* 109:116–17
57. Mellor, F. C., Fitzpatrick, R. E. 1951. Studies of virus diseases of strawberries in British Columbia. II. The separation of the component viruses of yellows. *Can. J. Bot.* 29:411–20
58. Murant, A. F., Goold, R. A. 1968. Purification, properties and transmission of parsnip yellow fleck, a semi-persistent, aphid-borne virus. *Ann. Appl. Biol.* 62:123–37
59. Murant, A. F., Goold, R. A., Roberts, I. M., Cathro, J. 1969. Carrot mottle—a persistent aphid-borne virus with unusual properties and particles. *J. Gen. Virol.* 4:329–41
60. Normand, R. A., Pirone, T. P. 1968. Differential transmission of strains of cucumber mosaic virus by aphids. *Virology* 36: 538–44
61. Novick, A., Szilard, L. 1951. Virus strains of identical phenotype but different genotype. *Science* 113:34–35
62. Okusanya, B. A. M., Watson, M. A. 1966. Host range and some properties of groundnut rosette virus. *Ann. Appl. Biol.* 58:377–87
63. Ossiannilsson, F. 1966. Insects in the epidemiology of plant viruses. *Ann. Rev. Entomol.* 11: 213–32
64. Peterson, J. F. 1971. *Possible formation of mixed or 'hybrid' virions in plants infected with brome mosaic and barley stripe mosaic viruses.* Ph.D. thesis. Univ. Nebraska, Lincoln. 109 pp.
65. Plumb, R. T. 1970. Yellow net virus (Y or V). *Rept. Rothamst. Expt. Sta., 1969, Part 1,* p. 144
66. Prentice, I. W. 1952. Resolution of strawberry virus complexes. V. Experiments with viruses 4 & 5. *Ann. Appl. Biol.* 39:487–94
67. Rochow, W. F. 1963. Variation within and among aphid vectors of plant viruses. *Ann. N.Y. Acad. Sci.* 105:713–29
68. Rochow, W. F. 1965. Apparent loss of vector specificity following double infection by two strains of barley yellow dwarf virus. *Phytopathology* 55:62–68
69. Rochow, W. F. 1965. Selective virus transmission by Rhopalosiphum padi exposed to two vector-specific strains of barley yellow dwarf virus. *Phytopathology* 55:1284–85 (Abstr.)
70. Rochow, W. F. 1969. Specificity in aphid transmission of a circulative plant virus. In *Viruses, Vectors, and Vegetation,* 175–98. Ed. K. Maramorosch. New York: Interscience
71. Rochow, W. F. 1969. Biological properties of four isolates of barley yellow dwarf virus. *Phytopathology* 59:1580–89
72. Rochow, W. F. 1970. Barley yellow dwarf virus: Phenotypic mixing and vector specificity. *Science* 167:875–78
73. Rochow, W. F., Aapola, A. I. E., Brakke, M. K., Carmichael, L. E. 1971. Purification and antigenicity of three isolates of barley yellow dwarf virus. *Virology* 46:117–26
74. Rochow, W. F., Jedlinski, H. 1970. Variants of barley yellow dwarf virus collected in New York and Illinois. *Phytopathology* 60:1030–35

75. Rochow, W. F., Muller, I. 1971. A fifth variant of barley yellow dwarf virus in New York. *Plant Dis. Reptr.* 55:874–77
76. Ross, A. F. 1957. Responses of plants to concurrent infection by two or more viruses. *Trans. N.Y. Acad. Sci.* 19:236–43
77. Ross, A. F. 1959. The interaction of viruses in the host. In *Plant Pathology, Problems & Progress 1908–58,* 511–20, chap. XLVI. Ed. C. S. Holton. Madison: Univ. Wisconsin Press
78. Sarkar, S. 1969. Evidence of phenotypic mixing between two strains of tobacco mosaic virus. *J. Mol. Gen. Genet.* 105:87–90
79. Schneider, I. R. 1971. Characteristics of a satellite-like virus of tobacco ringspot virus. *Virology* 45:108–22
80. Schroeder, W. T., Provvidenti, R., McEwen, F. L. 1959. Pea streaks naturally incited by combinations of viruses. *Plant Dis. Reptr.* 43:1219–26
81. Simons, J. N. 1958. Titers of three nonpersistent aphid-borne viruses affecting pepper in south Florida. *Phytopathology* 48: 265–68
82. Smith, K. M. 1931. On the composite nature of certain potato virus diseases of the mosaic group as revealed by the use of plant indicators and selective methods of transmission. *Proc. Roy. Soc. London B* 109:251–67
83. Smith, K. M. 1945. Transmission by insects of a plant virus complex. *Nature* 155:174
84. Smith, K. M. 1946. The transmission of a plant virus complex by aphids. *Parasitology* 37:131–34
85. Smith, K. M. 1965. Plant virus-vector relationships. *Advan. Virus Res.* 11:61–96
86. Stace-Smith, R. 1956. Studies on Rubus virus diseases in British Columbia. III. Separation of components of raspberry mosaic. *Can. J. Bot.* 34:435–42
87. Streisinger, G. 1956. Phenotypic mixing of host range and serological specificities in bacteriophages T2 and T4. *Virology* 2: 388–98
88. Swenson, K. G. 1963. Effects of insect and virus host plants on transmission of viruses by insects. *Ann. N.Y. Acad. Sci.* 105:730–40
89. Swenson, K. G. 1968. Role of aphids in the ecology of plant viruses. *Ann. Rev. Phytopathol.* 6:351–74
90. Sylvester, E. S. 1956. Beet yellows virus transmission by the green peach aphid. *J. Econ. Entomol.* 49:789–800
91. Trautman, R., Sutmoller, P. 1971. Detection and properties of a genomic masked viral particle consisting of foot-and-mouth disease virus nucleic acid in bovine enterovirus protein capsid. *Virology* 44:537–43
92. Valleau, W. D., Johnson, E. M. 1930. The relation of some tobacco viruses to potato degeneration. *Kentucky Agr. Exp. Sta. Res. Bull. no.* 309, 473–507
93. Watson, M. A. 1946. The transmission of beet mosaic and beet yellows viruses by aphides; a comparative study of a nonpersistent and a persistent virus having host plants and vectors in common. *Proc. Roy. Soc. London B* 133:200–19
94. Watson, M. A. 1956. The effect of different host plants of potato virus C in determining its transmission by aphids. *Ann. Appl. Biol.* 44:599–607
95. Watson, M. A. 1960. Evidence for interaction or genetic recombination between potato viruses Y and C in infected plants. *Virology* 10:211–32
96. Watson, M. A. 1961. Plant virus diseases. *Nature* 190:220–21
97. Watson, M. 1962. Yellow-net virus of sugar beet. I. Transmission and some properties. *Ann. Appl. Biol.* 50:451–60
98. Watson, M. A. 1967. Transmission of cucumber mosaic virus from plants infected with two strains. *Rept. Rothamst. Expt. Sta., 1966,* p. 120
99. Watson, M. A., Dunning, R. A., Serjeant, E. P., Lack, A. J. 1965. Carrot motley dwarf in the field in 1964. *Rept. Rothamst. Expt. Sta., 1964,* p. 128–29
100. Watson, M. A., Plumb, R. T. 1972. Transmission of plant pathogenic viruses by aphids. *Ann. Rev. Entomol.* 17:425–52

101. Watson, M., Serjeant, E. P., Lennon, E. A. 1964. Carrot motley dwarf and parsnip mottle viruses. *Ann. Appl. Biol.* 54:153–66
102. Whitcomb, R. F., Davis, R. E. 1970. Mycoplasma and phytarboviruses as plant pathogens persistently transmitted by insects. *Ann. Rev. Entomol.* 15: 405–64
103. Yamamoto, N., Anderson, T. F. 1961. Genomic masking and recombination between serologically unrelated phages P22 and P221. *Virology* 14:430–39
104. Zettler, F. W., Christie, R. G., Edwardson, J. R. 1967. Aphid transmission of virus from leaf sectors correlated with intracellular inclusions. *Virology* 33: 549–52
105. Zitter, T. A. 1970. Titers of two virus diseases of celery affecting field spread. *Phytopathology* 60:1321 (Abstr.)

PLANT VIRUSES WITH A DIVIDED GENOME 3546

A. VAN KAMMEN
Laboratory of Virology, State Agricultural University, Wageningen, The Netherlands

A considerable number of plant viruses produce in infected plants more than one nucleoprotein component. Typical examples are tobacco rattle virus (TRV), alfalfa mosaic virus (AMV), viruses of the cowpea mosaic virus (CPMV) group, tobacco streak virus (TSV) and brome mosaic virus (BMV). It has been shown in recent years for some of these multiparticulate viruses that the genetic information necessary for virus multiplication is divided among two or more nucleoprotein components.

Discovery of plant viruses with a divided genome has changed the concept of the nature of viruses. The classical concept implied that the mature virus, the ultimate phase of viral development, was a single particle, the virion, containing the genome of the virus. This is of course still true for many plant viruses, but no longer for all. Several important groups of plant viruses have been recognized to have a genome divided among two or more physically separable particles.

A certain heterogeneity occurs in practically all purified preparations of plant viruses. In preparations of tobacco mosaic virus (TMV) for example, rodshaped particles with a modal length of 300 mμ predominate, but smaller and longer rods occur also. The presence of such particles has been explained by possible fragmentation of 300 mμ particles during purification, faulty synthesis of particles, or, in case of longer particles, by aggregation. Such heterogeneity is quite normal in preparations of rodshaped and threadlike viruses. Similarly, preparations of spherical viruses may contain some particles with fragmented RNA. This type of particle we shall regard as defective; it consists of viral structural proteins and fragments of the viral genome. It is possible that these defective particles have some biological significance, and that they can interfere with virus replication. The infectivity of TMV, for example, is enhanced by the presence of short particles derived from the virus in the inoculum (23, 37). In these cases it is a question of occurrence of defective particles, besides nondefective homologous virus particles which contain a complete genome necessary for multiplication (36). Heterogeneity in such virus preparations does not reflect functional heterogeneity. This is different from the multiparticulate viruses we will discuss here.

Plant viruses with a divided genome are characterized by physical heterogeneity, and this goes with functional heterogeneity. Each of the physical components contains a well-defined part of the genome of the virus. The separate components are not defective in the sense described above (36). Particles with a complete genome do not occur, but there is a characteristic distribution of genetic properties of the virus among the components.

Plant viruses with a divided genome, as far as now known, can be subdivided into the following six groups, each with different characteristics.

1. Satellite Viruses

A satellite virus is unable to multiply by itself. It is dependent for multiplication on association with a virus which can multiply by itself. The famous example of this type of behavior is satellite virus (SV) (42, 43), which multiplies only in plants also infected with tobacco necrosis virus (TNV). TNV is fully autonomous. Purified preparations of this virus contain one type of particle with a sedimentation coefficient of 118 *S*. The particles are isometric, about 26 nm in diameter, and contain one single-stranded RNA with a molecular weight of ca 1.45×10^6, which constitutes about 19% of the particle weight. The survival of TNV does not depend on the presence of SV, but the continued existence of SV can only be guaranteed by TNV.

Purified preparations of SV also contain only one type of particle with a sedimentation coefficient of 50 *S*. The particles are isometric, 17 nm in diameter, and contain about 20% RNA. The RNA is single-stranded and has a molecular weight of about 0.4×10^6. The proteins of the capsids of SV and TNV are different. Each capsid contains one type of protein. The protein subunit of SV has a molecular weight of about 22,000 (62, 63), and that of TNV 33,500 (50). TNV and SV are serologically unrelated.

The association between SV and TNV is very specific (42). No plant viruses other than TNV are able to effect multiplication of SV. The presence of SV in TNV-infected plants affects the behavior of TNV in several ways. Preparations containing both TNV and SV produce two sizes of local lesions on infected leaves of *Nicotiana tabacum* and *Phaseolus vulgaris*. Large lesions contain only TNV, whereas small lesions contain both SV and TNV. If TNV and SV multiply together, the amount of TNV produced may be much less than when SV is absent.

Strains of TNV can be divided into two serological groups (44). They do not coincide with groupings of TNV based on different symptoms on tobacco and French beans, nor with those based on the ability of strains to support the growth of satellite-virus strains. However, if the ability to activate SV does not seem to be correlated with the serological affinity of TNV, there seems to be a correlation between the ability to activate SV and host range. Those TNV viruses that multiply readily in tobacco and French bean activate SV strains different from those activated by virus strains that multiply with difficulty in these plants.

The ability of TNV to activate SV is thus distinct from the coat-producing function of TNV. The relationship between TNV and SV, which allows the latter to multiply, is not known. As the protein of SV is synthesized only if SV is multiplying, the genetic information for SV protein apparently is coded into the SV-RNA. The RNA chain of SV has about 1200 nucleotides. Half of this number would be required to code for the 208 amino acid residues of the SV protein (62). The rest of the nucleotides might code for one other protein of a reasonable size; one can only speculate about the possible function of such a protein. It is obvious that SV needs replicase activity, which it lacks and TNV appears to possess. Viral RNA-replicases are known to be restrictive enzymes, which are only able to recognize and replicate very specific RNAs. It is therefore of interest that 5′-terminal nucleotide sequences of SV-RNA and TNV-RNA are identical (51). Both RNAs have the sequence ppApGpUp--- at the 5′terminus. The 5′-terminal sequence of the RNAs might play a role in the recognition by virus-specific RNA replicase.

Kassanis et al (45) found that wherever SV occurred in infected cells, TNV was nearby, as might be expected with SV dependent on TNV for its multiplication. The presence of two types of double-stranded RNA in tobacco leaves infected with TNV + SV has been demonstrated (46), one specific for SV-RNA and the other for TNV-RNA. This might indicate that SV-RNA and TNV-RNA are indeed synthesized on separate templates, and if so, the common recognition of the RNAs by a replicase might be important. It is plausible that the RNA of SV is too small to specify all the proteins needed for replication, for the molecular weight of the RNA of plant viruses capable of independent multiplication in a suitable host is far more than 1×10^6. An intriguing question remains whether SV-RNA contains information for another protein besides the cistron for SV coat protein, or whether the SV-RNA is monocistronic with the rest of the RNA required for structural and recognition purposes.

One might wonder if SV is a plant virus with a divided genome. The first requirement for SV multiplication is the presence of TNV, and SV thus appears to be more a satellite than a virus. If one considers SV as a virus of full value, as has been done by the editors of the *Descriptions of Plant Viruses* (43), the genetic information for SV multiplication is divided among two nucleoprotein components. Moreover, the relationship between SV and TNV is clarifying, as it is distinct from the relationship between components of the other multiparticulate viruses.

A satellite-like particle has been described by Schneider (65) to accompany tobacco ringspot virus (ToRSV). The satellite nucleoprotein is noninfective alone, but infective when mixed with ToRSV. Local lesions produced by mixtures of satellite-like particles and ToRSV are much smaller than these caused by ToRSV alone. The relative number of satellite-like particles increased in the mixture of ToRSV and satellite in inoculated and systemically invaded leaves of bean and cowpea plants. The satellite-like particles have the ability in such mixed infections to become the larger part of the population. The satellite-like particles thus need ToRSV for multiplication, and interfere specifically with the growth of ToRSV.

The satellite-like particles have the same size and shape as ToRSV and are serologically indistinguishable from it. Each protein shell of a satellite-like particle encloses many small nucleic acid strands. The RNA molecules are single-stranded, have a sedimentation coefficient of about 7 S and a molecular weight of about 86,000. Each particle should contain about 14 or more RNA strands of this size, to give the sedimentation coefficient of 122 S of the nucleoprotein particles.

The nature of the RNAs of the satellite-like particles is unknown. Since no gene products have been attributed to these RNAs, it is not known if they are translated at all. The RNAs in a satellite-like particle may be all one kind of RNA, but different strands cannot be ruled out. The RNAs could represent a class of replicating RNA molecules, which can compete successfully with the ToRSV-RNA for the viral RNA-replicating system.

The RNA of the satellite-like particles might equally well be derived from the viral RNA, and consist of parts of the viral genome. If so, they would meet the definition of defective viral particles given by Huang & Baltimore (36). According to these authors defective virus particles have the following properties: they contain normal viral structural protein; they contain part of the viral genome; they can reproduce in the presence of helper virus; they can interfere specifically with the intracellular replication of nondefective homologous virus. But the remarkable advantage of the satellite-like RNA over the viral RNA in the replication process is not thereby explained.

2. Tobacco Rattle Viruses

The second group of viruses is characterized by a genome divided among two nucleoprotein components: one can multiply by itself; the other component depends on the first for multiplication; only together the components produce complete virions.

The group consists of tobacco rattle viruses (TRV) (34) and pea early browning virus (PEBV). These are RNA-containing viruses with straight tubular particles of two predominant lengths.

Most of the research of this group of viruses has been done with TRV. The properties of these viruses have been reviewed recently by Lister (55).

The long particles of TRV have sedimentation coefficients of 300 *S*, and the short ones 155–243 *S*, depending on the isolate. The two types of particles are therefore easily separable by zonal density gradient centrifugation. The corresponding RNAs from the particles have S-values of 26 S and 12–20 S, respectively. The two nucleoprotein rods are serologically identical.

The studies of Lister (52, 54, 56) and Sänger (64) make it clear that the long particles are infective and induce in infected plants the synthesis of RNA of the long particles. They are, however, not capable of producing virus-coat protein. The product of infection by long particles is referred to as the unstable variant of TRV. It is considered to be "free" long-particle RNA, not coated by viral protein. The short-rod particles are noninfective. They do not multiply if inoculated alone. The short particles require the presence of long particles for multiplication. If the inoculum contains long and short particles or the corresponding RNAs, the stable variant of TRV is produced (i.e., long and short particles both encapsulated in viral protein). The short particles appear to carry the gene for the virus-coat protein. Two essential functions for the production of complete virions are therefore distributed among the two rods.

There is some resemblance between the short rod of TRV and the SV of TNV, described earlier. Both carry information for virus-coat protein, and depend on another nucleoprotein for their multiplication, whereas that nucleoprotein cannot multiply by itself. The clear difference is that the long particles of TRV are not autonomous. They require the short particle for the production of stable complete virions. Moreover, short and long particles are serologically identical, and have the same protein. The RNA of the long and short particles therefore appear to be specialized parts of the genome of one virus, TRV. This conception is supported by the results of many experiments.

There are many strains of TRV (13, 35, 56, 68). The interactions of the two types of particles have been investigated by making mixtures of short particles of one strain and long particles of another strain. The biological behavior of such mixtures has been studied to examine the distribution of functions between the two pieces of RNA or TRV. In this way it is hoped to determine the genetic functions present in each nucleoprotein component, and the nature of these functions.

However, not all mixtures of short particles of one isolate with long particles of another strain show interaction, i.e., give rise to a tobacco rattle virus of mixed phenotype. A rather close relationship appears necessary for strains in order readily to interact heterologously and to give hybrid strains. The nature of the relationship is still ill-defined.

If long particles or unstable variants of one strain and short particles of another were inoculated (56, 69) into tobacco plants, the length of short particles produced was the length of short particles used in the inoculum. This is evidence that the unstable variants do not contain uncoated short particle RNA, and that the short particles are not derived from the long ones. The long particle has no influence on the short particle produced. If the short particle can be replicated in combination with the long particle of any strain, it is replicated as such.

The short particles contain the cistron for viral-coat protein. The coat protein consists of about 218 amino-acid residues. This would require 654 nucleotides for coding. The short-particle RNA has about 2400 nucleotides, and can therefore easily accommodate information for more proteins. Evidence for the presence of cistrons in the short particles that influence symptomatology was obtained by Lister & Bracker (56). These authors performed experiments with three strains of TRV, which were antigenically closely related but could be differentiated by the length of their short particles and in causing mild (TRV-Mild), severe (TRV-Sev), and yellow (TRV-Yel) symptoms in a systemic host, *Nicotiana clevelandii* Gray. If *N. clevelandii* plants were inoculated with long particles of each strain, unstable variants arose, as might be expected. The symptoms of the unstable variants derived from each strain were similar, and distinct from those characterizing the respective stable forms. This indicates an effect of the short particles on the symptoms produced, but it does not yet show that there are cistrons on the short particles influencing the type of symptoms, different from the coat-protein cistron. Mixed inoculation experiments made to compare the products resulting from inoculating *N. clevelandii* with heterologous mixtures of the strains mentioned, showed that the type of symptom of the stable variant obtained was that of the strain from which the short particles were derived. The combination of unstable variant of either TRV-Mild or TRV-Sev with short particles of TRV-Yel resulted in stable variants, which caused yellow symptoms when inoculated on *N. clevelandii*. In the same way stable variants obtained by combining unstable variants from TRV-Yel with short particles of TRV-Mild, gave mild symptoms. As the coat proteins of the three strains are closely similar, these results suggest that there occur cistrons on the short particles influencing these symptoms, apart from those specifying coat protein (56).

It has been shown that long particles also accommodate information influencing symptoms in characteristic ways. For instance unstable variants derived from several strains caused distinctive lesions in *Nicotiana glutinosa* (54). Sänger (64) used strains from Germany (TRV-Ger) and the United States (TRV-USA), which differed in particle size, symptoms, and antigenicity. TRV-USA had particles 1950 and 1050 Å long and they caused solid necrotic spots in Xanthi-nc tobacco. TRV-Ger caused necrotic rings on inoculated leaves of the same tobacco species; this strain had particles 1800 and 700Å long. The unstable variants derived from each strain produced symptoms identical to those of the respective stable viruses, illustrating that long particles are responsible for the type of symptoms of these strains. Inocula consisting of unstable variants from TRV-Ger mixed with short particles from TRV-USA, caused ringlike lesions like TRV-Ger on Xanthi-nc tobacco. The lengths of particles of the stable hybrid virus were those appropriate for TRV-Ger long particles and TRV-USA short particles, and the virus protein was serologically of the TRV-USA type. The RNA of both kinds of particles was apparently coated with protein specified by the RNA of TRV-USA. This is direct proof that the short particle accommodates the cistron-specifying coat protein. Once more these experiments show that the short TRV-USA RNA, though noninfective when used alone is replicated in the presence of TRV-Ger long particles. The situation was reversed in the reciprocal experiments. The intriguing problem is what determines whether long and short will interact or not. This was further demonstrated by experiments of Lister (55). Mixed infections were achieved with closely related strains from the U.S. and Europe, but several preparations of components that interacted in this way failed to produce hybrids with an antigenically distinct strain from Brazil (TRV-Braz). Stable hybrid forms were produced by inoculating the systemic host, *N. clevelandii,* with a mixture of long particles from TRV-Braz and short particles of the serologically distinct TRV-Yel. The particles had protein coats of TRV-Yel, but the reciprocal experiment failed to induce the development of a stable form of virus.

It has been found further (55) that hybrids apparently do not multiply as well as parent strains. Only about one tenth as much viral nucleoprotein was purifiable from *N. clevelandii* plants infected with the hybrids as was purifiable from plants infected with either TRV-Yel or TRV-Braz. This low productivity is not due to any difficulty in the association of TRV-Braz-RNA with TRV-Yel coat protein (55).

It appears that interaction takes place less readily and efficiently between particles of some strains than between others. Closely related strains appear to interact more readily than others, but serological relationship is not an exclusive criterion. An element of recognition is involved, but its nature is not known.

There are very few reports on the behavior of PEBV. It was shown that inoculation with long particles of PEBV only gave rise to formation of unstable virus. Short particles alone were not infectious; inoculation with a mixture of long and short particles led to formation of stable virus (39). There was a symptomatological difference between some stable and unstable variants of PEBV (53).

In summary, it has been shown that the short particles of TRV control coat-protein production, symptoms, and their particle length; long particles control replication, symptoms, and their particle length. Some strains are hybridizable, some not, depending on some kind of recognition between the RNA components. Short particles are noninfective; long particles are infective, but produce no coat protein and therefore long particles produce unstable variants.

It is not known whether any genetic information is duplicated on both components. If the genetic information in both components is entirely different, the total molecular weight of the RNA genome of TRV would be about 2.3×10^6 (long particle RNA) $+ 0.6\text{–}1.3 \times 10^6$ (short particle RNA) $= 2.9\text{–}3.6 \times 10^6$. This is considerably larger than the size of the genome of TMV and other single-particle viruses. The question should be answered as to why the rattle viruses need so much information, and why the genome size may vary for different strains of this group.

3. Cowpea Mosaic Virus Group

The viruses of this group are characterized by a genome divided among two nucleoprotein components. None of the components can multiply by themselves. Both are necessary for virus multiplication. This is now the largest group. To it belong cowpea mosaic virus (CPMV) (79), bean pod mottle virus (BPMV) (2), squash mosaic virus (SMV) (15), red clover mottle virus (RCMV) (73), radish mosaic virus (14), broad bean stain virus (30),

Echtes Ackerbohne Mosaik-virus (31), to mention only those properly described and recognized to be distinct. Members of the group are not closely related; there are large differences in host ranges, and the serological relationship between them is very weak. Viruses of the cowpea mosaic virus group have a number of characteristics in common. They produce three types of particles, all icosahedral with a diameter of about 28 nm, but differing in RNA content.

There are two nucleoprotein components; the third component consists of empty particles—protein capsids devoid of RNA. The protein capsid of the three components of each virus is the same. The sedimentation coefficients and the RNA contents of the corresponding components of the different viruses are in the same range. These are 54–60 *S*, 91–100 *S*, and 112–127 *S* for the respective components, which contain 0%, 24–29%, and 33–37% RNA. Each nucleoprotein component contains a single RNA molecule; the molecular weights of these RNAs are about 1.45×10^6 and 2.5×10^6. The two RNAs differ distinctly in base composition (76, 81).

The three components can be separated by either rate-zonal or equilibrium-density gradient centrifugation. According to their position in the gradient the components are referred to as top (*T*), middle (*M*) and bottom (*B*) components, with increasing RNA content. The sedimentation coefficient of the *M* and *B* nucleoprotein components are close, hence it has been difficult to prepare the single components with a high degree of purity. This has left a mark on the history of these viruses. In the first descriptions it was thought that only the bottom component, with the highest RNA content, was infectious (2, 60, 76). At that time the origin and metabolic significance of the middle component were obscure.

It was found in further experiments that addition of middle component to bottom component increased the specific infectivity of the latter greatly, often by a factor 8 (3, 72, 88). It was assumed that there were two types of bottom component particles: normal, intact, infectious particles, and particles which were damaged or did not give visible infection for some other reason. Middle component should be able to repair damaged particles, e.g., by complementation, or to activate bottom component so that the infection became visible. The nature of the damage and the activation could only be speculative. It was clear however that interaction between the components required the RNAs. It was shown (88) that nucleic-acid-free top component did not affect the supposed infectivity of the bottom component, and the largest infectivity activation was found if intact middle component RNA and intact bottom component RNA were mixed (12).

Bruening & Agrawal (12) offered an alternative explanation for the infectivity of CPMV. According to these authors, most infections were initiated by a group of middle and bottom component particles rather than by single bottom particles or a bottom plus middle component particle pair; only undamaged RNAs were active. Their data, however, did not exclude that bottom component alone was infectious. It was shown by extensive purification of separated middle and bottom components by rate-zonal centrifugation in sucrose-density gradients and equilibrium-density-gradient centrifugation in CsCl that neither bottom component nor middle component alone were infectious (77, 78, 81); only mixtures of the two components were infectious. The ultimate infectivity of CPMV depended upon the proportion of both middle and bottom component in the inoculum and of the component present in the lowest amount. Both middle component and bottom component were equally essential for infectivity. This led to the conclusion that there is no question of activation as presumed before, but that middle and bottom component each contain part of the information necessary for virus multiplication (78). Although this situation has been shown to be true properly only for CPMV, it holds very probably also for the other viruses of the CPMV-group.

The conclusion that the genetic information for CPMV is spread over the two nucleoprotein components, was confirmed by the use of strains or mutants of CPMV, which differed characteristically in phenotypic properties. This made it possible to prepare homologous and heterologous mixtures of middle and bottom components and to compare the properties of the homologous virus with the heterologous hybrids.

Bruening (11) isolated two strains of CPMV, which differed in top component production. By mixing the separate components of these strains, it was shown that the degree of top-component production depended on the middle component present in the inoculum.

De Jager & Van Kammen (17) obtained a mutant of CPMV after treatment with nitrous acid, which differed from the parent virus in two respects: whereas the wild-type CPMV infected *Phaseolus vulgaris* L. var. 'Beka', systemically, systemic infection by the mutant was inhibited. Moreover the mutant produced much more top component than middle component, in contrast to the wild-type virus where the situation is the reverse. If bean plants were inoculated with homologous and heterologous combinations of the middle and bottom components of the wild type and the mutant strains, it appeared that the specific properties of the strains were found only in those combinations in which their middle component was a partner. This indicated that the middle component carries important genetic information for the systemic reaction in bean, and the production of top component.

The properties of the mutant were temperature sensitive. At 30°C, as contrasted with 22–24°, the mutant infected Beka beans systemically as easily as the parent strain, and the production of top component was also reduced to almost the same level as that of the parent virus. This suggested that inhibition of systemic infection in beans and formation of excess top component may have basic mechanisms in common.

De Jager (unpublished results) isolated another mutant of CPMV, which also failed to infect beans systemically. This mutant did not show excess production of top component. In this case it was found by recombination experiments that the mutation was located in the bottom component. Bottom and middle components may therefore carry information which affect the same phenotypic property.

De Jager (16) recently described the isolation of a variant strain from the wild-type CPMV. The wild type and the variant strain could be distinguished by the type of symptoms on *Vigna unguiculata* (L.) Walp. var. 'Early Red'. The wild-type virus caused typical local lesions on the inoculated primary leaves of this *Vigna,* whereas the variant strain caused initially small chlorotic spots, which shaded off into a superficial browning, visible only on the inoculated upper side of the leaf. The variant strain was further characterized by mosaic symptoms and distortion of systemically infected secondary leaves of the 'Early Red' variety of *Vigna,* which were lacking with the wild-type strain. This difference could also be ascribed to bottom component by mixing middle and bottom components homologously and heterologously.

It has been shown by these experiments that middle and bottom components indeed carry different genetic information. As with TRV, both components appear to influence symptom expression.

The strains and mutants of CPMV which up to now have been isolated, were serologically indistinguishable. It has not been possible to locate the cistrons for the protein coat. This is a problem of considerable interest as it has been shown recently that the protein capsid of CPMV and other members of the cowpea mosaic virus group is constructed of two different proteins (90). The strains of BPMV, which differ from the wild-type BPMV in serological characteristics, therefore deserve special attention (61). Homologous and heterologous mixtures of middle and bottom components of the two BPMV strains were infectious and it was found that the property responsible for the antigenic difference was carried by the middle component. The cistron of at least one of the capsid proteins is therefore probably located in the middle component.

The interaction between the components of the viruses of the CPMV group is virus specific. No infective mixtures have been obtained by combining middle and bottom component derived from different viruses of the group (78, 88). It appears necessary that the viruses should be closely related in order to enable the formation of infective hybrids.

The genetic relationship between middle and bottom component of CPMV also has been investigated in a different way. From CPMV-infected *Vigna* plants double-stranded RNA specific for CPMV has been isolated (75). It was demonstrated that there occurred two preferential sizes of double-strand RNA: one of the size to be expected for double-stranded middle component RNA; the other for double-stranded bottom component RNA (74, 75). Apparently middle- and bottom-component RNA are each synthesized on their own template. This CPMV-specific double-stranded RNA was used in molecular hybridization experiments with ^{32}P- or ^{3}H-labeled middle and bottom component RNA to compare the homology in nucleotide sequences of the two RNAs (80, 82). These experiments demonstrated that there is no competition between middle- and bottom-component RNA in the process of hybridization. The RNAs of middle and bottom components hybridize independently of each other with the double-stranded RNA. Common nucleotide sequences on middle and bottom component RNA do not occur, and the base sequence of the two RNAs are completely different. Genetic information on bottom component RNA is not duplicated on middle component RNA, and vice versa.

The size of the genome of CPMV now appears very large. With a molecular weight of middle-component RNA of 1.45×10^6, and of bottom-component RNA of 2.5×10^6, the total genome of CPMV becomes 4×10^6. Again it is striking that these viruses with a divided genome need such a large amount of genetic information. A possible reason why the genome of these viruses is distributed among two particles might be that it is too large to be contained in a single particle. The important question to be answered first is, why this group of multicomponent viruses needs so much more genetic information than the single-particle plant viruses like TNV, TMV, and TYMV.

Only a few experiments have been conducted in which components of different viruses of the CPMV group were mixed. There is no interaction between the nucleoprotein components of BPMV and those of CPMV (88; Van Kammen, unpublished), nor between the components of CPMV and a cowpea mosaic virus which has been referred to as a severe strain of CPMV (78). Apparently the genetical relationship between these viruses is not so close that mixing of components of different viruses results in infective preparations.

The genetical relationship among CPMV and other members of the CPMV group has also been compared by molecular hybridization experiments (82). It was shown that the RNAs of BPMV and RCMV did not interfere with the hybridization of labeled CPMV-RNA with double-stranded RNA from CPMV-infected *Vigna* leaves. Therefore, there is no extensive homology in nucleotide-sequences of the RNA of CPMV and the RNAs of BPMV and RCMV. This confirms the conclusion from the mixing experiments that if the viruses of the CPMV group are physically rather similar, there may be no close genetical relationship between the viruses.

There is yet no answer to the question why the components of CPMV and the other members of the family are thrown entirely on each other for multiplication. As both middle and bottom component are necessary for infectivity and for virus multiplication to occur, it is obvious that both components carry information for the viral RNA synthetase. It is known (40, 47) that the RNA-replicating enzyme induced by the RNA phage Qβ consists of a single phage-specific and three host-specific polypeptide chains. Presumably in CPMV, the virus should introduce two virus-specific polypeptide chains, each originating from one of the components. It seems of considerable interest to answer this question as it will give us an insight in the virus-host interaction.

4. Alfalfa Mosaic Virus

The genome of alfalfa mosaic virus (AMV) (10, 38) is divided among three nucleoprotein components necessary for infectivity of AMV.

Purified preparations of AMV are very complex. Bancroft & Kaesberg (6, 7) were the first to observe the multicomponent nature of the virus. Purified preparations reveal up to six components in the analytical ultracentrifuge (39). The components that sediment with sedimentation coefficients of 99, 89, 75, and 68 *S* are indicated as bottom, middle, top *b* and top *a* components. We shall limit our discussion to these four components, as they have been shown to be involved in the genetic system of the virus. The two or three additional components that can be found have lower sedimentation coefficients. They are not infective and cannot replace any of the larger components.

The four components top *a*, top *b*, middle, and bottom, are bacilliform particles. All contain single-stranded RNA, which represent about 10% of the weight of each particle. The four particles have lengths of 58, 49, 38, and 28 nm, respectively, and a diameter of about 18 nm.

The biological significance of the particle heterogeneity of AMV is a complicated problem. Recently however, due to improved methods of virus purification, and extensive purification of the separate components, some insight has been gained into this problem.

The question of which components contribute to the infectivity of AMV was first investigated by Bancroft (1). After fractionation of purified AMV in sucrose-density gradients, he found the virus infectivity in a position coinciding with that of bottom component, not with that of middle or any of the other components. His conclusion was that bottom was the only infectious component of AMV. These results were challenged by other research workers, who were not able to exclude the possible involvement of the other components, particularly middle in the infectivity of AMV (32).

Gillaspie & Bancroft (33) tried in a subsequent paper to correlate infectivity and sedimentation properties of the RNAs of AMV. Although they found that infectivity coincided best in sucrose-density gradients with the fastest sedimenting class of RNA corresponding to RNA from the bottom component, at that time the possibility was not completely excluded that RNA of the middle component might have a kind of activating effect on infectivity.

Van Vloten-Doting (84, 85) noticed that the specific infectivity of the separate bottom component RNA was much less than that of a mixture of AMV-RNAs. This might have been caused by the removal of one or more of the other RNAs. Van Vloten-Doting & Jaspars (85) therefore determined the infectivity of AMV-RNA in density-gradient fractions, and in all possible combinations of two fractions. Certain combinations of fractions, notably from the heavy region of the gradient, corresponding with the position of bottom-component RNA sedimenting with 27 *S*, and from the 14 *S* region of the gradient, were remarkably more infectious than any of the single fractions. It became clear from this study that the infectivity of AMV-RNA is not so much associated with a particular component as with a specific combination of two or more components; one of them necessarily should be the heaviest RNA, originating from bottom component. When subsequently(86) the RNAs from bottom component and top component *a* were further purified, it was found that each separate RNA was noninfectious, or had only very little infectivity, but mixtures of top *a* RNA and bottom RNA were highly infectious. It was concluded that bottom component RNA and top *a* RNA were necessary for infectivity, and that the infectivity of AMV depended both on bottom component and top component *a*.

Further proof that the infectivity of AMV did not depend on a single component, but on two or more types of particles was derived from the infectivity-dilution curve for AMV (86). It was evident that more than one particle was necessary at each infection site. The dilution curve had the shape of a multiparticle, rather than a single-particle infection curve.

Experiments by which purified bottom-component RNA and top *a* RNA from two different strains were mixed homologously and heterologously confirmed the conclusion that the genetic information of AMV is spread among its components (86).

Strain AMV 425 induced on tobacco chlorotic spots or circles which were followed by systemic chlorosis, and on beans it produced only local lesions but no systemic symptoms. The strain YSMV induced necrotic spots on tobacco, followed by systemic infection; bean plants developed chlorosis on inoculated and systemically infected leaves. The two strains differed serologically and also in component composition. AMV 425 had a relative high concentration of bottom component, and YSMV had relatively more top *a* and top *b* components.

The heterologous combination of bottom component RNA from AMV 425 and top *a* RNA from YSMV had properties of either initial strain. The symptoms of this "new" strain were a combination of those of the starting strains. It gave local lesions of the AMV 425 type on bean, and systemic necrosis on tobacco, like YSMV. The mixed strain had the protein coat of YSMV, and the component composition was also that of YSMV.

Further experiments (87) demonstrated that AMV was not so simple that only bottom component and top *a* component should be involved in its infectivity. Up to this stage most of the work on the infectivity of AMV had been done with the RNAs from AMV, because of the instability of nucleoprotein components, which could not stand a prolonged purification procedure. It was found, however, that the nucleoproteins could be stabilized by the addition of 0.001 M EDTA to the purified virus preparations. This enabled (87) the purification of each of the nucleoprotein components, top *a*, top *b*, middle, and bottom components to a level that they were noninfectious at the concentration level normally used for inoculation of AMV (0.015 mg/ml). The specific infectivity of a virus preparation could be restored by combining the three components bottom, middle, and top *b*. Other combinations of two or three components showed little or no infectivity, whereas addition of top *a* did not enhance the infectivity of the mixture of top *b*, middle, and bottom components. This led to the conclusion that the infectivity of AMV was not associated with a single component nor with two components, but with a particular combination of three components: bottom, middle, and top *b*. Each of the active components should contain some essential part of the virus genome not occurring in the others.

This was apparently in contradiction with former results (86), which had indicated that bottom-component RNA and top *a* RNA were necessary for infectivity. The experiments with AMV-RNA were therefore repeated (9), and at the same time the purity of the RNA fractions as they were used in previous experiments (86) was checked by electrophoresis in polyacrylamide gels (22, 58). This demonstrated that the RNA fraction previously referred to as "pure" bottom component RNA was in fact heterogeneous and consisted of a mixture of bottom, middle, and top *b* component RNA. This concerned the bottom component RNA fraction obtained after fractionation of AMV RNA on a sucrose-density gradient. The RNA referred to previously as top *a* RNA was indeed found to be only top *a* RNA. At the same time it was demonstrated (9) that the purified bottom, middle, and top *b* nucleoprotein components contained each only one type of RNA. These findings caused the remarkable situation that infectivity of AMV was obtained with a mixture of the nucleoprotein bottom, middle, and top *b* components (87), whereas a mixture of bottom, middle, and top *b* RNAs had been found (86) to be noninfective. This was confirmed in experiments that used the single RNAs from purified bottom, middle, and top *b* components. Such a mixture of three RNAs was noninfective. If RNA from top *a* component was added, however, then virus infectivity was restored.

As the only difference between the inoculum consisting of three RNAs and of these nucleoproteins appeared to be the presence of coat protein, the effect of coat protein on the infectivity was investigated (9). The capsid of AMV consists of a large number of identical protein subunits with a molecular weight of 24,500 daltons (48). Indeed it was found that the addition of a small amount of protein subunits was sufficient for restoring the infectivity of a mixture of bottom, middle, and top *b* component RNAs. The amount of protein necessary to induce the effect ranged between 1 and 16 protein subunits per RNA molecule, depending on the RNA concentration and the batch of protein. The effect was specific for AMV protein. Other proteins did not have an activating effect. Heat treatment, which denatured the protein but left the RNAs intact, reduced the effect of the coat protein on infectivity. Such treatment had no effect on activation by top *a* RNA. This indicated that the effect of the virus protein could not be ascribed to contamination of the virus protein with top *a* RNA. On the other hand, the activating effect of top *a* RNA on the mixture of bottom, middle, and top *b* component RNAs was not heat sensitive, but could be destroyed by ribonuclease.

The picture that has emerged is that, on the nucleoprotein level, a mixture of top *b*, middle, and bottom components is infectious. The RNAs of these three components represent therefore the complete genome of AMV. However, the mixture of the three RNAs of these three components is not infective. Such a mixture has to be activated either by top *a* RNA or by AMV coat protein. The effect was observed with tobacco (*Nicotiana tabacum var. Samson NN*) and with bean (*Phaseolus vulgaris var. Berna*) as the host plants.

A functional equivalence between AMV coat protein and top *a* component RNA becomes understandable in view of the indications that the coat protein is the translation product of top *a* RNA. It has been reported (83) that the product in a cell-free protein-synthesizing system from *Escherichia coli* under direction of top *a* RNA resembled the AMV coat protein. The effect of top *a* RNA could then be that it is translated in an early stage of infection so that coat protein becomes available. This would indicate some other function of the virus-coat protein in the infection and multiplication besides its function of protecting the RNA in the resting virus.

The fact that top *a* can be omitted from the inoculum consisting of top *b*, middle, and bottom nucleoprotein components, implies that its genetic information is also present in one of the other components, probably in top *b* (unpublished results of the Leyden group). But apparently this information cannot be expressed in the absence of coat protein, whereas that of top *a* RNA can.

An inoculum of top *b*, middle, and bottom nucleoprotein components produces top *a*. So top *a* RNA appears to be synthesized under the direction of the other three components. It is not clear how this occurs and what the significance of this is. We have mentioned the experiments by which top *a* component RNA and "bottom" component RNA (in fact a mixture of bottom, middle, and top *b* RNA) of two strain of AMV were mixed. The results of these experiments showed that the genetic information introduced by top *a* RNA could be replicated, for the hybrid strain showed several properties which could only have been introduced by top *a* RNA. The impression is created that in that hybrid the genetic information present in top *a* RNA dominated over the information of the same kind present in top *b*, middle, or bottom component RNA.

The total size of the genome of AMV is about 3.2×10^6 expressed in terms of the molecular weights of the RNAs present in *b*, middle, and bottom component. The question may be raised if top *a* RNA should be considered as part of the genome or not. On the basis of the foregoing I think it should be. The naturally occurring AMV consists of four components: top *a*, top *b*, middle, and bottom. These represent together the genome of AMV. The genome is at least partially diploid. The significance of this diploidy is not yet clear. It has just been recognized.

A few reports have been published (18, 59) which state that bottom component of AMV alone is infectious. These results should be reconsidered against the backround of the careful analysis of the infectivity of AMV summarized above.

5. Tobacco Streak Virus

Tobacco streak virus (TSV) (24, 26) is small, isometric, about 28 nm in diameter, and is composed of three major nucleoprotein components. The relationship between the components in regard to infectivity is not yet so clear as in the previous cases. This might be partly due to the degree of purity achieved for the separate components.

The three major nucleoprotein components have sedimentation coefficients of 90, 98, and 113 *S* (57). In this order the components have been termed T, *a* and *b* (25) or top, middle, and bottom (57). We shall adopt the notation T, *a* and *b* to prevent confusion with terms used for the other viruses described in this review. According to Fulton (25) there occur still other particle types: one less rapidly sedimenting than component T, and one or more that sediment more rapidly than *b*. These minor particle types make up a very small proportion of the total virus and, as no biological activity has been associated with them, they are not discussed here. Absorbancy measurements on the separate T, *a* and *b* components showed $A_{260/280}$ ratios of 1.56, 1.60, and 1.65 respectively (57) suggesting a reduced RNA content in components T and *a* as compared to *b* component.

The three components were separated and purified by rate-zonal centrifugation in sucrose-density gradients. Component T was found to be noninfective; components *a* and *b* were weakly infective as compared to the unfractionated virus (25, 57). The components *a* and *b* sediment were, however, so close that critical separation was probably not achieved. No stringent criteria for the purity of components were given. Mixtures of *a* and *b* components showed markedly enhanced infectivity. Adding component T to mixtures of *a* and *b* further enhanced infectivity (25). This pattern of infectivity is thought to be typical for TSV.

A number of variant strains of TSV are known (25, 27). Variants are often distinguishable by the type of lesion they produce on *Vigna cylindrica,* and by symptoms on other hosts. These strains were used to examine whether there is a distribution of genetic information among the components. When *a* and *b* components derived from strains differing in type of lesion on *Vigna cylindrica* were mixed, resulting lesions were predominantly typical of the strain supplying the *a* component. In three component mixtures, when *a* and *b* were derived from one strain and T from another, the increased number of lesions resembled the strain from which T was derived. Therefore both T and *a* components carry information for the lesion type.

Further results on heterologous mixtures of components of different strains of TSV were reported (27). A strain of TSV from Brazil had an antigen not found in two North American strains, WC and HF. Specific antisera were prepared that reacted only with the Brazilian or with the American isolates. If component *b* from the Brazilian strain was mixed with component *a* from the WC isolate, then the mixture was highly infectious. The reverse combination, component *b* from WC with component *a* from the Brazilian isolate, did not show enhancement of infectivity. This might be explained by incompatibility of the combination or by assuming that a piece of genetic information, a cis tron essential for multiplication, was lacking in the heterologous combination, whereas it was present in the homologous combination. The combination a_{WC} b_{Braz} produced lesions of the WC type, and 78% of the lesions carried the WC antigen. This indicated that component *a* was responsible for antigen and lesion type. In the combination a_{Braz} b_{HF}, lesions were predominantly of the Brazilian type, but most of the lesions contained virus with HF antigen. Antigenic determinants in this combination seemed to be carried by the *b* component. In a mixture of components of strain NR with those of strain HF, the lesions were for 85% of the HF type in the combination a_{NR} b_{HF}, and thus component *b* appeared to carry the determinant for lesion type. In the reciprocal combination a_{HF} b_{NR} the lesions were again like HF.

It has appeared very difficult to assign specific characters to either of the components *a* and *b*. In regard to lesion type, this phenotypic characteristic might result from cooperation of factors originating from both components, which could make the final outcome rather unpredictable.

The presence of antigenic determinants on both *a* and *b* components is remarkable. This might suggest that there is rather an arbitrary, and not a definite distribution of the genetic information among the components. The distribution also need not necessarily be the same for different strains of TSV.

It has been reported (28) that particle composition of TSV is different, depending on the time of infection, the host plant, and method of purification (57). This should be a warning to be very careful in using component composition as a characteristic in this genetic work.

Further chemical and biological characterization of the components appears necessary before a good comparison can be made of TSW with other plant viruses with a divided genome.

6. Brome Mosaic Virus Group

It recently has been recognized that brome mosaic virus (BMV) and cowpea chlorotic mottle virus (CCMV) (5, 49) are plant viruses with a divided genome. Although not yet shown, broad bean mottle virus (BBMV), which is physically similar to BMV and CCMV, probably also belongs to this group. There are indications that cucumber mosaic virus (CMV) has likewise a divided genome of the same type (41).

BMV (4) behaves as a single, apparently homogeneous component both in sedimentation and in electrophoresis. The molecular weight of the particle is 4.5×10^6 daltons and the RNA content 1×10^6. In fact BMV was thought to belong to the smallest independently replicating plant viruses.

The RNA extracted from such virus preparations had three components with molecular weights of 1.0, 0.7, and 0.3 daltons approximately. Virus infectivity was associated only with the largest RNA (8). The other two RNAs were considered biologically inactive, possibly arising from the largest component by a single cleavage of the RNA chain.

Similarly, it has been known for some years that CCMV, BBMV, and CMV sediment as single nucleoprotein components, and contain more than one sedimenting species of RNA. Infectivity was associated in all cases with the fastest sedimenting RNA. This picture now has radically changed.

If RNA from BMV was fractionated on 2.5% polyacrylamide, 0.5% agarose gels, there were four components of molecular weights 1.09, 0.99, 0.75, and 0.28×10^6 daltons (49) referred to as RNA components 1, 2, 3, and 4 respectively. The difference in molecular weights of the RNA components 1 and 2, together with the known similarity of the virus capsids, suggested that BMV should be heterogeneous in buoyant density. In equilibrium-density-gradient centrifugation carried out in 4.2 molar rubidium chloride a wide asymmetric band was found, indicating that buoyant density heterogeneity was present. It was shown that the largest RNA, component 1, was recovered mostly from BMV particles of highest density. RNA component 2 was recovered from particles of lowest density, and RNA components 3 and 4 were recovered from particles of intermediate densities. Since the ratio of RNA components 3 and 4 was independent of density, and since the RNA content of the intermediate-density particles is the sum of their molecular weights, they must be contained in the same virion. Therefore, BMV appeared to be a three-component virus: the three nucleoprotein components—one with RNA component 1, a second with RNA 2, and a third with RNA 3 and 4—have identical capsids but differ in RNA content. The difference is so small that BMV sediments as a homogeneous peak. The heterogeneity can only be detected by small differences in buoyant density among the three components.

Mixtures of heavy and light virus particles invariably had higher specific infectivities than isolated components when compared at equal concentrations. This indicated that more than one component might be required for infectivity. Equilibrium-density-gradient centrifugation by itself was unsatisfactory for isolation of biologically pure components. The biological role of the different RNA components by infectivity studies was therefore examined after separation of the RNAs on agarose-polyacrylamide gels (49, 58). It was shown that the specific infectivity of RNA of BMV was greatly decreased if any of the components 1, 2, or 3 was omitted from the inoculum. Addition or deletion of RNA component 4 had no effect. Purification of the

RNA components increased the effect of each on the specific infectivity of a mixture. This suggested that further purification of components would still further decrease the specific infectivity of incomplete mixtures. The data were therefore most simply interpreted if RNA components 1, 2, and 3 are all required for infectivity.

Lane & Kaesberg (49) further demonstrated that the genetic information of BMV is really distributed among the RNA components. They isolated a strain of BMV differing in electrophoretic mobility, which means that the isolate differed in the viral coat protein. They prepared mixtures in which two RNA components were derived from one strain, and the third RNA component was from the heterologous strain. In all cases the electrophoretic mobility of the progeny virus agreed with the source of RNA component 3. Thus BMV component 3 carries the genetic information for the viral-coat protein. The nature of the genetic information carried by components 1 and 2 remains to be discovered.

Lane & Kaesberg (49) demonstrated the presence in BMV-infected plants of replicative forms corresponding to all 4 components. The existence of these multiple replicative forms suggested that all single-stranded RNA components of BMV are replicated. Component 4 was not required for infectivity, but was produced if RNA components 1, 2, and 3 were inoculated.

The origin and function of component 4 is therefore puzzling. Perhaps it is necessary for the construction of the virion, which contains component 3. In vitro protein-synthesis studies suggested that RNA component 4 directed the synthesis of a portion of the virus-coat protein in an *E. coli* cell-free system (70). This implies that component 4 can be derived from component 3 during infection. If so, this would indicate a certain similarity with AMV top *a* RNA, which appeared to carry information for the virus-coat protein and might be derived from top component *b* during infection (9, 83). The difference is that top *a* RNA from AMV has an activating effect on the infectivity of AMV, which has not been demonstrated for RNA component 4 for BMV.

Bancroft (5) reported the significance of the multicomponent nature of CCMV-RNA. By polyacrylamide-gel electrophoresis it was shown that the RNA of CCMV consist also of four components, which have molecular weights of approximately 1.15, 1.0, 0.85, and 0.32 $\times$ 10^6 daltons. At least two RNA types, namely the 1.15 $\times$ 10^6 and the 1.0 $\times$ 10^6 component, were required for initiating infection. The 0.85 $\times$ 10^6 RNA component increased infectivity of the mixture of the two larger RNAs, but its presence in the inoculum was not absolutely necessary for the infectivity of CCMV. According to Bancroft (5) the 0.85 $\times$ 10^6 RNA increased the efficiency of infection.

Bancroft (5) also prepared heterologous mixtures of RNA components 1 and 2 from BMV with the 0.85 $\times$ 10^6 RNA component from CCMV. Whereas the mixture of the RNA components 1 and 2 of BMV alone was noninfective, lesions were produced on *Chenopodium hybridum* if 0.85 $\times$ 10^6

RNA of CCMV was added to the inoculum. In this mixture then, 0.85×10^6 RNA of CCMV was essential to establish infection and to start virus multiplication. It appears that 0.85×10^6 CCMV RNA can supply some genetic information necessary for infection, lacking on components 1 and 2 of BMV.

If the picture of the infectivity of CCMV is correct, only two RNA components, namely the 1.15×10^6 and the 1.0×10^6, are required for infectimn. The 0.85×10^6 RNA component contains genetic information, present also on the large RNA components of CCMV. CCMV therefore might represent another case of diploidy of a virus genome analogous to that of AMV.

BMV and CCMV are similar in physical properties, but these viruses are serologically unrelated and have different host ranges. They do not seem to be closely related genetically. It is striking that CCMV can furnish a part of the genetic information essential for infectivity of BMV-RNA; this might indicate some genetical relationship.

CMV (29) belongs to another group of plant viruses than BMV and CCMV. It corresponds with these viruses in sedimenting homogeneously as one single nucleoprotein, whereas the RNA extracted from the virion consists of at least four sedimenting components (20). It recently was reported (41) that for optimal infectivity of CMV-RNA the cooperation of more than one of its constituents is required. The exact relationship between these components has still to be worked out, but it appears reasonable that CMV will come out also as a virus with a divided genome.

Summary

It appears well established that a considerable number of plant viruses have a divided genome. We have placed the viruses in groups, each with some special features. A general pattern in the distribution of genetic information among the different RNA components has not yet been found, if it exists. The only marked agreement for all cases is that the genetic determinant for virus-coat protein is mostly found in the smaller RNA component, not the larger one. It is possible that TSV constitutes an exception to this rule.

There are many more plant viruses that produce more than one nucleoprotein component. Examples are: arabis mosaic virus, pea enation mosaic virus, prunus necrotic ringspot virus, raspberry ringspot virus, tobacco ringspot virus, tomato black ring virus, Tulare apple mosaic virus. There are probably more viruses of the type of BMV and CMV; tomato aspermy virus might be one. Wound tumor virus and rice dwarf virus have several pieces of double-stranded RNA in homogeneously sedimenting nucleoprotein particles (89); these might well represent divided genomes. These double-stranded RNA plant viruses are rather similar to the animal Reoviruses. Recently a statistical analysis of the recombination frequencies resulting from mixed infection with temperature-sensitive mutants of Reovirus type 3 was published

(21). The results were consistent with the biochemical data that the Reovirus genome is segmented.

One may conclude that a genome divided among two or more RNA molecules is widely occurring among plant viruses.

An important result of the work on these viruses is that the origin and significance of the different nucleoprotein or RNA component found with these viruses are explained. Formerly the occurrence of these RNA components either present in a separate nucleoprotein component, or in one homogeneously sedimenting particle, was thought to be caused by specific scissions of infectious RNA in accidents during viral RNA synthesis or in the mature virus particle. It is now clear that these RNA components may represent different parts of the virus genome. In a few cases, like CPMV and BMV, it has been shown that in infected plants replicative forms occur, corresponding with and specific for each of the RNAs. The RNA components seem therefore self replicating.

Top *a* RNAs of AMV, RNA component 4 of BMV and the 0.85×10^6 RNA component of CCMV develop even if they are not introduced into the plant. Apparently they are derived from any of the RNAs of respective AMV or BMV, which are required for infectivity. Smaller pieces of viral RNA therefore originate from the larger RNA molecules during infection. The origin of these RNAs is particularly intriguing in top *a* RNA of AMV and the 0.8×10^6 RNA component of CCMV, as the activating effect on the virus infectivity of these components is specially important in the initial phases of virus infection.

Tobacco ringspot virus (ToRV) is probably the best studied example of a virus that produced two nucleoprotein components, apparently without being a virus with a divided genome. ToRV nucleoproteins are isometrical particles with a diameter of about 28 nm and sedimentation coefficients of 91 *S* and 126 *S*. The 126 *S* particle appears to contain one infectious RNA molecule or two pieces of noninfective RNA strands. With increasing age of infection the proportion of 126 *S* particles with two noninfective strands increases (19, 66, 67); both pieces are about half the size of the infective RNA.

This system deserves more attention since the chemical relation between the RNA components, their biological significance, and their origin are not understood. Some of the multiparticulate viruses for which a divided genome has not been demonstrated, may be of the type of ToRS.

Now it has been recognized that the genetic information for a considerable number of plant viruses is distributed among two or more nucleoprotein components, it remains to be elucidated what kind of information is present on each component and how the components complement each other in the infection process. Experiments in which components of strains and mutants of these viruses are mixed may give insight into the number and nature of the properties located on each component. It also appears of great importance to identify the gene products of the different RNAs, and to determine their function in order to obtain an idea about the interaction of the components.

Recently (71) it has become possible to isolate tobacco leaf-cell protoplasts and to infect a high percentage of these protoplasts with TMV. The development of similar protoplast systems for multi-particulate plant viruses would be of interest. These systems offer the possibility of synchronous infection of a large number of cells. It might thereby become possible to determine biochemically whether the separate components have some biological activity, and how the components complement each other in the infection process.

LITERATURE CITED

1. Bancroft, J. B. 1961. Association of infectivity with alfalfa mosaic virus bottom component only. *Virology* 14:296–97
2. Bancroft, J. B. 1962. Purification and properties of bean pod mottle virus and associated centrifugal and electrophoretic components. *Virology* 16:419–27
3. Bancroft, J. B. 1968. Plant viruses: defectiveness and dependence. In: *The Molecular Biology of Viruses* Symp. Soc. Gen. Microbiol. Ed. J. Crawford, R. Stoker 229–47, London: Cambridge Univ. Press
4. Bancroft, J. B. 1970. Brome mosaic virus. Commonw. Mycol. Inst. *Descriptions of Plant Viruses* No. 3
5. Bancroft, J. B. 1971. The significance of the multicomponent nature of cowpea chlorotic mottle virus RNA. *Virology* 45:830–34
6. Bancroft, J. B., Kaesberg, P. 1958. Size and shape of alfalfa mosaic virus. *Nature* 181:720–21
7. Bancroft, J. B., Kaesberg, P. 1960. Macromolecular particles associated with alfalfa mosaic virus. *Biochim. Biophys. Acta* 39:519–28
8. Bockstahler, L. E., Kaesberg, P. 1965. Infectivity studies of brome grass virus RNA. *Virology* 27:418–25
9. Bol, J. F., Vloten-Doting, L. van, Jaspars, E. M. J. 1971. A functional equivalence of top component *a* RNA and coat protein in the initiation of infection by alfalfa mosaic virus. *Virology* 46:73–85
10. Bos, L., Jaspars, E. M. J. 1971. Alfalfa mosaic virus. Commonw. Mycol. Int. *Descriptions of Plant Viruses* No. 46
11. Bruening, G. 1969. The inheritance of top component formation in cowpea mosaic virus. *Virology* 37:577–84
12. Bruening, G., Agrawal, H. O. 1967. Infectivity of a mixture of cowpea mosaic virus nucleoprotein components. *Virology* 32:306–20
13. Cadman, C. H., Harrison, B. D. 1959. Studies on the properties of soil-borne viruses of the tobacco-rattle type occurring in Scotland. *Ann. Appl. Biol.* 47:542–56
14. Campbell, R. N. 1964. Radish mosaic virus, a crucifer virus serologically related to strains of bean pod mottle viruses and to squash mosaic virus. *Phytopathology* 54:1418–24
15. Campbell, R. N. 1971. Squash mosaic virus. Commonw. Mycol. Inst. *Description of Plant Viruses* No. 43
16. De Jager C. P. 1971. Genetical investigations on a multi-component virus. *Proc. 7th Conf. Czechoslovak Plant Virol. High-Tatras.* In press
17. De Jager, C. P., Kammen, A. van. 1970. The relationship between the components of cowpea mosaic virus. III Location of genetic information for two biological functions in the middle component of CPMV. *Virology* 41:281–87
18. Desjardins, P. R., Steere, R. L. 1969. Separation of top and bottom components of alfalfa mosaic virus by combined differential and density gradient centrifugation. *Arch. Ges. Virusforschung* 26:127–37
19. Diener, T. O., Schneider, I. R. 1966. The two components of tobacco ringspot virus nucleic acid: origin and properties. *Virology* 29:100–05
20. Diener, T. O., Scott, H. A., Kaper, J. M. 1964. Highly infectious nucleic acid from crude and purified preparations of cucumber mosaic virus (Y strain). *Virology* 22:131–41
21. Fields, B. N. 1971. Temperature-sensitive mutants of Reo virus type 3. Features of genetic recombination. *Virology* 46:142–48
22. Fowlks, E., Young, R. J. 1970. Detection of heterogeneity in plant viral RNA by polyacrylanide gel electrophoresis. *Virology* 42:548–50
23. Francki, R. I. B. 1966. Some factors affecting particle length distribution in tobacco mosaic virus preparations. *Virology* 30:388–96

24. Fulton, R. W. 1967. Purification and some properties of tobacco streak and Tulare apple mosaic viruses. *Virology* 32:153–62
25. Fulton, R. W. 1970. The role of particle heterogeneity in infection by tobacco streak virus. *Virology* 41:288–94
26. Fulton, R. W. 1971. Tobacco streak virus. Commonw. Mycol. Inst. *Descriptions of Plant Viruses* No. 44
27. Fulton, R. W. 1972. Tobacco streak virus as a genetic system. p. 235. *2nd Int. Congr. Virol.*, Budapest
28. Fulton, R. W., Potter, K. T. 1971. Factors affecting the proportion of nucleoprotein components of tobacco streak virus. *Virology* 45:736–38
29. Gibbs, A. J., Harrison, B. D. 1970. Cucumber mosaic virus. Commonw. Mycol. Inst. *Descriptions of Plant Viruses* No. 1
30. Gibbs, A. J. Smith, H. G. 1970. Broad bean strain virus. Commonw. Mycol. Inst. *Descriptions of Plant Viruses* No. 29
31. Gibbs, A. J., Paul, H. L. 1970. Echtes Ackerbohnemosaik-Virus. Commonw. Mycol. Inst. *Descriptions of Plant Viruses* No. 20
32. Gibbs, A. J., Nixon, H. L., Woods, R. D. 1963. Properties of purified preparations of lucerne mosaic virus. *Virology* 19:441–49
33. Gillaspie, A. G. Bancroft, J. B. 1965. Properties of ribonucleic acid from alfalfa mosaic virus and related components. *Virology* 27: 391–97
34. Harrison, B. D. 1970. Tobacco rattle virus. Commonw. Mycol. Inst. *Descriptions of Plant Viruses* No. 12
35. Harrison, B. D., Woods, R. D. 1966. Serotypes and particle dimensions of tobacco rattle viruses from Europe and America. *Virology* 28:610–20
36. Huang, A. S., Baltimore, D. 1970. Defective viral particles and viral disease processes. *Nature* 226: 325–27
37. Hulett, H. R., Loring, H. S. 1965. Effect of particle length distribution on infectivity of tobacco mosaic virus. *Virology* 25:418–30
38. Hull, R. 1969. Alfalfa mosaic virus. *Advan. Virus Res.* 15:365–428
39. Huttinga, H. 1969. Interaction between components of pea early-browning virus. *Netherl. J.: Plant Pathol.* 75:338–42
40. Kamen, R. 1970. Characterization of the subunits of Qβ-replicase. *Nature* 228:517–33
41. Kaper, J. M. 1971. *Separation and characterization of five RNA species obtained from cucumber mosaic virus (strain S).* Paper presented at 62nd Ann. Meet. Am. Soc. of Biol. Chemists. San Francisco, June 1971
42. Kassanis, B. 1968. Satellitism and related phenomena in plant and animal viruses. *Advan. Virus Res.* 13:147–80
43. Kassanis, B. 1970. Satellite virus. Commonw. Mycol. Inst. *Descriptions of Plant Viruses* No. 15
44. Kassanis, B., Phillips, M. P. 1970. Serological relationship of strains of tobacco necrosis virus and their ability to activate strains of satellite virus. *J. Gen. Virol.* 9: 119–26
45. Kassanis, B., Vince, D. A., Woods, R. D. 1970. Light and electron-microscopy of cells infected with tobacco necrosis and satellite viruses. *J. Gen. Virol.* 7:143–51
46. Klein, A., Reichmann, M. E. 1970. Isolation and characterization of two species of double-stranded RNA from tobacco leaves doubly infected with tobacco necrosis and satellite tobacco necrosis viruses. *Virology* 42:269–72
47. Kondo, M., Gallerani, R., Weissmann, C. 1970. Subunit structure of *Qβ*-replicase. Nature 228:525–27
48. Kruseman, J., Kraal, B., Jaspars, E. M. J., Bol, J. F., Brederode, F. T., Veldstra, H. 1971. Molecular weight of the coat protein of alfalfa mosaic virus. *Biochemistry* 10:447–54
49. Lane, L. C., Kaesberg, P. 1971. Multiple genetic components in bromegrass mosaic virus. *Nature* 232:40–43
50. Lesnaw, J. A., Reichmann, M. E. 1970. Determination of molecular weights of plant viral protein subunits by polyacrylamide gel electrophoresis. *Virology* 42: 724–31
51. Lesnaw, J. A., Reichmann, M. E.

1970. Identity of the 5′-terminal RNA nucleotide sequence of the satellite tobacco necrosis virus and its helper virus: possible role of the 5′-terminus in the recognition by virus-specific RNA replicase. *Proc. Nat. Acad. Sci.* 66: 140–45
52. Lister, R. M. 1966. Possible relationships of virus-specific products of tobacco rattle virus infection. *Virology* 28:350–53
53. Lister, R. M. 1967. A symptomatological difference between some unstable and stable variants of pea early browning virus. *Virology* 31:739–42
54. Lister, R. M. 1968. Functional relationships between virus-specific products of infection by viruses of the tobacco rattle type. *J. Gen. Virol.* 2:43–50
55. Lister, R. M. 1969. Tobacco rattle, NETU, viruses in relation to functional heterogeneity in plant viruses. *Proc. Fed. Am. Soc. Exp. Biol.* 28:1875–89
56. Lister, R. M., Bracker, C. E. 1969. Defectiveness and dependence in three related strains of tobacco rattle virus. *Virology* 37:262–75
57. Lister, R. M., Bancroft, J. B. 1970. Alterations of tobacco streak virus component ratios as influenced by host and extraction procedure. *Phytopathology* 60:689–94
58. Loening, U. E. 1967. The fractionation of high-molecular-weight ribonucleic acid by polyacrylamide gel electrophoresis. *Biochem. J.* 102:251–57
59. Majorana, G., Paul, H. L. 1969. The production of new types of symptoms by mixtures of different components of two strains of alfalfa mosaic virus. *Virology* 38:145–52
60. Mazzone, H. M., Incardona, N. L., Kaesberg, P. 1962. Biochemical and biophysical studies of squash mosaic virus and related macromolecules. *Biochim. Biophys. Acta* 55:164–75
61. Moore, B. J., Scott, H. A. 1971. Properties of a strain of bean pod mottle virus. *Phytopathology* 61:831–35
62. Rees, M. W., Short, M. N., Kassanis B. 1970. The amino acid composition, antigenicity and other characteristics of the satellite viruses of tobacco necrosis virus. *Virology* 40:448–61
63. Roy, D., Fraenkel-Conrat, H., Lesnaw, J., Reichmann, M. E. 1969. The protein subunit of the satellite tobacco necrosis virus. *Virology* 38:368–69
64. Sänger, H. L. 1969. Functions of the two particles of tobacco rattle virus. *J. Virol.* 3:304–12
65. Schneider, I. R. 1969. Satellite-like particles of tobacco ringspot virus that resemble tobacco ringspot virus. *Science* 166:1627–29
66. Schneider, I. R., Diener, T. O. 1966. The correlation between the properties of virus related products and the infectious component during the synthesis of tobacco ringspot virus. *Virology* 29:92–99
67. Schneider, I. R., Diener, T. O. 1968. In vivo and in vitro decline of specific infectivity of tobacco ringspot virus correlated with nucleic acid degradation. *Virology* 25:150–57
68. Semancik, J. S. 1970. Identity of structural protein from new isolates of TRV with different length of associated short particles. *Virology* 40:618–23
69. Semancik, J. S., Kajiyama, M. B. 1968. Enhancement of tobacco rattle virus. Stable form infections by heterologous short particles. *Virology* 34:170–71
70. Stubbs, J. D., Kaesberg, P. 1967. Amino acid incorporation in an *Escherichia coli* cell-free system directed by bromegrass mosaic virus ribonucleic acid. *Virology* 35:385–97
71. Takebe, I., Otsuki, Y. 1969. Infection of tobacco mesophyll protoplasts by tobacco mosaic virus. *Proc. Nat. Acad. Sci.* 64:843–48
72. Valenta, V., Marcinka, K. 1968. Enhanced infectivity of combined bottom and middle components of red clover mottle virus. *Acta Virol.* 12:288
73. Valenta, V., Marcinka, K. 1971. Red clover mottle virus. Commonw. Mycol. Inst. *Descriptions of Plant Viruses* No. 74
74. Van Griensven, L. J. L. D. 1970. *De zuivering en de eigenschappen van de replicatieve vorm van het RNA van cowpea mosaic virus.* Thesis, State Agricultural

Univ. Wageningen, The Netherlands

75. Van Griensven, L. J. L. D., Van Kammen, A. 1969. The isolation of ribonuclease-resistant RNA induced by cowpea mosaic virus: Evidence for two double stranded RNA components. *J. Gen. Virol.* 4:423–28
76. Van Kammen, A. 1967. Purification and properties of the components of cowpea mosaic virus. *Virology* 31:633–42
77. Van Kammen, A. 1968. The relationship between the components of cowpea mosaic virus. Proc. Symp. Biochemical Regulation in Diseased Plants or Injury. *Phytopathol. Soc Japan,* Tokyo
78. Van Kammen, A. 1968. The relationship between the components of cowpea mosaic virus. I. Two ribonucleoprotein particles necessary for the infectivity of CPMV. *Virology* 34:312–18
79. Van Kammen, A. 1971. Cowpea mosaic virus. Commonw. Mycol. Inst. *Descriptions of Plant Viruses* No. 47
80. Van Kammen, A. 1971. Cowpea mosaic virus, un virus au énome divisé. *Physiol. Vég.* 9:479–85
81. Van Kammen, A., Van Griensven, L. J. L. D. 1970. The relationship between the components of cowpea mosaic virus. II. Further characterization of the nucleoprotein components of CPMV. *Virology* 41:274–80
82. Van Kammen, A., Rezelman, G. 1972. A comparison of the RNA's from the nucleoprotein components of cowpea mosaic virus by hybridization experiments. Proc. Coll. Plant Viruses as Genetic System. p. 235–36. *2nd Int. Congr. Virol.* Budapest
83. Van Ravenswaaij Claasen, J. C., Van Leeuwen, J. C. J., Duyts, G. A. H., Bosch, L. 1967. In vitro translation of alfalfa mosaic virus RNA. *J. Mol. Biol.* 23:535–44
84. Van Vloten-Doting, L. 1968. *Verdeling van de genetische informatie over de natuurlijke componenten van een plantevirus, alfalfa mosaic virus.* Thesis, Univ. of Leyden
85. Van Vloten-Doting, L., Jaspars, E. M. J. 1967. Enhancement of infectivity by combination of two ribonucleic acid components from alfalfa mosaic virus. *Virology* 33:684–93
86. Van Vloten-Doting, L., Kruseman, J., Jaspars, E. M. J. 1968. The biological functions and mutual dependence of bottom component and top component *a* of alfalfa mosaic virus. *Virology* 34:728–37
87. Van Vloten-Doting, L. Dingjan-Versteegh, A., Jaspars, E. M. J. 1970. Three nucleoprotein components of alfalfa mosaic virus necessary for infectivity. *Virology* 40:419–30
88. Wood, H. A., Bancroft, J. B. 1965. Activation of a plant virus by related incomplete nucleoprotein particles. *Virology* 27:94–102
89. Wood, H. A., Streissle, Gert. 1970. Wound tumor virus: purification and fractionation of the double-stranded ribonucleic acid. *Virology* 40:329–34
90. Wu, G., Bruening, G. 1971. Two proteins from cowpea mosaic virus. *Virology* 46:596–612

PHYSIOLOGY OF FUNGAL HAUSTORIA[1] 3547

W. R. BUSHNELL

Plant Physiologist, U. S. Department of Agriculture, Cooperative Rust Laboratory,[2] University of Minnesota, St. Paul, Minnesota

INTRODUCTION

The position of the haustorium inside the host cell and the apparent needs of attached hyphae outside the cell suggested to De Bary and other early workers that the haustorium is primarily an organ of absorption. This concept remains credible today, but it has gone untested in spite of intensive investigation of haustoria by electron microscopy and increasing knowledge of the changes that are induced in host plants by haustorial fungi. Our understanding of absorption and other activities of the haustorium is generally poor because the small intracellular haustorium is difficult to study by conventional physiological techniques.

This paper presents selected topics related to the development and physiological activities of the haustorium as they are understood today. The haustoria of rust and powdery mildew fungi are emphasized, but other haustorial fungi are included, especially the lichen and mycorrhizal fungi which have been given little or no attention in earlier reviews of haustoria (44, 100, 101). Their inclusion provides a more complete concept of the haustorium than has been available heretofore, and serves to emphasize: (*a*) that many fungi other than obligate parasites produce haustoria; and (*b*) that some haustorial fungi are similar to certain types of biotrophic fungi which merely contact the walls of host cells. Specifically, this paper attempts to answer the following questions: What are haustoria and what types of fungi produce them? How do haustoria develop and how do host cells defend themselves against haustorial invasion? What is known about the physiological activities of the haustorium, particularly with respect to haustorial structure and the transfer of nutrients from host to fungus? Since space is limited, citations are representative instead of all-inclusive.

[1] Paper No. 7836, Misc. J. Series, Minnesota Agricultural Experiment Station.

[2] A laboratory operated cooperatively by the University of Minnesota; the Plant Science Research Division, Agricultural Research Service, U. S. Department of Agriculture; and the Plant Protection and Quarantine Programs, Animal and Plant Health Service, U. S. Department of Agriculture.

The Fungal Haustorium

Definition.—The fungal haustorium is a specialized organ which is formed inside a living host cell as a branch of an extracellular (or intercellular) hypha or thallus, which terminates in that host cell, and which probably has a role in the interchange of substances between host and fungus. To produce a true haustorium, the fungus must penetrate the wall of the host cell, not merely invaginate it as some lichen fungi do. The terminal characteristic of the haustorium helps to distinguish it from intracellular hyphae (like those of the smut fungi) which are otherwise similar to haustoria (52). The attachment to a hypha or thallus outside the host cell differentiates it from thalli of parasites that are entirely intracellular like *Plasmodiophora* spp. The multicellular "haustoria" of parasitic higher plants are excluded here, although they do have specialized "searching hyphae" which penetrate individual living host cells as do fungal haustoria (35). Also excluded are the haustorium-like structures produced intercellularly by the Laboulbeniales in epidermal tissues beneath the cuticle of insects (102). As defined here, the haustorium can be regarded as a specialized organ evolved in fungi to contend with the rigid thick wall of the plant cell, bringing the fungus into close contact with the host protoplast.

Haustorial fungus—host associations.—Haustoria are found in five types of fungus-host associations: (*a*) *Diseases of higher plants.* Haustoria are produced in the downy mildews, the white rusts, the powdery mildews, the rusts, and several miscellaneous diseases caused mostly by ascomycetes according to Rice (100). The fine structure of haustoria in these diseases has been reviewed recently by Bracker (14) and by Ehrlich & Ehrlich (44). (*b*) *Lichens.* Intracellular haustoria have been demonstrated in lichens by light microscopy (56, 57, 98, 124) and more recently by electron microscopy (53, 54, 74, 75, 103, 129). The lichens with haustoria are usually of the anatomically primitive crustose type. Fungi in more advanced lichens do not produce haustoria but, instead, either invaginate the algal wall without penetration (the intruding hypha is inappropriately termed an "intramembranous" structure) or produce simple wall-to-wall contacts without intruding into the cell (53, 98, 124). (*c*) *Endotrophic mycorrhizae.* Haustoria are produced in the endophytic mycorrhizae that occur on *Monotropa* sp., other nongreen higher plants, and orchids (64, 91). The fine structure of such haustoria has been described recently by Dörr & Kollmann (36) and by Hadley et al (60). Haustoria are also produced in the vesicular-arbuscular mycorrhizae that are found on most species of higher plants (58). Haustoria are sometimes produced in young ectotrophic mycorrhizae, but not when this type of mycorrhiza is well developed. (*d*) *Fungi parasitized by filamentous mycoparasites.* Haustoria are produced by several species of mycoparasites in the Piptocephalidaceae and the Dimargaritaceae (7–9, 13). These are species in the Mucorales which usually parasitize other species of Mucorales. The fine structure of this type of haustorium has been described by Armentrout & Wilson (5) and Manocha & Lee (87). (*e*) *Cells parasitized by nonfilamentous aquatic phycomycetes.* Haustoria are produced by the lower fungi that have the "Chytridium" type of thallus in which a monocentric body is formed on the outer surface of a host cell, and a haustorium is produced inside the host cell (115). The host may be a fungal or algal cell or the pollen grain of a higher plant. The haustorium may be simple or branched, and often has a fine, tapering, rhizoidal pattern of branching. When produced in a living host, such rhizoidal systems conform to the definition of a haustorium used here, but some workers have not designated such rhizoidal systems as haustoria, primarily because they resemble the rhizoids produced by aquatic phycomycetes on nonliving substrates (78, 115). The "Chytridium" type of haustorial fungi are found in the Phlyctidiaceae and Chytridiaceae (in the Chytridiales), the Rhizidiomycetaceae (Hyphochytriales), and the Thraustochytriaceae (Saprolegniales) (115). Haustoria of these fungi have not, to my knowledge, been examined by electron microscopy.

There are many biotrophic associations in which the cells of host and fungus are intimately associated, but in which true intracellular haustoria are not formed. These include the "cuticle" diseases of higher plants (55), many lichens (as mentioned earlier), the ectotrophic mycorrhizae (64), the contact mycoparasites (6, 13), and some aquatic phycomycetes (131). Each of these strictly extracellular associations resembles a haustorial counterpart. We do not know if the haustorium facilitates biochemical interactions between host and parasite that are not possible when the association is of the wall-to-wall, extracellular type. Direct comparisons, like those being made by Barnett and co-workers (6, 7) between contact and haustorial mycoparasites will be necessary to determine what aspects of host-parasite interaction, if any, are unique to haustorial fungus-host associations.

Culturability of haustorial fungi.—The endotrophic mycorrhizal fungi of the Pyrolaceae and Monotropaceae include *Armillaria* spp., *Rhizoctonia* spp., and other fungi that grow rapidly on artificial media, can digest cellulose and lignin, and compete as saprophytes in the soil (64). The endotrophic mycorrhizal fungi of the orchids also grow in artificial culture, but more slowly (64). Most lichen fungi grow in artificial culture, but may increase only 1–2 mm in diameter per year, a rate comparable to that of some lichen thalli (1). Lichen fungi frequently require biotin and thiamine (1) as do many nonhaustorial fungi. Both mycorrhizal and lichen fungi grow on malt extract, potato dextrose agar, or other media commonly used for saprophytic fungi, and thus do not have any known nutritional requirements that distinguish them from other types of fungi. The haustorial aquatic phycomycetes are also apparently readily cultured on artificial media since at least three such fungi [as judged from the descriptions of Sparrow (115)] have been cultured, including *Rhizidiomyces apophysatus, Rhizidium carpophilum,* and *Phlyctochytrium biporosum* (66, 130).

In contrast, several types of haustorial fungi are difficult or impossible to grow on artificial media by present techniques and for this reason, have been considered traditionally as obligate parasites. These include the vesicular-arbuscular mycorrhizal fungi (58), the rust, powdery mildew, and downy mildew fungi (16), and some of the haustorial mycoparasites (7, 8, 13). Although several of the haustorial mycoparasites (7), some rust fungi (65, 106, 125), and a downy mildew fungus (121) recently have been grown apart from living hosts, such fungi are "strictly obligate parasites" in the ecological sense originally intended by De Bary (29), in that they probably live only as parasites in the natural "spontaneous course of things." The nutrition of these fungi in relation to the posible role of haustoria is discussed later. However, there is neither structural nor physiological evidence to indicate that haustoria of obligate parasites differ from those of fungi with greater natural saprophytic capabilities.

Relations with the living host.—Many haustorial fungi coexist biotrophically for a long period in apparent harmony with their hosts. Rust and powdery mildew fungi actually tend to prolong the life of the host cell if host and parasite are highly compatible, maintaining or increasing the synthetic activities of host cytoplasm. The concentrations of ribosomes and endoplasmic reticulum may increase (44, 108), senescence of tissues around infection sites may be delayed, and host cells may enlarge and divide in some cases (101, 108). At sporulation of the fungus, the juvenescence can be partly reversed as the tissues tend to become chlorotic and to leak solutes. The role of the haustorium in these alterations is uncertain, although the fungus probably induces changes in the host by withdrawing or introducing chemical substances by way of the haustorium.

The biotrophic period may be extremely short in some cases. For example, the cells of *Mycotypha microspora* started to degenerate within 6 hr after the haustorium was formed by the mycoparasite *Piptocephalis virginiana* and became empty within 36 hr (5). The death of the host here was apparently a normal consequence of the host-parasite interaction, and one that apparently served to nourish the parasite. Likewise, the presence of dead or dying algal cells in the lichen thallus has been interpreted to mean that the haustorium caused death of the host protoplast and then possibly absorbed part of it (97, 129). Haustorial aquatic phycomycetes can cause rapid decline of host cells, especially if many thalli of the parasite become attached to a single host cell (78, 107). Lysosomes, possibly of haustorial origin, have been postulated to release hydrolytic enzymes which, in turn, could have a role in the degeneration of host cells in some fungus-host associations (5, 129).

Haustorial structure and terminology.—Haustoria can be filamentous, vesicular, or club-shaped, and may branch and coil in a variety of ways, sometimes filling the host cell (51, 64, 71, 98, 100, 124). Many haustoria have a thin neck that places the main haustorial body away from the host wall. The extrahaustorial structures that constitute the interface between the haustorium and the host protoplast are shown in Fig. 1 and include: (*a*) The *papilla* —a mound of material, laid down on the inside surface of the host wall, either around the haustorial neck or opposite the infection peg before wall penetration; (*b*) the *sheath*—material that forms a collar around the haustorial neck as an extension of the papilla, and can extend to the haustorial body and enclose the entire haustorium; (*c*) the *extrahaustorial membrane*—which borders the host cytoplasm in the region peripheral to the haustorium; and (*d*) the *extrahaustorial matrix*—the liquid or solid substances between the extrahaustorial membrane and the haustorial wall. The haustorium with all associated structures is termed the haustorial apparatus (14). Not all haustoria have all the features shown in Fig. 1, nor are the structures always clearly differentiated from one another. However, most haustoria have the following structures or layers through which substances probably move to make the journey from the cytoplasm of the host to the cytoplasm of the haustorium: (*a*) the extrahaustorial membrane; (*b*) the extrahaustorial matrix (this varies in thickness and probably also in composition); (*c*) the haustorial wall; and (*d*) the plasma membrane of the fungus.

The terms used here are those that most clearly express the concepts to be developed herein. "Papilla" follows the precedent set by Smith (113). "Extrahaustorial membrane" was used by Ehrlich & Ehrlich (44) although not as a formal term. "Extrahaustorial matrix," newly introduced here, corresponds to the "sheath matrix" of Bracker (14) and the "encapsulation" of Ehrlich & Ehrlich (41). "Sheath" is used here to designate a type of structure that can wall-off the haustorium. This is a common usage, especially among rust workers, although sheath has been used by some authors to designate the extrahaustorial matrix, or the extrahaustorial matrix and extrahaustorial membrane combined.

Formation of the Haustorium in the Host

Infection structures.—Haustorial aquatic phycomycetes penetrate host cells directly beneath spores, but other kinds of haustorial fungi produce special infection structures to bring the fungus from the site of spore germination on the plant surface to a site suitable for haustorium formation. Most powdery mildew fungi, some downy mildew fungi, and the lichen fungi each produce a short germ tube, an appressorium, and then an infection peg that penetrates the host cell directly, producing a haustorium in that cell. Some other downy mildew fungi produce an infection hypha that forces its way into the leaf along the middle lamella between adjacent epidermal cells, producing haustoria laterally in those cells (23, 29). Some rust fungi produce an infection hypha that penetrates an epidermal cell directly and then grows through one or two host cells before establishing an intercellular hypha which, in turn, produces a haustorium in a mesophyll cell (4, 136). Most rust and some downy mildew fungi enter stomates and grow in substomatal

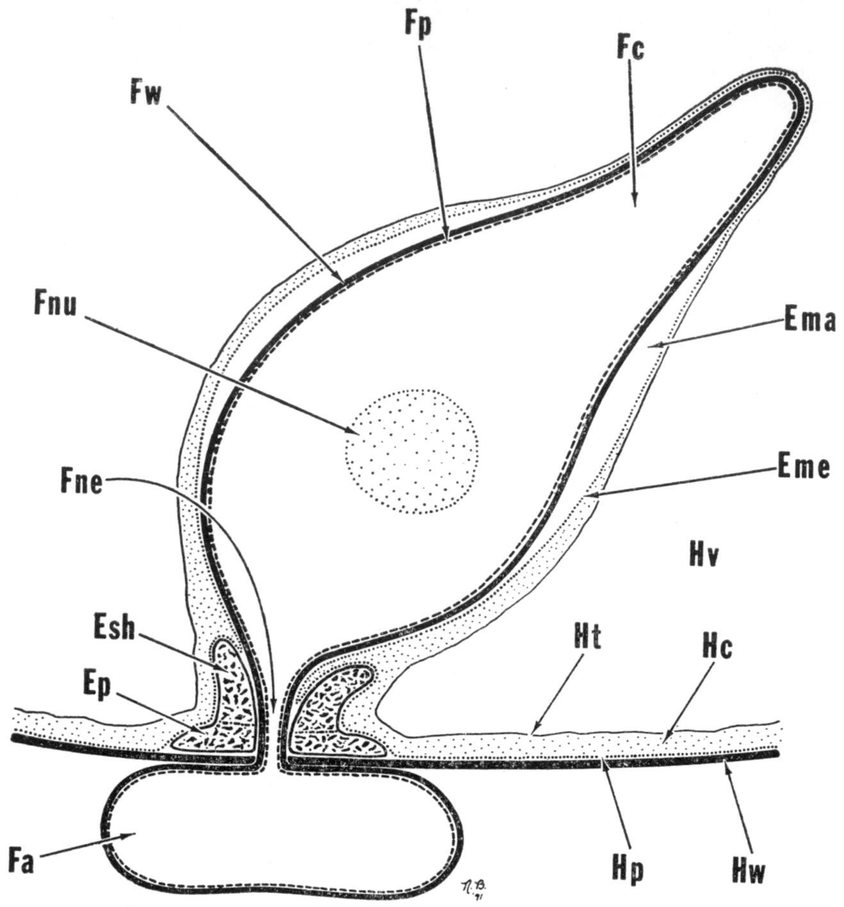

FIGURE 1. Diagrammatic cross-section of the fungal haustorium and associated structures in the cell of a higher plant. *Extrahaustorial components:* Ep, papilla; Esh, sheath; Ema, extrahaustorial matrix; Eme, extrahaustorial membrane; *Fungal components:* Fa, appressorium or haustorial mother cell; Fne, haustorial neck; Fw, haustorial wall; Fp, plasma membrane of fungus; Fnu, nucleus of haustorium; Fc, haustorial cytoplasm; *Host components:* Ht, host tonoplast, Hc, host cytoplasm; Hp, plasma membrane of host (thought to be continuous with the extrahaustorial membrane); Hw, host wall; Hv, host vacuole. Cytoplasmic organelles are not shown.

cavities by producing a complex set of infection structures consisting of an appressorium, substomatal vesicle, infection hyphae, and a haustorium mother cell from which a haustorium is produced in a mesophyll cell. The physiology of spore germination and the development of infection structures have been studied intensively for certain rust fungi (37, 117). Regardless of the mode of primary infection, subsequent haustoria arise from haustorial mother cells produced by ramifying intercellular hyphae, or, as in the case of ectoparasitic powdery mildew fungi, from appressoria produced by hyphae growing on the surfaces of epidermal cells.

Wall penetration.—Several types of evidence indicate that infection pegs of haustorial fungi penetrate the walls of host cells with the help of enzymatic digestion. A "halo" region on the host cell wall around the infection peg of powdery mildew fungi stains differently than does the normal host cell wall, whether the specimen is stained for light or electron microscopy. Similar, though less distinct, alterations are visible near infection pegs of rust fungi (44). Cytochemical tests indicate that the halo contains reduced amounts of cutin, polysaccharide (including cellulose), and pectin, and increased amounts of reducing sugar, pentose, and uronic acid—all indicating that the wall and cuticle have been partially degraded (80, 88). The wall in the halo region also has an altered fibrillar structure (88). However, the halo is distinct from the small zone immediately ahead of the infection peg where the wall appears to be completely dissolved. Here the peg usually advances without tearing or laterally displacing wall materials, as shown by Edwards & Allen (40) for *Erysiphe graminis.* In lichens there is a similar absence of visible displacement of wall materials around the infection peg (75, 129). In line with these morphological observations is the fact that rust spore germlings apart from a host can produce hemicellulase, β-glucosidase, and polygalacturonase (127). Regardless of the degree of dissolution of cuticle and wall, some mechanical force is required for penetration by the infection peg. This force sometimes produces a visible displacement of wall materials. Hardwick and co-workers (63) reported a "scuffing" of the wall around the infection peg of the bean rust fungus, a rare event in their material. With *Erysiphe graminis,* small portions of the cuticle (116) or of the cuticle and outer layers of the host wall (C. E. Bracker, Purdue Univ., personal communication) have been found to be pushed inward in the region immediately beside the infection peg. Akai and co-workers (2) observed a crack in the host wall intersecting the infection peg of *E. graminis,* which suggests that the peg had applied a large mechanical force.

Papillae and sheaths.—The host often deposits materials that form a papilla on the host wall outside the host protoplast, and the host may deposit additional materials that form a sheath around the neck and body of the haustorium, as shown in Fig. 1. Both structures are apparently attempts by the host to wall out the parasite. The composition and degree of rigidity of the sheath and papilla are not known, although some workers have suggested that these structures are gels or semi-liquids (23, 61). Both structures give positive staining reactions for callose (28, 40, 51, 63, 69, 95, 116), and both usually contain irregular areas of membranous substances and darkly stained particles as viewed by electron microscopy.

The papilla is deposited on the host wall during or after penetration as a mound-shaped structure, about 3–10μ in diameter, 0.5–2μ thick, located in a position centered on the infection peg (Fig. 1) (15, 23, 40, 69, 88, 116, 119). The papilla is distinct from the host cell wall even though the wall may be slightly thickened near the site of penetration (23, 88). Smith (113) provided the first detailed description of the papilla, showing by light microscopy that the papilla of *Erysiphe* spp. appeared to bar entrance of the penetration peg in some cases, but was penetrated by the peg in other cases, depending on the relative vigor of fungus and host. Recent electron micrographs show that a papilla is deposited before the penetration pegs of powdery mildew fungi complete penetration of the host wall (40, 88, 116). The "wall lesions" described by Hanchey & Wheeler (61) in cells attacked by *Phytophthora parasitica* var. *nicotiana,* and the lomasome-like vesicles in cells attacked by *Puccinia hordei* described by Calonge (21), closely resemble the papilla and are likewise produced ahead of wall penetration.

Preceding development of the visible papilla by two or more hours, rapidly moving cytoplasmic organelles aggregated on the host wall beneath the appressorium of *E. graminis* (18). The timing and location of these aggregates suggested that they were directly involved in deposition of the papilla. The formation of papillae in advance of the penetration peg suggests that their deposition is induced by chemical substances, perhaps materials related to the enzymatic dissolution of the wall. On the other hand, papilla-like deposits have been produced by a glass microprobe inserted into a living plant cell and left in place (95), and by other types of wounding (44).

The sheath, produced as an extension of the papilla, forms a collar around the haustorial neck (Fig. 1) or an encasement around all or part of the haustorial body. Sheaths occur in invaded algal cells of lichens (98, 103, 124), as well as in the cells of higher plants (10, 15, 23, 28, 63, 69, 85, 89, 126). Encasement of the entire haustorium occurs frequently around small aborted haustoria or aging dysfunctional ones (3, 10, 93). The small knob-like haustoria of *Phytophthora* spp. are often sheathed (12, 42), but Hanchey & Wheeler (61) found that the structure around haustoria of *P. parasitica* var. *nicotiana* was unlike the typical sheath in that it lacked inclusions and appeared to be continuous with the host cell wall. Structures analogous with the sheath have been termed lomasomes by Ehrlich et al (45) and primary encapsulations by Kajiwara (77).

The collar portion of the sheath may result from a displacement of a part of the papilla during penetration by the fungus (23, 40), but Heath & Heath (69) showed in a host resistant to cowpea rust that the sheath was deposited after the haustorium had formed. The plasma membrane of the host was highly convoluted adjacent to the sheath, and irregular portions of this membrane became trapped between the developing sheath and the haustorial wall.

Papillae and sheaths are so commonly found that they can be regarded as a normal response to haustorial fungi and to other plant pathogens if the host cell remains alive for a short time after penetration (21, 23, 55, 100). Thus, these reactions are probably nonspecific responses to wounding on the part of the host. It follows that the successful haustorial parasite must have a means to limit this response. The aggregation of cytoplasmic organelles associated with papilla deposition ahead of *E. graminis* dispersed abruptly as the haustorium enlarged inside the host cell (18). For reasons unknown the host quit trying to encase the haustorium soon after the young haustorium had started to grow toward the interior of the host cell. Thus, we need to determine how the deposition processes are turned off in this and other host-parasite systems as well as how they are turned on.

Invagination of the host's plasma membrane.—The initial stages of haustorium formation occur rapidly after wall penetration and tend to be obscured by the papilla or by complex configurations in the membranes and cytoplasm of the host so that the relation between the very young haustorium and host cytoplasm has rarely been seen clearly in either fixed or living specimens. For example, the young haustorium of *Erysiphe graminis* was shielded from view by unidentified, rapidly moving elliptical vesicles (5–10 μ long) which developed in the last few minutes preceding emergence of the haustorium (18). Nevertheless, the haustorium grows rapidly toward the interior of the host cell, the localized hyperactivity of host cytoplasm ceases (18), and the haustorium then appears to have invaginated the plasma membrane of the host cell (Fig. 1). To determine if the extrahaustorial membrane (Fig. 1) is the invaginated plasma membrane (or a derivative thereof), several workers have tried to trace continuities between these two membranes in the region of the haustorial neck. This has been accomplished most clearly by Littlefield & Bracker (84) in specimens of flax rust in which complexities in the neck region were relatively few. That the plasma membrane is invaginated during haustorium formation is also indicated by the experiments of Thatcher (120) and of Fraymouth (51) who were able to remove and then replace the host protoplast from its position around the haustorium by successively plasmolysing and deplasmolysing the host cell. Additional experiments of this type are needed since they were not described in detail and since Hoskin (see 18) was unable to do similar experiments with *Erysiphe graminis*.

Formation of the Haustorium on Artificial Substrates

Methods for inducing haustorium formation apart from the host have not been available until very recently. Williams (134) noted some haustorium-like knobs produced by the wheat stem rust fungus in artificial culture, but offered no detailed morphological or experimental evidence to support his conjecture that they were haustoria. Infection structures of both rust and powdery mildew fungi can be induced apart from the host on membranes, surfaces of liquids, or in liquid suspension (33, 37), but haustoria are not usually so induced. As the culmination of many years of work with artificial membranes, Dickinson (34) recently obtained infection structures and haustoria of a rust fungus, *Puccinia coronata.* The most effective membranes contained nitrocellulose and paraffin wax, poured, gelled, and chemically "stretched" in a precise fashion. On these membranes about 45% of the leaf hyphae produced haustoria, either at the tip or at the side of the hyphae. Although Dickinson offers only a diagrammatic sketch of these haustoria, their occurrence in the proper developmental sequence, their long necks, and their development at high frequencies all suggest that these structures are, in fact, haustoria.

Host Defenses Against the Haustorial Fungus

The formation of the haustorium is opposed by many chemical and physical barriers in the host, some preformed, others like the papilla and sheath, produced in response to attack. If the species under attack is not a natural host for the attacking pathogen (is a "nonhost species"), fungal development is generally stopped before or during formation of the first haustorium at the infection site (the primary haustorium). The early halt to fungus growth can be the consequence of a preformed defense or a product of rapid interactions between host and parasite. If the species under attack is a natural host (is a "host species"), the fungus grows without interruption whenever host and parasite are specifically compatible, but growth is stopped or retarded at various stages if they are incompatible. In the latter case, preformed defenses may play a role early in fungus development, but fungus development is often inhibited only after one or more haustoria are produced when products of host-parasite interaction apparently become effective. Within this framework, then, consider briefly the various ways in which the host plant defends itself against haustorial fungi.

No haustoria are formed.—The specialized infection structures that bring the fungus to a site of potential haustorium formation are subject to limiting factors of the environment and of the host, but with rust and powdery mildew fungi in the laboratory, 20% or more of applied spores usually produce a complete set of infection structures on a susceptible host. This success rate is reduced frequently when the plant under attack is a member of a "nonhost species" (59, 62, 81, 96), and less frequently when it is an incompatible member of a "host species" (24, 70, 96, 104). Thus, rust fungi often fail to enter the stomates of a "nonhost" freely, sometimes injuring guard cells, and upon gaining entrance, may produce abnormal substomatal structures and fail to produce haustorial mother cells. Leath & Rowell (82) found that the inhibited substomatal growth of the wheat stem rust fungus on a nonhost (*Zea mays*) was not caused by nutritional deficiencies, wall thickenings, or preformed toxins, but probably by phytoalexins which could be elicted by infiltrating leaves with exudates from germinating spores. It appeared that interactions between fungus and host led to toxin production which, in turn, prevented complete development of infection structures.

If the fungus arrives successfully at a site of potential haustorium formation, at least three factors can prevent formation of the haustorium: (*a*) The cuticle, the wall, or the combined cuticle and wall of the host cell may be too thick or hard for penetration. This can be a preformed condition (24, 86, 90) or a localized thickening induced at the site of potential wall penetration (68). Similar thickenings occur in response to intercellular hyphae or dying host cells (3, 23, 67, 69). (*b*) Excessive osmotic pressure in the host cell may prevent haustorium formation. Schnathorst (105) has shown that the osmotic pressure of lettuce sometimes exceeds that of a powdery mildew fungus, and that resistance to the fungus correlates consistently with the presence of high osmotic pressure in the host. Similar though less complete correlations occur in powdery mildew of apple (24). These are unusual cases since the osmotic pressure of most haustorial fungi is well above that of their hosts (120), a condition that allows water to move into the parasite for maintenance of turgor during growth. (*c*) The papilla, formed in advance of the penetration peg as described earlier, serves to prevent haustorium formation in both "host" and "nonhost" species according to several workers (26, 86, 113). However, in the host-parasite combinations in which the papilla has been examined by electron microscopy, the fungus has breached this structure (40, 88, 119). Furthermore, Edwards (39) found that the papilla stained with bromophenol blue, and that this stain spread throughout the host cell at the time that the papilla was probably being penetrated in an incompatible host-parasite combination. Edwards indicated that the papilla may have limited fungus development by serving as a source of chemical agents involved in resistance instead of by acting as a physical barrier.

The primary haustorium is partly or fully formed.—The primary haustorium may induce classic hypersensitive reactions in which a single host cell dies, and further fungal development is virtually stopped (69, 83, 104). The cell may die when the haustorium is starting to form (3), or after it is fully formed (83), but in either case the host cell usually dies before the fungus does. A sheath occasionally encases the young haustorium forming a protective barrier that delays or prevents the hypersensitive response (67, 69, 70, 83). Heath & Heath (69) found convolutions in the plasma membrane of the host that preceded either sheath formation or host cell necrosis. Membrane dissolution occurs early in the hypersensitive response, and may involve the extrahaustorial membrane (69), or the plasma and chloroplast membranes of the host cell (119). What triggers the hypersensitive reaction is not known. Zimmer (137) found crystal-containing microbodies in the perivascular ring cells of safflower and suggested that these bodies might be a source of lysosomes involved in the selective hypersensitivity of the ring cells. Hirata (71, 72) found a correlation between calcium content and susceptibility of barley epidermal cells to powdery mildew and speculated that the cell falls into an unstable condition unless calcium or other divalent ions are present in concentrations sufficient to allow host cytoplasm to adjust physically to the intruding haustorium.

Phytophthora spp. induce very rapid hypersensitive responses in incompatible host cells, usually within 3 hr after cells are penetrated (61, 122). Some host cells show signs of degeneration within a few minutes, as if the responses were triggered by contact between the fungus and the host protoplast. The fungus, on the other hand, may continue to live for several hours, apparently until phytoalexins are produced (122, 128) Although the intruding hypha does not usually form a haustorium in a hypersensitive cell, the hypersensitivity induced by the hypha of *Phytophthora* and that induced by haustoria of other fungi may well involve similar triggering mechanisms.

Several haustoria are formed.—The battle between a parasite and a member of a "host species" often persists indecisively for many days while the fungus produces several haustoria and a small hyphal colony (83, 110, 114, 132). The result is a *partial compatibility* expressed initially by retarded hyphal growth and poor haustorial development. The volume of the extrahaustorial matrix may become larger than normal (109). Although host cells may die at the colony center or elsewhere, Ogle & Brown (96) have shown that the amount of fungal development and necrosis are unrelated and that the fungus usually continues to live in spite of host degeneration. Haustoria may function poorly in this important type of partial compatibility, but we lack experimental evidence to support this possibility.

The host cell invaded by endophytic mycorrhizal fungi can mount a belated counterattack by partially digesting the haustorium. After a period of coexistence between fungus and host which can last several months, mycorrhizal haustoria lose viability, the cytoplasmic contents of haustoria disappear, and the undigested remnants of the haustorial wall and other materials form a permanent clump within the host cell (17, 58, 64, 91). New haustoria may form in the same host cell and again be digested. This phenomenon probably supplies nutrients to the host plant, an important benefit if the host is heterotrophic for carbon. Antifungal toxins were demonstrated in cells in which haustoria were being digested—toxins that probably kill the fungus preceding digestion (17). Gaumann and co-workers have shown that the phytoalexin orchinol is produced by the tissue of orchid, a common host to endophytic mycorrhizal fungi (27). Whether digestion of the haustorial contents in the mycorrhizae is autolytic or the result of enzymes secreted by the host is not known. Dörr & Kollmann (36 have shown that the parietal host cytoplasm in *Neottia nidus-avis* expands inward in response to invasion by a mycorrhizal fungus, forming an unusual network of cytoplasm in the host vacuole. Otherwise we have no indication why mycorrhizal host cells in roots and rhizomes digest haustoria whereas host cells in other types of fungus-host associations do not.

Multiple defenses.—A host may invoke several of the above defenses in series against the potential haustorium producer. Thus the resistance of a cowpea variety to an isolate of rust included thickening of the host cell wall (in a low percentage of cases), hypersensitive death of the host cell, and sheathing of the entire haustorium (67, 69). Ellingboe and co-workers (46, 110) have shown that a part of the attacking population of powdery mildew germlings fails to produce haustoria in barley or wheat, and that the remaining part goes on to produce limited amounts of colony growth depending on which genes are controlling the host-parasite interaction. Stanbridge et al (116) indicate that the two defense mechanisms are papilla formation and hypersensitivity. Pre-existing differences among cells of the host may contribute to such dual responses as Hirata has shown with respect to calcium content (71, 72). In any case, the successful haustorial parasite must bypass, suppress, tolerate, or avoid triggering the entire battery of defensive mechanisms that the host can muster. The investigator of defense must establish carefully which mechanisms are operating in the system he has under study.

The degree of compatibility between many haustorial fungi and members of "host species" is known to be controlled by corresponding genes in host and parasite as with flax rust, stem rust of wheat, and the powdery mildews of wheat and barley (49). The incompatibility conditioned by gene-for-gene interaction is most often expressed either as a hypersensitivity during formation of the first haustorium, or as a partial compatibility in which several haustoria may be formed. The specific physiological processes through which the gene-controlled interactions are expressed are virtually unknown (46, 49).

The Functional Haustorium

Structural aspects.—In compatible higher plants the developing haustorium appears to invaginate not only the plasma membrane, but also the thin parietal layer of cytoplasm and the tonoplast of the host, so that the haustorium intrudes into the vacuolar region of the host cell (Fig. 1). In algal hosts, which generally do not have large vacuoles, the haustorium intrudes into the cytoplasm of the host and often extends into the single large chloroplast of the algal cell. The fully grown haustorial cell has a normal complement of cytoplasmic organelles, including mitochondria, ribosomes, and endoplasmic reticulum, and usually is not vacuolate. Cytochemical tests for RNA, protein, dehydrogenase, and the enzymes of glycolysis (5, 118, 123, 133) indicate that the haustorium is a site of relatively high synthetic activity, but such tests have not revealed any specialized activities that distinguish the haustorium from other types of fungal cells.[3]

The extrahaustorial membrane, the boundary between the host cytoplasm and the haustorial apparatus (Fig. 1), closely resembles the plasma membrane of the host from which it is probably derived, but it sometimes becomes

[3] Note added in proof: Coffee et al (*Can. J. Bot.* 50:231–40, 1972) reported that the cristae and matrices of mitochondria were more highly developed in the haustorium than in the parasitic hyphae of two rust fungi, and therefore suggested that the hyphae may be dependent on products of primary metabolism that are produced in the haustorium.

thickened as the haustorium matures (15, 41, 126), and resistant to disruption by mechanical or chemical treatment (15, 32). Littlefield & Bracker (85) found that the extrahaustorial membrane of *Melampsora lini* lacked the particulate material seen on plasma membranes in freeze-etch specimens, and that the extrahaustorial membrane stained lightly whereas the plasma membrane of the host stained intensely with periodate-chromate-phosphotungstate. Associated with the extrahaustorial membrane of several fungi are vesicles postulated to be involved in movement of materials across the membrane from the host cytoplasm to the extrahaustorial matrix or in the opposite direction (44). For rust fungi several workers have concluded that the extrahaustorial membrane, unlike the plasma membrane of the host, is connected to the endoplasmic reticulum of the host (44). For powdery mildew fungi, small irregular invaginations extend from the extrahaustorial membrane into the extrahaustorial matrix (44). Bracker (14) and Ehrlich & Ehrlich (44) have reviewed these specialized configurations of the extrahaustorial membrane in more detail than space permits here.

The volume of the extrahaustorial matrix (the substances between the extrahaustorial membrane and the haustorial wall, Fig. 1) tends to increase as the haustorium matures. In powdery mildews, a large part of the extrahaustorial matrix is composed of an aqueous solution (the matrical solution) which may contain small amounts of suspended particulates, but in which all components tend to become uniform in concentration, probably by diffusion and convection. This is indicated for *E. graminis* and other powdery mildew fungi (15, 32, 89, 119) by the light, uniform stain present throughout most of the region occupied by the extrahaustorial matrix in specimens prepared for electron microscopy. In powdery mildews, the matrical solution often occupies a zone 1 micron or more thick, a zone easily seen by light microscopy (15, 32, 89, 119). In *E. graminis* the matrical solution is known to be in osmotic equilibrium with the cytoplasm of the host, a feature to be described later.

The extrahaustorial matrix also has components that do not intermix freely with the matrical solution, as indicated by the presence of differentially stained regions. Thus in *E. graminis* there is a distinct, lightly stained layer bordering the haustorial wall, generally no thicker than the wall itself (C. E. Bracker, Purdue Univ., personal communication). The physical and chemical properties of this layer are not known, but its failure to intermix with the matrical solution suggests that it is either not soluble in water (it may contain small amounts of matrical solution), or that it is separated from the matrical solution by structural barriers. The extrahaustorial membrane sometimes contacts the differentially stained layer, and sometimes is separated from it by the matrical solution.

For fungi other than the powdery mildews, the relative amounts of solution and other components in the extrahaustorial matrix are not known. Generally the distance between the extrahaustorial membrane and the haustorial wall is less than 0.5 microns, so that the extrahaustorial matrix is not readily visible by light microscopy. The matrix may stain lightly throughout when the haustorium is young (5, 69, 84, 85, 126) but it otherwise stains in varying intensities, sometimes in layers suggesting the presence of one or more solid or semi-solid layers. Usually no recognizable cytoplasmic organelles are found.

Studies of *Erysiphe graminis* from several laboratories have revealed several important properties of the extrahaustorial matrix and membrane, especially when the haustorium is young and functional. The primary haustorium of this fungus starts to form in an epidermal cell during the first day after inoculation, continues to elongate until the third day, and may add additional finger-like branches until the fifth or sixth day (18, 71). Hirata (71) has shown that during this period the hyphae that are attached to the primary haustorium elongate about 250 μ per day, branch, and produce additional haustoria, so that each colony has 100–300 haustoria by the time the primary haustorium is mature. The colonies start to sporulate at about 5 days, intensifying demands by the fungus for nutrients at a time when most haustoria are less than fully grown.

Hirata found a sac-like membrane around the haustorium in specimens of *E. graminis* fixed in FAA and examined by light microscopy. This membrane, which he assumed to be analogous with the extrahaustorial membrane, began to separate from the haustorial wall as the haustorium matured. The separation occurred earlier if the mycelium attached to the haustorium had developed poorly or had been removed or disabled. From these observations Hirata concluded that the haustorium is more functional before, than after, the extrahaustorial membrane separates from the haustorial wall and that the separation serves to protect the surviving host cell or haustorium when one or the other degenerates. Further, Hirata & Kojima (73) showed that the sac expanded and contracted with changes in osmotic pressure in unfixed specimens "pulled" from host cells.

That the extrahaustorial membrane of *E. graminis* is subject to large shifts in position was confirmed through observation of living specimens in my laboratory (18, 19). Here the sac as seen by light microscopy was judged to include the entire thin layer of host cytoplasm that surrounds the haustorium, a layer bordered on the haustorial side by the extrahaustorial membrane, and on the host side by the tonoplast of the host. The entire structure was so thin that it appeared to be a single membrane by light microscopy except for occasional protuberances attributed to cytoplasmic organelles. The extrahaustorial membrane moved as a part of this sac, much as Hirata had described, when changes in osmotica were introduced through incisions in the host cell wall. Thus the sac expanded to form a balloon-shaped structure around the central body of the haustorium on media hypotonic to the host, and shrank into a position close to the haustorial wall on media hypertonic to the host as viewed by light microscopy. Some unknown property of the interface between the extrahaustorial membrane and haustorium prevented expansion at the tips of the haustorial fingers. [The recent electron microscope study of C. E. Bracker (Purdue Univ., personal communication) suggests that the membrane was in contact with the differentially stained part of the matrix that borders the haustorial wall in these finger tip regions.] These experiments provide evidence for the following properties of the extrahaustorial membrane or matrix: (*a*) the membrane (together with the layer of host cytoplasm and tonoplast) is highly permeable to water, and less so to sucrose; (*b*) the matrical solution is normally in osmotic equilibrium with the host cytoplasm and vacuole; (*c*) the cavity that contains the matrical solution is sealed in the neck region of the haustorium and elsewhere so that no open channels exist between the cavity and regions either inside or outside the host cell; (*d*) the cavity that contains the matrical solution does not contain more than trace amounts of structural or solid material.

Further experiments showed that chemical fixatives, particularly $KMnO_4$, caused the sac to expand, as did the fairly common procedure of stripping pieces of epidermis from leaves and placing them in water for observation (18). To determine the position of the extrahaustorial membrane relative to the wall of functional, undisturbed haustoria, mildewed tissues were placed in special micro-culture mounts so that haustoria in living host cells could be examined while they were attached to growing hyphae. The primary haustorium was observed 2–3 days after inoculation, before secondary haustoria were produced. The sac was found to be closely appressed to the haustorium except for short distances along the wall of the central body and regions near the base of adjacent haustorial fingers in some specimens. Thus, the osmotic equilibrium minimized the volume of the extrahaustorial matrical solution, so that the extrahaustorial membrane was kept close to the haustorium, probably in contact with the thin differentially stained layer that borders the haustorial wall.

Indirect evidence indicates that hyphal growth was a valid sign that the young haustoria were serving as a source of water or nutrients for the fungus in these experiments. Hyphae of *E. graminis* do not grow from appressoria until the primary haustorium is partly formed in a host cell (46, 71), and they do not continue to grow (except at an extremely low rate) if the primary haustorium is removed before secondary haustoria are produced (18). Furthermore, Ellingboe and co-workers (94, 111) have shown that large amounts of ^{35}S or ^{32}P applied to the host moved into the fungal hyphae after the primary haustorium began to branch (16 hr after inoculation), compared to movement of very small amounts of labeled substances prior to that time. Such findings indicate strongly that the young haustorium was functional at the time when the extrahaustorial membrane was found to be close to the haustorial wall, i.e., when the volume of the matrical solution was small.

Does the haustorium remain functional as the volume of the extrahaustorial matrix increases during maturation of the haustorium? Although Hirata's observations with *E. graminis* indicate that the volume of the matrical solution is large when the haustorium is obviously dysfunctional, the relation between volume of the matrical solution and functional capacity has not been assessed when the haustorium first becomes fully grown (5–6 days after inoculation) or during the period immediately following. The development of secondary haustoria in experimental colonies has hampered efforts to locate and determine if the maturing primary haustorium remains functional. In fungi other than *E. graminis,* the relation between functional capacity and the amount and type of substances in the extrahaustorial matrix has not been determined for either young or mature haustoria. Therefore, we do not know if the extrahaustorial matrical materials serve as a barrier to movement of substances, facilitate such movement, or are unrelated to interchange of substances between host and parasite.

The microradioautographs of Ehrlich & Ehrlich (43) show that ^{14}C from labeled rust spores did not accumulate in the extrahaustorial matrix of *Puccinia graminis* in samples harvested 5 days after inoculation. Likewise, Jacobs & Ahmadjian (76) found no accumulation of radioisotope in the extrahaustorial region in a lichen fed $^{14}CO_2$ in the light. On the other hand, M. S. Manocha (Brock Univ., personal communication) found tritium from tritiated N-acetyl glucosamine in the extrahaustorial region of *Piptocephalis virginiana,* a mycoparasite of *Choanephora cucurbitarum.* In the noninfected fungus, the tritium was incorporated into the fungal wall, presumably into chitin. Thus host wall substances or their precursors may be excreted by the host into the extrahaustorial matrix, a possibility also suggested by the presence of pectin in the extrahaustorial matrix of a powdery mildew fungus (*Sphaerotheca pannosa*) in a higher plant host (22). Smith et al (112) have discussed at length the possibility that heterotrophic organisms may derive carbohydrate from autotrophic hosts by diverting the processes of wall synthesis.

Transfer of nutrients.—The presence of a haustorial fungus is usually associated with an increase in the amount of available nutrient in a host cell, as has been summarized by Smith et al (112) with respect to carbohydrates. Carbohydrate availability in the algal cell in the lichenized state is increased over that in the free living state by reduced algal growth, by conversion of insoluble carbohydrates to soluble ones, and by markedly increased excretion of carbohydrate. In rust and mildew diseases of higher plants, large amounts of organic and inorganic substances accumulate in infected tissue at the expense of the rest of the plant. Substances tend to stay at infection sites instead of being recirculated to growing parts of the plant, and long distance patterns of translocation apparently are altered so that the movement of organic substances toward the infection site is favored (38). Thatcher (120) provided evidence that the permeability of host cells was increased by rust infection, facilitating the movement of solutes to cells that contained haustoria. The mechanisms are not well understood, but the host generally tends to replenish the supply of nutrients in the host cell as they move into a haustorial fungus.

Smith et al (112) have, in addition, summarized the types of carbohydrates that are transferred from host cells into heterotrophic fungi. These carbohydrates have been determined principally by supplying $^{14}CO_2$ in the light to the photosynthetic host, dissecting apart the two organisms, and determining what carbohydrates are labeled in each. In a technique used with lichens, the intact thallus is immersed in carbohydrate solution to find what types of carbohydrate cause excretion of photosynthetically fixed ^{14}C into the medium, a specific property of that carbohydrate that is normally transferred. In general, a single soluble carbohydrate is transferred from the donor and then converted into a different soluble carbohydrate in the recipient fungus. Thus in lichens with green algal components (algalbionts), the donor produces erythritol, ribitol, or sorbitol, which is converted into arabitol or mannitol in the recipient fungal cell. In lichens with blue-green algalbionts, glucose is transferred and converted to mannitol. In higher plants, sucrose is transferred and converted into trehalose or a polyol (mannitol, arabitol, and erythritol) by the fungal parasite. The conversion is thought to reduce back flow to the donor since some donors are unable to utilize the product of the conversion.

Reisener and co-workers (99) have determined what types of amino acids and carbohydrates are transferred to rust spores from wheat leaf tissue by following changes in the ^{14}C labeling patterns of carbon skeletons. They conclude that glutamic acid, tyrosine, and hexose are taken up from the host by the fungus without alteration and that a part of fungal alanine also comes directly from the host. Furthermore, ^{14}C in the fungus was found first in free amino acids and later in protein, indicating that the bulk of fungal protein was synthesized by the fungus and did not enter in a ready-made form.

In summary, there is abundant evidence that small molecules move from the host cell to the mycelium of haustorial fungi. Unfortunately, there is little evidence to indicate if the transfers are made by way of the haustorium. The movement of a labeled substance from host to fungal hyphae is not in itself evidence of uptake by way of haustoria since substances such as glucose can move directly into hyphae of haustorial fungi (92) as they do into hyphae of nonhaustorial ectomycorrhizal fungi (112). Neutral red and tetrazolium salts applied to fresh tissue sections or epidermal cells entered haustoria, probably directly from host cells, but the possibility of entrance via extracellular contacts between hyphae and host was not ruled out (19, 123). Similarly, the radioautographs of Ehrlich & Ehrlich (43) and Jacobs & Ahmadjian (76) showed that ^{14}C moved from fungus to host or in the reverse direction, but not if the transfer was primarily through haustoria. Only in powdery mildews where the cuticle of the host obstructs the flow of nutrients betwen the host epidermal cell and the ectoparasitic hyphae does the haustorium appear to be essential for movement of nutrients into the fungus. The cuticle is not an absolute barrier, however. The best experimental evidence that substances move into fungi by way of haustoria comes from the developmental studies of Ellingboe and co-workers noted earlier (94, 111) in which a marked increase in the flow of ^{32}P and ^{35}S from host to parasite occurred at the time that the primary haustorium was forming.

Role of Haustoria in Obligate Parasites

The development and nutrition of germlings or hyphae of haustorial fungi not connected to haustoria can provide indirect indications of the possible functional role of the haustorium. This is especially true for the obligately parasitic haustorial fungi, which apparently have special nutritional requirements apart from hosts which may relate to their dependency on the haustorium in the host. Hyphae of obligately parasitic fungi have been studied in three situations in which the haustorium is absent: (*a*) the germling produced by a spore before the primary haustorium is formed; (*b*) hyphae which have been excised from leaves or cotyledons: (*c*) hyphae grown on artificial media apart from the host.

Germlings.—Spores of rust or powdery mildew fungi produce germ tubes and infection structures in the laboratory if appropriate environmental factors are provided, but usually produce little or no vegetative hyphae unless a haustorium is produced in a host cell. Germlings have a limited ability to produce protein (117), a deficiency thought to be the major reason for their limited ability to grow. This deficiency is apparently relieved about the time the first haustorium is produced in the host. The deficiency is also relieved on artificial nutrient media without formation of haustoria, but the transition takes a long time (5–7 days) as part of a poorly understood adaptation to the nonliving medium (79).

Excised hyphae.—The hyphae produced upon leaves maintain a part of their metabolic capabilities after they are excised from the leaf, i.e., after they are separated from haustoria and host cells. Hyphae of *Melampsora lini* excised from the surface of cotyledons continued to take up oxygen for several hours, to absorb amino acids, sugars, and sugar alcohols, and to convert glucose into other carbohydrates, probably trehalose and polyols (92, 135). Hyphae isolated from homogenates of rusted bean leaves accumulated label from tritiated orotic acid into an RNA fraction, although they had a very low rate of oxygen uptake and may have been injured during isolation (31). Powdery mildew hyphae (*Erysiphe graminis*) grew in trace amounts in a 2½ day period after haustoria were removed with microneedles, especially when supplied sugar exogenously (18, 19). These rust and mildew hyphae apparently maintained an ability to oxidize carbohydrate in amounts sufficient to maintain the fungal protoplasts in a living condition and to allow synthesis of small amounts of wall material in some cases. Like the germling, the hypha may have difficulty in synthesizing protein.

Hyphae grown in vitro.—Media rich in amino acids are required for growth of obligately parasitic haustorial fungi apart from host plants. The media commonly employed contain peptone, yeast extract, casein hydrolysate, or combinations thereof (20, 25, 50, 106, 121, 125). High concentrations (0.4–1.0%) of casein hydrolysate supported optimal growth of *Puccinia graminis* (20, 50) and a synthetic mixture of the 16 amino acids in the kinds and amounts found in a casein hydrolysate supported growth if glucose and a complete mineral solution were added (50). Preliminary trials indicate that *P. graminis* can grow when supplied aspartic acid and either cystine, cysteine, or glutathione (79) and that *Melampsora lini* can grow with alanine and methionine (25) (in combination with minerals and sugar). These results with rust fungi are closely paralleled by those obtained for *Tieghemiomyces parasiticus* and other haustorial mycoparasites in the laboratory of Barnett (7, 11). High concentrations of casein hydrolysate (1–4%) gave optimal growth. Cystein and methionine were required and the fungus was highly dependent on valine and leucine. It can be concluded tentatively that the haustorial obligate parasite requires a sulfur containing amino acid and one or more other amino acids, and that optimal growth occurs at concentrations of amino acids higher than are usually required by microorganisms.

The results with germlings, isolated hyphae, and in vitro cultures together suggest that the haustorium is needed for uptake or metabolism of amino acids in such a way that the hyphae are able to obtain or synthesize protein. Thus, the haustorium may provide a balanced, selective uptake of amino acids, perform metabolic conversions before the amino acids are transported to hyphae, or otherwise facilitate protein synthesis by the fungus.

For some parasites, the haustorium may also have a key role in carbohydrate uptake and metabolism. *Tieghemiomyces parasiticus* and several other haustorial mycoparasites do not grow on nonliving media when supplied glucose alone as the carbohydrate source, but do grow when supplied glycerol, or glucose combined with a surfactant (7, 11). Recent investigations by Fehrmann (47, 48) indicate that the oxidative metabolism of *Phytophthora infestans* is unusual in that it requires a high oxidation-reduction potential relative to that of cytochrome c. Oxidation of NADH was inhibited by small amounts of ATP, ADP, or AMP, or by substrate (NADH) in cell-free preparations from *P. infestans,* but not in preparations from saprophytic *P. erythroseptica.* Fehrmann regarded *P. infestans* as an ecologically obligate parasite, intermediate in metabolic capabilities between saprophytic species of *Phytophthora* and the obligately parasitic downy mildew fungi. Whether unusual NADH oxidation systems occur in other haustorial parasites is yet to be determined.

Epilogue

Several recently developed techniques will help bring our understanding of the haustorium more nearly in line with that of other types of fungal cells. Dekhuijzen (30, 32) developed a method of sieving haustoria from homogenates of mildewed cucumber leaves. The isolated haustoria were judged to be alive in that they reduced Janus green B and had a normal fine structure. Bushnell, Dueck, & Rowell (19) developed a technique for microsurgically removing host protoplasts from individual haustoria, although accumulation of extraneous cytoplasm on the haustorium has so far limited the usefulness of the procedure (18). Electron microscopy combined with high resolution radioautography (44, 76) will be an important tool for studying the movement of substances in and out of the haustorium. However, major progress will depend on the continued development of different and better techniques for study of the functioning haustorium. Dickinson's method of producing haustoria from leaf hyphae on membranes (34) would be especially useful if mycelium apart from a host could be grown on one side of a membrane while attached to haustoria on the other side.

Acknowledgments

For their generous help the writer wishes to thank P. G. Rothman, T. Kommedahl, and N. A. Anderson who critically reviewed the manuscript; L. J. Littlefield and C. E. Bracker who provided critical discussion of key concepts herein; and the several authors who provided manuscripts before publication, especially M. A. Ehrlich and H. G. Ehrlich. The writer also expresses appreciation to J. B. Rowell for help with this manuscript and for a decade of teaching and encouragement.

LITERATURE CITED

1. Ahmadjian, V. 1967. *The Lichen Symbiosis.* Waltham, Mass.: Blaisdell. 152 pp.
2. Akai, S., Kunoh, H., Fukutomi, M. 1968. Histochemical changes of the epidermal cell wall of barley leaves infected by *Erysiphe graminis hordei. Mycopathol. Mycol. Appl.* 35:175–80
3. Allen, R. F. 1923. A cytological study of infection of Baart and Kanred wheats by *Puccinia graminis tritici. J. Agr. Res.* 23: 131–52
4. Allen, R. F. 1932. A cytological study of heterothallism in *Puccinia coronata. J. Agr. Res.* 45: 513–41
5. Armentrout, V. N., Wilson, C. L. 1969. Haustorium-host interaction during mycoparasitism of *Mycotypha microspora* by *Piptocephalis virginiana. Phytopathology* 59:897–905
6. Barnett, H. L. 1963. The physiology of mycoparasitism. 65–90. *The Physiology of Fungi and Fungus Diseases.* West Virginia Agr. Exp. Sta. Bull. 488T. 106 pp.
7. Barnett, H. L. 1970. Nutritional requirements for axenic growth of some haustorial mycoparasites. *Mycologia* 62:750–60
8. Benjamin, R. K. 1959. The merosporangiferous Mucorales. *Aliso* 4:321–433
9. Benjamin, R. K. 1961. Addenda to the merosporangiferous Mucorales. *Aliso* 5:11–19
10. Berlin, J. D., Bowen, C. C. 1964. The host-parasite interface of *Albugo candida* on *Raphanus sativus. Am. J. Bot.* 51:445–52
11. Binder, F. L., Barnett, H. L. 1971. Nutrition and metabolism of the haustorial mycoparasite *Tieghemiomyces parasiticus* in axenic culture. *Phytopathology* 61:885
12. Blackwell, E. M. 1953. Haustoria of *Phytophthora infestans* and some other species. *Trans. Brit. Mycol. Soc.* 36:138–58
13. Boosalis, M. G. 1964. Hyperparasitism. *Ann. Rev. Phytopathol.* 2:363–76
14. Bracker, C. E. 1967. Ultrastructure of fungi. *Ann. Rev. Phytopathol.* 5:343–74
15. Bracker, C. E. 1968. Ultrastructure of the haustorial apparatus of *Erysiphe graminis* and its relationship to the epidermal cell of barley. *Phytopathology* 58: 12–30
16. Brian, P. W. 1967. Obligate parasitism in fungi. *Proc. Roy. Soc. Ser. B* 168:101–18

17. Burges, A. 1939. The defensive mechanism in orchid mycorrhiza. *New Phytol.* 38:273–83
18. Bushnell, W. R. 1971. The haustorium of *Erysiphe graminis:* An experimental study by light microscopy. *Morphological and Biochemical Events in Plant-Parasite Interaction,* ed. S. Akai, S. Ouchi, 229–54. Tokyo: Phytopathol. Soc. Japan. 415 pp.
19. Bushnell, W. R., Dueck, J., Rowell, J. B. 1967. Living haustoria and hyphae of *Erysiphe graminis* f. sp. *hordei* with intact and partly dissected host cells of *Hordeum vulgare. Can. J. Bot.* 45:1719–32
20. Bushnell, W. R., Rajendren, R. B. 1970. Casein hydrolysates and peptones for artificial culture of *Puccinia graminis* f. sp. *tritici. Phytopathology* 60:1287
21. Calonge, F. D. 1969. Ultrastructure of the haustoria or intracellular hyphae in four different fungi. *Arch. Mikrobiol.* 67: 209–25
22. Caporali, M. L. 1960. Sur la formation des sucoirs de *Sphaerotheca pannosa* (Wallr.) Lév. var. *rosae* dans les cellules épidermiques des folioles de *Rosa pouzini* Tratt. *C. R. Acad. Sci.* 250:2415–17
23. Chou, C. K. 1970. An electron-microscope study of host penetration and early stages of haustorium formation of *Peronospora parasitica* (Fr.) Tul. on cabbage cotyledons. *Ann. Bot.* 34:189–204
24. Cimanowski, J., Millikan, D. F. 1970. Resistance of apple to powdery mildew and its relationship to osmotic activity of the cell sap. *Phytopathology* 60: 1848–49
25. Coffey, M. D., Shaw, M. 1972. Nutritional studies with axenic cultures of the flax rust, *Melampsora lini. Physiol. Plant Pathol.* 2:37–46
26. Corner, E. J. H. 1935. Observations on resistance to powdery mildews. *New Phytol.* 34:180–200
27. Cruickshank, I. A. M. 1963. Phytoalexins. *Ann. Rev. Phytopathol.* 1:351–74
28. Davison, E. M. 1968. Cytochemistry and ultrastructure of hyphae and haustoria of *Peronospora parasitica* (Pers. ex Fr.) Fr. *Ann. Bot.* 32:613–21
29. De Bary, A. 1887. *Comparative Morphology and Biology of the Fungi, Mycetozoa, and Bacteria.* Transl. H. E. F. Garnsey. Rev. I. B. Balfour. Clarendon: Oxford. 525 pp.
30. Dekhuijzen, H. M. 1966. The isolation of haustoria from cucumber leaves infected with powdery mildew. *Neth. J. Plant Pathol.* 72:1–11
31. Dekhuijzen, H. M., Singh, H., Staples, R. C. 1967. Some properties of hyphae isolated from bean leaves infected with the bean rust fungus. *Contrib. Boyce Thompson Inst.* 23:367–72
32. Dekhuijzen, H. M., Van Der Scheer, C. 1969. The ultrastructure of powdery mildew, *Sphaerotheca fuliginea* isolated from cucumber leaves. *Neth. J. Plant Pathol.* 75:169–77
33. De Waard, M. A. 1971. Germination of powdery mildew conidia in vitro on cellulose membranes. *Neth. J. Plant Pathol.* 77:6–13
34. Dickinson, S. 1971. Studies in the physiology of obligate parasitism. VIII. An analysis of fungal responses to thigmotropic stimuli. *Phytopathol. Z.* 70:62–70
35. Dörr, I. 1969. Feinstruktur intrazellular wachsender Cuscuta-Hyphen. *Protoplasma* 67:123–37
36. Dörr, I., Kollmann, R. 1969. Fine structure of mycorrhiza in *Neottia nidus-avis* (L.) L. C. Rich. (Orchidaceae). *Planta* 89:372–75
37. Dunkle, L. D., Allen, P. J. 1971. Infection structure differentiation by wheat stem rust uredospores in suspension. *Phytopathology* 61:649–52
38. Durbin, R. D. 1967. Obligate parasites: effect on the movement of solutes and water. *The Dynamic Role of Molecular Constituents in Plant-Parasite Interaction.* ed. C. J. Mirocha, I. Uritani, 80–99. St. Paul: Am. Phytopathol. Soc. 372 pp.
39. Edwards, H. H. 1970. A basic staining material associated with the penetration process in

resistant and susceptible powdery mildewed barley. *New Phytol.* 69:299–301
40. Edwards, H. H., Allen, P. J. 1970. A fine-structure study of the primary infection process during infection of barley by *Erysiphe graminis* f. sp. *hordei. Phytopathology* 60:1504–09
41. Ehrlich, H. G., Ehrlich, M. A. 1963. Electron microscopy of the host-parasite relationships in stem rust of wheat. *Am. J. Bot.* 50:123–30
42. Ehrlich, M. A., Ehrlich, H. G. 1966. Ultrastructure of the hyphae and haustoria of *Phytophthora infestans* and hyphae of *P. parasitica. Can. J. Bot.* 44: 1495–1503
43. Ehrlich, M. A., Ehrlich, H. G. 1970. Electron microscope radioautography of ^{14}C transfer from rust uredospores to wheat host cells. *Phytopathology* 60:1850–51
44. Ehrlich, M. A., Ehrlich, H. G. 1971. Fine structure of the host-parasite interfaces in mycoparasitism. *Ann. Rev. Phytopathol.* 9:155–84
45. Ehrlich, M. A., Schafer, J. F., Ehrlich, H. G. 1968. Lomasomes in wheat leaves infected by *Puccinia graminis* and *P. recondita. Can. J. Bot.* 46:17–20
46. Ellingboe, A. H. 1968. Inoculum production and infection by foliage pathogens. *Ann. Rev. Phytopathol.* 6:317–30
47. Fehrmann, H. 1971. Obligater Parasitismus phytopathogener Pilze I. Vergleichende Wachstumsversuche mit *Phytophthora infestans* und verwandten Arten. *Phytopathol. Z.* 70:89–113
48. Fehrmann, H. 1971. Obligater Parasitismus phytopathogener Pilze: II. Vergleichende biochemische Untersuchungen zum Energiestoffwechsel von *Phytophthora*-Arten. *Phytopathol. Z.* 70:230–62
49. Flor, H. H. 1971. Current status of the gene-for-gene concept. *Ann. Rev. Phytopathol.* 9:275–96
50. Foudin, A. S. 1971. *Determination of a Defined Medium for In Vitro Growth of Puccinia graminis tritici.* M.S. Thesis. Univ. of Georgia. 40 pp.
51. Fraymouth, J. 1956. Haustoria of the Peronosporales. *Trans. Brit. Mycol. Soc.* 39:79–107
52. Fullerton, R. A. 1970. An electron microscope study of the intracellular hyphae of some smut fungi (Ustilaginales). *Aust. J. Bot.* 18:285–92
53. Galun, M., Paran, N., Ben-Shaul, Y. 1970. The fungus-alga association in the Lecanoraceae: An ultrastructural study. *New Phytol.* 69:599–603
54. Galun, M., Paran, N., Ben-Shaul, Y. 1970. An ultrastructural study of the fungus alga association in *Lecanora radiosa* growing under different environmental conditions. *J. Microscopie* 9:801–06
55. Gäumann, E. 1950. *Principles of Plant Infection.* Trans. W. B. Brierley. New York: Hafner. 543 pp.
56. Geitler, L. 1955. Gehäufte Haustorien bei einer Collematacee. *Öster. Bot. Z.* 102:317–21
57. Geitler, L. 1963. Über Haustorien bei Flechten und über *Myrmecia Biatorellae in Psora globifera. Öster. Bot. Z.* 110:270–80
58. Gerdemann, J. W. 1968. Vesicular-arbuscular mycorrhiza and plant growth. *Ann. Rev. Phytopathol.* 6:397–418
59. Gibson, C. M. 1904. Notes on infection experiments with various uredineae. *New Phytol.* 3: 184–91
60. Hadley, G., Johnson, R. P. C., John, D. A. 1971. Fine structure of the host-fungus interface in orchid mycorrhiza. *Planta* 100:191–99
61. Hanchey, P., Wheeler, H. 1971. Pathological changes in ultrastructure: tobacco roots infected with *Phytophthora parasitica* var. *nicotianae. Phytopathology* 61:33–39
62. Hanes, T. B. 1936. Observations on the results of inoculating cereals with the spores of cereal rusts which do not usually cause their infection. *Trans. Brit. Mycol. Soc.* 20:252–92
63. Hardwick, N. V., Greenwood, A. D., Wood, R. K. S. 1971. The fine structure of the haustorium of *Uromyces appendiculatus* in

Phaseolus vulgaris. Can. J. Bot. 49:383–90
64. Harley, J. L. 1969. *The Biology of Mycorrhiza.* London: Leonard Hill. 2nd ed. 334 pp.
65. Harvey, A. E., Grasham, J. L. 1970. Growth of *Cronartium ribicola* in the absence of physical contact with its host. *Can. J. Bot.* 48:71–73
66. Hasija, S. K., Miller, C. E. 1970. Pure culture studies of some Chytridiomycetes. *Mycologia* 62:1032–40
67. Heath, M. C. 1971. Haustorial sheath formation in cowpea leaves immune to rust infection. *Phytopathology* 61:383–88
68. Heath, M. C. 1972. Ultrastructure of host and non-host reactions to cowpea rust. *Phytopathology* 62:27–38
69. Heath, M. C., Heath, I. B. 1971. Ultrastructure of an immune and a susceptible reaction of cowpea leaves to rust infection. *Physiol. Plant Pathol.* 1:277–87
70. Hilu, H. M. 1965. Host-pathogen relationships of *Puccinia sorghi* in nearly isogenic resistant and susceptible seedling corn. *Phytopathology* 55:563–69
71. Hirata, K. 1967. Notes on haustoria, hyphae, and conidia of the powdery mildew fungus of barley, *Erysiphe graminis* f. sp. *hordei. Mem. Fac. Agr. Niigata Univ.* 6:207–59
72. Hirata, K. 1971. Calcium in relation to the susceptibility of primary barley leaves to powdery mildew. See Ref. 18, 207–28
73. Hirata, K., Kojima, M. 1962. On the structure and the sack of the haustorium of some powdery mildews, with some considerations on the significance of the sack. *Trans. Mycol. Soc. Japan* 3:43–46
74. Jacobs, J. B., Ahmadjian, V. 1969. The ultrastructure of lichens. I. A general survey. *J. Phycol.* 5: 227–40
75. Jacobs, J. B., Ahmadjian, V. 1971. The ultrastructure of lichens. II. *Cladonia cristatella:* The lichen and its symbionts. *J. Phycol.* 7: 71–82
76. Jacobs, J. B., Ahmadjian, V. 1971. The ultrastructure of lichens IV. Movement of carbon products from alga to fungus as demonstrated by high resolution radioautography. *New Phytol.* 70:47–50
77. Kajiwara, T. 1971. Structure and physiology of haustoria of various parasites. See Ref. 18, 255–77
78. Karling, J. S. 1932. Studies in the Chytridiales VII. The organization of the chytrid thallus. *Am. J. Bot.* 19:41–74
79. Kuhl, J. L., Maclean, D. J., Scott, K. J., Williams, P. G. 1971. The axenic culture of Puccinia species from uredospores: experiments on nutrition and variation. *Can. J. Bot.* 49:201–09
80. Kunoh, H., Akai, S. 1969. Histochemical observation of the halo on the epidermal cell wall of barley leaves attacked by *Erysiphe graminis hordei. Mycopathol. Mycol. Appl.* 37:113–18
81. Leath, K. T., Rowell, J. B. 1966. Histological study of the resistance of *Zea mays* to *Puccinia graminis. Phytopathology* 56: 1305–09
82. Leath, K. T., Rowell, J. B. 1970. Nutritional and inhibitory factors in the resistance of *Zea mays* to *Puccinia graminis. Phytopathology* 60:1097–1100
83. Littlefield, L. J., Aronson, S. J. 1969. Histological studies of *Melampsora lini* resistance in flax. *Can. J. Bot.* 47:1713–17
84. Littlefield, L. J., Bracker, C. E. 1970. Continuity of host plasma membrane around haustoria of *Melampsora lini. Mycologia* 62: 609–14
85. Littlefield, L. J., Bracker, C. E. 1972. Ultrastructural specialization at the host-pathogen interface in rust-infected flax. *Protoplasma:* In press
86. Lupton, F. G. H. 1956. Resistance mechanisms of species of *Triticum* and *Aegilops* and of amphidiploids between them, to *Erysiphe graminis* DC. *Trans. Brit. Mycol. Soc.* 39:51–59
87. Manocha, M. S., Lee, K. Y. 1971. Host-parasite relations in mycoparasite. I. Fine structure of host, parasite, and their interface. *Can. J. Bot.* 49:1677–81
88. McKeen, W. E., Smith, R., Bhattacharya, P. K. 1969. Alterations of the host wall surround-

ing the infection peg of powdery mildew fungi. *Can. J. Bot.* 47:701–06

89. McKeen, W. E., Smith, R., Mitchell, N. 1966. The haustorium of *Erysiphe cichoracearum* and the host-parasite interface on *Helianthus annuus. Can. J. Bot.* 44:1299–1306
90. Melander, L. W., Craigie, J. H. 1927. Nature of resistance of *Berberis* spp. to *Puccinia graminis. Phytopathology* 17:95–114
91. Meyer, F. H. 1966. Mycorrhiza and other plant symbioses. *Symbiosis,* ed. S. M. Henry, Vol. 1, 171–255. New York: Academic. 478 pp.
92. Mitchell, D., Shaw, M. 1968. Metabolism of glucose-^{14}C, pyruvate-^{14}C, and mannitol-^{14}C by *Melampsora lini.* II. Conversion to soluble products. *Can J. Bot.* 46:453–60
93. Moss, E. H. 1926. The uredo stage of Pucciniastrease. *Ann. Bot.* 40:813–47
94. Mount, M. S., Ellingboe, A. H. 1969. ^{32}P and ^{35}S transfer from susceptible wheat to *Erysiphe graminis* f. sp. *tritici* during primary infection. *Phytopathology* 59:235
95. Nims, R. C., Halliwell, R. S., Rosberg, D. W. 1967. Wound healing in cultured tobacco cells following microinjection. *Protoplasma* 64:305–14
96. Ogle, H. J., Brown, J. F. 1971. Quantitative studies of the post-penetration phase of infection by *Puccinia graminis tritici. Ann. Appl. Biol.* 67:309–19
97. Peveling, E. 1968. Elektronenoptische Untersuchungen an Flechten. I. Strukturveränderungen der Algenzellen von *Lecanora muralis* (Schreber) Rabenh. beim Eindringen von Pilzhyphen. *Z. Pflanzenphysiol.* 59:172–83
98. Plessel, A. 1963. Über die Beziehungen von Haustorientypus und Organisationshöhe bei Flechten. *Öster. Bot. Z.* 110:194–269
99. Reisener, H. J., Ziegler, E. 1970. Über den Stoffwechsel des parasitischen Mycels und dessen Beziehungen zum Wirt bei *Puccinia graminis* auf Weizen. *Angew. Bot.* 44:343–46
100. Rice, M. A. 1927. The haustoria of certain rusts and the relation between host and pathogene. *Bull. Torrey Bot. Club* 54:63–153
101. Rice, M. A. 1935. The cytology of host-parasite relations. *Bot. Rev.* 1:327–54
102. Richards, A. G., Smith, M. N. 1956. Infection of cockroaches with *Herpomyces* (Laboulbeniales) II. Histology and histopathology. *Ann. Entomol. Soc. Am.* 49:85–93
103. Roskin, P. 1970. Ultrastructure of the host-parasite interaction in the basidiolichen *Cora pavonia* (Web.) E. Fries. *Arch. Mikrobiol.* 70:176–82
104. Rothman, P. G. 1960. Host-parasite interactions of eight varieties of oats infected with race 202 of *Puccinia coronata* var. *avenae. Phytopathology* 50: 914–18
105. Schnathorst, W. C. 1959. Resistance in lettuce to powdery mildew related to osmotic value. *Phytopathology* 49:562–71
106. Scott, K. J., Maclean, D. J. 1969. Culturing of rust fungi. *Ann. Rev. Phytopathol.* 7:123–46
107. Seymour, R. L. 1971. Studies on mycoparasitic chytrids. I. The genus *Septosperma. Mycologia* 63:83–93
108. Shaw, M. 1967. Cell biological aspects of host-parasite relations of obligate fungal parasites. *Can. J. Bot.* 45:1205–20
109. Shaw, M., Manocha, M. S. 1965. The physiology of host-parasite relations. XV. Fine structure in rust-infected wheat leaves. *Can. J. Bot.* 43:1285–92
110. Slesinski, R. S., Ellingboe, A. H. 1969. The genetic control of primary infection of wheat by *Erysiphe graminis* f. sp. *tritici. Phytopathology* 59:1833–37
111. Slesinski, R. S., Ellingboe, A. H. 1971. Transfer of ^{35}S from wheat to the powdery mildew fungus with compatible and incompatible parasite/host genotypes. *Can. J. Bot.* 49:303–10
112. Smith, D., Muscatine, L., Lewis, D. 1969. Carbohydrate movement from autotrophs to heterotrophs in parasitic and mutualistic symbiosis. *Biol. Rev.* 44: 17–90
113. Smith, G. 1900. The haustoria of

the Erysipheae. *Bot. Gaz.* 29: 153–84

114. Sood, P. N., Sackston, W. E. 1970. Studies on sunflower rust. VI. Penetration and infection of sunflowers susceptible and resistant to *Puccinia helianthi* race I. *Can. J. Bot.* 48:2179–81
115. Sparrow, F. K. 1960. *Aquatic Phycomycetes.* 2nd ed. Ann Arbor: Univ. Mich. 1187 pp.
116. Stanbridge, B., Gay, J. L., Wood, R. K. S. 1971. Gross and fine structural changes in *Erysiphe graminis* and barley before and during infection. *Ecology of Leaf Surface Microorganisms,* ed. T. F. Preece, C. H. Dickenson, 367–79. London & New York: Academic
117. Staples, R. C., Yaniv, Z., Ramakrishnan, L., Lipetz, J. 1971. Properties of ribosomes from germinating uredospores. See Ref. 18, 59–90
118. Stavely, J. R., Hanson, E. W. 1966. Some basic differences in the reactions of resistant and susceptible *Trifolium pratense* to *Erysiphe polygoni. Phytopathology* 56:957–62
119. Stavely, J. R., Pillai, A., Hanson, E. W. 1969. Electron microscopy of the development of *Erysiphe polygoni* in resistant and susceptible *Trifolium pratense. Phytopathology* 59:1688–93
120. Thatcher, F. S. 1943. Cellular changes in relation to rust resistance. *Can. J. Res. Sect. C.* 21: 151–72
121. Tiwari, M. M., Arya, H. C. 1969. *Sclerospora graminicola* axenic culture. *Science* 163:291–93
122. Tomiyama, K. 1971. Cytological and biochemical studies of the hypersensitive reaction of potato cells to *Phytophthora infestans.* See Ref. 18, 387–401
123. Tschen, J., Fuchs, W. H. 1968. Endogene Aktivität der Enzyme in rostinfizierten Bohnenprimärblättern. *Phytopathol. Z.* 63: 187–92
124. Tschermak, E. 1941. Untersuchungen über die Beziehungen von Pilz und Alge im Flechtenthallus. *Öster. Bot. Z.* 90:233–307
125. Turel, F. L. M. 1969. Saprophytic development of the flax rust *Melampsora lini,* race No. 3. *Can. J. Bot.* 47:821–23
126. Van Dyke, C. G., Hooker, A. L. 1969. Ultrastructure of host and parasite in interactions of *Zea mays* with *Puccinia sorghi. Phytopathology* 59:1934–46
127. Van Sumere, C. F., Van Sumere-De Preter, C., Ledingham, G. A. 1957. Cell-wall-splitting enzymes of *Puccinia graminis* var. *tritici. Can. J. Microbiol.* 3: 761–70
128. Varns, J. L., Kuć, J. 1971. Suppression of rishitin and phytuberin accumulation and hypersensitive response in potato by compatible races of *Phytophthora infestans. Phytopathology* 61:178–81
129. Webber, M. M., Webber, P. J. 1970. Ultrastructure of lichen haustoria: symbiosis in *Parmelia sulcata. Can. J. Bot.* 48: 1521–24
130. Whiffen, A. J. 1941. Cellulose decomposition by the saprophytic chytrids. *J. Elisha Mitchell Sci. Soc.* 57:321–30
131. Whiffen, A. J. 1949. Two new chytrid genera. *Mycologia* 34: 543–57
132. White, N. H., Baker, E. P. 1954. Host pathogen relations in powdery mildew of barley. 1. Histology of tissue reactions. *Phytopathology* 44:657–62
133. Whitney, H. S., Shaw, M., Naylor, J. M. 1962. The physiology of host-parasite relations. XII. A cytophotometric study of the distribution of DNA and RNA in rust-infected leaves. *Can. J. Bot.* 40:1533–44
134. Williams, P. G. 1969. Haustoria-like branches in axenic cultures of *Puccinia graminis* f. sp. *tritici. Can. J. Bot.* 47:1816–17
135. Williams, P. G., Shaw, M. 1968. Metabolism of glucose-^{14}C, pyruvate-^{14}C, and mannitol-^{14}C by *Melampsora lini.* I. Uptake. *Can. J. Bot.* 46:435–40
136. Zimmer, D. E. 1965. Rust infection and histological response of susceptible and resistant safflower. *Phytopathology* 55: 296–301
137. Zimmer, D. E. 1970. Fine structure of *Puccinia carthami* and the ultrastructural nature of exclusionary seedling-rust resistance of safflower. *Phytopathology* 60:1157–63

LOCALIZATION AND INDUCED RESISTANCE IN VIRUS-INFECTED PLANTS[1,2]

3548

G. LOEBENSTEIN
The Volcani Institute of Agricultural Research, Bet Dagan, Israel

The interaction between two populations—that of the virus and that of the cells of the organized host—may vary between two extremes, from complete susceptibility to immunity. Within this range, various defense phenomena, resulting in different levels of resistance or tolerance, are known. In an immune plant, the virus is apparently unable to multiply, while in other cases the plant may be resistant to infection, i.e., to the introduction of the virus into the susceptible cell. This resistance to infection was found in tobacco T.I. 245 to be correlated with the presence of fewer ectodesmata in the epidermis (115). Partial resistance (tolerance), whereby virus multiplication is suppressed but not excluded, is known in many virus-host interactions (37), *inter alia* in CMV-resistant cucumber varieties. Plants infected with certain viruses, e.g. TRSV may "recover" from the acute disease symptoms and develop resistance to reinfection—*apparent recovery*. The restriction of virus movement from the local lesion to adjacent tissues in hypersensitive hosts—*localization*—is one of the most efficient resistance mechanisms. The tissues surrounding a necrotic lesion, as well as other noninfected parts of such a hypersensitive host, may also become resistant to reinfection—*localized and systemic induced (acquired) resistance.*

Although these resistance phenomena have been well known for many years their mechanism and physiology are not yet fully understood. Some defenses are considered to be of a passive kind prepared in advance of infection; e.g. mechanical barriers or lack of compatible infection sites may result in resistance to infection. Other resistance phenomena require the active metabolism of the host cell, and evidence is growing that some defense mechanisms become activated only after infection (76).

[1] Contribution from the Volcani Institute of Agricultural Research, Bet Dagan, Israel. 1971 Series; No. 1997-E.

[2] The following abbreviations are used: AMV (alfalfa mosaic virus), AVF (antiviral factor), CarMV (carnation mosaic virus), CMV (cucumber mosaic virus), HMP (hexose monophosphate), PPO (polyphenol oxidase), PVX (potato X virus), PVY (potato Y virus), SBMV (southern bean mosaic virus), TMV (tobacco mosaic virus), TNV (tobacco necrosis virus), TRSV (tobacco ringspot virus), TomRSV (tomato ringspot virus), TSW (tomato spotted wilt virus).

This review will be centered around localization, induced resistance, and the possible development of antiviral or interfering agents in the resistant tissue. The term 'induced resistance' will be used to describe (*a*) resistance developing in noninvaded tissue after viral or fungal infection of other parts of the plants, and (*b*) the phenomena whereby nonmultiplying subtances interfere via a host-mediated process, requiring the transcription mechanism of the cell, with the multiplication of a virus. Approaches that activate or suppress the defense reaction will be emphasized, in the hope of thus obtaining a better understanding of the mechanism.

A review on localization and induced resistance, as such, has not come to the author's attention, but these subjects have been treated in various reviews on related topics (see 16, 42, 68, 69, 86, 87).

Localization

Symptoms.—In a localized infection the virus moves into and multiplies in only a small group of cells around the point of entry, without further invasion of the plant. This is often associated with a strong response of the host, resulting in the death of the tissue in the immediate vicinity of the first infected cells, producing conspicuous *necrotic local lesions*. These may vary in color; e.g. dull red in cowpea infected by CMV, brown in *Nicotiana glutinosa* infected by TMV, and greyish with a red anthocyanin margin in PVX-infected *Gomphrena globosa*. In other localized responses, infected cells may lose chlorophyll and other pigments, giving rise to *chlorotic local lesions, i.e., Chenopodium amaranticolor* infected by PVY. Occasionally, ring-like patterns or ring spots that remain localized are observed, *e.g. Beta vulgaris* infected either by TSW or AMV, and *Tetragonia expansa* infected by TSW. Some localized infections produce no obvious lesions on the intact leaf, but when leaves are decolorized with ethanol, and stained with iodine, *starch lesions* become apparent. To obtain well-defined lesions in cucumber cotyledons or leaves, the latter are harvested 5–7 days after inoculation with TMV and stored for 24 hours in the dark in a moist chamber. Generally, plants must be kept in the dark long enough to free the uninfected but not the infected cells of starch, resulting in a better contrast between lesions and the rest of the leaf. Chlorophyll is then removed by heating in 70% ethanol at 80°C. Starch lesions are developed by placing the cotyledons or leaves in a solution of iodine in potassium iodide containing lactic acid (50), the latter apparently aiding in penetration of the iodine.

Certain host-virus interactions may result in microlesions not detectable macroscopically. Thus, under certain experimental conditions, the U_2 strain of TMV forms, on Pinto bean leaves, mainly necrotic microlesions, with a mean size of $1.1 \times 10^{-2}mm^2$. The U_1 strain, for comparison, causes under the same conditions macrolesions with a mean size of $5.85 \times 10^{-2}mm^2$ (31). The microlesions can be detected on a green leaf with transmitted light at magnifications higher than 1.33 ×, or as necrotic starch lesions on bleached iodine-stained leaves, at a magnification of 10 ×.

Infections may remain localized without showing any symptoms, e.g. both TMV and tobacco etch virus multiply locally in leaves of many plants without moving systemically (35). Subliminal symptomless infections not detectable as starch lesions, as in TMV infected cotton cotyledons, and where only limited virus synthesis occurs, may also be due to localization of virus (11).

Lesion growth and virus content.—Lesions may be self limiting, i.e., reach maximum size and then stop growing, or be of a nonlimited type. TMV is self-limiting in *Datura stramonium,* where lesions reach their maximum size 3 days after inoculation, whereas in beans the lesions caused by most strains of TNV continue to grow indefinitely. The radius of local lesions incited by TMV in *Nicotiana glutinosa,* kept at 25°C, is a function of time. Each of three strains examined (90) had its own characteristic spreading rate, with a range of about 0.13–0.33 mm per day. It is generally assumed that lesion size is correlated with extractable infectious material. Thus, the U_1 and U_2 strains of TMV yield about 10^{-5} mg TMV per mm^2 of lesion area in *N. glutinosa.* However, with the U_8 strain a decrease in infectivity was associated with larger lesion areas.

Studies of a developing local necrotic lesion in *Chenopodium amaranticolor* by electron microscopy revealed new virus particles in a few palisade cells 30 hours after heavy inoculation with TMV, though no necrosis was apparent until 15 hours later (72). In infections 50 hours old, a great number of particles in crystalline arrays was found in about one-third of the cells, but interspersed with these were other cells in which little or no virus material was detected. A similar pattern was found in *N. glutinosa,* though far fewer cells—only about 2%—harbor visible virus and much less is found in each cell (73). This discontinuous distribution of accumulated virus from cell to cell in local lesion hosts is distinct from the rather uniform distribution in a systemic host (104). Milne (73) suggested that this uneven distribution and the apparent resistance of many cells to virus multiplication (though not necrosis) may be connected with the confinement of the infection to the lesion.

The estimated number of TMV particles per cell, averaged over all cells in the infected area of *N. glutinosa,* is about 10^3 (73), 2–4 orders of magnitude lower than in a comparable systemic infection. An estimate of TMV particles per tobacco mesophyll cell 2 weeks after inoculation was 10^5 to 10^6, according to Harrison (27); Nixon reached a higher estimate of 6×10^7 particles in a single hair cell of tobacco (82). The number of TMV particles observed in cells along the periphery of a local lesion in *D. stramonium* is apparently much higher, forming relatively large aggregates (8).

Ultrastructural changes.—The cytopathic changes leading to collapse and necrosis of cells in a local lesion have been reviewed (68, 87). In brief, these start with a swelling of starch grains and distortion of the chlorophyll lamella 8 hours after infection (123). About 16 hours later, disintegration of cytoplasmic membranes, accompanied by a two- to threefold increase in the number of mitochondria, is observed. After that, chloroplasts disintegrate before necrosis takes place. At this time the nucleus still preserves its integrity. About 78 hours after inoculation the cells collapse and their walls tear away from adjacent cells. Ragetli (87) suggested that this disorganization of the cell is indicative of uncontrolled activity of hydrolases as, for example, activation of structure-bound latent phosphatases in chloroplasts (89). Although Ragetli suggested that this activation is a possible cause of chloroplast degradation, it could well be that the disruption of cell organelles results in enzyme release.

Around the core of necrotic cells of the lesion in the hypersensitive tobacco cv. Samsun NN, there is a zone of cells that stain darkly (39). Seven days after inoculation with TMV, the mesophyll cells surrounding the lesion showed structural features suggestive of metabolic activity. Compared with normal mesophyll cells, cells within this zone had smaller vacuoles, more cytoplasm and ribosomes, and a marked system of endoplasmic reticulum. The chloroplasts in this zone were more ameboid and had more ribosomes than those in normal cells, and were invested with large starch grains and lipid globules. Crystal-containing spherosomes were observed, and it was suggested these were identified with peroxisomes (95). These changes were most marked in cells nearest the lesion edge, but were evident also in cells about 10 cell diameters beyond the lesion, graduating into normal-appearing tissue. Apparently this zone of active cells is associated with the localizing mechanism.

In Samsun NN, Israel & Ross (39) visualized this zone as a cylinder of active metabolic cells, encircling a network of collapsed and disorganized cells, but in *N. glutinosa* the peripheral tissue was found to consist of necrotic cells intermingled with healthy ones and others showing various levels of disorganization (28). This region in *N. glutinosa* apparently also has enhanced metabolic activity, as numerous mitochondria were observed in the peripheral region of cells adjacent to a necrotic cell.

Membrane-bound vesicular bodies were observed in the surrounding uninfected zone of a TMV lesion in Pinto leaves between the plasmalemma and cell wall. Plasmodesmata in the cell wall adjacent to these paramural bodies appeared to be broken (111).

Ultrastructural changes have so far been investigated mainly in necrotic or chlororotic (43) local lesion hosts. It may be worthwhile to study the ultrastructure in plants where the infection remains localized without causing necrosis, as in a starch-lesion host or in a symptomless reaction. The observed changes in such a host could be more convincingly interpreted as related to the localizing process, without having to consider their possible association with the necrotic reaction.

Respiration.—The appearance of necrotic local lesions is fairly consistently accompanied by an enhancement of host respiration (70). The increase in oxygen uptake is proportional to lesion number, except at highest lesion densities, and seems generally to be related to the development of necrosis rather than to virus increase (84). The green tissue between necrotic local lesions also respires at a greater rate than healthy controls, but copper sulfate-induced necrosis does not lead to an increase in respiration (121).

As mentioned above, an increase in the number of mitochondria has been observed in both necrotic and peripheral cells (28, 123). However, Pierpoint did not find an increase in mitochondrial material accompanying the infection of *N. glutinosa* with TMV. Extracts from infected tissue contained similar amounts of mitochondrial protein N, slightly less cytochrome oxidase, and considerably more polyphenol oxidase. He suggested that the respiratory rise is not based on mitochondrial respiration, but on some terminal oxidase, other than cytochrome oxidase (85).

Activity of enzymes of the HMP pathway, in particular glucose-6-phosphate dehydrogenase and phosphogluconate dehydrogenase, were increased in the zone of yellow tissue surrounding TMV necrotic local lesions in *N. tabacum* (110). Hexokinase and glucose phosphate isomerase, enzymes of the glycolytic pathway, showed unaltered activity in local lesion hosts. That increased respiration in tissues where local necrotic lesions develop is at least partly due to the activation of the HMP shunt, is also supported by the lowering of the C_6/C_1 ratio observed in several local lesion hosts, but not in systemic infections. In beans infected with SBMV, the C_6/C_1 ratio decreased sharply when local lesions appeared, suggesting that more glucose was metabolized by the HMP pathway (4). A lowering of the C_6/C_1 ratio was also observed in *D. stramonium* infected with TMV (54). However, in Xanthi tobacco no change in the C_6/C_1 ratio was observed after infection with TMV, although the overall glucose metabolism increased (71).

A marked increase in the activity of polyphenol oxidase (PPO), as o-diphenol oxidase, has been reported for necrotic lesion-infected hosts (17, 84). PPO increased markedly about one day after lesions appeared in *N. glutinosa* or *D. stramonium* infected with TMV, and in *Vigna sinensis* and *N. glutinosa* infected by CMV and TNV, respectively. This was accompanied by a decrease in o-dihydric phenols, mainly chlorogenic acid. During the same period, PPO in systemic hosts was stimulated to a lesser extent, if at all (17). It has been suggested (17) that quinones, which accumulate simultaneously, cause necrosis and cell death, because feeding of ascorbic acid and other reducing compounds that hold phenols in a reduced state, inhibit lesion formation. In TMV-infected *N. glutinosa,* which did not develop necrosis when kept at 36°C, PPO was only slightly stimulated. This, and the accelerated appearance of lesions in leaves treated with polyphenols (84), may also be interpreted as additional evidence that oxidation of phenols to quinones causes necrosis. However, in Xanthi tobaccos, where PPO was activated before inoculation by stem cutting, TMV lesions appeared one day later than in control plants (40). Stem cutting activated PPO equally well at temperatures between 23° and 35°C, though plants kept at higher temperature did not develop necrosis. PPO activation may therefore be the result of decompartmentalization or injury and not the cause of cell death. Furthermore, treatment with ascorbic acid—which prevents necrosis without inhibiting PPO—does not lead to systemic movement of the virus from the infection center (17, 84), and the observed change in ascorbate concentration occurs after localization and lesion appearance (74). This, and other evidence discussed below, does not support the idea that increased PPO activity is responsible for necrosis and localization of the infection.

Necrotic lesion formation is accompanied by a marked increase in peroxidase activity, more so than in systemic infections (2, 55). This increase, apparent after the appearance of symptoms, seems to be caused by a quantitative change of isozymes (83), including those associated with senescence (2), and not by a new isozyme associated with necrobiosis (18). The precise function of peroxidase in infected plants is still unclear. It has been associated with lignification (118), as an oxidase of aromatic compounds produced via the HMP pathway (55), or said to promote reduced nicotinamide adenine dinucleotide (NADH) oxidation (96). It may also be that peroxidases bound to mature cell walls (41) are released during necrobiosis. Recently it was suggested that high peroxidase activity leads to early killing of infected cells, which in turn leads to enhanced changes in advance of infection forming a barrier to virus spread (108).

Whether increased respiration in local necrotic lesion hosts is due to a change in enzyme concentration or activity is open to question. As summarized by Merrett & Bayley (70), rate-limiting steps, i.e., increased concentration of adenine nucleotides and enhanced ATP turnover, may be the major cause of increased oxygen uptake.

Ethylene.—Production of ethylene was found to be enhanced concurrently with necrotic local lesion development, but not with systemic infections (80); the increase was proportional to the number and size of lesions, i.e., with necrotization of the tissue. Exogenously introduced ethylene caused a small increase of TMV multiplication in tobacco and lesion numbers on beans though it slightly inhibited peroxidase activity. In cucumber and cotton cotyledons, a starch lesion and subliminal host respectively, application of Ethrel markedly enhanced TMV content (12).

Accumulation of metabolites.—By feeding TMV-inoculated leaves of *N. glutinosa* and *D. stramonium* with $C^{14}O_2$ and preparing autoradiographs of the leaves, Thrower (116) showed that the young lesions acted as a "sink" for assimilated substances. The radius of the radioactive spot increased for 60–80 hours after appearance of the lesion; in *D. stramonium*, it was larger than the necrotic lesion.

Various fluorescent compounds accumulate around the necrotic lesion. Scopoletin can be detected in TMV-infected *N. glutinosa* before lesion formation (5). In bean or cowpea plants inoculated with TMV or TNV, the appearance of abnormal fluorescent compounds, some of a phenolic nature, coincided with incipient lesion formation (23). This is apparently a response to necrotization, nonspecific as to pathogen, because similar compounds accumulate also in rust-infected beans (23) and in tobacco with necrotic lesions caused by *Colletotrichum destructivum* (132).

Deposition of calcium in the walls of cells surrounding a TMV lesion for a radius of about 50 cells has been observed in *N. glutinosa*. These cells contained predominantly insoluble calcium pectate in the middle lamella, in contrast to healthy tissue which contained primarily pectic acid (122). The question was raised whether this change plays some part in localizing the infection, but unfortunately this approach did not receive further attention.

Recently, intense yellow fluorescence was observed around necrotic lesions produced by U_2 TMV in Pinto bean, stained with aniline blue, indicative of callose (127, 128). Wu et al observed a rapid deposition of callose in the walls of living epidermal and spongy parenchyma cells adjacent to the necrotic cells, but none in the palisade parenchyma. In *N. glutinosa,* where TMV lesions increase with time (90), little callose is deposited and that in only a small number of cells. It was suggested that callose forms a barrier that seals off the plasmodesmata between infected and noninfected cells, thereby preventing virus movement. However, it may be that callose is a response to necrotization and not necessarily responsible for localization, because in TMV-infected cucumber cotyledons, where infection is localized without necrotization, no callose deposition was observed (53). Furthermore, there is no evidence that yellow fluorescence after aniline blue staining is specific only for callose.

Genetics.—The genetic mechanism that controls localization is determined in many cases by one gene (36). Thus, the dominant N and L genes in *N. glutinosa* and Tobasco peppers, respectively, confer hypersensitivity to infection by TMV, and a single dominant gene in beans controls local lesion formation by SBMV. Inheritance of local lesion response to PVX in crosses between two species of *Capsicum* appears to depend on the female parent, indicating cytoplasmic control (78). However, as no data on reciprocal crosses were provided, other explanations cannot be ruled out (R. Frankel, personal communication).

Suppression of localization.—Treatments that suppress localization, resulting in enhanced virus multiplication and invasion of tissues outside the lesion, have been studied in several laboratories, thereby providing additional information on the localizing mechanism.

Heat.—Breakdown of localization at high temperatures and renewed spread of virus, leading to systemic infection, have been known since 1931. Samuel (97) pointed out that with increasing temperatures TMV lesions on *N. glutinosa* increase in size; above 28°C they tend to coalesce and at 35°C no necrosis occurs on the inoculated leaves, but systemic infection does. Rapid necrosis of the infected tissue occurs when such plants are transferred again to 21°C. Enlargement of lesions with increasing temperature is correlated with virus multiplication (20), and at 32°C numerous virus particles are observed (67). Heat activation causing larger lesions (129) seems also to be responsible for enlargement of very small microlesions (30). By exposing Samsun NN plants for 24 hours to 32°C at different days after inoculation, it was shown that lesions are most sensitive to high temperatures 48 hours after inoculation (95). At this time infection was well established but necrosis had not yet developed, and apparently localization was not yet complete. Exposing plants to 32°C at later time intervals after inoculation resulted in only slight increases in lesion size.

Brief immersions of inoculated leaves in water at 50°C also increased the size of TMV lesions in cucumber cotyledons and Pinto bean leaves (127, 129). Studies with Pinto beans showed that the increase in lesion size is associated with a proportional increase in infectivity per lesion and with a decrease in the rate and amount of callose deposition on the walls of adjacent nonnecrotic cells (127). Maximum effect on lesion size was obtained by a 40-second treatment given 20 hours after inoculation. The magnitude of activation energy required to increase lesion size by pre-inoculation hot-water treatment, was of the same order of magnitude as that for denaturation of protein (91).

In other host-virus combinations, hot-water treatments did not increase lesion size; immersion of bean leaves in hot water 20 or 43 hours after inoculation decreased the size of SBMV lesions (94). In Samsun NN leaves, hot-water treatments applied at times ranging from 18 hours before to 36 hours after inoculation, markedly inhibited development of TMV lesions. The same treatment applied after lesion appearance enhanced lesion development by collapsing a ring of tissue around each lesion, but without increasing virus multiplication (95). It may be that in the Pinto bean and cucumber cotyledon-TMV systems, the hot-water treatment selectively inactivates the localizing mechanism, leaving the cells around the lesion in a state capable of supporting virus multiplication. As a result, both lesion size and virus multiplication increase. In the Samsun NN-TMV system, however, the hot-water treatment causes severe damage to the cells around the lesion, leading to a rapid collapse of cells. This area, which is apparently more sensitive to heat in Samsun NN than in Pinto beans or cucumber cotyledons, is therefore unable to support additional virus multiplication.

The effect of temperature on the hypersensitive reaction in TMV-infected Xanthi plants has been studied in relation to PPO (7). Maximum PPO activity was observed 3 days after the appearance of lesions, when plants were kept at 20°C. Exposing plants to 30°C resulted in systemic infection. Keeping plants at 30°C for 68 hours after inoculation and then exposing them to 20°C resulted in lower PPO activity than in comparable uninoculated plants at the time of lesion appearance. Therefore, PPO activation seems to be a secondary effect, and not the cause, of lesion formation. Furthermore, in similarly treated plants, changes in phenolic compounds were detected only 36–

72 hours after transferring the plants to 20°C, whereas necrotic lesions appeared after 12 hours (114).

Light.—Light intensity often affects symptom expression, and pre-inoculation darkening or shading is used routinely to increase the susceptibility of test plants. Thus, TNV causes complete necrosis in England during winter, but only relatively few lesions are produced in summer. This is correlated with virus yields, and winter plants produced ten times more virus than autumn- or spring-grown plants, while purification during the summer does not yield a satisfactory preparation. Studying the effect of light intensities, Bawden & Roberts (3) observed that shading leaves during the summer sensitized the plants, so that many more and larger lesions were produced and virus titer was significantly higher than in unshaded plants. The susceptibility of *Phaseolus vulgaris* to TNV infection was increased by keeping plants for a 24-hour period before inoculation at light intensities below 250 lumen/ft^2, and maximum susceptibility was detected in plants kept in complete darkness (44). Shading of TMV-infected *N. glutinosa* plants also increased virus titers.

The explanation of the effect of shading is still obscure, and it is not known if this is related to suppression of the localizing mechanism. It should also be emphasized that in the above-mentioned experiments the plants were shaded before inoculation, and specific experiments relating postinoculation shading to lesion size or virus titer are few. Thus, postinoculation treatments generally resulted in larger AMV-induced lesions on bean (15). Darkness after inoculation inhibited lesion formation in detached leaves of *G. globosa* inoculated with PVX, though the virus titer was three times higher than in illuminated leaves (38). Limited movement occurred from inoculated regions in darkened leaves. If glucose was added to the nutrient medium, lesions formed on darkened leaves, and infectivity titers decreased. Postinoculation darkness also reduced lesion number in TNV-inoculated beans, but virus concentration decreased simultaneously (33). It has been suggested that a product(s) of photosynthesis interferes with virus multiplication (3, 38). Virus multiplication is thus allowed to proceed more smoothly in shaded plants. It may also be that shaded plants are deficient in high-energy compounds, produced by photosynthesis, which are required for the operation of the localizing mechanism. Studies with light at different wavelengths, especially in the blue, red, and far-red regions, may help to evaluate the causal relation between carbohydrate and protein synthesis and localization.

Leaf detachment.—Detaching bean leaves just after inoculation with TMV caused a two- and fourfold increase in lesion diameter and infectivity per unit lesion size, respectively (79). The number and size of TMV packets observed by electron microscopy increased greatly in *N. glutinosa* when leaves were detached after inoculation, and kept at nearly 100% humidity (124). Concentration assays resulted in a 30-fold increase of virus. Keeping detached leaves under normal atmospheric conditions resulted in only a 1.7-fold increase in TMV concentration, and in leaves excised 96 hours before inoculation less virus was produced than in comparable attached controls (88). It is probable that under conditions of moderate senescence, when compounds required for localization are depleted, the localizing mechanism—but not virus synthesis—is suppressed; subsequent advance of cell degeneration interferes with virus production.

Cytokinins and 2,4-D.—Appearance of TMV lesions on detached leaves of *N. glutinosa* was completely inhibited by five cytokinins applied immediately after inoculation, although virus production was substantially stimulated—more than in systemic hosts (75). However, both lesion formation and virus production per leaf were inhibited in Pinto beans, although N^6-benzyladenine caused a threefold increase in lesion size. As no correlation between virus production and chlorophyll retention was noted, their effect in *N. glutinosa* does not seem to be by delaying senescence. Apparently, cytokinins at low concentrations affect the equilibrium between synthetic processes required for localization and those for virus multiplication, favoring the latter. In other cases kinetin is known to inhibit virus multiplication; a marked increase in lesion size is also observed on plants sprayed with 2,4-dichlorophenoxyacetic acid (2,4-D), within 24 hours after inoculation (106). It is not known whether 2,4-D specifically suppresses the development of the localizing mechanism or increases lesion size by its herbicidal action, causing a general disorganization of the host. Also 2,4-D stimulated the low TMV-replicating capacity of cotton (12).

Metabolic inhibitors and ultraviolet irradiation.—Antimetabolites that selectively suppress host-mediated processes, but not virus multiplication, have recently been used as tools to study localization. Thus actinomycin D—known to inhibit DNA-dependent m-RNA synthesis, and chloramphenicol—which inhibits protein synthesis, partially suppressed the localizing mechanism in several hypersensitive hosts when applied close to the time of inoculation (60, 99). The average diameter of lesions in Pinto bean leaves injected with actinomycin D prior to TMV inoculation was 2½ times that of controls. Virus concentration in cucumber cotyledons increased 11 times when cotyledons were injected with 10 $\mu g/ml$ actinomycin D one day after inoculation, even though the number of visible starch lesions was reduced. Controls with mannitol, at concentrations equal to those present in the actinomycin D preparations, had no effect (unpublished results). When the antibiotic was applied at later intervals, the effect on lesion size and virus concentration decreased; it was negligible when applied at the third day or later after inoculation. Comparable results were obtained with chloramphenicol, which apparently does not enter the host cell nucleus, where TMV or TMV-RNA is thought to be synthesized (99). Virus concentration increased 4 times in cucumber cotyledons injected with 200 $\mu g/ml$ chloramphenicol within 1 day of inoculation, but not when applied after 3 days. Lesion size in *N. glutinosa* leaves injected with the antibiotic 24 hours after inoculation was more than three times greater than that of controls. Applications of chloramphenicol made later than 24 hours after inoculation were less effective. The increase in size of lesions in *N. glutinosa* was correlated with their extractable infectivity. Application of the antibiotic to Pinto beans for 6 successive days before inoculation increased lesion size significantly (46). Pretreatment of tobacco Xanthi leaf discs with chloramphenicol or actinomycin D also caused an earlier appearance of TMV lesions (98). In systemic hosts it was observed that either no increase in virus multiplication followed application of one of these antibiotics (60, 99), or that the increase—sometimes temporary—was one or two orders of magnitude lower than in hypersensitive hosts (98, 103).

2-Thiouracil, assumed to interfere with host RNA metabolism, suppresses the localization of cowpea chlorotic mottle virus and causes systemic movement from one side of a cowpea leaf to the other (49). Treatments within 24 hours of inoculation increase both lesion size and sap infectivity very markedly, but applications made after 48 hours had no effect. With other viruses, 2-thiouracil is recognized as a potent inhibitor.

Ultraviolet irradiation, which affects a variety of morphogenetic processes in plants (47), may also be used to suppress localization. Ultraviolet irradiation also inhibits induced enzyme synthesis and may directly inactivate m-RNA synthesis. However, suppressing localization with ultraviolet, consequently enhancing virus multiplication, becomes evident only if ultraviolet (at certain dosages) hits the localizing mechanism more effectively than the viral nucleic acid. Thus, short-wave ultraviolet irradiation (2537 Å) significantly enhanced TMV multiplication in cucumber cotyledons and lesion size in *N. glutinosa* plants if applied 24 hours after inoculation (53). Short-wave radiation applied 2–4 days after inoculation, or irradiation with a long-wave (3660 Å) lamp one day after inoculation, did not increase virus concentration or lesion size. Larger lesions on primary leaves of Pinto beans, irradiated one day after inoculation and incubated afterwards for 3 days in the dark, were also observed (128). Callose deposition around the lesions is prevented in irradiated leaves. It was suggested that callose deposition by noninfected cells may act as an effective barrier to viral movement, by sealing the plasmodesmata. However, the localizing mechanism in cucumber cotyledons does not seem to be associated with callose deposition, as no fluorescence was observed in relation to virus lesions (53). Because short-wave ultraviolet radiation was more effective in suppressing the localizing mechanism than long-wave ultraviolet, it seems that the targets are nucleic acid-like in their adsorption, and are probably located in the cellular DNA. Ultraviolet probably produces thymine dimers, which prevent the template activity of DNA in DNA-dependent RNA synthesis. It might also be mentioned that in irradiated rat embryo cells, interferon yields decrease, whereas the size of Sindbis virus plaques increases (62).

It is suggested that localization requires transcription of cellular DNA to

RNA and subsequent protein synthesis. This mechanism becomes activated after infection, perhaps by a de-repression process, and can be suppressed by actinomycin D, chloramphenicol, or ultraviolet radiation applied within 24–48 hours after inoculation. Applying these treatments 3–4 days after inoculation, when the localizing mechanism has been operating for several days, does not increase virus multiplication or lesion size. During this period a substance(s), probably a protein, is produced which is responsible for the localization of the virus. This protein is probably produced in the extranuclear parts of the cell. The possibility that a protein may be involved in the localization is supported by the results obtained when tannic acid was infiltrated into cucumber cotyledons and brushed for 2 weeks on the new developing leaves (13). Tannic acid, which is known to complex with proteins, interfered with the mechanism that limits infection within a starch lesion and permitted systemic development from the cotyledons to the leaves.

Summary of hypotheses explaining localization.—Several hypotheses have been advanced to explain why a virus is confined to a lesion. It has been assumed that movement is prevented by the death of the host cell due to necrosis. This explanation is not adequate, even in necrotic local lesion hosts; *a fortiori* in a chlorotic- or starch-lesion host, as virus particles are found outside the necrotic area in viable cells (73). Furthermore, the virus resumes its spread if plants with fully developed lesions are placed at temperatures above 30°C. It has been suggested that the increase in PPO activity observed in necrotic local lesion hosts is responsible for necrosis and for localizing the infection, due to an accumulation of quinones. However, PPO activation seems to be the result of cellular injury and necrosis, and not the cause of cell death. Deposition of other barrier substances, such as calcium pectate (122) and callose (127, 128), in cells surrounding necrotic local lesions, thereby sealing off the plasmodesmata between infected and noninfected cells, has been mentioned as a possible factor preventing virus movement. However, callose deposition has not been observed in TMV-infected cucumber cotyledons, where the infection remains localized without necrotization, and data on calcium pectate deposition have come only from one necrotic local-lesion host. It may therefore be that these are responses to necrotization and not necessarily responsible for localization.

Based on electron microscope examination of the region surrounding a necrotic lesion, Ross & Israel (95) recently postulated that structural changes induced in advance of infection leading to cell collapse, may be instrumental in virus localization. These changes may progress to a level that causes collapse of a narrow ring of cells not yet invaded by virus, resulting in a barrier to further movement of virus. This may be the result of high peroxidase activity (108), though activation of peroxidase is also observed in plants kept at 100% humidity, thereby inhibiting the necrotic reaction (81). Spencer & Kimmins (111) suggested that the spread of virus may be inhibited because of a lack of cytoplasmic connections between cells in the resistant zone, arising perhaps from the accumulation of paramural bodies. Blocking of plasmodesmata could also result from thickening of cell walls, observed in the periphery of lesions incited by potato virus M on beans (117). However, as long as no comparable data from chlorotic- and starch-lesion hosts are available, these structural changes may well be due to movement of some product of necrosis. Furthermore, the above arguments, i.e., the presence of virus particles outside the necrotic area and resumed spread at higher temperatures, seem not to favor these hypotheses.

The author prefers the following hypothesis. In a local lesion host the virus and possibly other agents (see below) activate the localizing mechanism by a de-repression process. Necrogenesis is a separate process, which may retard virus multiplication but is not the major factor in localization. The localizing mechanism requires transcription of cellular DNA to RNA and subsequent protein synthesis. A substance(s)—probably a protein—is produced, which 24–48 hours after inoculation reaches a level sufficient to depress (nonlimited lesions as in *N. glutinosa*) or inhibit completely (limited lesions as in *D. stramonium*) further virus multiplication. The mechanism requires energy and precursors to produce this substance, and is presumably less efficient when plants are placed in the dark or leaves are detached. The localizing mechanism does not work at high temperatures, and it may be that at these temperatures the localizing substance is not produced, is inhibited or metabolized, or both. Production of the localizing substance may cease, either because of insufficient supply of high-energy compounds or because of a shift in synthetic processes (protein synthesis?) necessary for localization to those required for virus multiplication or other host functions.

The nature of the localizing substance, produced in the zone surrounding the lesion, is still a matter for speculation. The analogy to an "interferon-like" protein, that interferes with virus multiplication, has been proposed (60), and White (126) suggested that an antiviral factor, presumably RNA (102), acts as a barrier to virus spread. The development of such materials will be discussed later. However, it should be emphasized that our understanding of localization is still fragmentary, and it may well be that in different hosts the virus is confined to the lesion by different mechanisms.

Induced Resistance

This section will be centered around resistance developing in noninvaded tissue after other parts of a plant have been infected with a virus, as well as on activation of resistance by various nonviral agents.

Induced by viruses.—The first observation that infection in one part of the plant induces resistance in other noninvaded tissues was reported by Gilpatrick & Weintraub (25). When the lower leaves of two clones of *Dianthus barbatus* were inoculated with CarMV, primary local lesions developed, whereas in the upper noninoculated leaves the presence of a virus could not be demonstrated. However, these upper leaves were resistant to infection, as only 5–25% as many lesions developed after inoculation with CarMV as in control plants, where the lower leaves had not been infected. However, after 2½ years of continuous operation the protection mechanism ceased abruptly, perhaps due to a change in the plant clones.

Resistance has been reported to develop in various hypersensitive hosts, both in tissues near the primary local lesions and in more distant parts of the plant. The term "localized induced (acquired) resistance" is used in the first case, whereas the resistance developing in other parts is termed "systemic induced (acquired) resistance." This distinction seems to be one of convenience rather than a basic one, as it depends only on the distance of the resistant tissue from the primary infection. Thus, around primary TMV lesions on beans (130) and tobacco NN (92), a zone of uninfected tissue became resistant to a second inoculation with TMV. In this area no lesions, or only a few minute ones, developed. The resistant zone in tobacco NN increased in size and degree of protection for about 6–8 days after the first inoculation. Development of the resistant zone was not affected when the plants were kept at 100% humidity, thereby negating the possibility that the resistance is due to a drying-out effect, resulting from abnormal loss of water from the tissues adjacent to necrotic areas. Resistance was highest when tests were carried out at 20–24°C; no resistance could be detected at temperatures close to 30°C.

Further studies showed that this resistance is not restricted to the areas around lesions—where it may be most pronounced—but develops also in other more distant tissues of the plant. Thus, inoculation of half-leaves of tobacco NN plants with TMV induced a high level of resistance to TMV in the opposite half-leaves (93). Similarly, resistance could be induced in the basal part of a leaf by inoculating the apical half. Challenge inoculation with TMV of the resistant half-leaves 7 days after the first inoculation resulted in limited lesion formation; the lesions were consistently only one-fifth to one-third as large in diameter as lesions in susceptible half-leaves, and they were usually fewer in number. Inoculation of the lower leaves on a plant induced resistance in the upper noninoculated leaves. Before the challenge inoculation both the upper resistant leaves, or the uninoculated half-leaves, were free of virus. Resistance could be detected 2 or 3 days after the inducing inoculation; it reached a maximum in about 7 days, persisted for at least 20 days, and was highest when the plants were kept at 20°C. The resistance induced by TMV in tobacco NN is not specific for TMV, and leaves with TMV-induced resistance were also resistant to TNV, TRSV, and TomRSV. Similarly, other viruses, such as TNV, which induce necrotic local lesions in tobacco NN plants, will also induce both local and systemic resistance against TMV. However, in plants infected by PVX, no resistance around TMV lesions was detected, and the level of TMV-induced systemic resistance was lower than in plants that were free of PVX. Systemic induced resistance was also demonstrated in Pinto beans and in cowpeas (*Vigna sinensis* var. Blackeye) (94). The following viruses, in various combinations, were used both for the first resistance-inducing inoculation and for the challenge: AMV, TMV, TNV, and SBMV with Pinto beans; and AMV, CMV, and TNV with cowpeas. Resistance developed in the opposite primary leaf regardless of the order in which any virus pair was used. Lesions developing in the resistant tissues were consistently smaller (about one-half the diameter) than lesions in the appropriate control leaves. Lesion numbers were reduced, but the effect on size was more consistent. However, at high levels of resistance, as measured by the effect on size, lesion numbers were also reduced considerably. Thus, systemic acquired resistance, as measured by a reduction in lesion numbers, could be induced by TMV in *D. stramonium* and by PVX in *Gomphrena globosa* plants (52).

It seems possible that the effects on size and number are caused by the same factor and that in highly resistant tissues the infection does not progress far enough to result in visible lesions. This might be the case also with localized induced resistance, developing near or around primary local lesions. Resistance there seems to be high and, as a result, lesion numbers, and not only size are reduced considerably. Ross's data (94) on the distribution of lesion size indicate that the progressive decrease in lesion size results in a reduction in the number of visible lesions in highly resistant tissue. In nonresistant tissue a normal distribution curve of lesion diameter was observed, ranging from 0.2–3 mm, with a mean of 1.6 mm, when SBMV lesions were measured. In resistant tissue, however, the distribution curve of lesion size was markedly skewed, ranging from 0.03–1.43 mm, with the mode of the curve at the 0.03–0.2 mm group. No lesions smaller than the mode were detected. Due to the resistance, lesions smaller than the mode could not progress further to become visible, the result being also a reduction in lesion number. As virus content is generally correlated with lesion size, the decreased size in resistant tissue seems therefore to be associated with a reduction in virus multiplication.

The development of resistance has so far been reported only for viruses that induce necrotic local lesions, and no systemic resistance has become evident in cucumber cotyledons with starch lesions (unpublished results). On the other hand, no resistance was induced by mechanically or chemically induced necroses (93). However, in *Capsicum pendulum,* heat-inactivated PVX induced systemic resistance, affecting both lesion number and size (77).

To elucidate the metabolic basis of induced resistance, several enzyme systems have been investigated. In the yellow ring of living tissue surrounding virus-induced lesions, the activity of a number of respiratory enzymes increased, compared with normal green areas of the same leaf; mainly the HMP shunt dehydrogenases were activated and sugars, mostly pentoses, accumulated. This accumulation of sugars in tissues surrounding the lesion could be explained if the later stages of the HMP shunt could not cope with the highly activated first two steps of the cycle. It was suggested as a possibility that the increased operation of the HMP pathway is associated with accumulation of phenols (110). As induced resistance in uninoculated tissues may have similarities to the mechanism responsible for localizing the infection within a lesion, polyphenoloxidase in resistant tissues was also studied. A twofold increase in PPO activity was observed in the basal parts of tobacco NN leaves, 2-3 days after inoculating the upper parts of the leaves (119). However, removing the inoculated part of the leaves 7 hours after inoculation was sufficient to induce an increase in PPO activity in the remaining tissue, but insufficient to induce resistance (6), for which at least 3 days are required (6, 107). No increase in PPO activity was observed in extracts of resistant leaves when a manometric (6) instead of spectrophotometric (119) method was employed. It seems, therefore that PPO does not play a major role in the resistance phenomenon. In addition, it might be mentioned that although accumulation of polyphenols is characteristic for many diseased tissues, it may also be induced by simple chemical damage which, however, does not induce resistance.

In cells peripheral to lesions the amount of protein increases (110), and recently it was reported that peroxidase and catalase activities increased parallel with the development of systemic resistance, and remained at high levels (107, 108). The induction of resistance was not accompanied by a permanent change in oxygen uptake or by changes in the activities of selected enzymes participating in the glycolytic, HMP, and tricarboxylic pathways. Likewise selected terminal oxidases were not affected. It was also observed that lesions appeared sooner in resistant than in nonresistant leaves. Simons & Ross (108) suggested that inoculation with a lesion-inducing virus elicits high peroxidase activity in resistant leaves, which leads to early appearance of necrosis. The lateral movement of some product(s) of necrosis or of reactions preceding necrosis, then causes an early limitation of lesion size. Although injection of commercial peroxidase into tobacco leaves increased resistance to *Pseudomonas tabaci* (61), it did not induce resistance to TMV (unpublished results).

The possibility that resistance is associated with a gaseous substance diffusing from the inoculated leaves can be rejected, because control plants placed close by do not develop resistance. By interrupting the supposed pathway, it has been demonstrated (94) that resistance is associated with the movement of substances from or to the resistant tissue. In tobacco NN plants, where the three lower leaves were inoculated with TMV, cutting the midvein of an upper leaf prevented development of resistance distal to the cut. Killing a section of the petiole of each inoculated leaf with boiling water also prevented development of resistance in upper leaves.

Two hypotheses have been advanced to explain induced resistance:

1. *Depletion theory*. Due to the increased metabolic reaction in the initially infected cells, acting as a metabolic sink, a depletion of metabolites occurs in the uninfected adjacent tissues. These tissues are therefore less able to support virus synthesis, and become "resistant". According to Silberschmidt & Caner (105), resistance is due to a depletion of sugars in the leaf as a consequence of the preceding infection. Systemic resistance was induced in the basal parts of *G. globosa* leaves, after inoculating the apical part of the leaf with PVX.

"Ringing" the shoots, which is considered to prevent the downward movement of sugars from the leaves, prevented the development of induced resistance. They suggested that in the ringed plants the leaves retained a sugar supply sufficient for lesion development, which was not depleted even by the first "resistance inducing" infection. Although ringing may prevent the sugar transport, this does not necessarily mean that resistance is due to a depletion of sugars. It may be that "ringing" prevented the production or movement of a precursor necessary for the development of resistance.

Most experimental data favor the second hypothesis:

2. *Production of a resistance inducing substance(s).* After the initial local necrotic infection a substance(s) is produced that induces resistance in uninfected tissues. When the inoculated leaves of tobacco NN plants were removed 7 days after inoculation, high resistance was induced in leaves that developed during a period when there was no localized infection present elsewhere in the plant (6). Thus, it is difficult to explain how drainage of metabolites could affect leaves challenged 19–25 days after removal of the lower inoculated leaves. Development of localized induced resistance to TMV in tobacco NN and Pinto bean plants was also partially or completely inhibited in the presence of actinomycin D (58). In tobacco plants challenged 5 days after the inducing inoculation, actinomycin D injected intercellularly 3 days before the challenge almost completely abolished resistance. However, no inhibition resulted when actinomycin D was injected close in time to the challenge. Resistance in Pinto bean plants was also significantly reduced in the presence of actinomycin D.

These results suggest that development of localized induced resistance depends on the transcription mechanism of the cell from DNA to RNA—a key reaction to synthetic processes, thereby producing a resistance-inducing substance. If actinomycin D is injected close in time to the inducing inoculation, when the hypothetical resistance-inducing agent is beginning to be produced, resistance is inhibited almost completely. However, when actinomycin D is injected at a later stage, close to the time of challenge, when such a substance has already been produced in larger quantities, induced resistance is not affected. The sensitivity of induced resistance to actinomycin D therefore negates the hypothesis that induced resistance is caused by drainage of metabolites. These results also strengthen the idea that localization of a virus and induced resistance may be caused by similar mechanisms (60, 94).

It is interesting to note that resistance in tobacco NN develops rapidly from the second or third day after the inducing inoculation onward, at a time when the localization of the virus is almost complete. It seems plausible that in a hypersensitive host a substance is synthesized after the initial virus infection that is responsible for the localization. Later, when it is produced in larger quantities, it diffuses from the primary infection area and induces resistance in adjacent tissues. It may also be that a substance associated with necrotization, or released by it, diffuses into uninfected tissue and there activates—before virus introduction—the localizing mechanism, perhaps by a de-repression process. The fact that induced resistance has so far been observed only in host-virus combinations that react with necrotic lesions, strengthens this suggestion. Simons & Ross proposed the possible role of peroxidase, which enhances the localizing process in resistant tissue (108).

Certain analogies between induced resistance and the interferon system in animals have been pointed out (52, 58, 60, 109). Interferon is a protein that develops in animal tissues previously inoculated with viable or inactivated viruses. It is nonspecific in its action, i.e., induced by one virus and acting against other nonrelated viruses. Interferon may also be induced by other nonviral agents such as nucleic acids, phytohemagglutinin, and synthetic polyanions. It prevents the synthesis of new viral nucleic acid. The production of interferon by the animal cell is inhibited in the presence of actinomycin D, puromycin, and ultraviolet radiation, and the host cell RNA and protein syntheses are also required for the antiviral action of exogenous interferon. When compared with the interferon system, induced resistance is also not virus specific, but is induced by various agents such as viruses, fungi, and yeast-RNA (see below), and its induction is sensitive to actinomycin D. Since induced resistance may be due to the same mechanism that is responsible for localization, it may be worthwhile to study whether localization might not also be associated with an interferon-like substance. However, interferons in animal virology have been defined as to their chemical nature, and their mode of action has been studied. Conclusions about similar mechanisms in plants can therefore be drawn only after obtaining more data on their mode of action and association with a chemically defined substance.

Fungi.—Systemic resistance to TMV was induced in tobacco Xanthi-nc leaves by injecting spore suspensions of the fungus *Peronospora tabacina* into the stems of the plants (65). The plants were challenge-inoculated 3 weeks later, to allow time for the development of the mycelium in the injected stems. TMV lesions in the fungus pre-inoculated plants were smaller and fewer than in the controls. Hecht & Bateman (29) also demonstrated the induction of systemic resistance against TMV and TNV by infection of the stem or leaf tissues with the fungus *Thielaviopsis basicola,* which causes a localized reaction. Virus lesions in the resistant tissue were markedly smaller than those in the comparable control. In *Datura stramonium* plants also, *T. basicola* induced systemic resistance affecting both the size and number of TMV lesions (unpublished results). It is not known whether resistance is activated either by a product of the fungus-induced necrosis, or directly by a pathotoxin of the fungus. Thus, bean leaves treated by stem adsorption with victorin, a pathotoxic product of *Helminthosporium victoriae,* were rendered highly resistant to TMV and AMV, at concentrations that did not cause necrosis (125). However, higher concentrations of victorin caused necrosis, and it may therefore be that the initiation of necrosis activated resistance. Fungal necrosis may also evoke resistance that suppresses virus multiplication in a systemic host.

Thus, the content of PVX in side shoots of potatoes was greatly reduced when spores of *Phytophthora infestans* were injected into the stem (32).

The decrease in lesion size or number in the preceding cases was associated with a reduction in virus multiplication, but the suppression of TMV lesions in rust-infected bean leaves was not. When TMV was inoculated 26 hours or more after rust inoculation, virus lesions were markedly suppressed, but virus amount was enhanced (24). It seems, therefore, that the rust infection inhibits the hypersensitive reaction of the host, but does not induce a resistance mechanism.

Development of resistance has been associated with necrosis, induced either by a virus or fungus that produces necrotic lesions. The following sections, however, will deal with resistance or interference, induced without visible necrotization of the host. Resistance in most of these cases is localized and not systemic; it develops only in tissues directly treated with the inducer, which activates a host-mediated process.

Heat-killed bacteria and bacterial fractions.—Heat-killed cells of *Pseudomonas syringae* (56) and *P. fluorescens* (48) injected into the intercellular spaces of tobacco NN or Xanthi-nc leaves, respectively, 0–7 days before inoculation with TMV, markedly reduced lesion numbers. Furthermore, heat-killed bacteria suppressed lesion formation even when applied 5–72 hours after inoculation, when the infection had become established and virus multiplication had started. Injection of a bacterial preparation 24 hours after inoculation or later, resulted in a decreased effect on lesion number, but a marked reduction in lesion size was still evident. Injection of casein or egg-albumin after inoculation had no effect. Preliminary experiments with actinomycin D showed that interference was inhibited in the presence of the antibiotic. The effect on lesion size, together with sensitivity to actinomycin D, suggests that the bacterial cell, or one of its components, activates or induces a resistance mechanism in the plant. The interfering agent, supposedly responsible for the suppression of lesion formation when applied close in time (5 hours) after TMV inoculation, was shown to be present in the cytoplasm of the bacteria and not in the cell wall. Heating the bacterial cytoplasm for 10 minutes at 100°C resulted in complete disappearance of its inhibitory potency (M. Halevy and G. Loebenstein, unpublished results).

Injection of an extract from *Escherichia coli* into Xanthi or Samsun tobacco leaves, 1 or 4 hours after inoculation, reduced TMV lesion number and virus concentration (1). Isopropanol-extracts from *Bacillus uniflagellatus* reduced the number of local lesions when added to the soil or sprayed on the leaves of Xanthi tobacco prior to inoculation with TMV (66). Cultures of the bacterium added to the soil also reduced the number of lesions on Xanthi and the viral content in a systemic host. However, further evidence is still required that the bacterial fractions in all these cases activate the host resistance mechanism or, alternatively, directly inhibit viral infection or multiplication. It is also possible that more than one mechanism is involved, a direct and an induced one, possibly but not necessarily brought about by different components of the bacterial preparations.

Foreign nucleic acids.—A high degree of resistance or interference can be induced when yeast-RNA (5 mg/ml) is injected intercellularly into leaves of tobacco NN. Yeast-RNA, when applied together with TMV, acts as an inhibitor of infection, but it also induces interference when injected at least 3 days before inoculation. Interference was highest when TMV was inoculated 5–6 days after RNA injection, and remained high even after 20 days. In the resistant tissue injected with RNA, TMV lesion number decreased by 80–90%, compared with that developing in water-injected control tissue. However, no decrease in lesion size was observed. The development of this localized interference was significantly inhibited when actinomycin D or puromycin was injected shortly after the RNA induction (22). The fact that a time interval of at least 3 days between RNA injection and TMV challenge is needed for detecting resistance, and that its development is sensitive to metabolic inhibitors, strongly suggests that a resistance-inducing agent is formed in the plant after RNA injection. Some local lesion hosts also became resistant to TMV after being infiltrated or sprayed with RNA isolated from plants different from the test species. This response was not elicited by RNA isolated from plants of the test species (14).

Recently it was also reported that injection of the synthetic double-stranded RNA polyinosinic-polycytidilic acid (poly I. poly C), in microgram quantities, into tobacco NN and *Datura stramonium* leaves, induced resistance to TMV (112). No competitive inhibition was observed even at concentrations three times higher than those necessary to induce interference; and a time interval between injecting poly I. poly C and challenge inoculation was necessary for the development of resistance, suggesting that poly I. poly C activates a resistance mechanism in the plant. In this connection it should be mentioned that foreign nucleic acids and poly I. poly C have been reported to induce the production of interferon in animal tissues (19).

Miscellaneous compounds.—Partial resistance can be induced by treating leaves of *N. glutinosa* or *D. stramonium* with TMV coat protein, and inoculating with TMV 4 days later (51). The treatment of leaves with TMV protein, 1 day or less before the challenge inoculation, did not reduce the number of lesions. Therefore, a direct inhibiting effect of the protein does not seem to be the cause. Reduced susceptibility of upper untreated leaves of plants whose lower leaves had been treated with TMV protein, showed that this type of protection was systemic. TMV protein protected against PVX but not against CMV. A similar effect was not obtained by several other proteins. Induction of resistance was also observed in bean leaves after inoculation with TMV protein (34) or ultraviolet-inactivated TMV (131).

Phytic acid (inositol hexaphosphoric acid) has been reported to reduce lesion number when applied to tobacco 24 hours before inoculation with TMV (64). The authors suggest that phytic acid may affect the host metabolism and confer on the plants a certain resistance to infection. From *D. stramonium* and tobacco plants an antiviral factor could be extracted 6–8 days after treatment with the sodium salt of phytic acid (63). This factor, which seems to be a protein, inhibited TMV lesions when inoculated with the virus and caused a slight reduction in the virus content, when discs systemically infected with TMV were floated on the medium 48 hours after inoculation. According to Maïa, this supports the suggestion that phytic acid confers resistance to the treated plants.

To summarize this section, it seems that activation of resistance by different nonmultiplying substances which do not cause necrosis, is in most cases limited to the tissue directly treated with the inducer, whereas necrosis-evoking viruses and fungi also activate resistance in noninvaded tissues. The fact that, in a local-lesion host, resistance can be activated by various agents, whereas attempts to induce resistance in a systemic host with several nonviral agents give negative results (unpublished results), suggests that these agents probably activate the localizing mechanism before the challenge inoculation. It could be speculated that certain macromolecules may initiate a de-repression process, by combining or removing histone repressors from cellular DNA responsible for coding the resistance substance(s).

Inhibitory Substances in Infected and Resistant Tissues

Antiviral or interfering agents have been extracted from both infected- and uninfected-resistant tissues, although their connection with localization, induced resistance, or recovery is still an open question. An antiviral factor (AVF) developing in local and systemically infected plants has been reported by Sela and coworkers (100, 102). Sap extracted from PVY- or TMV-infected tissue was treated with hydrated calcium phosphate, resulting in a noninfective solution. This solution, when mixed with TMV or PVY, was more inhibitory than the comparable control from healthy plants. The agent was not virus specific, and apparently does not inactivate the virus in vitro. The active principle, purified by column chromatography and paper electrophoresis from *N. glutinosa* plants, 2 days after inoculation with TMV, was sensitive to RNAse and therefore considered to be RNA. This conclusion, however, was questioned by Solymosy (109), because ultraviolet absorption and electrophoretic migration of purified AVF were not typical for RNA. A suppression of infectivity was observed when AVF was applied to leaves of 4 species previously infected with TMV or CMV (101). Young leaves from plants infected for 10 or 2 months, respectively, were used, and after treatment with AVF for 24–28 hours, they were assayed for infectivity. The authors suggest that the reduced infectivity provides evidence that the site of action is inside the cell. However, it is difficult to visualize this agent as affecting virus particles already present. On the other hand, if this agent affects virus multiplication it would seem advisable to test it under controlled conditions, shortly after inoculation of healthy tissue during the exponential period of virus multiplication and not in young leaves of old infected plants, where variation may be considerable. Association of AVF with induced resistance is unknown, as it is extracted and purified from infected tissues and not from resistant uninfected parts.

Another antiviral principle was isolated from tomato plants systemically infected with TMV, showing no signs of necrosis, also by employing hydrated calcium phosphate clarification. The active principle reduces infectivity when mixed with TMV and may directly affect the virus in vitro. After adsorption by plant tissues it is translocated to new sites, where it either inhibits infection or multiplication (9). The development of this agent was concomitant with the decline of virus. The production of the agent reached a maximum at 26°C, and was maintained at high levels at 32°C—a temperature supraoptimal for virus multiplication (10).

A substance that inhibited PVX was found in stems and leaves of potato plants, following inoculation with the fungus *Phytophthora infestans* (32). Sap from within or around blight lesions markedly inhibited PVX lesion production, but no inhibition was noted when sap from beyond the lesion was used. It seems reasonable to assume that the inhibition is caused by a substance in the fungal mycelium acting as an inhibitor of infection. An inhibitor of TMV was also extracted from *Nicotiana glutinosa* infected with *Thielaviopsis basicola* (26). The inhibitor is probably also of fungal origin, though extracts of tissue not invaded by the fungus possessed a low inhibitory capacity.

Inhibitory substances have also been extracted from resistant uninfected tissues. Thus, a crude extract from the uninfected apical halves of leaves of *Datura stramonium* previously inoculated on their basal halves with TMV or TNV, consistently interfered with infection by TMV (when mixed in the inoculum) to a much greater extent than did juice from apical halves of control leaves (59). After partial purification, the active protein-like substance resembled interferon in many of its properties. It was stable at pH 2.5 and in perchloric acid, precipitated by zinc acetate, acetone, and ammonium sulfate, sensitive to trypsin, and had an apparent molecular weight of approximately 28,000. However, its role in induced resistance has yet to be determined (57, and unpublished).

Increased inhibitory effects were found in saps from resistant parts of bean leaves (131) and whole leaves of *Capsicum pendulum* (77), after inoculating other tissues with TMV or PVX, respectively, and in extracts from bean leaves treated with ultraviolet-irradiated TMV (131). Kimmins also found that crude extracts from resistant tissues of *D. stramonium, N. glutinosa, Phaseolus vulgaris,* and *Chenopodium amaranticolor,* after inoculation of other parts with TMV or TNV, were more inhibitory than control extracts (45). The active principle was characterized as a low molecular weight RNA (3 S). However, yeast RNA has been observed to inhibit infection when present in the inoculum and also to induce resistance when applied 3 or more days before inoculation (22). Therefore, it has yet to be shown that the low molecular weight RNA isolated by Kimmins is directly responsible for the development of systemic resistance. Evidence for such a causal relationship has to be presented for all the above-mentioned substances. In most of these reports, the agents were assayed as inhibitors of infection and it has yet to be shown convincingly that they affect virus multiplication. It should also be emphasized that plant juices themselves are inhibitory, and age of plants, environment, or other stresses affect their inhibitory capacity. Testing for qualitative or quantitative differences in infected or resistant tissues should therefore be carried out under controlled conditions with plants carefully selected for uniformity.

Another approach recently employed is the study of protein profiles in infected hypersensitive and in noninfected resistant tissues. Using disc electrophoresis, a new band was observed in extracts from TMV-infected Xanthi tobacco kept at 20°C, which was absent from extracts of similar plants kept at 30°C, as well as from noninoculated controls (21). This band may, however, be related to TMV coat protein. In Samsun NN plants, four new bands were observed one week after infection (120). These were not related to TMV coat protein, and were not new isoenzymes of six enzymes tested. These four bands were also present in the noninfected, young, developing leaves of inoculated plants, and their amount was reduced by treatment with actinomycin D. They were not found in Samsun plants systemically infected with TMV. None of these four new protein components was found to be identical to one of the three protein bands, which appear or are greatly increased after infection of *N. glutinosa.* Van Loon & van Kammen suggest that the new components may help to limit virus multiplication or spread, but are not a product of the N gene responsible for the necrotic lesion formation. It now has to be seen whether these proteins affect virus multiplication.

Concluding Remarks

Our knowledge regarding localization and induced resistance is still fragmentary, and many uncertainties continue to exist in the interpretation of their mechanisms. Strong indications suggest that localization and necrogenesis are separate processes. Care has therefore to be taken when relating metabolic processes occurring in a necrotic local lesion host to the localizing process, as they may well be associated with necrotization.

Both localization and induced resistance appear to be active processes involving transcription of cellular DNA to RNA, and they may be caused by similar mechanisms. Different viral or nonviral agents may induce resistance (activation of the localizing process?), presumably by activating the information contained in the host's DNA, perhaps by a de-repression process. The host's genome carries, therefore, the information necessary for localization or resistance in a repressed state, and different agents may initiate de-repression. Certain analogies have been suggested between the resistance induced by viruses or nonviral agents in plants and the interferon system in animals. However, conclusions cannot be drawn until the mode of action of a chemically defined agent has been established.

Difficulties arise when a causal relationship has to be established between substances developing in the host after infection and localization or resistance. Inhibition of infection, the most often used technique, provides only very preliminary evidence. Evaluation of these substances as inhibitors of viral multiplication in the intact plant is also often difficult, because external application does not ensure penetration into the cell. New techniques, such as the use of plant protoplasts (113), may open additional possibilities for the study of these substances, as well as inhibitors of virus increase in general.

Acknowledgments

I wish to thank Dr. M. Smookler for criticisms of the manuscript and Miss Nilly Bachar for aid in its preparation.

LITERATURE CITED

1. Albouy, J., Lapierre, H., Maury, Y., Staron, T. 1969. Mise en évidence de l'effet inhibiteur d'une extrait d'*Escherichia coli* sur la production du virus de la mosaïque du tobac. *Ann. Phytopathol.* 1:251–55
2. Bates, D. C., Chant, S. R. 1970. Alterations in peroxidase activity and peroxidase isozymes in virus-infected plants. *Ann. Appl. Biol.* 65:105–10
3. Bawden, F. C., Roberts, F. M. 1947. The influence of light intensity on the susceptibility of plants to certain viruses. *Ann. Appl. Biol.* 34:286–96
4. Bell, A. A. 1964. Respiratory metabolism of *Phaseolus vulgaris* infected with alfalfa mosaic and southern bean mosaic viruses. *Phytopathology* 54:914–22
5. Best, R. J. 1944. Studies on a fluorescent substance present in plants. 2. Isolation of the substance in a pure state and its identification as 6-methoxy-7-hydroxy 1:2 benzo-pyrone. *Aust. J. Exp. Biol. Med.* 22: 251–55
6. Bozarth, R. F., Ross, A. F. 1964. Systemic resistance induced by localized virus infections: extent of changes in uninfected plant parts. *Virology* 24:446–55
7. Cabanne, F., Scalla, R., Martin, C. 1968. Activité de la polyphénoloxydase au cours de la réaction d'hypersensibilité chez *Nicotiana Xanthi* n.c. infecté par le virus de la mosaïque du tabac. *Ann. Physiol. Vég.,* 10:199–208
8. Carroll, T. W., Shalla, T. A. 1965. Visualization of tobacco mosaic virus in local lesions of *Datura stramonium. Phytopathology* 55:928–29
9. Chadha, K. C., MacNeill, B. H. 1969. An antiviral principle from tomatoes systemically infected with tobacco mosaic virus. *Can. J. Bot.* 47:513–18
10. Chadha, K. C., MacNeill, B. H. 1969. Influence of temperature upon levels of virus and an induced antiviral principle in tomato plants systemically infected with tobacco mosaic virus. *Can. J. Microbiol.* 15: 1469–71
11. Cheo, P. C. 1970. Subliminal infection of cotton by tobacco mosaic virus. *Phytopathology* 60:41–46
12. Cheo, P. C. 1971. Effect of plant hormones on virus-replicating capacity of cotton infected with tobacco mosaic virus. *Phytopathology* 61:869–72
13. Cheo, P. C., Lindner, R. C. 1964. *In vitro* and *in vivo* effects of commercial tannic acid and geranium tannin on tobacco mosaic virus. *Virology* 24:414–25
14. Cheo, P. C., Lindner, R. C., McRitchie, J. J. 1968. Effect of foreign RNA on tobacco mosaic virus local lesion formation. *Virology* 35:82–86
15. Desjardins, P. R. 1969. Alfalfa mosaic virus-induced lesions on bean: effect of light and temperature. *Plant Dis. Reptr.* 53: 30–33
16. Diener, T. O. 1963. Physiology of virus-infected plants. *Ann. Rev. Phytopathol.* 1:197–218
17. Farkas, G. L., Király, Z., Solymosy, F. 1960. Role of oxidative metabolism in the localization of plant viruses. *Virology* 12:408–21
18. Farkas, G. L., Stahmann, M. A. 1966. On the nature of changes in peroxidase isoenzymes in bean leaves infected by southern bean mosaic virus. *Phytopathology* 56:669–77
19. Field, A. K., Tytell, A. A., Lampson, G. P., Hilleman, M. R. 1967. Inducers of interferon and host resistance, II. Multistranded synthetic polynucleotide complexes. *Proc. Nat. Acad. Sci.* 58:1004–10
20. Gáborjányi, R., El Hammady, M. 1969. Effect of temperature on size of local lesions induced by TMV in tobacco plants. *Acta Phytopathol. Hung.* 4:125–29
21. Gianinazzi, S., Vallée, J. C., Martin, C. 1969. Hypersensibilité aux virus, température et protéines solubles chez le *Nicotiana Xanthi n.c. C.R. Acad. Sci. Paris* 268:800–02
22. Gicherman, G., Loebenstein, G.

1968. Competitive inhibition by foreign nucleic acids and induced interference by yeast-RNA with the infection of tobacco mosaic virus. *Phytopathology* 58:405–09
23. Gill, C. C. 1965. Fluorescent metabolites in virus- or rust-infected bean leaves. *Can. J. Bot.* 43:201–15
24. Gill, C. C. 1965. Suppression of virus lesions by rust infection. *Virology* 26:590–95
25. Gilpatrick, J. D., Weintraub, M. 1952. An unusual type of protection with the carnation mosaic virus. *Science* 115:701–02
26. Harpaz, I., Bar-Joseph, M., Sela, I. 1969. Inhibition of tobacco mosaic virus infectivity by the fungus *Thielaviopsis basicola* (Berk. & Br.) Ferr. *Ann. Appl. Biol.* 64:57–64
27. Harrison, B. D. 1955. *Studies on virus multiplication in inoculated leaves.* Ph.D. Thesis, London Univ.
28. Hayashi, T., Matsui, C. 1965. Fine structure of lesion periphery produced by tobacco mosaic virus. *Phytopathology* 55:387–92
29. Hecht, E. I., Bateman, D. F. 1964. Nonspecific acquired resistance to pathogens resulting from localized infections by *Thielaviopsis basicola* or viruses in tobacco leaves. *Phytopathology* 54:523–30
30. Helms, K. 1965. Role of temperature and light in lesion development of tobacco mosaic virus. *Nature* 205:421–22
31. Helms, K., McIntyre, G. A. 1962. Studies on size of lesions of tobacco mosaic virus on Pinto bean. *Virology* 18:535–45
32. Hodgson, W. A., Munro, J. 1966. An inhibitor of potato virus X in potato plants infected with *Phytophthora infestans*. *Phytopathology* 56:560–61
33. Hofferek, H. 1967. Untersuchungen über den Stoffwechsel viruskranker Pflanzen. V. Beeinflussung der Stoffwechselveränderungen bei *Phaseolus vulgaris* L. nach Infektion mit dem Tabaknekrose-Virus (TNV) durch Lichtmangel. *Phytopathol. Z.* 58:357–66.
34. Hofferek, H., Proll, E. 1969. Influence of noninfectious virus components on virus infection. In *Plant Virology,* Proc. 6th Conf. Czechoslovak Plant Virologists Olomouc 1967, 77–84, Academia Prague. 346 pp.
35. Holmes, F. O. 1946. A comparison of the experimental host ranges of tobacco-etch and tobacco-mosaic viruses. *Phytopathology* 36:643–59
36. Holmes, F. O. 1954. Inheritance of resistance to viral diseases in plants. *Advan. Virus Res.* 2:1–30
37. Holmes, F. O. 1965. Genetics of pathogenicity in viruses and of resistance in host plants. *Advan. Virus Res.* 11:139–61
38. Huguelet, J. E., Hooker, W. J. 1966. Latent infection of *Gomphrena globosa* by potato virus X. *Phytopathology* 56:431–37
39. Israel, H. W., Ross, A. F. 1967. The fine structure of local lesions induced by tobacco mosaic virus in tobacco. *Virology* 33:272–86
40. Jockusch, H. 1966. The role of host genes, temperature and polyphenoloxidase in the necrotization of TMV infected tobacco tissue. *Phytopathol. Z.* 55:185–92
41. de Jong, D. W. 1967. An investigation of the role of plant peroxidase in cell wall development by the histochemical method. *J. Histochem. Cytochem.* 15:335–46
42. Kassanis, B. 1963. Interactions of viruses in plants. *Advan. Virus Res.* 10:219–55
43. Kim, K. S. 1970. Subcellular responses to localized infection of *Chenopodium quinoa* by pokeweed mosaic virus. *Virology* 41:179–83
44. Kimmins, W. C. 1967. The effect of darkening on the susceptibility of French bean to tobacco necrosis virus. *Can. J. Bot.* 45:543–53
45. Kimmins, W. C. 1969. Isolation of a virus inhibitor from plants with localized infections, *Can. J. Bot.* 47:1879–86
46. Király, Z., El Hammady, M., Pozsár, B. I. 1968. Susceptibility of tobacco mosaic virus in relation to RNA and protein synthesis in tobacco and bean plants. *Phyto-*

pathol. Z. 63:47–63
47. Klein, R. M. 1967. Effect of ultraviolet radiation on auxin-controlled abscission. *Ann. N.Y. Acad. Sci.* 144:146–53
48. Klement, Z., Király, Z., Pozsár, B. I. 1966. Suppression of virus multiplication and local lesion production in tobacco following inoculation with a saprophytic bacterium. *Acta Phytopathol. Hung.* 1:11–18
49. Kuhn, C. W. 1971. Cowpea chlorotic mottle virus local lesion area and infectivity increased by 2-thiouracil. *Virology* 43: 101–09
50. Lindner, R. C., Kirkpatrick, H. C., Weeks, T. E. 1959. Some factors affecting the susceptibility of cucumber cotyledons to infection by tobacco mosaic virus. *Phytopathology* 49:78–88
51. Loebenstein, G. 1962. Inducing partial protection in the host plant with native virus protein. *Virology* 17:574–81
52. Loebenstein, G. 1963. Further evidence on systemic resistance induced by localized necrotic virus infections in plants. *Phytopathology* 53:306–08
53. Loebenstein, G., Chazan, R., Eisenberg, M. 1970. Partial suppression of the localizing mechanism to tobacco mosaic virus by uv irradiation. *Virology* 41: 373–76
54. Loebenstein, G., Linsey, N. 1963. Effect of virus infection on peroxidase activity and C_6/C_1 ratios. *Phytopathology* 53:350
55. Loebenstein, G., Linsey, N. 1966. Alterations of peroxidase activity associated with disease symptoms, in virus-infected plants. *Israel J. Bot.* 15:163–67
56. Loebenstein, G., Lovrekovich, L. 1966. Interference with tobacco mosaic virus local lesion formation in tobacco by injecting heat-killed cells of *Pseudomonas syringae*. *Virology* 30:587–91
57. Loebenstein, G., Rabina, S., van Praagh, T. 1966. Induced interference phenomena in virus infections. In *Viruses in Plants,* eds. A. B. R. Beemster, J. Dijkstra, 151–57. North Holland Publishing Co., Amsterdam, Holland, 342 p.
58. Loebenstein, G., Rabina, S., van Praagh, T. 1968. Sensitivity of induced localized acquired resistance to actinomycin D. *Virology* 34:264–68
59. Loebenstein, G., Ross, A. F. 1963. An extractable agent, induced in uninfected tissues by localized virus infections, that interferes with infection by tobacco mosaic virus. *Virology* 20:507–17
60. Loebenstein, G., Sela, B., van Praagh, T. 1969. Increase of tobacco mosaic local lesion size and virus multiplication in hypersensitive hosts in the presence of actinomycin D. *Virology* 37:42–48
61. Lovrekovich, L., Lovrekovich, H., Stahmann, M. A. 1968. The importance of peroxidase in the wildfire disease. *Phytopathology* 58:193–98
62. de Maeyer-Guignard, J., de Maeyer, E. 1965. Inhibition of interferon synthesis and stimulation of virus plaque development in mammalian cell cultures after ultra-violet irradiation. *Nature* 205:985–87
63. Maïa, E. 1966. Induction de synthèse d'un facteur antiviral chez le Tabac et la *Datura* par application d'acide phytique. *C.R. Acad. Sci. Paris* 262:2099–2101
64. Maïa, E., Morel, G. 1965. Action de l'acide phytique sur l'installation et la multiplication du virus de la mosaïque du Tabac. *C.R. Acad. Sci. Paris* 261:2727–29
65. Mandryk, M. 1963. Acquired systemic resistance to tobacco mosaic virus in *Nicotiana tabacum* evoked by stem injection with *Peronospora tabacina* Adam. *Aust. J. Agr. Res.* 14:315–18
66. Mann, E. W. 1969. Inhibition of tobacco mosaic virus by a bacterial extract. *Phytopathology* 59:658–62
67. Martin, C., Scalla, R., Meignoz, R., Gallet, M. 1969. Etude au microscope électronique, chez le *Nicotiana Xanthi* n.c., des cellules vivantes situées au voisinage d'une lésion nécrotique d'hypersensibilité au virus de la mosaïque du Tabac. *C.R. Acad. Sci. Paris* 286:2183–85
68. Matsui, C., Yamaguchi, A. 1966. Some aspects of plant viruses *in*

situ. *Advan. Virus Res.* 12: 127–74
69. Matthews, R. E. F. 1970. *Plant Virology*. Academic Press, New York & London. 778 pp.
70. Merrett, M. J., Bayley, J. 1969. The respiration of tissues infected by virus. *Bot. Rev.* 35: 372–92
71. Merrett, M. J., Sunderland, D. W. 1967. The metabolism of carbon-14 specifically labelled glucose in leaves showing tobacco mosaic virus induced local necrotic lesions. *Physiol. Plant.* 20:593–99
72. Milne, R. G. 1966. Electron microscopy of tobacco mosaic virus in leaves of *Chenopodium amaranticolor*. *Virology* 28: 520–26
73. Milne, R. G. 1966. Electron microscopy of tobacco mosaic virus in leaves of *Nicotiana glutinosa*. *Virology* 28:527–32
74. Milo, G. E., Santilli, V. 1967. Changes in the ascorbate concentration of pinto bean leaves accompanying the formation of TMV-induced local lesions. *Virology* 31:197–206
75. Milo, G. E., Srivastava, B. I. S. 1969. Effect of cytokinins on tobacco mosaic virus production in local-lesion and systemic hosts. *Virology* 38:26–31
76. Nachman, I., Loebenstein, G., van Praagh, T., Zelcer, A. 1971. Increased multiplication of cucumber mosaic virus in a resistant cucumber cultivar caused by actinomycin D. *Physiol. Plant Pathol.* 1:67–71
77. Nagaich, B. B., Singh, S. 1970. An antiviral principle induced by potato virus X inoculation in *Capsicum pendulum* Willd. *Virology* 40:267–71
78. Nagaich, B. B., Upadhya, M. D., Prakash, O., Singh, S. J. 1968. Cytoplasmically determined expression of symptoms of potato virus X crosses between species of *Capsicum*. *Nature* 220:1341–42
79. Nakagaki, Y., Hirai, T. 1971. Effect of detached leaf treatment on tobacco mosaic virus multiplication in tobacco and bean leaves. *Phytopathology* 61:22–27
80. Nakagaki, Y., Hirai, T., Stahmann, M. A. 1970. Ethylene production by detached leaves infected with tobacco mosaic virus. *Virology* 40:1–9
81. Nienhaus, F., Hoogen, H. 1970. Stoffwechselphysiologische Veränderungen in der Pflanze nach Virusinfektion unter Einfluss von Wundreiz. II. Enzymaktivitätsbestimmungen nach Acrylamidgelelektrophorese. *Phytopathol. Z.* 69:38–48
82. Nixon, H. L. 1956. An estimate of the number of tobacco mosaic virus particles in a single hair cell. *Virology* 2:126–28
83. Novacky, A., Hampton, R. E. 1968. Peroxidase isozymes in virus-infected plants. *Phytopathology* 58:301–05
84. Parish, C. L., Zaitlin, M., Siegel, A. 1965. A study of necrotic lesion formation by tobacco mosaic virus. *Virology* 26:413–18
85. Pierpoint, W. S. 1968. Cytochrome oxidase and mitochondrial protein in extracts of leaves of *Nicotiana glutinosa* L. infected with tobacco mosaic virus. *J. Exp. Bot.* 19:264–75
86. Price, W. C. 1964. Strains, mutation, acquired immunity, and interference. In *Plant Virology*, eds. M. K. Corbett, H. D. Sisler. 93–117. Univ. Florida Press, Gainesville 527 pp.
87. Ragetli, H. W. J. 1967. Virus-host interactions, with emphasis on certain cytopathic phenomena. *Can. J. Bot.* 45:1221–34
88. Ragetli, H. W. J., Weintraub, M., Lo, E. 1970. Degeneration of leaf cells resulting from starvation after excision. II. Correlation with water movement and effect on virus synthesis. *Can. J. Bot.* 48:1923–29
89. Ragetli, H. W. J., Weintraub, M., Rink, U. M. 1966. Latent acid phosphatase in chloroplasts. *Can. J. Bot.* 44:1723–25
90. Rappaport, I., Wildman, S. G. 1957. A kinetic study of local lesion growth on *Nicotiana glutinosa* resulting from tobacco mosaic virus infection. *Virology* 4:265–74
91. Resconich, E. C. 1961. Heat-induced susceptibility to tobacco mosaic virus and thermal injury in bean. *Virology* 13:338–47
92. Ross, A. F. 1961. Localized ac-

quired resistance to plant virus infection in hypersensitive hosts. *Virology* 14:329–39

93. Ross, A. F. 1961. Systemic acquired resistance induced by localized virus infections in plants. *Virology* 14:340–58
94. Ross, A. F. 1966. Systemic effects of local lesion formation. In *Viruses of Plants,* eds. A. B. R. Beemster, J. Dijkstra, 127–50. North Holland Publishing Co., Amsterdam, Holland. 342 pp.
95. Ross, A. F., Israel, H. W. 1970. Use of heat treatments in the study of acquired resistance to tobacco mosaic virus in hypersensitive tobacco. *Phytopathology* 60:755–70
96. Rubin, B. A., Ivanova, T. M. 1963. On the oxidase function of plant peroxidase. *Life Sci.* 2: 281–89
97. Samuel, G. 1931. Some experiments on inoculating methods with plant viruses, and on local lesions. *Ann. Appl. Biol.* 18: 494–507
98. Sander, E. 1969. Stimulation of biosynthesis of tobacco mosaic virus by antimetabolites. *J. Gen. Virol.* 4:235–44
99. Sela, B., Loebenstein, G., van Praagh, T. 1969. Increase of tobacco mosaic virus multiplication and lesion size in hypersensitive hosts in the presence of chloramphenicol. *Virology* 39: 260–64
100. Sela, I., Applebaum, S. W. 1962. Occurrence of antiviral factor in virus-infected plants. *Virology* 17:543–48
101. Sela, I., Harpaz, I., Birk, Y. 1965. Suppression of virus infectivity in diseased plant tissue following treatment with an antiviral factor from virus-infected plants. *Virology* 25:80–82
102. Sela, I., Harpaz, I., Birk, Y. 1966. Identification of the active component of an antiviral factor isolated from virus-infected plants. *Virology* 28:71–78
103. Semal, J. 1967. Effects of actinomycin D in plant virology. *Phytopathol. Z.* 59:55–71
104. Shalla, T. A. 1964. Assembly and aggregation of tobacco mosaic virus in tomato leaflets. *J. Cell Biol.* 21:253–64
105. Silberschmidt, K., Caner, J. 1967. Physiological factors responsible for a specific form of resistance against plant virus infection. *Arq. Inst. Biol. S. Paulo* 34:127–33
106. Simons, T. J., Ross, A. F. 1965. Effect of 2,4-dichlorophenoxyacetic acid on size of tobacco mosaic virus lesions in hypersensitive tobacco. *Phytopathology* 55:1076–77
107. Simons, T. J., Ross, A. F. 1970. Enhanced peroxidase activity associated with induction of resistance to tobacco mosaic virus in hypersensitive tobacco. *Phytopathology* 60:383–84
108. Simons, T. J., Ross, A. F. 1971. Metabolic changes associated with systemic induced resistance to tobacco mosaic virus in Samsun NN tobacco. *Phytopathology* 61:293–300
109. Solymosy, F. 1970. Biochemical aspects of hypersensitivity to virus infection in plants. *Acta Phytopathol. Hung.* 5:55–63
110. Solymosy, F., Farkas, G. L. 1963. Metabolic characteristics at the enzymatic level of tobacco tissues exhibiting localized acquired resistance to virus infection. *Virology* 21:210–21
111. Spencer, D. F., Kimmins, W. C. 1971. Ultrastructure of tobacco mosaic virus lesions and surrounding tissue in *Phaseolus vulgaris* var Pinto. *Can. J. Bot.* 49:417–21
112. Stein, A., Loebenstein, G. 1970. Induction of resistance to tobacco mosaic virus by poly I. poly C in plants. *Nature* 226: 363–64
113. Takebe, I., Otsuki, Y. 1969. Infection of tobacco mesophyll protoplasts by tobacco mosaic virus. *Proc. Nat. Acad. Sci.* 64: 843–48
114. Tanguy, J. 1971. Quelques aspects du métabolisme des composés phénoliques chez les *Nicotiana* hypersensibles au virus de la mosaïque du Tobac souche commune (V.M.T.). *Physiol. Vég.* 9:169–87
115. Thomas, P. E., Fulton, R. W. 1968. Correlation of ectodesmata number with nonspecific resistance to initial virus infection. *Virology* 34:459–69
116. Thrower, L. B. 1965. A radioauto-

graphic study of the formation of local lesions by tobacco mosaic virus. *Phytopathology* 55: 558–62

117. Tu, J. C., Hiruki, C. 1971. Electron microscopy of cell wall thickening in local lesions of potato virus-M infected red kidney bean. *Phytopathology* 61: 862–68

118. van Fleet, D. S. 1962. Histochemistry of enzymes in plant tissues. In *Handbuch der Histochemie* Vol. 7, part II, eds. W. Graumann, K. Neumann. 1–38. G. Fischer Verlag, Stuttgart. 386 pp.

119. van Kammen, A., Brouwer, D. 1964. Increase of polyphenoloxidase activity by a local virus infection in uninoculated parts of leaves. *Virology* 22:9–14

120. van Loon, L. C., van Kammen, A. 1970. Polyacrylamide disc electrophoresis of the soluble leaf proteins from *Nicotiana tabacum* var. 'Samsun' and 'Samsun NN.' II. Changes in protein constitution after infection with tobacco mosaic virus. *Virology* 40:199–211

121. Weintraub, M., Kemp, W. G., Ragetli, H. W. J. 1960. Studies on the metabolism of leaves with localized virus infections. *Can. J. Microbiol.* 6:407–15

122. Weintraub, M., Ragetli, H. W. J. 1961. Cell wall composition of leaves with a localized virus infection. *Phytopathology* 51: 215–19

123. Weintraub, M., Ragetli, H. W. J. 1964. An electron microscope study of tobacco mosaic virus lesions in *Nicotiana glutinosa* L. *J. Cell Biol.* 23:499–509

124. Weintraub, M., Ragetli, H. W. J., John, V. T. 1967. Some conditions affecting the intracellular arrangement and concentration of tobacco mosaic virus particles in local lesions. *J. Cell Biol.* 35:183–92

125. Wheeler, H., Pirone, T. P. 1969. Pathotoxin-induced disease resistance in plants. *Science* 166: 1415–17

126. White, N. H. 1969. The local lesion reaction and resistance in plants to systemic virus disease. *Aust. J. Sci.* 31:223–25

127. Wu, J. H., Blakely, L. M., Dimitman, J. E. 1969. Inactivation of a host resistance mechanism as an explanation for heat activation of TMV-infected bean leaves. *Virology* 37:658–66

128. Wu, J. H., Dimitman, J. E. 1970. Leaf structure and callose formation as determinants of TMV movement in bean leaves as revealed by uv irradiation studies. *Virology* 40:820–27

129. Yarwood, C. E. 1958. Heat activation of virus infections. *Phytopathology* 48:39–46

130. Yarwood, C. E. 1960. Localized acquired resistance to tobacco mosaic virus. *Phytopathology* 50:741–44

131. Yoshizaki, T. 1966. On the inhibitory effect induced by tobacco mosaic virus in *Phaseolus vulgaris* plants. *Japan J. Microbiol.* 10:85–91

132. Yu, L. M., Hampton, R. E. 1964. Biochemical changes in tobacco infected with *Colletotrichum destructivum*—I. Fluorescent compounds, phenols, and some associated enzymes. *Phytochemistry* 3:269–72

PHYTOALEXINS[1]

3549

J. Kuć

Departments of Biochemistry, & Botany and Plant Pathology, Purdue University, Lafayette, Indiana

Introduction

Phytoalexins have been an object of continuing debate by plant scientists for more than three decades. Some of these debates have been mainly semantic; others have involved basic concepts of the nature of injury and disease; still others have been caused by controversy about the importance of the Phytoalexin Theory and the effectiveness of the protection these compounds provide against disease. This review certainly will not end these debates, but I hope it will facilitate more complete understanding of past researches on phytoalexins and stimulate future researches on the nature and role of these compounds.

Müller & Börger (125) originally defined phytoalexin in a restricted sense —a chemical compound produced only when the living cells of the host are invaded by a parasite and undergo necrobiosis. In 1956 Müller (123) redefined phytoalexins as "antibiotics which are produced as a result of the interaction of two different metabolic systems, the host and parasite, and which inhibit the growth of microorganisms pathogenic to plants." This review has been written according to my concept that the term phytoalexins should serve as an umbrella under which chemical compounds contributing to disease resistance can be classified whether they are formed in response to injury, physiological stimuli, the presence of infectious agents or are the products of such agents.

The Irish Potato

Phytophtohora infestans is the pathogen commonly used in studies of phytoalexins of potato. Chlorogenic and caffeic acid, scopolin, α-solanine, α-chaconine, solanidine, rishitin, and phytuberin all accumulate after infection and, with the exception of rishitin and phytuberin, are present in the peel of tubers at levels equal to or greater than those produced after infection of peeled slices. Both resistance controlled by major (R) genes and susceptibility to races of *P. infestans* are clearly expressed and can be differentiated in tuber slices both with respect to visible symptoms and biochemical altera-

[1] Journal paper no. 4638 of the Purdue Agricultural Experiment Station Lafayette, Indiana 47907.

tions. These responses can be differentiated from the wounding response qualitatively and quantitatively by the use of appropriate controls of uninoculated slices.

Chlorogenic and caffeic acids.—The association of chlorogenic and caffeic acids and related compounds with the disease resistance of potato tubers has been reviewed (42, 104, 107–110, 181, 202), but doubt still exists regarding their role in disease resistance. Chlorogenic and caffeic acids are present in all parts of the potato plant. They are produced in the potato tuber in response to injury and infection with a broad spectrum of pathogens and nonpathogens, and may accumulate or are present in some healthy tissues at levels which are inhibitory to some microorganisms. Injury or infection generally causes their rapid oxidation and the oxidation products are also toxic to a broad spectrum of microorganisms. The ubiquitous nature of the compound and its repeated association with the response of plants to stress & infection have led many investigators to assign chlorogenic acid or its oxidation products a role in containing the invading microorganisms.

Rubin & Aksenova (157) demonstrated twice as much chlorogenic acid in tubers of a cultivar resistant to *P. infestans* as compared to that in a susceptible cultivar; polyphenoloxidase activity was much greater and persisted longer in resistant tubers after infection. They suggested the importance of quinones as chemical barriers to the development of the pathogen. Oxidation products of chlorogenic acid, but not the acid, are toxic to *P. infestans* (69). Sokolova, Savel'eva & Rubin (177) reported less chlorogenic acid and more caffeic acid in resistant than in susceptible tubers after inoculation; they suggested mobilization of chlorogenic acid to sites of infection and its rapid oxidation to quinones in resistant tubers. Polyphenols accumulated during early stages of infection following inoculation of tubers with incompatible races of *P. infestans* (186–188).

Thickness of infected slices was important in determining phenol accumulation, and differences between compatible and incompatible interactions were evident only with thick slices (185). Chlorogenic acid is the principle phenol accumulating in cut tissue or tissue inoculated with incompatible races of *P. infestans* (159). Though a 25-fold increase in chlorogenic acid occurred after inoculation with an incompatible race, this was only one third that found in aged cut slices. From earlier reports (187, 188) it appears that compatible races prevented or more markedly reduced accumulation than incompatible races. The tissue zone where the metabolism of phenols was accelerated by infection was 10–15 cells in thickness (158). During the first 24 hr after inoculation with an incompatible race, the content of total phenols increased more rapidly in this zone than in the corresponding cut tissue. Tomiyama et al (185) suggested that oxidized products of chlorogenic and ascorbic acid are inhibitors of aldolase and hence shift respiration to the pentose pathway leading to increased synthesis of phenols.

Scopolin.—Scopolin, a derivative of ferrulic acid, accumulates in potato tubers infected with *P. infestans* (92). Accumulation occurs in the vacuoles of living cells and possibly in cell walls of necrotic tissue, and greater accumulation was observed in susceptible than resistant cultivars (36). A time study of scopolin accumulation after infection with compatible and incompatible races of the fungus is needed.

Other phenols.—Cultures of suspended tuber cells susceptible to *P. infestans* supported vigorous growth of the fungus, whereas cultures of the resistant cells inhibited the fungus (154, 155). Several days after inoculation, the suspension liquor from the resistant but not susceptible cultivar caused lysis of *P. infestans* zoospores. Liquors from uninoculated tissue were not inhibitory to the fungus. The toxic liquor from the resistant culture contained p-hydroxybenzoic, vanillic and salicylic acids and these acids could account for most of the inhibitory activity. Tomiyama and coworkers (161) and Metlitskii & Ozeretskovskaya (120) still consider oxidation products of phenols as having a role in the disease resistance of potato to *P. infestans*. Tomiyama et al state (161)

> Apparently, oxidized phenolics played an important role in slowing growth, but not causing the death, of hyphae. Therefore it is suggested that the infecting hyphae may be arrested at various stages such as that of cell wall browning and of rishitin formation depending upon the degree of aggressiveness of the parasite and the defense activity of the host tissue.

Isoprenoid derivatives.—Rishitin, a fungitoxic terpenoid has been isolated (28, 189, 192) from tubers infected with *P. infestans*. The compound is a bicyclic norsesquiterpene alcohol ($C_{14}H_{22}O_2$). Its structure was established by infrared, ultraviolet, nuclear magnetic resonance and mass spectral data (98). Rishitin could not be detected in fresh tuber tissue or in culture filtrates and extracts of the fungus. A trace of rishitin was found in sliced non inoculated potato tissue and in compatible interactions of host and pathogen, whereas 120 mg/kg fr wt of tissue was detected in the incompatible interaction.

The development of *P. infestans* in compatible and incompatible tuber or leaf petiole tissue and the rate of rishitin accumulation suggest that rishitin has a role in disease resistance (161, 162). Rishitin was first detected when reduction in growth of the intracellular hyphae began and reached levels sufficient almost completely to inhibit hyphal growth when development of the lesion was about to stop. The close association of rishitin accumulation with host cell death and browning suggested that cell death may be a trigger for the synthesis of rishitin. Browning and restricted cell death induced by dry ice, chlorogenic acid, L-dihydroxy-phenylalanine, catechol, dimethylsulfoxide, mercuric chloride, cupric chloride and tyrosinase, however, did not induce rishitin accumulation, and tuber slices treated with boiled cell-free sonicates of *P. infestans* accumulated high levels of rishitin with little browning (193).

Rishitin and other isoprenoid derivates also accumulate in response to inoculation with numerous nonpathogens of potato (189, 192, 193). Sato et al (163) demonstrated that rishitin accumulated in incompatible interactions of four cultivars (R_1,R_2,R_3,R_4) inoculated with race 0 of the fungus. A cultivar susceptible to all the known races of the fungus accumulated rishitin when dipped into a cell-free homogenate of the fungus. Varns, Kuć & Williams (195) and Varns (192) demonstrated a consistent response of tubers to infection by incompatible races of *P. infestans* with eleven cultivars and three races of the fungus. The response included rapid necrosis and the accumulation of rishitin and phytuberin, a new aliphatic, unsaturated, sesquiterpene acetate ($C_{17}H_{26}O_4$). Two additional terpenoids, rishitinol $C_{15}H_{22}O_2$ (97) and lyubimin (121), have been isolated from tuber tissue infected by incompatible races of *P. infestans*. The potential for resistance appears to exist in completely susceptible cultivars but this potential is not expressed. This is consistent with disease resistance mechanisms found in animals and is the basis of protection by immunization. Rishitin does not appear to have a role in the establishment of specificity between host and pathogen (184).

Rishitin also influences the metabolism of plants (189), and may have a role in conditioning a resistance environment in addition to its role as an inhibitor of fungal growth (94).

Tomiyama (182) suggested that a compatible race of the fungus delayed the hypersensitive collapse of cells in the presence of an incompatible race. Varns & Kuć (194) demonstrated that a compatible interaction suppressed both necrosis and the accumulation of rishitin and phytuberin in the tuber. An alteration of cellular response in the host during the compatible interaction suppressed the ability of the host to respond normally to a subsequent infection by an incompatible race. Once the host was inoculated with an incompatible race, suppression from a subsequent inoculation with a compatible race did not occur. An incubation period of 12 hr was sufficient either to establish suppression or elicit the hypersensitive response. The races of the fungus used were not mutually antagonistic in vitro. A compatible interaction also suppressed the hypersensitive host response to sonicated cell-free homogenates of the fungus.

It appears that the hypersensitive response to incompatible races and its suppression by compatible races are active metabolic processes. In at least the potato—*P. infestans* interaction, susceptibility may be determined by the suppression of a biochemical response that is quite general for a broad spectrum of fungi. The "turning off" or preventing of a general resistance or immune response apparently is a rare occurrence in nature, and this may account for the fact that susceptibility is the exception rather than the rule. The trigger for terpenoid accumulation may be a fungal cell wall component with lipophylic properties (183, Currier & Kuć, unpublished data), and it may influence IAA levels. Fehrman & Dimond (67, 68) reported the IAA content of sliced tubers inoculated with compatible and incompatible races of *P. infestans* increased 5–10 and 10–20 fold, respectively.

At least two other isoprenoids, the steroid-glycoalkaloids α-solanine and α-chaconine, may also be associated with resistance. They have been reported in potato tubers and foliage (4, 5, 113, 119, 134, 146, 201), and appear localized around sites of injury in tubers (119). The steroid glycoalkaloids are largely restricted to the peel of whole tubers (5). The major antifungal components in extracts of potato peel are α-solanine and α-chaconine (5). Locci & Kuć (113) reported many isoprenoid derivatives accumulated in tubers of two cultivars inoculated with either *H. carbonum,* a pathogen of corn, or two incompatible races of *P. infestans.* The authors suggested that the accumulation of the compounds was due to physiological stress induced by the microorganisms.

Tomiyama et al (184) verified that cutting induced tubers to accumulate high concentrations of α-solanine and α-chaconine in tissue close to the cut surface. They reported 538 mg of steroid-glycoalkaloids (including α-solanine and α-chaconine)/kg fr wt of tissue 48 hr after slicing. In the incompatible and compatible interactions the quantities of steroid-glycoalkaloids were 3 and 72 mg/kg fr wt of tissue, respectively. Rishitin was not detected in fresh tuber or sprouts and occurred in trace amounts in cut slices. Approximately 120 mg of rishitin/kg fr wt of tissue accumulated in the incompatible and 0.44 mg in the compatible interaction.

It appears that infection of tubers with *P. infestans* diverts biosynthesis from α-solanine and α-chaconine to rishitin in the resistant reaction, whereas accumulation of the steroid-glycoalkaloids and rishitin are markedly inhibited in the susceptible interaction. Shih & Kuć (unpublished data) inoculated two R_1 cultivars with compatible and incompatible races of *P. infestans* and verified Tomiyama's data (184) that inoculation suppresses the accumulation of steroid-glycoalkaloids in potato slices. Steroid glycoalkaloid accumulation in inoculated and control slices was approximately equal during the first 16–24 hr in the top mm of tissue. Accumulation continued to increase sharply in noninoculated tissue for an additional 48 hr, but it was markedly reduced in inoculated tissue. Determinations of rishitin in the top mm of incompatible interactions indicated a marked increase 24-32 hr after inoculation which lasted for 36–48 hr. Thus, the time when marked suppression of steroid glycoalkaloid accumulation became evident approximately coincided with the time when rapid rishitin accumulation was initiated.

Some evidence argues against the importance of rishitin in all tissues of the potato. Potato sprouts contain very large amounts of α-solanine and α-chaconine, rishitin accumulation was detected only in compatible interactions of sprouts and *P. infestans* (193). Rishitin and phytuberin also could not be detected in extracts of leaf blade tissue after inoculations with compatible or incompatible races (Shih & Kuć, unpublished data).

Rishitin has been detected in tomato fruit inoculated with *P. infestans* (164); however, it is not clear whether the variety used was susceptible or resistant to the fungus. Leaf blade tissue of potato is very high in α-solanine and α-chaconine and that of tomato contains tomatine. Though rishitin has recently been reported in infection droplets on leaves inoculated with incompatible races of *P. infestans* (122), further work is necessary to establish whether rishitin and phytuberin accumulate only in tissues containing little or no α-solanine and α-chaconine, e.g., freshly sliced tubers inoculated with fungi, or whether its accumulation is part of the resistance mechanism of the entire plant.

Another consideration complicating the assignment of a role in disease resistance to a compound found at fungitoxic amounts in plants is evident when considering the amounts of α-solanine and α-chaconine in potato foliage. The amount of these compounds in susceptible and resistant varieties is apparently more than sufficient to inhibit completely spore germination and hyphal growth of *P. infestans*. The steroid alkaloids apparently are present in foliage in a form that is not inhibitory or they are compartmentalized in areas unavailable to the fungus (Shih & Kuć, unpublished data).

Amici & Locci (9) obtained strains of *H. carbonum* resistant to α-solanine by growing the fungus in media containing a high concentration of the compound. Such strains would be useful in determining whether α-solanine and α-chaconine have a role in the resistance of potato to *H. carbonum*. Similarly, the development of strains of *P. infestans* resistant to rishitin and phytuberin would help clarify the role of these compounds in resistance.

Summary.—The primary event in determining a compatible or incompatible reaction of potato to *P. infestans* is the recognition of the race of the fungus by the host and/or recognition of the host by the fungus (192, 194). This event occurs within hours (182, 192, 194) and probably seconds after penetration and is responsible for establishing the nature of the disease reaction. All metabolic alterations including the accumulation of phytoalexins are the result of the initial recognition event. A top priority for research in this area should be assigned to determining the nature of this recognition and the structure of the compound(s) triggering the hypersensitive host response or its suppression.

The Garden Pea

The technique for studying the accumulation of post-infectional compounds in green bean pods (123, 124) was employed by Cruickshank & Perrin (43) to detect and isolate pisatin from the pea, *Pisum sativum* L. Endocarps of detached open pea pods were inoculated with spore suspensions, and the drops of liquid remaining on the tissue, containing diffusate, were recovered after suitable incubation. The fungitoxic material was extracted from the crude aqueous diffusate with light petroleum ether. The structure of pisatin was established as 3-hydroxy-7-methoxy-4′,5′-methylenedioxy-chromanocoumaran (140). Pisatin is a weak antibiotic with a broad biological spectrum (40). Mycelial growth of *Monolinia fructicola* is three times more sensitive to pisatin than is spore germination. Fungi pathogenic to pea are relatively insensitive to the amounts of pisatin accumulating after infection, whereas, nonpathogens of pea are generally sensitive.

Pisatin is not formed as a result of gross mechanical injury (44) although a low concentration of pisatin was later recorded in uninoculated and incubated open pea pods (46). Pisatin accumulation was stimulated by nineteen fungi- and spore-free germination fluids (44). Bacteria did not stimulate accumulation of pisatin. A broad spectrum of metabolic inhibitors (140) and ethylene (31) also stimulate pisatin accumulation. Pisatin did not accumulate in pods inoculated immediately after treatment at 45°C for 2 hr, and the pod tissue was susceptible to nonpathogens (48). Exposure of pods to anaerobic storage for 6, 9, or 12 days prior to inoculation also resulted in reduced pisatin accumulation and susceptibility to *M. fructicola* (49).

Cruickshank & Perrin reported pisatin is stable in infected pea pods and is not degraded by the pea pathogen *A. pisi* (48). Several workers (57, 58, 127, 129), however, reported pisatin is degraded by a number of pathogens of pea but not appreciably by nonpathogens. They suggest that a mechanism for the detoxication of pisatin may determine whether an organism can parasitize pea. Cruickshank, Biggs & Perrin (42) take issue with this suggestion. A subsequent study (59) demonstrated that degradation of pisatin by two pea pathogens in vitro was influenced by the carbohydrate source in the nutrient medium. Synthesis of pisatin-degrading enzymes may be subject to catabolite repression and sugar levels in plants could, therefore, indirectly influence phytoalexin concentration.

The formation of pisatin by 58 cultivars, 9 numbered lines of *P. sativum* and *P. arvense,* and 3 other species of *Pisum,* following inoculation by *M. fructicola,* is further evidence supporting a role for pisatin in the disease resistance mechanism of pea pods (47). Heath & Wood (79) reported an important quantitative study of the reaction of pea leaflets to fungi. It appears that pisatin has a role in the limitation and the size of lesions in two susceptible host-pathogen interactions. Bailey (11) reported a sterile culture filtrate of *Penicillium expansum* induced pisatin accumulation in pea leaf discs. The amount of pisatin produced by leaves was shown to decrease as they senescenced.

Biosynthesis of pisatin appears to require participation of the acetate-malonate pathway and the shikimic acid pathway (71, 72). The activity of phenylalanine deaminase, an enzyme that may be in the pathway of pisatin biosynthesis, was found to increase ten-fold when spore suspensions of pathogenic and nonpathogenic fungi were applied to detached pea pods (73, 75). The significance of increased phenylalanine deaminase activity is somewhat obscure. It would appear the increase in activity is closely associated with a stress response of the plant, but its significance as a control point in phytoalexin biosynthesis would only be apparent if the activity of the enzyme was a limiting factor in the biosynthetic pathway. Pisatin accumulation is stimulated by ultra-violet radiation (77), DNA intercalating compounds with planar triple ring systems (74), and many microbial metabolites including well known antibiotics (165, 166). Schwochau & Hadwiger (76, 167) presented an induction hypothesis for host resistance based on the Jacob-Monod model for gene activation in bacteria. They propose that any one of several microbial metabolites, in low concentrations, powerfully induces high metabolic activity and phytoalexin biosynthesis in the host resulting in and from increased protein synthesis. The induction occurs via the derepression of certain genes by selectively interfering with the negative gene control mechanism. Cruickshank (42) cites the need for further clarification of certain aspects of the induction hypothesis.

The induction hypothesis is not a new explanation for disease resistance and many reports support it (34, 39, 41, 42, 46, 62, 80, 105–110, 123, 125). Kuć (109) postulated that immunity and varietal resistance are dependent on the expression rather than the presence or absence of structural genes and this concept may prove important in developing new methods of disease control. The final proof for the role of a phytoalexin in immunity or varietal resistance must be based on tissue normally used by the infectious agent for entrance and development.

Green Bean

Müller (124) reported the presence of antifungal activity in extracellular fluids collected from seed cavities of green bean pods inoculated with *M. fructicola.* Cruickshank & Perrin (45) and Perrin (139) isolated phaseollin from detached bean pods inoculated with *M. fructicola* and established its structure as 7-hydroxy-3′,4′-dimethylchromanocoumaran. Evidence is offered for its role in the disease resistance mechanism of bean pods (51).

Unlike pisatin production in pea, phaseollin is formed by bean endocarp (51) and foliage (179) in response to inoculation with some but not all bacteria. Many bacteria, however, are insensitive to high levels of the compound. Phaseollin and another unidentified compound, apparently phenolic and more water soluble than phaseollin, were reported produced by green bean seedlings in response to infection by the pathogen *Rhizoctonia solani* (142, 144). Both compounds were toxic to *R. solani* in vitro and the authors suggested the compounds are associated with a mechanism responsible for restricting the size of lesions in susceptible plants. The compounds also have been found in bean seedlings inoculated with other fungi (142). Subsequent investigations (55, 143) substantiated the production of a hydrophylic antifungal compound.

The first report associating phaseollin in a mechanism for varietal resistance was by Rahe et al (152). A resistant cultivar showed a fleck response 60–72 hr after inoculation with *Colletotrichum lindemuthanum,* and the level of phaseollin in the tissue at this time was much higher than in a susceptible cultivar. Accumulation of phaseollin, however, is not always linked to necrosis on bean foliage (152, 179). A number of unidentified compounds accumulate in infected tissue in addition to phaseollin (55, 142–144, 151–153, 176), and this supports the concept that more than a single compound may contribute to resistance. Studies of induced resistance (151–153) also suggest that phaseollin and other phenols associated with visible cell collapse are not the only sources of protection. Bean plants inoculated with two nonpathogens or a cultivar nonpathogenic race were protected against a cultivar pathogenic race of *C. lindemuthianum* (153). Protection against *C. lindemuthianum* was also demonstrated by infecting seedlings with a cultivar pathogenic race of fungus that was heat inactivated in the plant after a period of incubation (151).

The concept of induced plant protection is not new (34) and has been suggested by many workers. Works of Rahe et al (151–153), Elliston et al (61), and Berard et al (18) help establish the validity of the concept with fungi. Elliston et al (61) reported that resistance induced in green bean hypocotyls by a varietal nonpathogenic race of *C. lindemuthianum* was exhibited by cells distant from the inducing interaction. Induced resistance was expressed after the varietal pathogenic race had penetrated, and was microscopically indistinguishable from the resistant reaction to a varietal nonpathogenic race. Induced resistance involved effects upon the cells of the plant rather than upon the spores of the challenge inoculum on the surface of the plant. It would appear that protected tissue recognized a varietal pathogenic race as a varietal nonpathogenic race.

Further substantiation of the importance of gene expression in varietal resistance was offered by the same authors (unpublished data). They demonstrated that protection against anthracnose could be induced in bean cultivars susceptible to all four races of *C. lindemuthianum* by using *Colletotrichum* sp. nonpathogenic on bean. Berard et al (18) reported a factor, produced by the interaction of bean hypocotyls and cultivar nonpathogenic races of *C. lindemuthianum,* induced resistance to cultivar pathogenic races. The factor diffused into water droplets at the sites of interaction prior to symptom expression. The factor was not obtained with interactions of bean and cultivar pathogenic races, and it was not inhibitory to the germination or growth of the fungus.

Many chemical reagents including low concentrations of metabolic inhibitors induce phaseollin accumulation in bean pods (51, 75, 82, 141). Monilicolin A, a peptide from the mycelium of *M. fructicola,* stimulated phaseollin accumulation at $2.5 \times 10^{-9} M$ (50). Monilicolin A does not appear to be fungitoxic or phytotoxic at levels which stimulate phaseollin production. Phaseollin appears synthesized via joint participation of the shikimic acid and acetate-malonate pathways (83).

The Carrot

The metabolic response of carrot root to infection or injury results in accumulation of a number of compounds including chlorogenic acid and 6-methoxymellein (MM) (37–39, 160). MM and chlorogenic acid reach fungitoxic levels around infected sites within 24 hr after inoculation with several fungi nonpathogenic to carrot. The resistance of carrot root slices to some but not all fungi nonpathogenic to carrots is lost by holding slices at 43–45°C for 3 hr immediately prior to inoculation, and the treatment reduces the accumulation of chlorogenic acid and MM to nonfungitoxic levels (160). Allowing heat-treated carrots to remain at room temperature for 2–3 days prior to inoculation restored resistance and the production of chlorogenic acid and MM. Inoculation of intact carrot roots with *C. fimbriata* resulted in the accumulation of 1×10^{-3} to 1×10^{-2} molal MM in the peel 72 hours after inoculation (81). Some mechanisms of disease resistance in carrot root, however, do not depend upon MM or chlorogenic acid accumulation. Spores of *C. fagacearum, C. pilifera,* and *C. coerulescens* do not germinate on carrot slices and these fungi did not stimulate the accumulation of chlorogenic acid or MM (81). The accumulation of MM has been reported with 7 isolates of *C. fimbriata* as well as 5 cultivars of carrot (81).

The accumulation of MM can also be induced by chemicals (39), cold treatment (60, 178), and ethylene (30). The compound is responsible for the condition of bitter carrot, a physiological disorder of carrots. Accumulation of MM is not a specific response to infection but rather a response to stress. Biosynthesis of MM is not established but it appears to proceed via the acetate-malonate pathway (39).

A controversy exists concerning whether MM is synthesized de novo by the carrot in the host-pathogen interaction and the role of the compound as a phytoalexin. MM has been reported produced by two *Sporormia* (10, 118). Traces of a compound, 8-hydroxy-6-methoxy-3-methyl-isocoumarin, differing from MM by the presence of a double bond, has been reported produced by *C. fimbriata* in liquid culture (52, 180). Curtis (52) indicated 18 mg of the isocoumarin were isolated from an ether extract of 7 l of culture medium after 11 days of incubation. If we compare this quantity of isocoumarin to the amount of MM produced in the fungus-carrot interaction, 2.3 g per kg infected carrot tissue 3 days after inoculation, it appears highly unlikely that the fungus is supplying the immediate precursor for MM. The amount produced in response to ethylene (29, 30) is equal to the amount produced in response to fungal infection, and it appears improbable that the quantities of MM accumulating in carrot discs after chemical treatment or ethylene exposure could be accounted for as Stoessl (180) states, "to low levels of fungal infection which may have eluded observation."

Recent reports establish MM and other dihydroisocoumarins (70) as well as the isocoumarin arthemidin (117) as constituents of higher plants. Stoessl's speculation (180, 181) that isocoumarins are not synthesized by higher plants does not appear valid. The work of Chalutz & coworkers (30) suggests that the accumulation of MM in carrot tissue is dependent on the quantity of the ethylene produced by the fungus and that the compound is not responsible for containing the growth of the fungus. Unfortunately, their data do not contain a time study of MM accumulation after inoculation.

The accumulation of MM is not the only determinant of disease resistance in carrot root. I believe, however, that the accumulation of compounds to fungitoxic levels at the sites of fungus penetration and the complex biochemical alterations associated with the synthesis of these compounds are part of the defense response and should not be ignored.

The Soybean

Uehara (190) and Nonaka et al (128) reported the accumulation of fungitoxic compounds in open soybean pods inoculated with four nonpathogens of soybean. Studies of phytoalexin accumulation in intact plants are rare, but a technique for doing so was reported by Klarman & Gerdemann (101, 102). Extracts were leached from soybean plants via strings through wounds inoculated with the pathogen *Phytophtora sojae* or two Phytophtora nonpathogenic to soybean. A soybean variety resistant to *P. sojae* became susceptible when compounds were leached from the wound. Nonpathogens of soybean or a cultivar nonpathogenic race of *P. sojae* induced the production of a fungitoxic compound, but a cultivar pathogenic race of *P. sojae* did not. Plants susceptible to *P. sojae* appeared protected against the pathogen by a fungitoxic compound produced in a resistant inoculated cultivar (33). Soybean plants heated before inoculation became susceptible to severe nonpathogens (32).

It appears the accumulation of fungitoxic compounds is more important to resistance in 0–2 week-old bean plants that in older plants (136). Biehn et al (19) reported phenols accumulated in etiolated seedlings of soybean, green bean, and lima bean inoculated with the pathogen of corn *H. carbonum,* and the plants showed a hypersensitive reaction to infection. Phenols accumulating after inoculation of lima bean and green bean were different from those produced by soybean. The same phenols accumulated when soybean seedlings were inoculated with *H. carbonum, M. fructicola, Trichoderma viride, Cercospora sojiiana* race 1 or *Alternaria* sp., but the quantity accumulated depended upon the fungus used for inoculation. Twenty four-twenty nine hr after inoculation with *H. carbonum,* total phenols increased to levels 4–5 times above those in uninoculated seedlings, and the initial increase in phenols accompanied a marked increase in phenylalanine deaminase. One of the major phenols accumulated to levels which inhibited the growth of fungi 70–90% (20).

On the basis of spectrophotometric and solubility data, Klarman & Sanford (103) suggested the inhibitor accumulating in soybean was closely related to phaseollin and pisatin. Early reports (102, 135) associated phytoalexin activity with a yellow-fluorescing compound produced by soybean after infection, but this association has not been confirmed (33).

Keen et al (99, 100, 173) confirmed the earlier work (103) and in a thorough study they identified 6 a-hydroxyphaseollin as a phytoalexin in soybean hypocotyls. On the basis of chromatographic and spectral data, the compound characterized by Keen et al appears to be the same as the major fungitoxic substance reported earlier by Biehn et al (19, 20). Keen et al reported that soybean cultivars carrying either of two allelomorphic resistant genes accumulated 6 a-hydroxy-phaseollin from 10–100 times faster when challenged with incompatible *P. megasperma var. sojae* races than cultivars inoculated with compatible races. In incompatible host-parasite combinations the levels of 6 a-hydroxyphaseollin obtained were 100 to 400 times the ED_{50} concentration for inhibition of mycelial growth of the fungus; however, 6 a-hydroxyphaseollin accumulated to only one to four times the ED_{50} concentration in compatible host-parasite combinations. Hydroxyphaseollin was not detected in nonwounded, unchallenged soybean hypocotyls.

Sweet Potato Root

Infection, injury, and treatment with numerous chemical agents (2, 191) all lead to the accumulation of chlorogenic acid, isochlorogenic acid, caffeic acid, scopoletin, esculetin, umbelliferone, and ipomeamarone in sweet potato root. The peel of sweet potato contains all of the above at levels equivalent to or greater than that produced by infected peeled tissue (Betaincourt & Kuć, unpublished data). Uritani (191) and Akazawa & Wada (2) suggested ipomeamarone may be associated with the disease resistance of sweet potato to *C. fimbriata*. The compound accumulated in inoculated resistant roots in excess of 1% and 0.1% markedly inhibited growth of the fungus in vitro. Ipomeamarone has at least some selective toxicity towards fungi, and is highly toxic to several nonpathogens of sweet potato but only slightly toxic to several pathogens (130). Doubt was cast on the role of ipomeamarone in disease resistance by Weber & Stahmann (199). They reported that inoculation of a susceptible variety of sweet potato with a nonpathogenic isolate induced resistance to a pathogenic isolate. Ipomeamarone, however, was not detected in the protected tissue.

Hyodo, Uritani & Akai (93) inoculated sweet potato roots with *C. fimbriata* isolates from sweet potato, prune, or coffee. Prune and coffee isolates invaded and browned a limited number of cells of sweet potato roots, whereas the potato isolate expanded continuously into inner tissues. The invasion of a resistant variety by the sweet potato isolate was continuous but considerably slower. On the basis of the amount of furano-terpenoids accumulated after infection by each isolate, it appears that furano-terpenoids may not be associated with disease resistance.

The status of ipomeamarone and related furano-terpenoids, including ipomeamaronol (96, 203), is not clear. The mechanism for immunity of sweet potato root to nonpathogenic isolates of *C. fimbriata* may be different from the mechanism for cultivar resistance to pathogenic isolates.

Cotton

Introduction of conidia of *Verticullium albo-atrum* into boll cavities or xylem vessels of excised stems of *Gossypium hirsutum* or *G. barbatense* induced the accumulation of fungitoxic levels of ether-soluble phenols, and the predominant phenol was identified as gossypol (13). Gossypol is normally found in glands distributed throughout leaves, stems, and root cortices of most cotton varieties (1). Several days after inoculating stems and intact plants, the rate of accumulation of gossypol and other related compounds was directly related to host resistance and inversely related to virulence of the pathogen (14). The influence of temperature on the resistance of varieties of *Gossypium* spp to a defoliating strain of *V. albo-atrum* has been studied in detail (16). The speed of gossypol accumulation relative to the speed of secondary colonization by *V. albo-atrum* appears to be important in wilt resistance. Further work (17) demonstrated that avirulent strains, heat-killed conidia or conidia inhibited by heat stimulated gossypol accumulation and increased resistance. It appears that genes for resistance enhance speed of production or the number of compounds produced after infection or stress as suggested by various investigators (76, 109, 167, 204).

Cellular integrity appears necessary for the synthesis of stress metabolites. Irreversible damage to membranes by high levels of fungal toxin or sensitivity to toxin may prevent the accumulation of stress metabolites resulting in susceptibility, whereas low levels of toxin or insensitivity to toxin may induce synthesis and accumulation of the compounds resulting in resistance. This may explain the effect of host-specific toxins.

Broad Bean

Several antifungal compounds have been reported in seedlings of broad bean *Vicia faba* L. (63). Purkayastha & Deverall (147, 148) reported production of phytoalexins in response to inoculation with the pathogens *Botrytis cinerea* or *B. fabae*. Lesions caused by *B. cinerea* remain small and limited, whereas those produced by *B. fabae* spread rapidly throughout the tissue and lead to widespread necrosis. The germ-tube development of *B. cinerea* was markedly inhibited on the leaf surface, but germ-tube growth of *B. fabae* was not inhibited and the fungus spread throughout the leaf (54, 55, 148). Large quantities of phytoalexin accumulated in healthy tissue surrounding lesions of *B. fabae*, and only trace amounts occurred in the centers of the spreading lesions; however, with *B. cinerea*, the largest quantity of phytoalexin accumulated in tissue underlying the infection droplets (56). These results suggest that the difference in pathogenicity between the two fungi depends on their abilities to degrade phytoalexin.

A pre-formed fungitoxic compound has been isolated from *V. faba* and identified as the acetylenic furanoid keto-ester, wyerone (65, 66, 198). Though the compound is highly fungitoxic to some nonpathogens, it is less active against *B. cinerea* (64). In subsequent work Letcher et al (111) demonstrated that wyerone was the methyl-ester of the broad bean phytoalexin, wyerone acid, which is probably responsible for the marked inhibition of *B. cinerea* in broad bean tissues. Infection of broad bean leaves by *B. fabae* led to a 400–500-fold increase in wyerone 3–4 days after inoculation. The observations that *B. cinerea* produced spreading lesions on broad bean if pollen grains of broad bean or strawberry were incorporated into spore suspensions of the fungus has been reported (35). These lesions were indistinguishable from those caused by *B. fabae*.

Cruickshank (41) also reported that an uncharterized phytoalexin named viciatin was produced by pod tissues after inoculation with a conidial suspension of *M. fructicola*.

Alfalfa

An antifungal compound accumulated in spore suspensions of two nonpathogens of alfalfa (*Helminthosporium turcicum* and *Colletotrichum phomoides*) on detached leaves of alfalfa (85). Only a small amount of inhibitor appeared to be present in infection droplets of *Stemphyllium loti*, a weak pathogen of alfalfa; none could be detected in those of the pathogen *S. botryosum* (86, 87, 97). Spore-free culture filtrates of both species of *Stemphyllium* induced phytoalexin formation. These results suggested that a phytoalexin was formed in response to infection with all the fungi tested but was degraded by *S. botryosum* and to a lesser extent by *S. loti*. *S. loti* and *C. phomoides* were also capable of degrading the phytoalexin, but in both cases the resulting compounds were inhibitory to the fungi (88).

One alfalfa phytoalexin has been characterized as (—)-3-hydroxy-9-methoxypterocarpan (175), a naturally occurring compound that has also been isolated from the heartwood of several tropical trees (78, 116, 132). The accumulation of formononetin, daidzein, 7,4′-dihydroxyflavone, 7,3′,4′-trihydroxyflavone, coumesterol, 7-hydroxyl-11, 12-dimethoxycoumesterol, 4′0-methyl coumesterol, sativol, medicagol, and some of the corresponding glycosides have been reported in leaves infected with *Aschochyta imperfecta* (131). Infection with the pathogens *Cylindrocladium scoparium, Colletotrichum trifolii,* and *Uromyces striatus,* also caused accumulation of coumesterol in alfalfa (171).

Red Clover

Several compounds inhibitory to the pathogen *Sclerotinia trifoloiorum* were detected in the leaves of red clover (196, 197). They included the isoflavones formononetin (12), and biochanin A (23). Trifolirhizin (84), a compound similar in spectral properties to pterocarpin (24, 26) was also isolated from red clover roots. Unlike pterocarpin, trifolirhizin is a glucoside; the aglycon was isolated from crushed roots (25) and identified as 3-hydroxy-8,9-methylenedioxypterocarpan (maackiain) (24, 89).

Another phytoalexin produced by red clover leaves inoculated with *Helminthosporum turcicum* was recently identified as 3-hydroxy-9-methoxypterocarpan (medicarpan). Medicarpin is also produced by alfalfa.

Other Phytoalexins

The literature concerning phytoalexins in the *Venturia inaequalis*-apple interaction has recently been reviewed (22, 200). It is doubtful that phloridzin and phloretin oxidation products are the only factors responsible for containing the growth of *V. inaequalis* in hypersensitive tissue (22). Pellizzari (138) found that antibiotics infused into apple shoots, in amounts too small to inhibit growth of *V. inaequalis,* markedly decreased protein synthesis and delayed the browning associated with the hypersensitive reaction in the host. These compounds did not interfere with cell collapse or enhance development of the fungus. Culture filtrates of *V. inaequalis* caused wilting and necrosis of the leaves of 42 resistant but not 70 susceptible plants from the cross Antonovka 34–20 × Golden Delicious (149, 150). Culture filtrates also caused necrosis and wilting of a resistant but not a susceptible crab apple (172). Necrosis of apple foliage led to the accumulation of fungitoxic oxidation products (150). A general mechanism of disease resistance was proposed based on the toxicity of extracellular fungal products to resistant but not susceptible plants (150, 172). The above mechanism was not verified by Boone (22) and Nicholson, Kuć & Williams (unpublished data) using commercial apple varieties, race differentials, and several races of the fungus.

The classic work of Gäumann and associates with orchinol and related compounds in orchids has also been thoroughly reviewed (41, 46, 107). Scopoletin and scoplin markedly increased in tobacco plants infected with *Pseudomonas solanacearum* (169). This was accompanied by a selective increase in the aromatic amino acids phenylalanine, tyrosine, and tryptophane (137). Heat-killed cells of *P. solanacearum* infiltrated into tobacco leaves protected them against the pathogen (115). The factor responsible for the protective response was probably nonspecific and common to several, but not all, bacteria. The protection factor apparently has its effect at some distance from the treated areas (see 18, 61, 114, 115, 151, 153, 156).

Other recent reports of phytoalexins include polyacetylenes from safflower (3, 6–8, 21, 95), benzoic acid from immature apples infected by *Nectria galligena* (27), 4-hydroxy benzaledhyde and vanillin formed in leaf wound sap of *Phaseolus lunatus* (15), unidentified phytoalexins from the interaction of corn and *Helminthosporium turcicum* (112), unidentified oxidation products of arbutin in the interaction of pear and *Erwinia amylovora* (145), phytoalexin-like compounds from the interaction of strawberry and *Phytophtohora fragariae* (126), hydroxymatairesinol in the interaction of Norway spruce and *Fomes annosus* (170), and phytoalexins from infected pods of 14 legume species (174).

Summary

The interactions between many plants and infectious agents have been investigated with regard to phytoalexin synthesis, accumulation, and the role of phytoalexins as inhibitors of the growth and development of infectious agents. The scientists in this area have been extremely ingenious in their use of tissues to study interactions, but many choice tissues remain for study. Avocado and muskmelon halves offer great promise for the collection of large quantities of diffusate. Nevertheless, progress in studies of phytoalexins would have been much slower were it not for the early work with pods, tubers, and other tissues that lend themselves to controlled biochemical experimentation. I believe it is time to deemphasize immunity experiments and tissues of convenience and to emphasize research to explain varietal resistance using tissues normally penetrated and affected by the pathogen.

Plants can synthesize and accumulate compounds in response to many stimuli as varied as polluted air (91) and the infectious agents and chemicals discussed in this review. It would be incredible if these compounds did not influence the growth and development of infectious agents in tissues where they are produced or accumulate. Even the repeated association of a metabolite or enzyme with a resistance reaction, however, does not prove it is responsible for resistance, since it is difficult to differentiate between cause and effect (53, 168).

Many factors probably interplay in determining the outcome of a host-pathogen interaction. The nutritive requirements of the microorganism, presence of suitable substrates for the induction of microbial enzymes, microbial toxins, factors which prevent production of toxins or inactivate them, influence of temperature and light on the plant and microbial metabolism, and the rapidity and magnitude of phytoalexin synthesis and degradation by host and pathogen, are only some of the variables that must be considered. Any alteration in the delicate metabolic balance that exists between a host and successful pathogen, especially during the initial period of infection, could stress the host and result in increased phytoalexin accumulation and possibly resistance. Hence, susceptibility is the exception resulting from a specific adaptation and resistance is the rule in nature. Ozeretskovskaya, Vasyukova & Metlitskii (133) state ". . . it is not likely that the defensive role of antibiotic substances in the phytoimmunity phenomenon is restricted to the formation of only one substance, since the fungitoxicity of individual phytoalexins on the whole is not very high, so that the parasite may adapt to it."

Phytoalexins reported to date generally accumulate for a limited period of time and their amounts then decline. It is just as vital to the plant to "turn on" its synthetic machinery for protection as it is to "shut it off," and the mechanism for the "shutting off" warrants more attention. Research on the complex biochemistry of disease resistance and the biosynthesis and degradation of phytoalexins must have a high priority in the future. Hirai (90) presents some interesting thoughts concerning the nature of the control mechanism in his comparison of biochemical changes associated with fungus and virus infections. He states: "Germination of fungus spores is accompanied by mRNA synthesis and subsequent changes in the ribosomal activity. This RNA seems to correspond to viral RNA, which shows the infectivity. Therefore, interactions between host and parasite can be understood as those between host genes and pathogenic or infective RNA. Locally infected tissues both in fungus and virus diseases had a common shift in the metabolic pattern including respiratory increase, accumulation of phenolic compounds, peroxidase activity and ethylene production."

The phytoalexins are with us to stay; they have already initiated an entire new chapter for biochemistry and plant pathology. Induced plant protection is a tool in studying mechanisms for disease resistance and its application for practical control of disease also deserves much more attention and support. Many investigators have emphasized repeatedly that expression of genetic information is the key to disease resistance. Science has progressed far in this field in the past 10 years, and since knowledge is autocatalytic, we can expect much more progress in the future.

LITERATURE CITED

1. Adams, R., Geissman, T., Edwards, J. 1960. Gossypol a pigment of cottonseed. *Chem. Rev.* 60:555–74
2. Akazawa, T., Wada, K. 1961. Analytical study of ipomeamarone and chlorogenic acid alterations in sweet potato roots infected by *Ceratocystis fimbriata. Plant Physiol.* 36:139–44
3. Aldwinckle, H. 1969. Phytoalexin-like activity in diffusates from safflower leaves inoculated with *Phytophthora drechsleri. Phytopathology* 59:1015
4. Allen, E., Kuć, J. 1964. Steroid alkaloids in the disease resistance of white potato tubers. *Phytopathology* 54:886
5. Allen, E., Kuć, J. 1968. α-Solanine and α-chaconine as fungitoxic compounds in extracts of Irish potato tubers. *Phytopathology* 58:776–81
6. Allen, E., Thomas, C. 1971. Time course of safynol accumulation in resistant and susceptible safflower infected with *Phytophthora drechsleri. Physiological Plant Pathol.* 1:235–40
7. Allen, E., Thomas, C. 1971. A second antifungal polyacetylene compound from *Phytophthora*-infected safflower. *Phytopathology* 61:1107–09
8. Allen, E., Thomas, C. 1971. Trans-trans-3,11-tridecadiene-5,7,9-triyne-1,2-diol, an antifungal polyacetylene from diseased safflower (*Carthamus tinctorius*). *Phytochemistry* 10:1579–82
9. Amici, A., Locci, R. 1968. Possible phytopathological implications of the behaviour of *Helminthosporium carbonum* in the presence of α-solanine. *Rivista Di Patologia Vegetale* IV:51–62
10. Aue, R., Mauli, R., Sigg, H. 1966. Production of 6-methoxy mellein by *Sporormia bipartis* Cain. *Experientia* 22:575
11. Bailey, J. 1969. Phytoalexin production by leaves of *Pisum sativum* in relation to senescence. *Ann. Appl. Biol.* 64:315–24
12. Bate-Smith, E. Swing, T., Pope, G. 1953. The isolation of 7-hydroxy-4′-methoxyisoflavone (formononetin) from red clover (*Trifolium pratense*) and a note on the identity of pratol. *Chem. Ind.* 1127
13. Bell, A. 1967. Formation of gossypol in infected or chemically irritated tissues of *Gossypium* species. *Phytopathology* 57: 759–64
14. Bell, A. 1969. Phytoalexin production and *Verticillium* wilt resistance in cotton. *Phytopathology* 59:1119–27
15. Bell, A. 1970. 4-Hydroxybenzaldehyde and vanillin as toxins formed in leaf wound sap of *Phaseolus lunatus. Phytopathology* 60:161–65
16. Bell, A., Presley, J. 1969. Temperature effects upon resistance and phytoalexin synthesis in cotton inoculated with *Verticillium alboatrum. Phytopathology* 59:1141–46
17. Bell, A., Presley, J. 1969. Heat-inhibited or heat-killed conidia of *Verticillium albo-atrum* induce disease resistance and phytoalexin synthesis in cotton. *Phytopathology* 59:1147–51
18. Berard, D., Kuć, J., Williams, E. 1972. A cultivar-specific protection factor from incompatible interactions of green bean with *Colletotrichum lindemuthianum.*

Physiological Plant Pathol. In press
19. Biehn, W., Kuć, J. Williams, E. 1968. Accumulation of phenols in resistant plant-fungi interactions. *Phytopathology* 58:1255–60
20. Biehn, W., Williams, E., Kuć, J. 1968. Fungitoxicity of phenols accumulating in *Glycine max*-fungi interactions. *Phytopathology* 58:1261–64
21. Bohlmann, F., Kohn, S., Arndt, C. 1966. Polyacetylenverbindungen CXIV. Die polyine der gattung *Carthamus* L. *Chem. Ber.* 99:3433–36
22. Boone, D. 1971. Genetics of *Venturia inaequalis. Ann. Rev. Phytopathol.* 9:297–318
23. Bredenberg, J. 1961. Identification of an antifungal factor in red clover as biochanin A. *Acta Chem. Fenn.* 34B:23
24. Bredenberg, J., Hietala, P. 1961. Investigation of the structure of trifolirhizin, an antifungal compound from *Trifolium pratense* L. *Acta Chem. Scand.* 15:696–99
25. Bredenberg, J., Hietala, P. 1961. Confirmation of the structure of trifolirhizin. *Acta Chem. Scand.* 15:936–37
26. Bredenberg, J., Skoolery, J. 1961. A revised structure for pterocarpin. *Tetrahedron Lett.* 285–88
27. Brown, A., Swinburne, T. 1972. Benzoic acid: an antifungal compound formed in Bramley's seedling apple fruits following infection by *Nectria galligena* Bres. *Physiological Plant Pathol.* 1:469–75
28. Chalova, La., Vasyukova, N., Ozeretskovskaya, O., Metlitskii, L. 1971. Chemical identification of one of the potato phytoalexins. *Prikl. Biokhim. Microbiol.* 7:55–58
29. Chalutz, E., DeVay, J. 1969. Production of ethylene in vitro and in vivo by *Ceratocystis fimbriata* in relation to disease development. *Phytopathology* 59:750–55
30. Chalutz, E., DeVay, J., Maxie, E. 1969. Ethylene induced isocoumarin formation in carrot root tissue. *Plant Physiol.* 44:235–41
31. Chalutz, E., Stahmann, M. 1969. Induction of pisatin by ethylene. *Phytopathology* 59:1972–73
32. Chamberlain, D., Gerdemann, J. 1966. Heat induced susceptibility of soybeans to *Phytophthora megasperma* var. *sojae, Phytophthors cactorum* and *Helminthosporium sativum. Phytopathology* 56:70–73
33. Chamberlain, D., Paxton, J. 1968. Protection of soybean plants by phytoalexin. *Phytopathology* 58:1349–50
34. Chester, K. 1933. The problem of acquired physiological immunity in plants. *Quarterly Rev. Biol.* 8:129–54, 275–324
35. Chou, M., Preece, T. 1968. The effect of pollen grains on infections caused by *Botrytis cinerea. Ann. Appl. Biol.* 62:11–22
36. Clarke, D. 1969. The accumulation of scopolin in potato tuber tissue after infection by *Phytophthora infestans* and its role in pathogenesis. *Phytochemistry* 8:7
37. Condon, P., Kuć, J. 1960. Isolation of a fungitoxic compound from carrot root tissue inoculated with *Ceratocystis fimbriata. Phytopathology* 50:267–70
38. Condon, P., Kuć, J. 1962. Confirmation of the identity of a fungitoxic compound produced by carrot root tissue. *Phytopathology* 52:182–83
39. Condon, P. Kuć, J., Draudt, H. 1963. Production of 3-methyl-6-methoxy-8-hydroxy-3,4-dihydroisocoumarin. *Phytopathology* 53:1244–50
40. Cruickshank, I. 1962. Studies on phytoalexins. IV. The antimicrobial spectrum of pisatin. *Aust. J. Biol. Sci.* 15:147–59
41. Cruickshank, I. 1963. Phytoalexins. *Ann. Rev. Phytopathol.* 1:351–74
42. Cruickshank, I., Biggs, D., Perrin, D. 1971. Phytoalexins as determinants of disease reaction in plants. *J. Indian Bot. Soc.* 50A:1–11
43. Cruickshank, I., Perrin, D. 1960. Isolation of a phytoalexin from *Pisum sativum* L. *Nature* 187:799–800
44. Cruickshank, I., Perrin, D. 1963. Studies on phytoalexins. VI.

The effect of some factors on its formation in *Pisum sativum* L. and the significance of pisatin in disease resistance. *Aust. J. Biol. Sci.* 16:111–28

45. Cruickshank, I., Perrin, D. 1963. Phytoalexins of the Leguminosae. Phaseollin from *Phaseolus vulgaris* L. *Life Sci.* 2:680–82
46. Cruickshank, I., Perrin, D. 1964. Pathological function of phenolic compounds in plants. *Biochemistry of Phenolic Compounds* ed. J. Harborne, 511–544, Academic Press, N.Y. 618 pp.
47. Cruickshank, I., Perrin, D. 1965. Studies on phytoalexins. IX. Pisatin formation by cultivars of *Pisum sativum* L. and other Pisum species. *Aust. J. Biol. Sci.* 18:829–35
48. Cruickshank, I., Perrin, D. 1965. Studies on phytoalexins. VIII. The effect of some further factors on the formation, stability and localization of pisatin in vivo. *Aust. J. Biol. Sci.* 18:817–28
49. Cruickshank, I., Perrin, D. 1967. Studies on phytoalexins. X. Effect of oxygen tension on the biosynthesis of pisatin and phaseollin. *Phytopathol. Z.* 60: 335–42
50. Cruickshank, I., Perrin, D. 1968. The isolation and partial characterization of monilicolin A, a polypeptide with phaseollin-inducing activity from *Monilinia fructicola. Life Sci.* 7:449–58
51. Cruickshank, I., Perrin, D. 1971. Studies on phytoalexins. XI. The induction, antimicrobial spectrum and chemical assay of phaseollin. *Phytopathol. Z.* 70: 209–29
52. Curtis, R. 1968. 6-Methoxy mellein as a phytoalexin. *Experientia* 24:1187–88
53. Daly, J., Seevers, P., Ludden, P. 1970. Studies on wheat stem rust resistance controlled at the Sr 6 locus. III. Ethylene and disease reactions. *Phytopathology* 60:1648–52
54. Deverall, B. 1967. Biochemical changes in infection droplets containing spores of *Botrytis* spp. incubated in the seed cavities of pods of bean (*Vicia faba* L.) *Ann. Appl. Biol.* 59:375–87
55. Deverall, B., Smith, I., Markis, S. 1968. Disease resistance in *Vicia faba* and *Phaseolus vulgaris. Neth. J. Plant Pathol.* 74 (Suppl. No. 1):137–148
56. Deverall, B., Vessey, J. 1969. Role of a phytoalexin in controlling lesion development in leaves of *Vicia faba* after infection by *Botrytis* spp. *Ann. Appl. Biol.* 63:449–58
57. DeWit-Elshove, A. 1968. Breakdown of pisatin by some fungi pathogenic to *Pisum sativum. Neth. J. Plant Pathol.* 74:44–47
58. DeWit-Elshove, A. 1969. The role of pisatin in the resistance of pea plants—some further experiments on the breakdown of pisatin. Neth. J. Plant Pathol. 75:164–68
59. DeWit-Elshove, A., Fuchs, A. 1971. The influence of the carbohydrate source on pisatin breakdown by fungi pathogenic to pea (*Pisum sativum*). *Physiol. Plant Pathol.* 1:17–24
60. Dodson, A., Fukui, N., Ball, C., Carolus, R., Sell, H. 1965. Occurrence of a bitter principle in carrots. *Science* 124:984–85
61. Elliston, J., Kuć, J., Williams, E. 1971. Induced resistance to bean anthracnose at a distance from the site of the inducing interaction. *Phytopathology* 61:1110–12
62. Farkas, G. L., Kiraly, Z. 1962. Role of phenolic compounds in the physiology of plant diseases and disease resistance. *Phytopathol. Z.* 44:105–50
63. Fawcett, C. 1961. Antifungal compounds in seedlings of *Vicia faba*. III Chemical results. *Soc. Chem. Ind. Monographs* 15: 119–31
64. Fawcett, C., Spencer, D., Wain, R. 1969. The isolation and properties of a fungicidal compound present in the seedlings of *Vicia faba. Neth. J. Plant Pathol.* 75: 72–81
65. Fawcett, C. et al. 1968. Natural acetylenes Part XXVII. An antifungal acetylenic furanoid keto-ester (wyerone) from shoots of the broad bean (*Vicia faba* L.; Fam. Papilionaceae). *J. Chem. Soc.* (c) 2455–62

66. Fawcett, C. et al. 1965. An antifungal acetylenic keto-ester from a plant of the papilionaceae family. *Chem. Comm.* 422–23
67. Fehrmann, H., Dimond, A. 1967. Studies on auxins in the *Phytophthora* disease of the potato. I. Role of indoleacetic acid in pathogenesis. *Phytopathol. Z.* 59:83–100
68. Fehrmann, H., Dimond, A. 1967. Studies on auxins in the *Phytophthora* disease of the potato tuber. II. Relation of indoleacetic acid to some physiological processes in pathogenesis. *Phytopathol. Z.* 59:105–21
69. Fehrmann, H., Dimond, A. 1967. Peroxidase activity and *Phytophthora* resistance in different organs of the potato plant. *Phytopathology* 57:69–72
70. Govindachari, T., Patankar, S., Viswanathan, N. 1971. Isolation and structure of two new dihydroisocoumarins from *Kigelia pinnata. Phytochemistry* 10: 1603–06
71. Hadwiger, L. 1966. The biosynthesis of pisatin. *Phytochemistry* 5:523–25
72. Hadwiger, L. 1967. Changes in phenylalanine metabolism associated with pisatin production. *Phytopathology* 57:1258–59
73. Hadwiger, L. 1968. Changes in plant metabolism associated with phytoalexin production. *Neth. J. Plant Pathol.* 74:163–69
74. Hadwiger, L. 1971. Specificity of DNA intercalating compounds in the control of phenylalanine ammonia lyase and pisatin levels. *Plant Physiol.* 47:346–51
75. Hadwiger, L., Hess, S., von Broembsen, S. 1970. Stimulation of phenylalanine ammonia liase activity and phytoalexin production. *Phytopathology* 60: 332–36
76. Hadwiger, L., Schwochau, M. 1967. Host resistance responses —an induction hypothesis. *Phytopathology* 59:223–27
77. Hadwiger, L., Schwochau, M. 1971. Ultraviolet light-induced formation of pisatin and phenylalanine ammonia lyase. *Plant Physiol.* 47:588–90
78. Harper, S., Kemp, A., Underwood, W., Campbell, R. 1969. Pterocarpanoid constituents of the heartwoods of *Pericopsis angolensis* and *Swartzia madagascariensis. J. Chem. Soc.* (c) 1109–16
79. Heath, M., Wood, R. 1971. Role of inhibitors of fungal growth in the limitation of leaf spots caused by *Ascochyta pisi* and *Mycosphaerella pinodes. Ann. Bot.* 35:475–91
80. Heitefuss, R. 1964. Sauerstoffaufnahme, phosphat- und uncleinsauertoffwechsel vonweizenpflanzen in beziehung zurumweltgesteuerten resistenz gegen *Puccinia graminis* tritici. *Biochemische Probleme der Kranken Pflanze. Symp. Proc. Inst. Phytopathol.* Aschersleben, Deutsche Demokratische Republik 227–39
81. Herndon, B., Kuć, J., Williams, E. 1966. The role of 3-methyl-6-methoxy-8-hydroxy-3,4-dihydroisocoumarin in the resistance of carrot root to *Ceratocystis fimbriata. Phytopathology* 56:187–91
82. Hess, S., Hadwiger, L. 1971. The induction of phenylalanine ammonia lyase and phaseollin by 9-aminoacridine and other deoxyribonucleic acid intercalating compounds. *Plant Physiol.* 48:197–202
83. Hess, S., Hadwiger, L., Schwochau, M. 1971. Studies on biosynthesis of phaseollin in excised pods of *Phaseolus vulgaris. Phytopathology* 61:79–82
84. Hietala, P. 1960. A countercurrent distribution method for separation of chemical compounds. *Ann. Acad. Sci. Fenn.* Ser. A. II. Chemica 100:1–69
85. Higgins, V., Millar, R. 1968. Phytoalexin production by alfalfa in response to infection by *Colletotrichum phomoides, Helminothosporium turcicum, Stemphylium loti* and *S. botryosum. Phytopathology* 58:1377–83
86. Higgins, V., Millar, R. 1969. Comparative abilities of *Stemphylium botryosum* and *Helminthosporium turcicum* to induce and degrade a phytoa-

lexin from alfalfa. *Phytopathology* 59:1493–99

87. Higgins, V., Millar, R. 1969. Degradation of alfalfa phytoalexin by *Stemphylium botryosum. Phytopathology* 59:1500–06
88. Higgins, V., Millar, R. 1970. Degradation of alfalfa phytoalexin by *Stemphylium loti* and *Colletotrichum phomoides. Phytopathology* 60:269–71
89. Higgins, V., Smith, D. 1972. Separation and identification of two pterocarpanoid phytoalexins produced by red clover leaves. *Phytopathology* In press
90. Hirai, T. 1970. Comparison of the biochemical changes due to fungus infection versus virus infection. *Phytopathol. Z.* 69: 256–66
91. Howell, R. 1970. Influence of air pollution on quantities of caffeic acid isolated from leaves of *Phaseolus vulgaris. Phytopathology* 60:1626–29
92. Hughes, J., Swain, T. 1960. Scopolin production in potato tubers infected with *Phytophthora* infestans. *Phytopathology* 50: 398–400
93. Hyodo, H., Uritani, I., Akai, S. 1969. Production of furanoterpenoids and other compounds in sweet potato root tissue in response to infection by various isolates of *Ceratocystis fimbriata. Phytopathol. Z.* 65: 332–40
94. Ishizaka, N., Tomiyama, K., Katsui, N., Murai, A., Masamune, T. 1969. Biological activities of rishitin, an antifungal compound isolated from diseased potato tubers and its derivatives. *Plant Cell Physiol.* 10: 183–92
95. Johnson, L. 1970. Influence of infection by *Phytophthora drechsleri* on inhibitory materials in resistant and susceptible safflower hypocotyls. *Phytopathology* 60:1000–04
96. Kato, N., Imaseki, H., Nakashima, N., Uritani, I. 1971. Structure of a new sesquiterpenoid, ipomeamaronol. *Tetrahedron Lett.* No. 13:843–46
97. Katsui, N., Matsunaga, A., Imaizumi, K., Masamune, T., Tomiyama, K. 1971. The structure and synthesis of rishitinol, a new sesquiterpene alcohol from diseased potato tubers. *Tetrahedron Lett.* No. 2, 83–86
98. Katsui, N., et al 1968. The structure of rishitin, a new antifungal compound from diseased potato tubers. *Chem. Com.* 43–44
99. Keen, N. 1971. Hydroxyphaseollin production by soybeans resistant and susceptible to *Phytophthora megasper*ma var. *sojae. Physiol. Plant Pathol.* 1:265–74
100. Keen, N., Sims, J., Erwin, D., Rice, E. Partridge, J. 1971. 6a-Hydroxyphaseollin: an antifungal chemical induced in soybean hypocotyls by *Phytophthora megasperma* var. *sojae. Phytopathology* 61:1084–89
101. Klarman, W., Gerdemann, J. 1963. Induced susceptibility in soybean plants genetically resistant to *Phytophthora sojae. Phytopathology* 53:863–64
102. Klarman, W., Gerdemann, J. 1963. Resistance of soybean to three *Phytophthora* species due to the production of a phytoalexin. *Phytopathology* 53: 1317–20
103. Klarman, W., Sanford, J. 1968. Isolation and purification of an anti-fungal principle from infected soybeans. *Life Sci.* 7: 1095–1103
104. Kosuge, T. 1969. The role of phenolics in host response to infection. *Ann. Rev. Phytopathol.* 7: 195–222
105. Kuć, J. 1963. The role of phenolic compounds in disease resistance. *Connecticut Agr. Exp. Sta. Bull.* 663:20–30
106. Kuć, J. 1964. Phenolic compounds and disease resistance in plants *Phenolics in Normal and Diseased Fruits and Vegetables* ed. V. Runeckles, 63–81 Imperial Tobacco Co., Montreal 102 pp.
107. Kuć, J. 1966. Resistance of plants to infectious agents. *Ann. Rev. Microbiol.* 20:337–70
108. Kuć, J. 1967. Shifts in oxidative metabolism during pathogenesis. *The Dynamic Role of Molecular Constituents in Plant-Parasite Interaction* ed. C. J. Mirocha, I. Uritani, 183–202, Bruce Publ. Co., St. Paul. 372 pp.
109. Kuć, J. 1968. Biochemical control

of disease resistance in plants. *World Rev. Pest Control* 7:42–55

110. Kuć, J. 1971. Compounds accumulating in plants after infection. *Microbial toxins* VIII ed. S. J. Ajl, G. Weinbaum, S. Kadis, 211–47, Academic Press, N.Y.
111. Letcher, R., Widdowson, B., Deverall, B., Mansfield, J. 1970. Identification and activity of wyerone acid as a phytoalexin in broad bean (*Vicia Faba*) after infection by *Botrytis. Phytochem.* 9:249–52
112. Lim, S., Hooker, A., Paxton, J. 1970. Isolation of phytoalexins from corn with monogenic resistance to *Helminthosporium turcicum. Phytopathology* 60:1071–75
113. Locci, R., Kuć, J. 1967. Steroid alkaloids as compounds produced by potato tubers under stress. *Phytopathology* 57:1272–73
114. Lovrekovich, L., Farkas, G. 1965. Induced protection against wildfire disease in tobacco leaves treated with heat-killed bacteria. *Nature* 205:823–24
115. Lozano, J., Sequeira, L. 1970. Prevention of the hypersensitive reaction in tobacco leaves by heat-killed bacterial cells. *Phytopathology* 60:875–79
116. Maekawa, E., Kitao, K. 1970. Isolation of pterocarpanoid compounds as heartwood constitutents of *Maackia amurensis* var. *Buergeri. Wood Res.* 50:29–35
117. Mallabaev, A., Yagudaev, M., Saitbaeva, I., Sidyakin, G. 1970. Isocoumarin artemidin from *Artemisia dracunculus. Khim. Prir. Soedin.* 6:467–68
118. McGahren, W., Mitscher, L. 1968. Dihydroisocoumarins from a *Spororomia* fungus. *J. Org. Chem.* 33:1577–80
119. McKee, R. 1955. Host parasite relationships in the dry-rot disease of potatoes. *Ann. Appl. Biol.* 43:147–48
120. Metlitskii, L., Ozeretskovskaya, O. 1968. *Plant Immunity,* Plenum Press, N. Y. 114 pp.
121. Metlitskii, L., Ozeretskovskaya, O. 1970. Phytoncides and phytoalexins and their role in plant immunity. *Mikol Fitopatol.* 4:146–55
122. Metlitskii, L., et al. 1970. Potato resistance to *Phytophthora infestans* as related to leaf phytoalexin activity. *Prikl. Biokhim. Microbiol.* 6:568–73
123. Müller, K. 1956. Einige einfache versuchenzum nachweis von phytoalexinen. *Phytopathol. Z.* 27:237–54
124. Müller, K. 1958. The formation and the immunological significance of phytoalexin produced by *Phaseolus vulgaris* in responses to infections with *Sclerotinia fructicola* and *phytophthora infestans. Aust. J. Biol. Sci.* 11:275–300
125. Müller, K., Börger, H. 1940. Experimentelle Untersuchungen uber die *Phytophthora*-resistenz der kartoffel. *Arb. Biol. Reichsanstat. Land-u Forstwirtsch.* Berlin 23:189–231
126. Mussell, H., Staples, R. 1971. Phytoalexin-like compounds apparently involved in strawberry resistance to *Phytophthora fragariae. Phytopathology* 61:515–17
127. Nonaka, F. 1967. Inactivation of pisatin by pathogenic fungi. *Agr. Bull. Saga Univ.* Japan 24:109–21
128. Nonaka, F., Isayama, S., Furukawa, H. 1966. On the phytoalexin produced by the results of the interaction between soybean pods and phytopathogens. *Agr. Bull. Saga Univ.* Japan 22:51–63
129. Nonaka, F., Kawakami, K. 1970. Inactivation of pisatin by pathogenic fungi. *Proc. Assoc. Plant Prot. Kyushu* 16:114–16
130. Nonaka, F., Kazuomi, Y. 1966. On the selective toxicity of ipomeamarone towards the phytopathogens. *Agr. Bull. Saga Univ.* Japan 22:39–49
131. Olah, A., Sherwood, R. 1971. Flavones, isoflavones and coumestans in alfalfa infected by *Ascochyta imperfecta. Phytopathology* 61:65–69
132. Ollis, W. 1966. The neoflavanoids, a new class of natural products. *Experientia* 22:777–83
133. Ozeretskovskaya, O., Vasyukova, N., Metlitskii, L. 1969. Study of potato phytoalexins. *Doklady Bot. Sci.* 189:158–60
134. Paseshnichenko, V. 1957. Content

of solanine and chaconine in the potato during the vegetation period. *Biokhimya* 22:929–31
135. Paxton, J., Chamberlain, D. 1967. Acquired local resistance of soybean plants to *Phytophthora* spp. *Phytopathology* 57:352–53
136. Paxton, J., Chamberlain, D. 1969. Phytoalexin production and disease resistance in soybean as related to age. *Phytopathology* 59:775–77
137. Pegg, G., Sequeira, L. 1968. Stimulation of aromatic biosynthesis in tobacco plants infected by *Pseudomonas solanacearum. Phytopathology* 58:476–83
138. Pellizzari, E. 1970. *A Biochemical Study on the Defense Mechanisms of* Venturia inaequalis *(cke) Wint.* Ph.D. Thesis, Purdue Univ., Lafayette 135 pp.
139. Perrin, D. 1964. The structure of Phaseollin. *Tetrahedron Lett.* No. 1, 29–35
140. Perrin, D., Bottomley, W. 1962. Studies on phytoalexins. V. The structure of pisatin from *Pisum sativum* L. *J. Am. Chem. Soc.* 84:1919–22
141. Perrin, D., Cruickshank, I. 1965. Studies on phytoalexins. VII. Chemical stimulation of pisatin formation in *Pisum sativum* L. *Aust. J. Biol. Sci.* 18:803–16
142. Pierre, R. 1966. *Histopathology and Phytoalexin in Beans Resistant or Susceptible to Fusarium and Thielaviopsis.* Ph.D. Thesis Cornell Univ., Ithaca, N. Y. 155 pp.
143. Pierre, R. 1971. Phytoalexin induction in beans resistant or susceptible to *Fusarium* and *Thielaviopsis. Phytopathology* 61:322–27
144. Pierre, R., Bateman, D. 1967. Induction and distribution of phytoalexins in *Rhizoctonia*-infected bean hypocotyls. *Phytopathology* 57:1154–60
145. Powell, C., Hildebrand, D. 1970. Fire blight resistance in *Pyrus:* Involvement of arbutin oxidation. *Phytopathology* 60:337–40
146. Prelog, V. Jeger, O. 1960. Steroid alkaloids: the solanum group *The Alkaloids-Chemistry and Physiology* ed. R. Manske, 7: 343–61 Academic Press, N. Y. 559 pp.
147. Purkayastha, R., Deverall, B. 1964. A phytoalexin type of reaction in the *Botrytis infection* of leaves of bean (*Vicia faba* L.) *Nature* 201:938–39
148. Purkayastha, R., Deverall, B. 1965. The growth of *Botrytis fabae* and *B. cinerea* into leaves of bean (*Vicia faba* L.). *Ann. Appl. Biol.* 56:139–47
149. Raa, J. 1968. *Natural resistance of apple plants to* Venturia inaequalis. Ph.D. Thesis. Univ. Utrecht, Utrecht, The Netherlands. 100 pp.
150. Raa, J., Sijpesteijn, A. 1968. A biochemical mechanism of natural resistance of apple to *Venturia inaequalis. Neth. J. Plant Pathol.* 74:229–31
151. Rahe, J., Kuć, J. 1970. Metabolic nature of the infection—limiting effect of heat on bean anthracnose. *Phytopathology* 60:1005–09
152. Rahe, J., Kuć, J., Chuang, C., Williams, E. 1969. Correlation of phenolic metabolism with histological changes in *phasolus vulgaris* inoculated with fungi. *Neth. J. Plant Pathol.* 75:58–71
153. Rahe, J., Kuć, J., Chuang, C., Williams, E. 1969. Induced resistance in *Phaseolus vulgaris* to bean anthracnose. *Phytopathology* 59:1641–45
154. Robertson, N., Friend, J., Aveyard, M. 1969. Production of phenolic acids by potato tissue culture after infection by *Phytophthora infestans. Phytochemistry* 8:7
155. Robertson, N. et al. 1968. The accumulation of phenolic acids in tissue culture pathogen combinations of *Solanum tuberosum* and *Phytophthora infestans. J. Gen. Microbiol.* 54:261–68
156. Ross, A. 1965. Systemic effects of local lesion formation. *Viruses of Plants. Int. Conf. Plant Viruses Proc.* Wageningen, John Wiley & Sons, N. Y. 329 pp.
157. Rubin, B., Aksenova, V. 1957. Participation of polyphenolase system in defense reactions of potato against *Phytophthora infestans. Biokhimiya* (Engl. Transl.) 22:191–97
158. Sakai, R., Tomiyama, K., Ishizaka, N., Sato, N. 1967. Phenol metabolism in relation to dis-

ease resistance of potato tubers. *Ann. Phytopathol. Soc. Japan* 33:216–22

159. Sakuma, T., Tomiyama, K. 1967. The role of phenolic compounds in the resistance of potato tuber tissue to infection by *Phytophthora infestans*. *Ann. Phytopathol. Soc. Japan* 33:48–58
160. Sandstedt, K. 1967. *The accumulation of 6-methoxy-mellein and chlorogenic acid in carrots under stress.* M.S. Thesis Purdue Univ. Lafayette 48 pp.
161. Sato, N., Kitazawa, K., Tomiyama, K. 1971. The role of rishitin in localizing the invading hyphae of *Phytophthora infestans* in infection sites at the cut surface of potato tubers. *Physiological Plant Pathol.* 1:289–95
162. Sato, N., Tomiyama, K. 1969. Localized accumulation of rishitin in the potato tuber tissue infected by an incompatible race of *Phytophthora infestans*. *Ann. Phytopathol. Soc. Japan* 35:202–07
163. Sato, N., Tomiyama, K., Katsui, N., Masamune, T. 1968. Isolation of rishitin from tubers of interspecific potato varieties containing different late-blight resistance genes. *Ann. Phytopathol. Soc. Japan* 34:140–42
164. Sato, N., Tomiyama, K., Katsui, N., Masamune, T. 1968. Isolation of rishitin from tomato plants. *Ann. Phytopathol. Soc. Japan* 34:344–45
165. Schwochau, M., Hadwiger, L. 1968. Stimulation of pisatin production in *Pisum sativum* by actinomycin D and other compounds. *Arch. Biochem. Biophys.* 126:731–33
166. Schwochau, M., Hadwiger, L. 1969. Regulation of gene expression by actinomycin D and other compounds which change the conformation of DNA. *Arch. Biochem. Biophys.* 134:34–41
167. Schwochau, M., Hadwiger, L. 1970. Induced host resistance—a hypothesis derived from studies of phytoalexin production. *Recent Advan. Phytochem.* 3:181–89
168. Seevers, P., Daly, J. 1970. Studies on wheat stem rust resistance controlled at the Sr6 locus. II. Peroxidase activities. *Phytopathology* 60:1642–47
169. Sequeira, L. 1969. Synthesis of scopolin and scopoletin in tobacco plants infected by *Pseudomonas solanacearum*. *Phytopathology* 59:473–78
170. Shain, L., Hillis, W. 1971. Phenolic extractives in Norway spruce and their effects on *Fomes annosus*. *Phytopathology* 61:841–45
171. Sherwood, R., Olah, A., Oleson, W., Jones, E. 1970. Effect of disease and injury on accumulation of a flavonoid estrogen, coumesterol, in alfalfa. *Phytopathology* 60:684–88
172. Sijpesteijn, A. 1969. Aspects of natural disease resistance. *Meded. Rijksfak. Landbouw-Wetensch. Gent.* 34:379–91
173. Sims, J., Keen, N., Howard, V. 1971. The structure of hydroxyphaseollin, and induced antifungal compound from soybeans. *Phytochemistry*. In press
174. Smith, I. 1971. The induction of antifungal inhibitors in pods of tropical legumes. *Physiological Plant Pathol.* 1:85–94
175. Smith, D. G., McInnes, A., Higgins, V., Millar, R. 1971. Nature of the phytoalexin produced by alfalfa in response to fungal infection. *Physiological Plant Pathol.* 1:41–44
176. Smith, D. A., VanEtten, D., Bateman, D. 1971. Isolation of substance II, an antifungal compound from *Rhizoctonia solani*-infected bean tissue. *Phytopathology* 61:912
177. Sokolova, V., Savel'eva, O., Rubin, B. 1958. The nature of the changes in chlorogenic acid of potato tubers attacked by *Phytophthora infestans*. *Doklady Akad. Nauk. SSR* (Engl. Translation) 123:251–53
178. Sondheimer, E. 1957. The isolation of 3-methyl-6-methoxy-8-hydroxy-3,4-dihydroisocoumarin from carrots. *J. Am. Chem. Soc.* 79:5036–39
179. Stholasula, P., Bailey, J., Severin, V., Deverall, B. 1971. Effect of bacterial inoculation of bean and pea leaves on the accumulation of phaseollin and pisatin.

Physiological Plant Pathol. 1: 177–83

180. Stoessl, A. 1969. 8-Hydroxy-6-methoxy-3-methylisocoumarin and other metabolites of *Ceratocystis fimbriata*. *Biochem. Biophys. Res. Comm.* 35:186–91
181. Stoessl, A. 1970. Antifungal compounds produced by higher plants. *Rec. Advan. Phytochem.* 3:143–80
182. Tomiyama, K. 1966. Double infection by an incompatible race of *Phytophthora infestans* of a potato cell which has previously been infected by a compatible race. *Ann. Phytopathol. Soc.* Japan 32:181–85
183. Tomiyama, K. 1970. Cytological and biochemical studies of the hypersensitive reaction of potato cells to *Phytophthora infestans*. *Morphological and Related Biochemical Events in Host-Parasite Interaction*. ed. S. Akai, S. Ouchi, 387–401, Denki Shoen, Tokyo 415 pp.
184. Tomiyama, K., Ishizaka, N., Sato, N., Masammine, T., Katsui, N. 1968. Rishitin, a phytoalexin-like substance its role in the defense reaction of potato tubers to infection. *Biochemical Regulation in Diseased Plants or Injury*. Phytopathological Soc. Japan, Tokyo 287–92
185. Tomiyama, K., Sakai, R., Otani, Y., Takemori, T. 1967. Phenol metabolism in relation to disease resistance of potato tuber. *Plant Cell Physiol.* 8:1–13
186. Tomiyama, K., Takakuwa, M., Takase, N. 1958. The metabolic activity in healthy tissue neighboring the infected cells in relation to resistance to *Phytophthora infestans* (Mont.) De Bary in potatoes. *Phytopathol. Z.* 31:237–50
187. Tomiyama, K., Takase, N., Sakai, R., Takakuwa, M. 1955. Physiological studies on the defense reaction of potato plant to infection by *Phytophthora infestans* II. Changes in the physiology of potato tuber induced by the infection of different strains of *P. infestans*. *Ann. Phytopathol. Soc. Japan* 20:59–64
188. Tomiyama, K., Takase, N., Sakai, R. Takakuwa, M. 1956. Physiological studies on the defense reaction of the potato plant to the infection of *Phytophthora infestans* and their varietal differences (I). *Res. Bull. Hokkaido Agr. Exp. St.* 71:32–50
189. Tomiyama, K., et al. 1968. A new antifungal substance isolated from resistant potato tuber tissue infected by pathogens. *Phytopathology* 58:115–16
190. Uehara, K. 1958. On the phytoalexin production of the soybean pod in reaction to *Fusarium* sp. the causal fungus of pod blight. *Ann. Phytopathol. Soc. Japan* 23:225–29
191. Uritani, I. 1963. The biochemical basis of disease resistance induced by infection. *Connecticut Agr. Exp. Sta. Bull.* 663:4–19
192. Varns, J. 1970. *Biochemical Response and its Control in the Irish Potato Tuber* (Solanum tuberosum L.) *-Phytophthora infestans* Interactions. Ph.D. Thesis Purdue Univ., Lafayette 148 pp.
193. Varns, J., Currier, W., Kuć, J. 1971. Specificity of rishitin and phytuberin accumulation by potato. *Phytopathology* 61:968–71
194. Varns, J., Kuć, J. 1971. Suppression of rishitin and phytuberin accumulation and hypersensitive response in potato by compatible races of *Phytophthora infestans*. *Phytopathology* 61:178–81
195. Varns, J., Kuć, J., Williams, E. 1971. Terpenoid accumulation as a biochemical response of the potato tuber to *Phytophthora infestans*. *Phytopathology* 61:174–77
196. Virtanen, A., Hietala, P. 1957. Additional information on the antifungal factor in red clover. *Acta Chem. Fenn.* 30B:99
197. Virtanen, A., Hietala, P. 1958. Isolation of an anti-Sclerotinia factor, 7-hydroxy-4′-methoxyisoflavone from red clover. *Acta Chem. Scand.* 12:597
198. Wain, R., Spencer, D., Fawcett, C. 1961. Antifungal compounds in seedlings of Vicia faba. *Fungicides in Agriculture and Horticulture. Soc. Chem. Ind. Monograph* No. 15 109–31
199. Weber, D., Stahmann, M. 1966. Induced immunity to *Ceratocystis* infection in sweet potato

root tissue. *Phytopathology* 56: 1066–70

200. Williams, E., Kuć, J. 1969. Resistance in *Malus* to *Venturia inaequalis*. *Ann. Rev. Phytopathol.* 7:223–46
201. Wolf, M., Duggar, B. 1946. Estimation and physiological role of solanine in the potato. *J. Agr. Res.* 73:1–32
202. Wood, R. 1967. *Physiological Plant Pathology* John Wiley & Sons, N. Y. 388 pp.
203. Yang, D., Wilson, B., Harris, T. 1971. The structure of ipomeamaronal: A new toxic furanosesquiterpene from moldy sweet potatoes. *Phytochemistry* 10:1653–54
204. Yarwood, C., Hatiro, I., Batra, K. 1969. Heat induced anthocyanin, polysaccharide and transpiration. *Phytopathology* 59: 596–98

EXPRESSION OF RESISTANCE IN PLANTS TO NEMATODES

3550

Richard A. Rohde

*Department of Plant Pathology, University of Massachusetts,
Amherst, Massachusetts*

Introduction

As in all areas of plant pathology, any information related to the development of a host-parasite interaction contributes to our understanding of resistance. The amount of literature on nematode-host interactions is immense (10, 33, 48, 70). Thus, it is the purpose of this paper to deal primarily with unfavorable host-parasite relationships that lead to resistance. Recent reviews on the breeding of nematode-resistant plants (25, 31) and the genetics of plant-parasitic nematodes (58, 61) are available or in preparation.

Resistance mechanisms that occur both before and after establishment of a parasitic relationship will be discussed. Because most resistance becomes apparent after infection, plant-parasitic nematodes are grouped into somewhat arbitrary categories based on increasing complexity of interactions. As complexity increases, resistance becomes more common and the bulk of the literature on resistance is devoted to three genera, *Ditylenchus, Meloidogyne,* and *Heterodera,* somewhat in keeping with their economic importance.

Concepts of Resistance

Wingard (73), in the 1953 Yearbook of Agriculture, defines resistance as the ability of a plant to withstand, oppose, lessen, or overcome the attack of a pathogen, but plant nematologists tend to place emphasis on the development of nematode populations, with only secondary attention to host injury. In general, a resistant plant is one on which nematodes reproduce poorly, and a tolerant plant is one that shows little injury, even under attack by large populations of nematodes.

Authors generally agree on a broad definition of resistance and often introduce new words to denote various concepts of host-parasite interactions. The source of the problem is that two factors, host growth and parasite growth, vary continuously from 0 to 100% and each host-parasite interaction can be represented as a single point on a graph (Figure 1). It can be seen that resistance varies from slight to absolute. An immune plant presumbly suffers no injury.

In Massachusetts, hay crops are tolerant to lesion nematodes and support

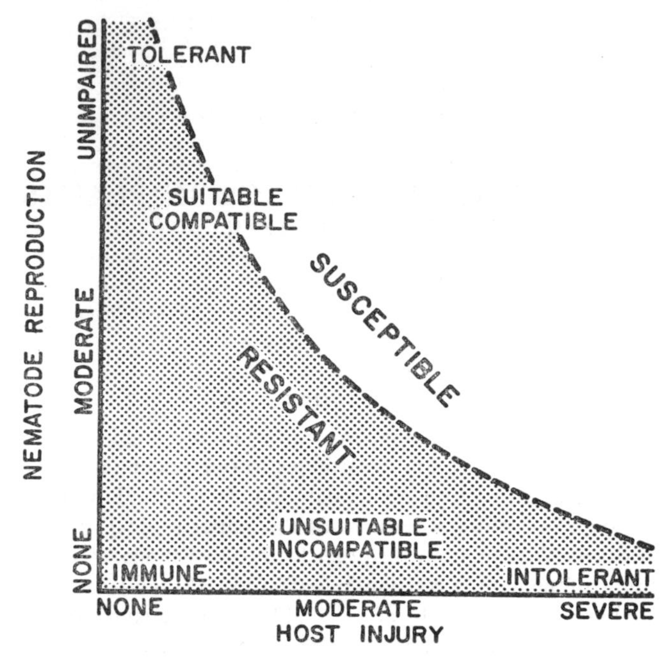

FIGURE 1. Concepts of resistance used by nematologists. The main criterion is usually failure of the parasite to reproduce, but lack of host injury is also of practical use.

large populations with little apparent injury. Tobacco planted in rotation in the same fields shows severe injury from lesion nematodes, but is an unsuitable host and populations may dwindle to the point that there is little injury to tobacco the second year. Tolerance that leaves a legacy of high populations to attack succeeding annual crops would be of limited usefulness, whereas tolerance may be of great practical significance in rootstocks of perennial plants (16).

Ample evidence exists to show that nematode species vary considerably in number of hosts and that these hosts differ in the size of populations they support and the amount of injury expressed. It is not uncommon, however, for a plant species or variety to be resistant in one locality and susceptible in another. Before it was shown in 1949 (6) that the root-knot nematode was actually several species, conflicting reports on host ranges were difficult to resolve, and probably similar situations exist today with other genera. Nematodes are not carried through the air like fungus spores and populations tend to show variation associated with geographic isolation. A further source of variation in local populations is that some nematode species reproduce only asexually. A better understanding of both the taxonomy and the cytogenetics of nematodes would be of great help in clarifying concepts of resistance.

Environmental Modification of Resistance.—Climatic and soil factors influence the development of host, parasite, and their interaction; inhibition of the parasite or of symptom development gives the appearance of resistance.

Temperature greatly affects rate of development of most nematodes and life cycles are prolonged at lower temperatures. Epicure potatoes grow well at 4° C and when planted early, they largely escape attack by *H. rostochiensis* (20). Lahontan alfalfa on the other hand loses much of its resistance to *Ditylenchus dipsaci* at temperatures above 12–15°C (23).

Many nematodes, particularly *Meloidogyne* and *Heterodera,* require large amounts of potassium and addition of this nutrient to soils may markedly alleviate symptoms of injury, although the parasite is stimulated as well (40). Crittenden (9) found that soybean varieties tolerant to low levels of potassium were resistant to root-knot, but only when grown in potassium deficient soils.

Failure to Establish a Parasitic Relationship

Interactions in the soil.—Root secretions from growing plants are an important feature of the environment of nematodes, although the exact relationship is not very well understood. Carbohydrates, amino acids, and a variety of nutrients are released by roots, but it has not been established that nematodes use them for food. These root secretions do, however, support a rhizosphere microflora that, in turn, produces metabolites which may influence nematodes more than the plants themselves.

Attraction of many species of plant parasites, hatching of *Heterodera* and *Meloidogyne,* and moulting of pre-adult *Paratylenchus* (45, 55) are influenced by root exudates and similar responses are probably very common. Viglierchio (67) found that *M. hapla* and *H. schachtii* are attracted to hosts more strongly than to nonhosts and proposed that oriented movement was governed by a combination of attractive and repellant stimuli. Several examples exist of failure of nonhosts to attract parasites, but it is doubtful that this is ever a deciding factor in host selection since random movement results in nematode accumulation and entry into unattractive hosts (21).

Spontaneous hatch from cysts of *Heterodera* spp. results in infection of plants that do not produce hatching factors, and lack of a hatching factor per se would not result in resistance. A number of compounds have been found to inhibit hatch of cyst nematodes, including a catechin in the cyst wall itself (30). Other phenols, such as p-cresol, are potent inhibitors of hatching (55). Although such compounds are common constituents of plants, particularly injured plants, they have not been shown to influence hatch under natural conditions.

Resistance to penetration.—Numerous examples can be found of resistant plants that are entered by larvae as readily as are susceptible plants. Less often, resistant plants are penetrated in small numbers and occasionally a plant is not attacked. Plant-parasitic nematodes are equipped with stylets and most secrete cell-wall-degrading enzymes and are usually able to complete the initial stages of penetration even if development stops at that stage. Reynolds, et al (44) have found that *Meloidogyne incognita acrita* larvae readily penetrate resistant alfalfa, but leave the root after 3 or 4 days. If such behavior proves common for other nematode species, it would help to explain conflicting reports of resistance to penetration.

Meloidogyne hapla seems to be an exception to the general pattern in that larvae fail to penetrate resistant hosts or do so in reduced numbers. Sasser (51) showed that *M. hapla* larvae do not penetrate the nonhosts oat and rye; decreased penetration has been found in resistant varieties of soybeans (12) and alfalfa (22).

Toxic Plants.—Juices extracted from a number of plants contain compounds toxic to nematodes and have been related to resistance. In no case have toxic compounds been found to be the sole mechanism of resistance, rather, they work in conjunction with other factors. Plants containing toxins are attacked and injured, but development of the nematodes is retarded and populations rapidly decline.

Asparagus (*Asparagus officinalis*) is resistant to *Trichodorus christiei* and contains a highly toxic glycoside in the roots, stems, and leaves (50). Populations of *T. christiei* die rather quickly in the root zone of asparagus and the toxic effect spreads through soil, protecting adjacent susceptible plants from attack. Concentrations of the glycoside equivalent to those encountered in roots are toxic to *T. christiei* in aqueous solutions, although feeding in soil may continue for several weeks.

Dutch bulb-growers have known for years that tulips and daffodils grow much better in fields that have been planted with marigolds the previous year (39). More recently, organic gardeners and commercial nurserymen have found that interplantings of marigolds improve plant growth. Marigolds (*Tagetes patula* and *T. erecta*) have been found to suppress populations of *Meloidogyne, Pratylenchus,* and several other nematode genera. Nematode populations decline around marigold roots although roots are attacked, often with considerable plant injury. Occasionally a root-knot female reaches maturity, but usually penetration results in necrosis and death of the larva. Penetration by lesion nematodes is usually infrequent, and reproduction, if any, is poor (74). As with asparagus, the soil becomes nematicidal and adjacent plants are partially protected. Thienyl compounds, particularly α-terthienyl and 5-(3-buten-l-ynyl)-2,2′-bithienyl, naturally occurring in marigold roots at nematicidal concentrations, are assumed to be the major factor in resistance (64).

Scheffer et al (52) concluded that resistance of *Eragrostis curvula* to 4 species of *Meloidogyne* was due to high concentrations of pyrocatechol in root exudates and found that pretreatment of larvae with pyrocatechol at 1:10^8 dilution prevented infection, possibly because pectic enzymes were inactivated.

Van Gundy & Kirkpatrick (66) found a positive correlation between toxicity of root juices and resistance of a number of citrus rootstocks to *Tylenchulus semipenetrans,* but concluded that toxicity was only partly responsible for overall resistance.

Winoto Suatmadji (74) has proposed that inhibitory compounds from plants that suppress nematode populations are quite common and that a compatible host compensates for this suppression by supporting reproduction. Soil from around apple, a compatible host, suppressed *Pratylenchus* populations more than soil from around marigolds. Addition of plant material to soil often reduces nematode populations and he speculates that the inhibitory effect may be caused by products from decomposition of crop residues.

Compounds derived from plants have been used since ancient times for the treatment of nematode parasites of man and domestic animals (28, 57). Santonin, from Levant wormseed (*Artemisea maratima*), ascaridole, the chief constituent of American wormseed (*Chenopodium ambrosioides*), *filicinic* acid, from the male fern (*Dryopteris filix mas*), and embelin from *Embelia ribes,* are widely used throughout the world (Fig. 2). Many of the compounds from plants that are toxic to animal-parasitic nematodes are phenols similar to those related to disease resistance in plants (Fig. 3).

Physostigmine, obtained from Calabar Bean (*Physostigma venenosum*), is the only known naturally occurring compound with an N–methyl carbamate group and is a potent cholinesterase inhibitor—as is the asparagus glycoside and many of the newer commercial nematicides. Physostigmine itself is not toxic to nematodes in vitro but it is systemically active in peas infected with *Ditylenchus dipsaci* (71). The assay procedures used by most investigators looking for toxic compounds in plant juices would fail to find compounds such as physostigmine.

Presumably toxic compounds are not uniformly distributed throughout all plant cells and variation in toxicity to different nematode species could be accounted for on the basis of the particular plant tissue fed upon. Toxicity of extracted juice is unsatisfactory evidence that the parasite actually encounters the toxin within the host.

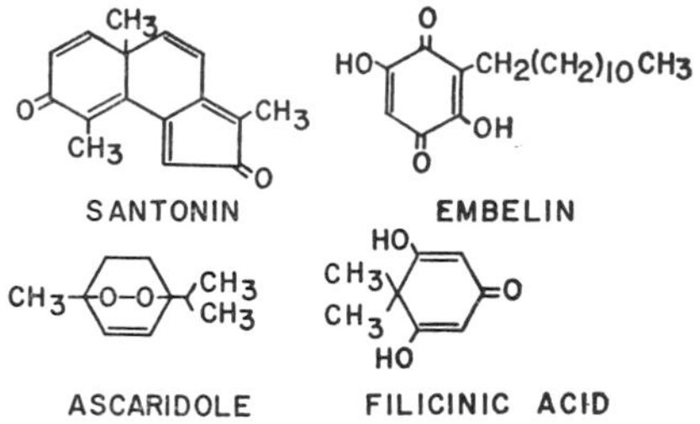

SANTONIN EMBELIN

ASCARIDOLE FILICINIC ACID

FIGURE 2. Anthelmintics obtained from plants used to control animal parasitic nematodes.

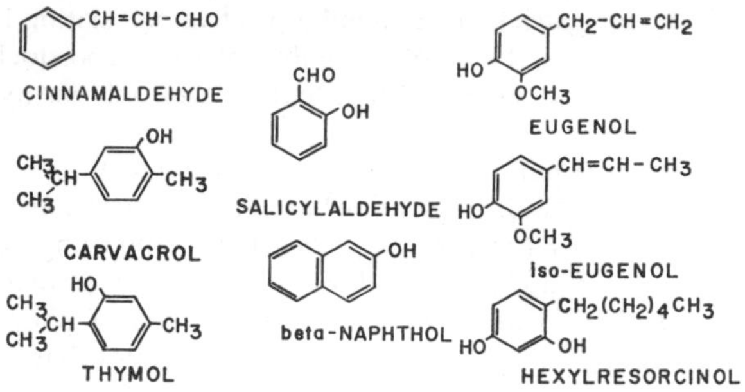
CH=CH-CHO
CINNAMALDEHYDE
OH
CH3
CH
CH3
CH3
CARVACROL
HO
CH3
CH
CH3
CH3
THYMOL
CHO
OH
SALICYLALDEHYDE
OH
beta-NAPHTHOL
CH2-CH=CH2
HO
OCH3
EUGENOL
CH=CH-CH3
HO
OCH3
Iso-EUGENOL
CH2(CH2)4CH3
HO
OH
HEXYLRESORCINOL

FIGURE 3. Phenolic compounds toxic to nematodes. All are of plant origin, except hexylresorcinol, which is a synthetic anthelmintic (Caprokol).

UNSUITABLE HOSTS

Parasitology has been described as that branch of ecology in which the host provides the environment of the parasite. With endoparasites, at least, the host provides nutrition and the proper physical environment, and possibly stimuli necessary for feeding, moulting, reproduction, and other functions. Failure to provide for the requirements of the parasite would result in an unsatisfactory or a resistant host. Many examples exist of hosts in which nematodes develop normally, but in small numbers or at a slow rate, perhaps because the host tissues are unsuitable.

Recent successes in the culture of the stem rust of wheat fungus (*Puccinia graminis tritici*) have placed renewed emphasis on the view that obligate parasitism may be primarily nutritional (53). The complexity of the problem is illustrated by the fact that no plant-parasitic nematode has been cultured on a defined medium. Several species of Tylenchida have been cultured on either excised roots or callus tissue, but plant tissues under these conditions are not basically different from those in the intact host (32).

None of the ectoparasites of the superfamilies Criconematoidea (Tylenchida) or Dorylaimoidea (Dorylaimida) have been successfully cultured, even on excised roots, indicating the probable importance of the physical environment of the soil in addition to nutritional requirements. Under natural conditions, however, these nematodes have very wide host ranges and resistance is relatively rare.

UNFAVORABLE HOST REACTIONS

Infection sets off a series of interactions that change both host and parasite to such a degree that every cell in each of the organisms is influenced. The conclusion derived from study of plant infections caused by a variety of pathogens, including nematodes, is that cellular responses in proper sequence are necessary for the survival of the parasite and delay or failure of the plant to respond properly results in resistance.

In addition to local pathology at the feeding site there are systemic biochemical and physiological alterations influencing both host and parasite. For example, the infected root is changed in its ability to absorb and transport water and nutrients leading to a reduction in photosynthesis and perhaps early maturity and senescence. Each of these changes in turn affects development of the parasite.

Migratory ectoparasites.—There are few reports in the literature concerning resistance to genera such as *Tylenchus, Tetylenchus, Trichodorus, Xiphinema,* and *Longidorus*. These nematodes feed at the root surface, usually for relatively short periods at a given cell, moving from place to place on the same or another root. Host response to attack by these wandering feeders is usually evident as cessation of growth of root apical meristems or swelling of root tips. It has not been established that any of these changes are essential for the survival of the parasite. Quite often, there is no visible evidence of feeding. In other hosts, brown surface lesions are formed such as with *Trichodorus proximus* on St. Augustine grass (46) and *T. viruliferous* on apple (42), but the nematodes do not remain at these lesions.

Species in this group have characteristically wide host ranges and very few nonhosts. *Trichodorus,* for instance, has several hundred known hosts, and the only reported nonhosts are toxic plants such as asparagus (49). There is, however, wide variation in the size of populations supported by different hosts and in the degree of tolerance. Grape rootstocks vary widely in injury from *Xiphinema index* and in the number of nematodes supported (34).

Nematodes in this group do not establish an intimate, long-lasting association with a host. The ability to move on to a more favorable area is never lost, which perhaps explains the relative scarcity of resistant hosts.

Sedentary ectoparasites.—Species of *Paratylenchus, Hemicycliophora, Criconemoides,* and similar genera remain attached to their hosts by their stylets for relatively long time periods. There is little, if any, injury to the cells fed upon, but infection may be associated with galling or change in growth characteristics of the root system (7, 45).

Nematodes in this group establish a closer relationship with their hosts than the migratory ectoparasites do, and they are not found in as wide a variety of habitats. Perhaps for these reasons, their host ranges tend to be smaller. There are reports of nonhosts or poor hosts, for instance in one greenhouse study of 101 plant species, *Paratylenchus projectus* failed to reproduce on 12 of them (8). Little else is known about resistance to nematodes in this group.

Migratory endoparasites.—Species of *Pratylenchus, Radopholus, Hoplolaimus,* and related forms attack the root cortex, breaking open cells and causing extensive brown lesions. *Aphelenchoides* spp. cause similar necrosis in the leaf mesophyll. The size, intensity of browning, and speed of formation of lesions varies widely according to the host plant.

The general metabolic response of plants to infection appears to be similar whether the infecting agent is a virus, bacterium, fungus, or nematode. Mechanical and chemical agents may produce similar responses. The compounds most commonly produced, and in the largest amounts, are the phenols and they have been the subject of a number of recent books and reviews (15, 24).

Phenols accumulate in infected tissues either from direct synthesis through altered respiratory pathways of the host or by release from precursors such as glycosides. Along with phenol accumulation is the buildup of oxidases. These enzymes oxidize the phenols to quinones which in turn polymerize, forming a number of complex products such as tannins, lignins, and melanins. These pigments give necrotic tissue its characteristic brown color.

The accumulation of phenols has been shown to be characteristic of *Pratylenchus penetrans* injury to a large number of plant species (1, 41, 60). The rapidity and intensity of the browning reaction varies from rapid and severe in Havana Seed tobacco to none in Kentucky bluegrass (62). In general, plant species with the largest lesions on the roots are the most severely stunted, whereas, those species that do not exhibit brown lesions are more tolerant. A very rapid, "hypersensitive" browning reaction, however, may result in death of the nematode, although if many nematodes are present, the host plant is severely injured.

Mountain & Patrick (37) found that in peach roots, the glycoside amygdalin is hydrolized by a nematode-produced β-glucosidase, releasing hydrocyanic acid and benzaldehyde, much to the detriment of both host and parasite. While this is an exceptional case, it does illustrate the types of reactions that lead to a hypersensitive response. Similar responses in which compounds are produced that are inhibitory to the parasite, while not as spectacular as those in peach, are probably a major form of resistance.

Several workers have noted that the root tissue that responds most markedly to the presence of lesion nematodes is the endodermis. Histochemical tests usually show a dark brown endodermis with a high concentration of phenols, even at early stages of infection when this tissue has not been fed upon. Lesion nematodes are confined to the cortex and do not reach vascular tissues until the later stages of infection. While there is no evidence that the endodermis protects plants from lesion nematode injury, it will be shown later that this type of reaction is important in plant resistance to nematodes that must feed on vascular tissues in order to survive.

Tolerance to *Radopholus similis,* a parasite similar in many respects to *P. penetrans,* might be related to failure to penetrate the endodermis. This species has a very wide host range, causing moderate damage to the cortical parenchyma of many crops, but in citrus, the endodermis is penetrated and the result is the devastating disease, spreading decline (13).

Chrysanthemum varieties resistant to *Aphelenchoides ritzemabosi* are characterized by a rapid, extensive browning of the leaf mesophyll associated with poor or no reproduction of the nematode (68). Susceptible varieties brown more slowly and there is progressive spread of the brown area as the nematodes multiply. The phenolic compounds and associated enzymes responsible for the browning reaction are the same and in similar concentrations in both resistant and susceptible varieties. The rapid browning of resistant varieties occurs during early stages of infection when nematodes are more active and damage more cells in their effort to find a suitable food supply. Wallace (68) attributed resistance to an absence of some nutritional factor rather than being directly related to the browning reaction. However, he did show that chlorogenic acid, the major phenol found, and its components, caffeic and quinic acids, increased uncoordinated movement of nematodes, although they did not kill them and did not prevent their migration. It is possible that there are large differences in phenol content of the cells actually fed upon which would not be detected in extracts of leaf discs. Wallace assumed that cell damage was essentially mechanical, although more recent evidence has shown that browning of fig leaves by a closely related species (*A. fragariae*) results from polymerization of oxidized phenols released from glycosides by a nematode-produced β-glucosidase (35). If, in the susceptible variety, this enzyme were somehow inactivated or perhaps if reducing substances such as ascorbic acid were present in large enough amounts to counteract oxidation, the browning reaction would be delayed.

Studying strawberry varieties resistant and susceptible to *Aphelenchoides fragariae*, Szczygiel & Giebel (59) found higher concentrations of phenols, particularly flavanols, in susceptible varieties. In this host, in contrast to chrysanthemum, the main symptom in susceptible varieties is distortion of leaves attributed to indole acetic acid (IAA) accumulation in injured areas. They hypothesized that flavanols in susceptible leaves inhibited enzymatic oxidation of IAA, whereas in resistant varieties IAA was destroyed. This was supported by their observations that extracts of susceptible leaves prevented destruction of IAA by horseradish peroxidase. Possibly related to this hypothesis is the report by Webster (69) that IAA promotes reproduction of *A. ritzemabosi.*

The very different symptoms produced on these two hosts result in part from a difference in feeding. In chrysanthemum, *A. ritzemabosi* is an endoparasite, migrating through mature leaf tissue, breaking open cells, releasing substrates and enzymes, and producing a lesion. In strawberry, this species feeds on the surface of leaves folded inside developing buds, and does not cause appreciable cell breakdown at this stage.

Oxidized phenols have been found to reduce significantly respiration of *P. penetrans* and to act as a repellant. Nematodes placed on agar plates near bits of necrotic tomato root or drops of oxidized chlorogenic acid, the major phenol in tomato, will immediately move in the opposite direction (5). It is interesting to note that chlorogenic acid itself attracts nematodes, whereas the oxidized form is repellant. Complete oxidation and polymerization again results in an inactive product so there appears to be only a certain time period when inhibitory products are present, and in the plant, speed of the necrotic reaction would be of extreme importance. Migratory forms might move away to a more favorable environment if the reaction were not too rapid.

Sedentary endoparasites.—The genus *Ditylenchus* represents an intermediate stage in the rather arbitrary separation of nematodes into groups based on feeding behavior. Within the genus, for those species that attack higher plants, injury ranges from extensive necrosis (*D. destructor*) to gall formation (*D. dipsaci*).

More typical of the sedentary endoparasites are genera such as *Meloidogyne, Heterodera, Tylenchulus,* and *Rotylenchus* in which the nematode induces a specialized feeding site within the host and develops to maturity at that point.

Two interrelated types of resistance are evident. Either the host fails to respond properly and the feeding site is not adequate for optimum development of the parasite, or else the host reaction is detrimental to the parasite. Typically, the detrimental reaction is necrosis.

Ditylenchus dipsaci can be divided into 20 or more morphologically-indistinguishable races based on host preference (58). The number of hosts may be wide or limited, depending on the race. The red clover race does not reproduce on alfalfa and the alfalfa race does not reproduce on red clover, and similarly each of the races may have both separate and common hosts. Within each race isolates or populations differ in ability to attack certain hosts or in the kind of symptoms produced. The Raleigh, North Carolina population of the alfalfa race caused galling in Wando peas, whereas the Waynesville, N.C. population caused necrosis on this variety (2). Both populations attacked alfalfa equally. Varieties of crop plants that are normally hosts vary in resistance so that Lahontan alfalfa is resistant to Swedish populations of the alfalfa race, but shows much more injury than it does when infected by populations from the United States (3).

From the above discussion it can be seen that a wide variety of interactions are possible. In a susceptible reaction, nematodes enter stem and leaf tissue causing swellings and various growth malformations. Internally, some parenchyma cells are destroyed and others become hypertrophied and their chloroplasts disintegrate. In the area of feeding, cells are separated from each other through dissolution of the middle lamella; Seinhorst (54) states that this maceration is necessary for proper development of the nematode.

Pectic enzymes are produced by *D. dipsaci* in large amounts and there is good evidence that they are primarily responsible for tissue maceration. Riedel & Mai (47) found nematode-produced endo-polygalacturonase in macerated host tissues and stressed its importance in pathogenicity. Other workers have not been able to demonstrate pectin removal and pectic enzymes in galls even though tissues were macerated. Riedel & Mai suggest that observations should be made at very early stages of infection because pectic enzyme activity may be diminished at later stages. They also suggest the possibility of enzyme inhibitors in host tissue extracts.

A number of workers have described resistance to *D. dipsaci* as hypersensitivity characterized by rapid necrosis of cells around entering larvae. Barker & Sasser (2) showed an intense staining with safranin in the necrotic area, but no detailed histochemical analyses have been made on necrotic tissue. Muse et al (38), found no correlation between the hypersensitive and susceptible responses of Wando pea to two populations of *D. Dipsaci* and the levels of pectolytic and cellulolytic enzymes in either nematode homogenates or infected tissues.

Not enough evidence is available to explain why, in a susceptible host, *A. ritzemabosi* causes browning of leaf mesophyll cells, but under similar conditions, *D. dipsaci* does not. Such browning usually does not inhibit *A. ritzemabosi,* whereas *D. dipsaci* will not develop in necrotic areas. A possible explanation might be based on the observation that *A. ritzemabosi* produces a β-glucosidase and *D. dipsaci* does not (38), but then, the necrosis produced by *D. dipsaci* in resistant plants must be accounted for. Soft-rot bacteria produce dehydrogenases that prevent oxidation of phenols (56). A similar reaction could occur in susceptible hosts, but not in resistant ones.

Another type of resistance is shown by hosts that apparently fail to provide adequate conditions for normal parasite development. Alaska 14A peas become infected and show initial symptom development similar to susceptible varieties, but later development is impaired (27). Epidermal cells in this variety browned rapidly, although penetration was not prevented and necrosis of underlying tissues became evident only after 5 days. Manod oats show an even milder form of resistance (4). Cavities are formed, although cell separation and hypertrophy are limited and there is no necrosis. Nematodes reproduce, but take twice as long to develop and are smaller compared with those from susceptible oats.

Resistance in Lahontan alfalfa is expressed as necrosis, reduced gall formation, and reduced reproduction of nematodes. Huisingh & Sherwood (26) working with this variety, found that inoculation with *D. dipsaci* resulted in accumulation of divalent cations at normal levels of nutrition. Plants grown at low calcium levels did not accumulate divalent cations and they also lost some of their resistance. These authors speculated that resistance might be related to plant tissue changes such as crosslinking of pectic substances of the middle lamella with divalent cations making them more resistant to enzymatic maceration. They pointed out, however, that although calcium levels modify the expression of resistance, their influence is not sufficient to explain the hypersensitive response of Lahontan alfalfa.

Resistance to root-knot and cyst nematodes is in general similar to resistance to the stem and bulb nematode. Hypersensitivity or necrosis is characteristic of some resistant hosts and in others, development of the parasite is retarded. The resistant reaction is variable, even in the same root, and both types of resistance may occur together, although usually they do not.

The complexity of the host-parasite interactions of *Meloidogyne* are diagrammatically represented in Figure 4. *Heterodera* species usually do not cause cortical hypertrophy, but otherwise, the actions and reactions are similar. It is probable that interference with any of these many interactions would impart some degree of unsuitability to the host.

The hypersensitive reaction of *Nicotiana repanda* (36) and Nemared tomato (41) to *M. incognita acrita* have been attributed to the rapid accumulation of oxidation products of chlorogenic acid. On a smaller scale, these lesions are similar to those produced by *Pratylenchus,* except that invading larvae of *Meloidogyne* are walled off by a group of necrotic cells and do not progress through the cortex. Nemared tomato has more chlorogenic acid in root tissue than other tomato varieties that are genetically similar, except for root-knot resistance, and early infection by *M. incognita* larvae leads to a rapid browning of cortical tissues, particularly the endodermis (41).

Heterodera larvae typically enter the cortex by penetrating through cells rather than intercellularly as do *Meloidogyne* larvae, so there is usually some necrosis and browning, even in susceptible varieties. Extensive necrosis, however, prevents normal development. Necrotic areas in potato varieties resistant to *H. rostochiensis* contain large amounts of β-glucosidase, peroxidase, and oxidized phenols which lead to lignification (18). These enzymes and substrates are reduced or lacking in susceptible varieties.

The necrotic reaction in the cortex either prevents the larva from reaching a suitable feeding site inside the cortex or else it inactivates the stimulus to produce giant cells, and if necrosis is prevented, resistance is lost. Biotype A of *H. rostochiensis* successfully attacks potato varieties resistant to Biotype B. The two biotypes differ in that Biotype A does not secrete a β-glucosidase and the phenols present in resistant cells are not released allowing larvae of this type to penetrate the cortex without causing necrosis (72). The hypersensitive reaction of tomatoes resistant to *M. incognita* can be partially blocked by applications of cytokinins (11) or by increasing temperatures above 28°C (10), in which case, a high percentage of larvae are able to produce galls.

An unusual type of resistance is shown in African Alfalfa resistant to *M. incognita acrita* and *M. javanica.* This variety is readily attacked and larvae migrate through the cortex without eliciting hypersensitivity, then most of them leave the root (44). As with larvae dissected from necrotic lesions, they retain their ability to develop on susceptible hosts.

Giant cell development is necessary for development of adult females capable of producing eggs. Dropkin & Nelson (12) classified 19 soybean varieties into four categories based on their reaction to *M. incognita.* Reaction Type I was hypersensitivity. Types II and III were incomplete development of giant cells and Type IV was a fully developed giant cell filled with dense cytoplasm. Rapid development of females with egg mass production occurred only with Type IV giant Cells, although Types II and III were found in all varieties.

INTERACTIONS IN THE LIFE CYCLE OF
MELOIDOGYNE spp.

PARASITE CHANGES STIMULATED BY HOST	TIME IN DAYS	HOST CHANGES STIMULATED BY PARASITE
	-5	
HATCH		
MOVE THROUGH SOIL (ATTRACTION)		
SURFACE FEEDING-ACTIVATION OF SUBVENTRAL ESOPHAGEAL GLANDS	0	CORTICAL HYPERTROPHY BEGINS
PENETRATION OF CORTEX		
ENTRY INTO STELE		
ACTIVATION OF DORSAL ESOPHAGEAL GLAND		SWELLING OF VASCULAR CELLS NEAR LARVA HEAD-DEVELOPMENT OF POLYPLOID CONDITION OF FUTURE GIANT CELL NUCLEI
ESTABLISHMENT OF FEEDING SITE	5	
2L BECOMES SWOLLEN AND SEDENTARY		SIMULTANEOUS MITOSES OF GIANT CELL NUCLEI
	10	INCREASE IN SIZE OF GIANT CELLS
		GIANT CELLS REACH FULL SIZE - VACUOLATE CYTOPLASM
MOULT II → 3L	15	
MOULT III → 4L		
MOULT IV → ADULT		INCREASE IN DENSITY
	20	OF GIANT CELL CYTOPLASM
MALES LEAVE ROOTS		(♂ GIANT CELLS DISINTEGRATE)
		FULLY DEVELOPED GIANT
	25	CELLS
RECTAL GLANDS SECRETE EGG MASS		
EGGS ARE LAID	30	GIANT CELLS BEGIN TO DISINTEGRATE

FIGURE 4. Development of a species of root-knot nematode. Each action and reaction is a potential source of resistance.

The stimuli that induce giant cells are exceedingly complex and not well understood. Development involves rapid synthesis of a wide variety of compounds, and lack of any of a large number of substrates and co-factors would impair proper formation. One of the important hormones involved is IAA, but its action is apparently modified by other growth factors.

Giebel & Wilski (19) have proposed that amounts of IAA are greater in susceptible potatoes infected with *H. rostochiensis* because polyphenols released by enzymatic breakdown of glycosides inhibit the peroxidases that oxidize IAA. In resistant potatoes, IAA oxidation is unimpaired or increased by the same nematode stimulus. Their evidence for this is that extracts from susceptible varieties inhibit horseradish peroxidase, but do not inactivate added IAA, whereas extracts from resistant potatoes do the reverse. Peroxidase inhibition, however, occurs only in extracts hydrolyzed by β-glucosidase. In a later study, Giebel (17) measured phenols in hydrolyzed root extracts of 21 species of Solanaceae and the effect of these extracts on IAA oxidase. Resistant species on which cysts failed to develop were characterized by a high ratio of monophenols to polyphenols and little IAA-oxidase inhibition.

An abnormally large number of larvae of *Meloidogyne* and *Heterodera* associated with plants under poor growing conditions, or in multiple infections where food is reduced, develop into males. The same thing often happens in resistant plants, although a high proportion of males is probably a symptom of resistance rather than a cause. The addition of tyrosine to susceptible tomato roots results in a necrotic type of response to invasion by *H. rostochiensis* and the development of larvae into males associated with poorly developed giant cells (63). It would be interesting to see if other monophenols would provoke a similar response.

Endo (14) found that *H. glycines* larvae that entered the stele opposite protophloem points initiated small giant cells and became males, but larvae near protoxylem points initiated larger giant cells incorporating more host cells and these larvae developed into females. This is an indication of variation even within a susceptible host of the ability of specific cells to give rise to optimum giant cells and could serve to explain variation of giant cell types in any host.

Little information is available about resistance to other sedentary endoparasites. *Rotylenchulus reniformis* produces both a hypersensitive necrotic response and poorly developed giant cells on resistant hosts (43). Soybean varieties react similarly to both *R. reniformis* and *H. glycines* and resistance to them may be controlled by the same gene or linked genes.

Citrus rootstocks resistant to *Tylenchulus semipenetrans* undergo a hypersensitive reaction in the outer cortex with an accumulation of safranin-staining materials in the collapsed cells (66). Cortical cells underlying the necrotic area divide to form a layer that separates the lesion from the rest of the cortex and nematodes do not establish a feeding site. In moderately resistant rootstocks, necrosis is reduced and a few nematodes are able to feed in the cortex and reach maturity. Roots of resistant varieties contain toxins that Van Gundy & Kirkpatrick (66) suggest might prevent larvae from giving the proper stimulus necessary for feeding site development.

The Control of Nematodes Using Resistant Plants

The use of resistant plants or nonhosts in crop rotation is widespread, either by design to keep nematode populations small or as a hidden benefit from good agricultural practice. Resistant varieties of many crops are available and are often used in areas where nematode problems are severe. The present situation in this regard has been extensively reviewed and evaluated (25, 31). Kehr concluded that "despite the fact that breeding research to control nematodes started nearly 50 years ago, there are virtually limitless unexploited opportunities in the field of effective biological control of nematodes through the use of resistant crop plants."

Presently, chemical control of nematodes is limited primarily to crops of high cash value because of the relatively high cost of fumigation and the specialized equipment necessary. The trend toward development of insecticide-type materials may introduce problems of residues and high mammalian toxicity. Rotation, on the other hand, is limited when a profitable market does not exist for alternate crops or when expensive planting, harvesting, or packaging machinery must be justified by constant use. All these trends in modern agriculture place additional emphasis on the use of resistant varieties of plants.

Van der Plank (65) has proposed the terms vertical and horizontal to describe plant resistance to fungi and bacteria, and examples can be found of resistance to nematodes of each type. Most nematode-resistant varieties of crop plants appear to have vertical resistance, which shows up as hypersensitivity or necrosis controlled by one or two genes. Variation in resistance of this type to different populations of the parasite is common, and resistance-breaking biotypes are to be expected. Plants with vertical resistance support lower parasite populations and are often easy to recognize. Root-knot-resistant plants of this type do not form galls and are readily separated from susceptible lines.

Resistance in potato to *H. rostochiensis* demonstrates vertical resistance. Jones & Parrott (29) have proposed that resistance involves at least two different gene pairs (A & B) and that larvae can be separated into pathotypes on a gene-for-gene basis. Resistance from *Solanum tuberosum andigena* (Ab) is overcome by females of pathotype 1 (aa) and resistance from *S. multidissectum* (aB) is overcome by females of pathotype 2 (bb). Pathotype 1, 2 females (aabb) mature on varieties with both types of resistance (AB) and Pathotype O females that are not homozygous recessive for either gene pair, are unable to attack varieties with either gene for resistance (A or B) (Table 1).

Horizontal resistance is polygenic and is generally effective against all races of the pathogen. Plants are only partially resistant, but the resistance is

TABLE 1. Reaction of potato varieties to pathotypes of *H. rostochiensis* Adapted from Jones & Parrott (29)

Host Resistance	Pathotype			
	0(A–B–)	1(aaB–)	2(A–bb)	1, 2(aabb)
S. tuberosum (ab)	+	+	+	+
X *S. tuberosum andigenum* (Ab)	–	+	–	+
X *S. multidissectum* (aB)	–	–	+	+
X *andigenum* X *multidissectum* (AB)	–	–	–	+

+ = ability to form cysts.
– = only males are able to develop.

quite stable and breakdown is unlikely. Poor hosts of root-knot and cyst nematodes are probably of this type. Although this type of resistance is much to be preferred, it is much more difficult to introduce into a breeding program because several genes are involved and because it is more difficult to recognize and assess.

Summary

1. Failure of nematodes to reproduce well (resistance) is usually accompanied by decreased plant injury, although some highly resistant plants may show severe injury (hypersensitivity) and some susceptible plants show little damage (tolerance).

2. Most plants attract and are fed upon by nematodes, and resistance does not become apparent until after the initial stages of infection.

3. Some plants contain compounds directly toxic to nematodes or contribute to poor nematode development.

4. Plant nematodes are obligate parasites, requiring nutritional, developmental, and environmental contributions from the host. Failure of a host to produce a favorable feeding site results in poor parasite development.

5. As complexity of the host-parasite interaction increases, resistance becomes more common. Ectoparasites apparently require little host modification and resistance is uncommon.

6. Endoparasites often induce necrosis. Migratory forms are usually able to move to new tissues, enlarging the necrotic area. When necrosis is rapid they may become trapped. Sedentary forms trapped by rapid necrosis (hypersensitivity) are unable to induce further host changes necessary for development.

7. Hypersensitive reactions are usually controlled by one or a few genes and response to different populations is variable (vertical resistance). Poor feeding site development is a separate response which is less clear cut, but may be more stable toward various populations (horizontal resistance).

LITERATURE CITED

1. Acedo, J. R., Rohde, R. A. 1971. Histochemical root pathology of *Brassica oleracea capitata* L. infected by *Pratylenchus penetrans* (Cobb) Filipjev and Schuurmans Stekhoven (Nematoda: Tylenchidae). *J. Nematol.* 3:62–68
2. Barker, K. R., Sasser, J. N. 1959. Biology and control of the stem nematode, *Ditylenchus dipsaci*. *Phytopathology* 49:664–70
3. Bingefors, S. 1961. Stem nematode in Lucerne in Sweden. II. Resistance in Lucerne against stem nematode. *Kungl. Lantbrukshögskolans Ann.* 27:385–98
4. Blake, C. D. 1962. The etiology of tulip-root disease in susceptible and resistant varieties of oats infested by the stem nematode *Ditylenchus dipsaci* (Kühn) Filipjev. II. Histopathology of tulip-root and development of the nematode. *Ann. Appl. Biol.* 50: 713–22
5. Chang, L. M., Rohde, R. A. 1969. The repellent effect of necrotic tissues on the nematode *Pratylenchus penetrans*. *Phytopathology* 69:398 (Abstr.)
6. Chitwood, B. G. 1949. Root-knot nematodes. I. A revision of the genus *Meloidogyne* Goeldi, 1887. *Proc. Helminthol. Soc. Wash.* 16:90–104
7. Coursen, B. W., Jenkins, W. R. 1958. Host-parasite relationships of the pin nematode, *Paratylenchus projectus,* on tobacco and tall fescue. *Plant Dis. Reptr.* 42: 865–72
8. Coursen, B. W., Rohde, R. A., Jenkins, W. R. 1958. Additions to the host lists of the nematodes *Paratylenchus projectus* and *Trichodorus christiei*. *Plant Dis. Reptr.* 42:456–60
9. Crittenden, H. W. 1954. Factors associated with root-knot nematode resistance in soybeans. *Phytopathology* 44:388 (Abstr.)
10. Dropkin, V. H. 1969. Cellular responses of plants to nematode infections. *Ann. Rev. Phytopathol.* 7:101–22
11. Dropkin, V. H., Helgesen, J. P., Upper, C. D. 1969. The hypersensitivity reaction of tomatoes resistant to *Meloidogyne incognita:* Reversal by cytokinins. *J. Nematol.* 1:55–61
12. Dropkin, V. H., Nelson, P. E. 1960. The histopathology of root-knot nematode infections in soybeans. *Phytopathology* 50:442–47
13. DuCharme, E. P. 1959. Morphogenesis and histopathology of lesions induced on citrus roots by *Radopholus similis*. *Phytopathology* 49:388–95
14. Endo, B. Y. 1965. Histological responses of resistant and susceptible soybean varieties and backcross progeny to entry and development of *Heterodera glycines.* *Phytopathology* 55:375–81
15. Farkas, G. L., Kiraly, Z. 1962. Role of phenolic compounds in the physiology of plant diseases and disease resistance. *Phytopathol. Z.* 44:105–50
16. Ford, H. W., Feder, W. A. 1969. Development and use of citrus rootstocks resistant to the burrowing nematode, *Radopholus similis*. *Proc. 1st Int. Citrus Symp.* 2:941–48
17. Giebel, J. 1970. Phenolic content in roots of some solanaceae and its influence on IAA-oxidase activity as an indicator of resistance to *Heterodera rostochiensis*. *Nematologica* 16:22–32
18. Giebel, J., Krenz, J., Wilski, A. 1971. Localization of some enzymes in roots of susceptible and resistant potatoes infected with *Heterodera rostochiensis*. *Nematologica* 17:29–33
19. Giebel, J., Wilski, A. 1970. The role of IAA-oxidase in potato resistance to *Heterodera rostochiensis* Woll. *Proc. IX Int. Nematol. Symp., Warsaw, 1967*. 239–45.
20. Grainger, J. 1964. Factors affecting the control of eelworm diseases. *Nematologica* 10:5–20
21. Green, C. D. 1971. Mating and host finding behavior of plant nematodes. *Plant Parasitic Nematodes*. eds. Zuckerman, B. M., Mai, W. F., Rohde, R. A. 2: Chap. 24, 247–66. New York: Academic. 345 pp.
22. Griffin, G. D., Waite, W. W. 1971. Attraction of *Ditylenchus dipsaci* and *Meloidogyne hapla* by resis-

tant and susceptible alfalfa seedlings. *J. Nematol.* 3:215–19

23. Grundbacher, F. J., Stanford, E. H. 1962. Effect of temperature on resistance of alfalfa to stem nematode (*Ditylenchus dipsaci*). *Phytopathology* 52:791–94
24. Harborne, J. B., ed. 1964. *Biochemistry of Phenolic Compounds*. New York: Academic. 618 pp.
25. Hare, W. W. 1965. The inheritance of resistance of plants to nematodes. *Phytopathology* 55:1162–67
26. Huisingh, D., Sherwood, R. T. 1968. The role of calcium in resistance of alfalfa to *Ditylenchus dipsaci*. *Nematologica* 14:8–9
27. Hussey, R. S., Krusberg, L. R. 1968. Histopathology of resistant reactions in Alaska pea seedlings to two populations of *Ditylenchus dipsaci*. *Phytopathology* 58:1305–10
28. Jenkins, G. L., Hartung, W. H., Hamlin, K. E. Jr., Data, J. B. 1957. *The Chemistry of Organic Medicinal Products*. New York: Wiley. 569 pp.
29. Jones, F. G. W., Parrott, D. M. 1965. The genetic relationships of pathotypes of *Heterodera rostochiensis* Woll. which reproduce on hybrid potatoes with genes for resistance. *Ann. Appl. Biol.* 56:27–36
30. Kaul, R. 1962. Untersuchungen über einen aus Zysten des Kartoffelnematoden (*Heterodera rostochiensis* Woll.) isolierten phenolischen Komplex. *Nematologica* 8:288–92
31. Kehr, A. E. 1966. Current status and opportunities for the control of nematodes by plant breeding. *Pest Control by Chem. Biol. Genet. & Phys. Means,* 126–38. Washington: U.S. Dep. Agr. 214 pp.
32. Krusberg, L. R. 1961. Studies on the culturing and parasitism of plant parasitic nematodes, in particular *Ditylenchus dipsaci* and *Aphelenchoides ritzemabosi* on alfalfa tissues. *Nematologica* 6:181–200
33. Krusberg, L. R. 1963. Host response to nematode infection. *Ann. Rev. Phytopathol.* 1:219–40
34. Kunde, R. M., Lider, L. A., Schmitt, R. V. 1968. A test of Vitis resistance to *Xiphinema index*. *Am. J. Enology Viticulture* 19:30–36
35. Maeseneer, J. de. 1964. Leaf browning of *Ficus* spp., new host plants of *Aphelenchoides fragariae* (Ritzema Bos). *Nematologica* 10:403–08
36. Milne, D. L., Boshoff, D. N., Buchan, P. W. W. 1965. The nature of resistance of *Nicotiana repanda* to the root-knot nematode, *Meloidogyne javanica*. *S. Afr. J. Agr. Sci.* 8:557–67
37. Mountain, W. B., Patrick, Z. A. 1959. The peach replant problem in Ontario. 7. The pathogenicity of *Pratylenchus penetrans* (Cobb, 1917) Filip. & Stek., 1941. *Can. J. Bot.* 37:459–70
38. Muse, B. D., Moore, L. D., Muse, R. R., Williams, A. S. 1970. Pectolytic and cellulolytic enzymes of two populations of *Ditylenchus dipsaci* on Wando pea (*Pisum sativum* L.). *J. Nematol.* 2:118–24
39. Oostenbrink, M., Kuiper, K., S'Jacob, J. J. 1957. Tagetes als fiendpflanzen von *Pratylenchus*-arten. *Nematologica Suppl.* 2:424S–33S
40. Otiefa, B. A. 1953. Development of the root-knot nematode, *Meloidogyne incognita,* as affected by potassium nutrition of the host. *Phytopathology* 43:171–74
41. Pi, C. L., Rohde, R. A. 1967. Phenolic compounds and host reaction in tomato to injury caused by root-knot and lesion nematodes. *Phytopathology* 57:344 (Abstr.)
42. Pitcher, R. S. 1967. The host-parasite relations and ecology of *Trichodorus viruliferus* on apple roots, as observed from an underground laboratory. *Nematologica* 13:547–57
43. Rebois, R. V., Epps, J. M., Hartwig, E. E. 1970. Correlation of resistance in soybeans to *Heterodera glycines* and *Rotylenchulus reniformis*. *Phytopathology* 60:695–700
44. Reynolds, H. W., Carter, W. W., O'Bannon, J. H. 1970. Symptomless resistance of alfalfa to *Meloidogyne incognita acrita*. *J. Nematol.* 2:131–34
45. Rhoades, H. L., Linford, M. B.

1961. A study of the parasitic habit of *Paratylenchus projectus* and *P. dianthus*. *Proc. Helminthol. Soc. Wash.* 28:185–90

46. Rhoades, H. L. 1965. Parasitism and pathogenicity of *Trichodorus proximus* to St. Augustine grass. *Plant Dis. Reptr.* 49:259–62
47. Riedel, R. M., Mai, W. F. 1971. A comparison of pectinases from *Ditylenchus dipsaci* and *Allium cepa* callus tissue. *J. Nematol.* 3: 174–78
48. Rohde, R. A. 1965. The nature of resistance in plants to nematodes. *Phytopathology* 55:1159–62
49. Rohde, R. A., Jenkins, W. R. 1957. Host range of a species of *Trichodorus* and its host-parasite relationships on tomato. *Phytopathology* 47:295–98
50. Rohde, R. A., Jenkins, W. R. 1958. Basis for resistance of *Asparagus officinalis* var. *altilis* L. to the stubby-root nematode *Trichodorus christiei* Allen 1957. *Md. Agr. Expt. Sta. Bull.* A-97, 19 pp.
51. Sasser, J. N. 1954. Identification and host-parasite relationships of certain root-knot nematodes (*Meliodogyne* spp.). *Md. Agr. Expt. Sta. Bull.* A-77, 31 pp.
52. Scheffer, F., Kickuth, R., Visser, J. H. 1962. Die Wurtzelauscheidungen von *Eragrostis curvula* (Schrad.) Nees und ihr Einfluss auf Wurtzel-knoten Nematoden *Z. für Pflanzenernährung Düngung Bodenkunde* 98:114–20
53. Scott, K. J., Maclean, D. J. 1969. Culturing of rust fungi. *Ann. Rev. Phytopathol* 7:123–46
54. Seinhorst, J. W. 1957. *Ditylenchus;* races, pathogenicity, and ecology. *Proc. S-19 Workshop Phytonem, Nashville,* 1–7
55. Shepherd, A. M., Clarke, A. J. 1971. Molting and hatching stimuli. *Plant Parasitic Nematodes.* eds. Zuckerman, B. M., Mai, W. F., Rohde, R. A. 2: Chap. 25, 267–87. New York: Academic, 345 pp.
56. Stahmann, M. A., Woodbury, W., Lovrekovitch, L., Macko, V. 1968. The role of enzymes in regulation of disease resistance and host-pathogen specificity. *Proc. Int. Symp. Plant Biochem. Reg. Viral Other Dis. Inj.* 263–74 Tokyo: Phytopathol. Soc. Japan. 350 pp.
57. Stecher, P. G., et al. eds. 1968. *The Merck Index.* Rahway: Merck, 1713 pp.
58. Sturhan, D. 1971. Biological races. *Plant Parasitic Nematodes.* eds. Zuckerman, B. M., Mai, W. F., Rohde, R. A. 2: Chap. 15, 51–72. New York: Academic. 345 pp.
59. Szczygiel, A., Giebel, J. 1970. Phenols in the leaf buds of two strawberry varieties resistant and susceptible to *Aphelenchoides fragariae* (Ritzema Bos) Christie. *Proc. IX Int. Nematol. Symp., Warsaw,* 1967. 247–53
60. Townshend, J. L. 1963. The pathogenicity of *Pratylenchus penetrans* to strawberry. *Can. J. Plant Sci.* 43:75–78
61. Triantaphyllou, A. C. *Ann. Rev. Phytopathol.* 11, In preparation
62. Troll, J., Rohde, R. A. 1966. Pathogenicity of *Pratylenchus penetrans* and *Tylenchorhynchus claytoni* on turfgrasses. *Phytopathology* 56:995–98
63. Trudgill, D. L. Personal communication
64. Uhlenbroek, J. H., Bijloo, J. D. 1959. Isolation and structure of a nematicidal principle occurring in *Tagetes* roots. *Proc. IV Int. Cong. Crop Protect., Hamburg, 1957.* 579–81
65. Van der Plank, J. E. 1968. *Disease Resistance in Plants.* New York: Academic. 206 pp.
66. Van Gundy, S. D., Kirkpatrick, J. D. 1964. Nature of resistance in certain citrus rootstocks to citrus nematode. *Phytopathology* 54: 419–27
67. Viglierchio, D. R. 1961. Attraction of parasitic nematodes by root emanations. *Phytopathology* 51: 136–42
68. Wallace, H. R. 1961. The nature of resistance in Chrysanthemum varieties to *Aphelenchoides ritzemabosi. Nematologica* 6:49–58
69. Webster, J. M. 1967. The influence of plant growth substances and their inhibitors on the host parasite relationships of *Aphelenchoides ritzemabosi* in culture. *Nematologica* 13:256–62
70. Webster, J. M. 1969. The host-parasite relationships of plant-parasitic nematodes. *Advan. Parasitol.* 7:1–40

71. Welle, H. B. A., Bijloo, J. D. 1965. The systemic nematicidal activity of physostigmine and structural analogues. *Meded. Landbouwhogesch. Opzoekingsstations Gent.* 30:1417–28
72. Wilski, A., Giebel, J. 1966. Beta-glucosidase in *Heterodera rostochiensis* and its significance in resistance of potato to this nematode. *Nematologica* 12:219–24
73. Wingard, S. A. 1953. The nature of resistance to disease. *Plant Diseases, The Yearbook of Agriculture* 165–73 Washington: U.S. Dep. Agr. 940 pp.
74. Winoto Suatmadji, R. 1969. *Studies on the Effect of Tagetes Species on Plant Parasitic Nematodes.* Sticht. Fonds. Landbouw. Export Bur. Publ. 47. Veenman and Zonen N.V., Wageningen, Netherlands, 132 pp.

METHODOLOGY OF EPIDEMIOLOGICAL RESEARCH

3551

J. C. ZADOKS
Laboratory for Phytopathology, Agricultural University, Wageningen, The Netherlands

No one can be a good observer unless he is a good theorizer.
Charles Darwin (quoted from 85).

Methodology is distinct from but shares some common ground with logic, mathematics (especially mathematical statistics), and systems analysis. The distinction is due in part to the behavioral aspect of methodology: it deals as much with the scientist and his behavior as with the science and its procedures. See references 27, 43, 48, 51, 52, 77, 83, 86, 113. Epidemiology is the science of disease in populations (100). "Botanical epidemiology" (121) is used to differentiate the subject of this paper from medical and veterinary epidemiology. The emphasis is on the epidemiology of foliar pathogens.

Scientific methods may be technical or logical, or both. The technical methods refer to the manipulation of phenomena, the setting of conditions, the techniques of measurement (114). Publications of a normative nature on the technology of epidemiology are rare and incomplete (24, 38, 58, 59, 73, 83); several are instruction books for practicals (26, 76, 77, 82, 83, 86). The logical methods refer to the way of reasoning, of drawing inferences from conditions, and of interpretation.

THE EMPIRICAL CYCLE

Phase 1. Observation.—Observation, including collection and classification of data, is the first systematic and purposeful activity of the researcher. A real genius is a keener and more original observer than the average scientist. Genius may err: Eriksson & Henning (36) studying the overwintering of *Puccinia striiformis* (= *P. glumarum*) found many sporulating leaves in the fall but few in the spring. Rarely, they found mycelium in the leaves during winter. These observations led Eriksson to the hypothesis that the rust overwintered as a "mycoplasma" (34). Later, he "substantiated" his hypothesis by a microscopical-anatomical study (35); he concluded that small bodies, now known as haustoria, were the transition forms between mycoplasma and mycelium. Ward (108), however, showed that the anatomical argument was invalid, because Eriksson had not observed the penetration hyphae connecting intercellular mycelium and intracellular haustoria owing to the use of glycerine as a medium to mount leaf sections. Having made the correct epidemiological observations—it was the period of merging insight in the role of mycoplasms (medicine) and viruses (phytopathology)—but without insight into the population dynamics of a fungus (116), Eriksson tried to substantiate a false epidemiological hypothesis with invalid microscopical evidence, because of his erroneous *expectation.*

Phase 2. Induction.—Induction, or the generalization of pre-digested observation, leads to a hypothesis: "The similar responses of seven isolates of *P. herbarum* indicate that the need for a cold treatment for maturation of protoperithecia is probably characteristic of this species" (66). Methodologically, a hypothesis is acceptable only when prediction, to be verified by experimentation, can be deduced from it. Inductive reasoning leading to abstraction and formalization, and especially the use of mathematics, has been rare in botanical epidemiology (43, 56). Van der Plank's (100, 103) work is still unique. Induction seems to be more frequent in medical epidemiology (5, 74). The difference may be due to the fact that the botanical epidemiologist has access to experimentation, whereas the medical epidemiologist often has to rely upon statistical investigations.

Phase 3. Deduction.—"If that hypothesis is true in general, then this prediction can be made in this special instance." The natural sciences often use the hypothetico-deductive approach (85). In the methodological sense, deduction implies even more (27). It means explicitation of the concepts used, operationalization of hypothesis and definitions, so that the prediction can be put to an experimental test. The fructification period (43) or latent period (5, 100) and its measurement may serve as an example. The abstract definition: "the period from inoculation to reproduction" is too vague for application in an experiment. An operational definition used in work on *P. striiformis* is "the period in days from the day of inoculation until the day of observation of the first open pustule" (116). The operational definition uses the day as the time unit, acknowledges the fact that not all pustules open at the same time, and allows for the possible overlooking of the very first open pustule during one or more days.

Phase 4. Testing.—A hypothesis is said to be true when it accurately predicts the condition of fresh materials, of new elements not used in designing the hypothesis. This testing is done by experimentation.

Phase 5. Evaluation.—As a result of the test the hypothesis is accepted or rejected. What are the consequences of the decision with respect to other related hypotheses? What is the practical value or utility of the result? The result of the test is an observation to be subjected to interpretation that may lead to additional, complementing or modifying hypotheses. Therewith, the empirical cycle is closed. An experiment on the effect of temperature on latent period may lead to a question on the interaction of temperature with leaf wetness (*Septoria nodorum*) (95).

Koch's Postulates

Most of the methodological instruments in phytopathology are derived from statistics, e.g. the analysis of variance, and these fall outside the scope of this paper. A methodological tool typical for pathology in general, including phytopathology, is Koch's Postulates (61, 99):

1. The organism should be found in all cases of the disease in question, and its distribution in the body should be in accordance with the lesions observed.

2. The organism must be shown to be a living thing, and must be cultivated outside the body of the original host, in pure culture, for several generations.

3. The organism, so isolated, must reproduce the disease in other susceptible animals.

These postulates are quite versatile. They can be applied to bacteria, fungi, viruses, nematodes, and even insects. The normative value of Koch's Postulates is great. The fact that, 80 years after their formulation, Koch's Postulates are still a major tool in phytopathological methodology, shows their excellence as well as the poor "state of the art" of epidemiology. The excellence, by the way, is limited, at least in medicine: there are several instances where the pathogen is present in the host either without causing disease (cholera, diphtheria, polyomyelitis, viral hepatitis) or after recovery (typhoid, cholera El-Tor) (42, 53, 99). Typically, Koch's Postulates have a qualitative and not a quantitative character.

Quantitative Methods

A respectable science goes quantitative. Around 1960, botanical epidemiology turned to quantitative methods under the incentive of Van der Plank (100), though there have been precursors (23, 63). Medical epidemiology was far ahead at that time (5, 42).

There seems to be no quantitative equivalent of Koch's Postulates. Gäumann's statement (43), that each epidemic has its own *genius epidemicus,* served as a starting point for the formulation of a set of rules proposed for testing the validity of quantitative evidence on epidemics.

1. The source(s) of inoculum at the onset of the epidemic must be known, and the amount of the pathogen in the source(s) must be expressed quantitatively.

2. The effect of environmental conditions on the development of an epidemic of the pathogen must be known in terms of quantitative relations between independent (usually abiotic) variables and dependent (biotic) variables.

3. The rate of development of the epidemic under the prevailing conditions must be calculated.

4. The successive levels of the epidemic must be calculated from the

TABLE 1. A comparison between the "rules for testing the validity of quantitative evidence on epidemics" and a few comprehensive epidemiological studies

Pathogen	*Phytophthora infestans*	*Puccinia striiformis*	*Puccinia striiformis*	*Alternaria solani*	*Puccinia recondita*
Host	*Solanum tuberosum*	*Triticum aestivum*	*Triticum aestivum*	*Solanum Lycopersicum*	*Triticum aestivum*
Country	Netherlands	Netherlands	Netherlands	U.S.A.	U.S.A.
Reference	(104)	(116)	(126)	(107)	(37)
Rule 1	+*	+	±	±	o
Rule 2	±	+	±	+	+
Rule 3	o g	±	+	+	+
Rule 3	o	o	+	+	±
Rule 5	o	o	±	±	+

* Explanation: rules observed (+), rules not observed (o), rules observed in part (±), and guestimates available (g).

known amount of pathogen in the source(s) and the calculated rates of development of the epidemic.

5. The calculated terminal level and the calculated intermediate levels of the epidemic must be equal to the observed terminal and intermediate levels. The rules, especially rule 5, provide checks for error. Furthermore, they have heuristic value; they may serve to discover unexpected relations.

It is an interesting exercise to test a few comprehensive epidemiological studies against the rules (Table 1). Such a test may help to locate areas of future research.

MEASURING

Host frequency.—Host frequency can be expressed as the number of susceptible host individuals (suscepts) per unit area, the percentage of susceptible leaf area in the canopy, the fraction of susceptible biomass (see below), etc. The larger the distance between suscepts, the slower the speed of the epidemic. According to Van der Plank's (100) "threshold theorem," if $iR_c < 1$, no epidemic will start, and low host frequency keeps R_c low. For similar reasons, outbreaks of insect pests in tropical rain forests are rare. Drawing from animal ecology (60), we may speak of "underpopulation" of the pathogen caused by low host frequency (this is true only for specialized pathogens). Underpopulation may partly explain why serious epidemics in natural mixed vegetation are rare: a topic waiting for investigation. Anderson's (3) remark, that diseases were rare in native Guatemalan gardens with typical mixed cropping, is suggestive. The principle of reduced frequency of suscepts is applied in the composite or multiline cultivars of cereals (17), where development of foci is suppressed (115, 120). The same principle is involved when an epidemic comes to a stop before 100% of the suscepts are killed (5, 101).

Amount of host.—The total amount of host per unit area of soil can be measured as fresh weight or dry weight, both giving yield per unit area; total green area (65), giving a Green Area Index (44); or leaf area, giving the well-known Leaf Area Index, LAI (111). Techniques for measuring leaf area were summarized by Marshall (70). Ecologists and crop physiologists are much interested in such data, epidemiologists rarely. In the epidemiology of *P. striiformis* on wheat in North West Europe, however, LAI is a key factor because its annual fluctuation can be as much as ten millionfold (116, 121). Quantitative assessment of the host as a part of an epidemiological study is, alas, a rare phenomenon, though Van der Plank (100) demonstrated the effect of the changing host quantity on the speed of the epidemic.

Vectors.—Vectors of plant diseases are man (119) and animals, usually insects, sometimes birds, nematodes, etc. Available knowledge is mainly qualitative. With respect to the effects of densities and movements of men and insects, medical epidemiology may be well in advance of botanical epidemiology (5).

Number of pathogens.—Counting individuals is usually not practical, because there are too many (bacteria and viruses), or individuals cannot be distinguished (fungi). The solution is to count diseased plants in the case of systemic diseases, or to count local lesions in the case of local lesion diseases. The latter is a tedious procedure, mostly avoided by estimating the amount of disease. Techniques of disease assessment have been developed recently (24, 38, 58, 62, 64, 82). An unsolved problem is the weighting of various symptoms; e.g. in *Piricularia oryzae* a leaf and a neck lesion are comparable in size, but their relative weights with respect to the progress of the epidemic (spore production, etc.) and to the loss induced, are completely different.

Number of propagules.—An alternative for or a complement to disease assessment is the counting of "propagules" or "viable counts." Most of the sporulating foliage pathogens produce one or a few types of propagules, usually spores. Within each type, the size distribution is unimodal, the variance is small, and the mean values are relatively stable. Such propagules are said to be monodisperse. In soil fungi, one has often to be content with "viable counts" on agar plates. Each viable count represents at least one living and germinable propagule, a spore or a mycelial fragment. Mycelial fragments can break into small but still infective units. Size distribution needs no longer to be unimodel, the mean is not constant, and the variance is high; these propagules may be called polydisperse. The justification for propagule counting is partly logical, partly technical. The logical justification is the consideration that one propagule may cause one infection. A one to one relation is rare in practice, however. With single-uredospore inoculations of *Puccinia recondita* on wheat the highest propagule/lesion ration found was ca 0.5 (Zadoks, unpublished). In the field, a 0.01 ratio is high (55). The technical justification is that propagule counting, especially in the case of wind-borne spores, is a fast, accurate, and nondestructive method that does not disturb the crop and its microclimate. In mathematical analyses, cumulative spore counts are analogous to disease severity (32, 37, 88).

Biomass.—Ecologists measure biomass in terms of dry weight or energy. Biomass is the net production (accumulation of chemical energy) per individual, species, trophic level, or ecosystem (78). Productivity is expressed as biomass per unit area and unit time. The annual production of mycelium in one square meter of forest soil was calculated to be ca 200,000 km hyphal length, equal to ca 150 grams dry weight (75). Similar data on plant-parasitic fungi are rare or absent. Considerations in terms of biomass might be fruitful, especially with respect to crop loss problems. The following approximative calculation illustrates the point. A wheat crop produces per hectare a final biomass of 6×10^3 kg grain + 5×10^3 kg straw + 2×10^3 kg roots = 13×10^3 kg. After a severe rust epidemic, the yield is 3×10^3 kg grain + 3.5×10^3 kg straw + 1.5×10^3 kg roots = 8×10^3 kg. A leaf produces up to its own dry weight in uredospores (72), and contains at any time about 10% of its dry weight in active mycelium. When the upper two leaf layers are completely rusted they will produce an estimated 1.25×10^3 kg spores + 0.13×10^3 kg mycelium = 1.38×10^3 kg rust biomass. Alternatively: a severely rusted field produces up to 10^{13} spores/ha/day (55) or 40 kg/ha/day, one spore having a biomass of ca 4×10^{-12} kg. Such peak production does not last longer than one month, so that the total rust production can be estimated at up to 1200 kg/ha/season in fresh weight, or ca $1. \times 10^3$ kg/ha/season in dry weight, excluding mycelium. We lost 5×10^3 kg wheat biomass and we gained 1 to 1.4×10^3 kg rust biomass. What happened to the difference of 3.6×10^3 kg biomass per ha per season? Has it ever been formed, or has it been dissipated? This imaginary case exemplifies an ecologist's approach to an epidemiological problem.

Ecologists also measure the complexities of nature in a more abstract way using binary units of "information" (67–69), but this will not be discussed here.

Experimentation

The diachronic approach.—Experimentation is an essential element in the empirical cycle but is relatively new in botanical epidemiology. The real world outside is so complicated that only keen observers dared to formulate associations between selected phenomena and the outbreaks of disease. A classical example is the Van Everdingen rules for outbreaks of *Phytophthora infestans* (105).

The diachronic approach studies a situation in its historical development. An example is the history of the *Puccinia graminis* epidemics in North America caused by race 15 B (98). Epidemics of successive years are studied in relation to environmental factors, and associations between key weather factors and disease severities formulated. Typical diachronic research is Grainger's work on "disease phenology plots" (49). Becker (9) found that high precipitation in June and July was crucial for a severe outbreak of *S. nodorum* on wheat in the Netherlands. The association or even correlation can lead to a possible explanation when detailed observations on the disease are made simultaneously. The migration path of wheat stem rust in North America was found in that way. The spore counting work provided part of the evidence (97, 98). In the case of *P. striiformis* on wheat in the Netherlands, diachronic studies with detailed quantitative observations on both host and parasite provided the basis for an analysis of cause and effect (116) but an explanation could be given only with the help of experiments designed for that purpose.

The experimental approach.—The complexities of nature make it hard to sort out the effects of innumerable factors often interacting or confounded. Experiments serve to simplify nature to an extent that the scientist can test a hypothesis under controlled conditions. The experiments should be quantitative, using types of measuring as discussed above.

All experiments have a common design. There are one or more independent variables (experimental factors) and one or more dependent variables. The independent variable is usually offered in two or more discrete dosages (treatments; experimental levels; stimuli). It may be a biotic variable like host susceptibility or inoculum density, or an abiotic one, as fertilizer treatment or light regime. The dependent variable is usually a biotic factor, e.g. terminal severity of a disease. Its response is quantified and related to the independent variable. Statistical techniques are used to establish the correct relationship and its confidence limits.

In a unifactorial experiment one dependent variable is related to one independent variable. A bifactorial experiment has two independent variables, and their effects and the effect of their interaction on the dependent variable is studied. Multifactorial experiments can be designed. In complex and costly multifactorial experiments, more than one dependent variable can be measured, like spore cell enlargement + formation of hyphae + elongation of hyphae + swelling of hyphal tips + stomatal penetration (10), or latent period + pustulation rate + final number of pustules + sporulation rate + sporulation period (72).

This could be called a "multiple output experiment." Four classes of experiments can be distinguished:

a. single input —single output experiment,
b. single input —multiple output experiment,

c. multiple input—single output experiment,
d. multiple input—multiple output experiment.

Classes *a* and *c* are treated in the ordinary statistical textbooks. Experiments of class *b* and, increasingly, of class *d* are common in epidemiological studies. The outputs (dependent variables) may not be correlated.

Experiments are usually performed under the implicit assumption *ceteris paribus*. The independent variables are manipulated but the remainder is constant. This remainder, the sum total of the circumstances under which the experiment is performed, is not always specified adequately. The circumstantial factors as distinct from the experimental factors, often remain implicit or even subconscious. The lack of explicitness and the assumption of constancy are dangerous, because they may invalidate the generalization of a perfectly correct inference from the experiment.

Another danger is overlooking a causal factor either because available evidence is overwhelming though incomplete or even invalid, or because of lack of imagination. The former danger is exemplified by the criticism on the early aerobiological research that demonstrated the wind-borne nature of wheat stem rust epidemics in North America. Recent research emphasizes the importance of early rain-borne inoculum (87, 91; Nagarajan, p.c.), a factor overlooked before. The latter danger, lack of imagination, is exemplified by the experience of the author. Studying the oversummering of *P. striiformis* in a temperate oceanic climate (116), it never occurred to him that dry uredospores might oversummer in dust or dirt in a dry and hot semi-desert climate (99a). The last example illustrates a point of wider interest—the frame of reference of the individual researcher. This is determined by such factors as his training, job, technology, the country in which he works, and the historical period in which he lives (see Eriksson's mycoplasm hypothesis). These circumstantial factors influence the design of the experiments, choice of the experimental factors, and interpretation of the results.

The technology gradient.—The experimenter tries to reduce those effects of the circumstantial factors that increase the variance of the responses, so that the observed variance can be attributed exclusively to the experimenal factors. The reduction is obtained by eliminating circumstantial factors, or by keeping them constant throughout the experiment(s). Simplification of the natural field situation follows a general pattern (Table 2). The words diachronic and synchronic stand for "over the years" and "within a year." Admittedly, the choice of the time unit is arbitrary. Epidemiological studies covering a sequence of recurrent infection cycles have been called "polycyclic"; those restricted to one infection cycle "monocyclic" (124). In both cases, a dependent variable can be chosen at one point in time, thus leading to "cross-sectional" (19) analysis, e.g. yield or terminal severity of an epidemic. Alternatively, the dependent variable can be monitored throughout the whole duration of the monocyclic or polycyclic experiment—"longitudinal" analysis (19).

TABLE 2. Types of epidemiological studies ranked along the technology gradient

Analysis	Years	Field etc.	Manipulation of experimental factors	
			Biotic	Abiotic
a Diachronic	Between	—	No	Yes
b Synchronic	Within	Between	No	No
c Synchronic	Within	Within	Yes	No
d Synchronic	Within	Within	Yes	Yes
e Greenhouse	Within	Within	Yes	Yes
f Growth chamber	Within	Within	Yes	Yes

Table 2 illustrates types of epidemiological studies ranked along the technology gradient. Type *a* analyses lead to generalizations like: *P. infestans* on potatoes usually occurs in wet years. It leads to peculiar contradictions like: *P. striiformis* is a disease of dry springs in Ireland (71), whereas it is usually regarded as a disease of wet years. *Alternaria solani* can be dangerous to tomatoes under wet and under semi-arid conditions (107). Type *b* analyses often lead to conclusions on agronomic aspects of the problem: cultivar, soil type, Ph of soil, water supply, fertilizers, etc. Type *c* analyses are experiments in the sense of this paper. The biotic factors manipulated can be host genome (cultivar), physiologic races, inoculum dosage, time of inoculation, etc. Type *d* experiments are not unlike those of type *c*. Fertilizer, fungicide, and irrigation experiments belong to this class. An unusual experiment is the heat treatment of the buds of sultana vines with incandescent lamps (7). Recent epidemiological experiments study the effect of duration of leaf wetness, either by interception of dew (90), thus shortening the duration, or by overhead irrigation at dawn prolonging the duration (89; Shearer, p.c.). Types *e* and *f*, where all experimental factors are manipulated, differ mainly in the degree of control: e.g. variable sunlight is replaced by nearly constant artificial irradiation.

In all types, longitudinal and cross-sectional studies can be made, but longitudinal studies are rare in types *a, e,* and *f*. Only very recently, typical longitudinal studies, covering a number of consecutive infection cycles, have been performed in growth chambers (25). Longitudinal studies of type *c*, without manipulation, attracted new interest since automated data acquisition coupled with computer-assisted data analysis became possible. Longitudinal field studies of types *b* and *c* can be analyzed statistically by (multiple) correlation and regression techniques, whereas cross-sectional studies are usually based on a factorial design.

From *a* to *f*, the researcher climbs the technology gradient, but he is often unaware of accompanying change in perspective. At each step, he chooses the experimental factors to be studied and the circumstantial factors to be eliminated. His choice may be wrong. In European rust identification work, some differentials yielded reproducible (lack of reproducibility is also reproducible) and therefore publishable results. Unfortunately, much of this information, though correct, had little or no explanatory value with respect to rust epidemiology (118). This unfortunate result was due to a wrong choice. First, differentials of non-European origin do not necessarily contain genes that differentiate European rust races (118). Second, seedlings at 15°C were used, whereas mature plants under field conditions differentiate better because of their genetically determined mature plant resistance (116, 117).

Going up along the technology gradient, the size of the plots is reduced tremendously, from whole fields (type *b*) to a few or even single plants (type *f*). This reduction could be obtained while maintaining or even improving the desired variance distribution, low variance within and high variance between treatments. This permits a lower number of replications, or, alternatively, a larger number of independent variables. The smaller the experimental units, the less the time spent in the preparation and maintenance of the experiment. The time saved becomes available to study more dependent variables in more detail (often longitudinally). Sophistication in technology works in favor of the multiple output experiments.

The disadvantages of sophisticated technology are not always evident. First, there is a steep increase in costs in the later stages of sophistication, from glasshouse to growth chamber, from constant conditions to profiling (temperature, humidity, and light variations according to any desired pattern), from moderate control (e.g. ± 2°C around set temperature) to strict control (± 0.2°C). The increase in costs of building, running, and servicing growth chambers should be weighed against the increase of efficiency as indicated above. Who has ever dared to make the necessary calculations? They may be very disappointing!

Second, a psychological aspect is involved. The experimenter tends to demand accuracy where he can get it easiest, not where he needs it most. High accuracy can be obtained in physical measurements, whereas biotic responses always have a relatively large variance. Environmental control can be effected with great precision, whereas the biotic materials, the plants, tend to show superposition of genetical, physiological, and morphological variability. Few scientists are able to avoid the error of *misplaced precision* (19) by adjusting the required degree of control over the abiotic variables to the inevitable variance of the biotic variables. An exercise in the optimization of experimental design (in the methodological sense) with a view to both financial plus labor input and scientific progress output needs to be done.

Third is the rarely mentioned fact that the elimination of some circumstantial factors may induce new ones. Many growth chambers abruptly switch the light on and off. This eliminates one circumstantial factor that is variable and difficult to evaluate: dawn and dusk. The abrupt switching may introduce another circumstantial factor: physiological shock due to abruptness. Constancy of environment is in itself a circumstantial factor foreign to nature; it may induce a lack of hardening against extreme conditions which can lead to unexpected and undesirable changes in the susceptibility of the host. New circumstantial factors are created when growth chambers have obvious imperfections like inadequate light intensity or spectral distribution. In one instance, *P. recondita* on mature wheat plants grew in stripes not unlike *P. striiformis* (Zadoks, unpublished).

Fourth, growth-chamber experiments usually are monocyclic (124), i.e. they cover a period of one infection cycle, from infection to subsequent sporulation. Epidemics usually develop over a much longer period consisting of a sequence of recurrent infection cycles: epidemics are polycyclic. The epidemiologist working at the upper side of the technology gradient exposes himself to the risk of becoming myopic, missing the perspective of the epidemic as a whole.

Fifth, the uphill movement along the technology gradient is a semi-autonomous process. It takes place because it is possible as much as because it is necessary. Especially when financial resources are limited, the researcher must be aware of the fact that good research can be done at any position along the technology gradient, when the researcher provides himself with good methodological guidance.

The Factor Time and Sequential Analysis

In operational research, interest in time is usually directed towards the timing of the first chemical treatment, and the spacing of the successive applications of chemicals. In theoretical research, the concepts of latent period and infection period were firmly established (5, 100). The abstract concept of e.g. latent period has to be operationalized by a working definition adapted to the technical limitations of the researcher.

Definition.—Sequential analysis in epidemiology is here defined as the quantitative analysis of a sequence of events by means of a combination of longitudinal and cross-sectional studies. In a monocyclic experiment, the events can be operationally defined phases (54), like the changes from one morphological stage to the next: ungerminated, ripe uredospores germinate to give uredospores with germ tubes which penetrate host cells and give rise to substomatal vesicles, etc.

Sequential analysis implies longitudinal studies of at least two consecutive and operationally defined phases, like spore storage and spore germination (94, 125). It also implies cross-sectional studies in which, for each of the phases involved, at least one experimental factor is explored in at least two levels. In (94) and (125) the experimental factors were hydration during the spore storage phase (levels: with and without) and light during the spore germination phase (levels: on and off). In (125) the spore production phase was added, with two experimental factors, light (levels: on and off) and humidity (levels: high and low).

Sequential analysis shows that a process (here the epidemiological sequence) is governed by different independent variables at different phases within the process. The output of the process can be measured as the response of a dependent variable, e.g. the terminal percentage of spores germinated. Statistical analysis of the response data may indicate interaction between phases and treatments. A few examples are: (*a*) The sequence of morphologic changes from ripe uredospore of *P. graminis* to substomatal vesicle has been produced in vitro by carefully selecting one particular treatment in every phase (33). Other treatments than those selected lead to lower vesicle/spore ratios, but this does not necessarily suggest interaction. (*b*) Uredospore germination in *P. striiformis* is greatly enhanced by hydration during preceding spore storage (94, 125). (*c*) Spores produced in light are very sensitive to light during germination when hydrated during storage but not when unhydrated, irrespective of humidity during spore production. Spores produced in the dark behave differently from those produced in light, but respond to hydration treatments only when produced under dry conditions (125). These data definitely suggest sequential *interaction* between treatments of a different nature given in different phases of the process. (*d*) Ascospore release in *Venturia inaequalis* shows a typical sequential effect (15). The two phases studied were a conditioning phase preceding release, and the release phase itself. High temperature or high humidity during the conditioning phase favored subsequent release, which was inhibited by absence of red light (16). (*e*) In *Pleospora herbarum,* induction, maturation, and terminal incubation of protoperithecia are sequentially interrelated phases in the production of ripe perithecia (66).

Within-phase longitudinal studies.—Within-phase longitudinal studies are essential to sequential analysis, but the results need to be adapted to the purpose. Longitudinal studies within a phase typically describe a response-time relationship. A few examples: (*a*) Postulation—the number of opening pustules per day (72), or the number of newly sporulating leaves per day (116). (*b*) Sporulation—the number of spores produced per day (72). (*c*) Germination—the number of spores germinated per hour (125).

These response-time relations yield many data, from which one representative parameter is to be selected for cross-sectional studies in sequential analysis. The curves representing cumulative response versus time usually are sigmoid curves. A representative parameter could be the terminal number of reacting units (as usual in spore germination studies), or the time at which the first unit reacted (e.g., to mark the end of the latent period, or the beginning of germination). The derivative of the response plotted in the sigmoid curve represents a rate, expressed in numbers per unit time, which can serve as a representative parameter (125). Suitable transformation of the sigmoid curve into a straight line permits the calculation of a constant rate and of an accurate mid-way time, the time at which 50% of the process is completed.

The latent periods of *P. striiformis* (116) and *P. recondita* (72) are stochastic variables. The same is true for the germination period of *P. recondita* uredospores, the period between seeding the spores on agar and their germination as indicated by a germ tube at least equal to the diameter of the spore (125). The probability distribution of these periods is near-normal, slightly skewed with a steep slope and a definite minimum value at the lower end, and a tail tapering off indefinitely at the upper end. The sigmoid curve representing the cumulative percentage of reacting individuals against time is transformed to a straight line, the probit-line (116), by the probit transformation (39). The slope of the probit line serves as an estimate of the rate of the process (spore germination, the opening of pustules, etc.), once it has begun. The standard deviation σ of the reaction period distribution is the inverse of the tangent of the slope α of the probit line. Though we can register the shortest observed response period, e.g. latent period, statistical considerations show that this is not necessarily the shortest possible response period. Reasons are the limited sample size and the lack of knowledge about the exact shape of the probability distribution. The mid-way time or t_{50} can be determined with great accuracy, however. t_{50} and $tg\ \alpha = \sigma^{-1}$ characterize the time course of a morphological development between two operationally defined phenological stages, under the specific circumstantial conditions of the experiment. The effect of quality of inoculum, of environmental conditions and of host resistance can be measured accurately using these parameters.

Validity of generalizations.—Many studies at the monocyclic level are, in the present terminology, cross-sectional studies of a single phase. The conclusions of such studies are not necessarily relevant to epidemiology. The conclusions may be true under the circumstantial conditions of the experiment, but the generalization of the conclusions may be invalid when the possible sequential interaction with preceding phases has been neglected. Other studies, still at the monocyclic level, are typical longitudinal studies at constant or continuously cycling conditions (10). For each phase, optima of various experimental factors are found, but again the possible effects of sequential interaction are neglected. Once more, generalization of conclusions may be invalid, because the circumstantial factor of constancy does not occur in nature.

Theorem.—A sequence of phases, under optimum conditions for the sequence as a whole, takes longer and is less efficient than a sequence of individual phases each under its own optimum conditions. Equation (1) symbolizes the theorem for the time-temperature relation. There are n phases, each lasting a finite period Δt_i, where $1 \leq i \leq n$. In the left part of the equation, the duration of Δt_i is minimal when function f_i (relating Δt_i to temperature T_i), specific for phase i, is optimal. In the right part of the equation, the duration of the sequence as a whole $\Sigma\Delta t_i$ is minimal when f_c (relating $\Sigma\Delta t_i$ to temperature T_c, which is constant throughout the sequence) is optimal.

Temperature can be replaced by other environmental factors, such as light intensity. Equation (2) states that the product Π of the ratios of success r_i at their respective optimum values of $f_i(T)$ is higher than the maximal Πr_i when the sequence is produced at its optimal constant temperature T_c. Examples for r_i are: number of germinated spores/number of total spores applied; number of appressoria formed/number of germinated spores; etc. The best evidence, though still incomplete, for this theorem comes from studies on the infection process of *P. graminis* uredospores (55, 122).

$$\sum_{i=1}^{n} \Delta t_i \cdot f_i(T_i) < f_c(T_c) \cdot \sum_{i=1}^{n} \Delta t_i \qquad 1.$$

$$\prod_{i=1}^{n} r_i \cdot f_i(T_i) > f_c(T_c) \cdot \prod_{i=1}^{n} r_i \qquad 2.$$

Prospects.—The prospects of sequential analyses cannot yet be fully envisaged, but three points can be made: (*a*) a rigorous sequential analysis is very laborious; (*b*) a complete sequential analysis, yielding an orthogonal matrix of experimental results, may lead to interesting mathematical exercises, and (*c*) sequential analysis is a prerequisite to the building of simulation models.

Procedure in sequential analysis.—Sequential analysis of monocyclic experiments opens the way to rather sophisticated interpretation. The procedure goes through various steps of data collection and data processing (at these two levels the ever present interpretation may be implicit or even subconscious), cross-sectional analyses, longitudinal analysis, and final analysis. At each level, the appropriate statistical tests and comparisons with existing literature must be made before decisions can be taken about how to proceed. Collateral to these analyses, a cost-benefit analysis and project evaluation has to be made. An experiment usually results in a relation of the type $y = f(x)$. A relation, which if true, is not necessarily valuable: see below, relevance and sensitivity. Cost-benefit analyses and project evaluations seem to be rarely made and never published. If this impression is true, it signals an evident omission that hampers real progress.

Relaxation Time and Integration Level

Biological processes can be studied on the molecular level, the cellular level, the plant level, and the crop level. Each of these levels has its own scale of space and time. Cybernetics uses the term "relaxation time" for the time needed by a system to recover from a sudden disturbance. When recovery proceeds logarithmically, the relaxation time can be expressed as the time needed to reduce the effect of the disturbance to e^{-1} times its original value, e being Euler's constant: 2.71 (21). The relaxation time needed to adapt a population of forest trees (*Pinus monticola*) to the introduction of a new pathogen (*Cronartium ribicola*) was estimated at 500 years (124). For annual crops, like *Zea mais* in West-Africa disturbed by *Puccinia polysora,* it was about 3 years, without the aid of professional breeders (100). Southern corn leaf blight, *Helminthosporium maydis,* caused a similar disturbance in the U.S.A. At the cellular level, the relaxation time needed to bring a forcibly ejected ascospore to a stop by the drag of the air is a few milliseconds (21). The epidemiologist needs some suppleness of mind.

The researcher studying molecular processes with relaxation times calculated in seconds works at a completely different integration level (31) than the forester who calculates in years. Between these two integration levels are several others, characterized by corresponding time spans, such as:

molecular conversions —seconds
cellular phenomena —minutes
plant physiology —hours
recurrent infection cycle —days
development of annual crops —weeks
development of perennial crops—years

De Wit (31) stated that the individual research worker can survey three integration levels at most. For the epidemiologist the most interesting integration levels seem to be: (*a*) Phases within recurrent infection cycle—monocyclic. (*b*) Recurrent infection cycles within crop growing season—polycyclic. (*c*) Effect of control measures over many years—diachronic.

In general, quantitative deductions from the lower to the higher integration level are considered to be valid, from the higher integration level to the lower one invalid.

Models and Systems

The researcher's mind switches back and forth between part and whole. The entirety can be understood only when details have been clarified; details are meaningful only when their position in the whole is evident. Splitting a whole into its component parts (analysis) and fitting parts together into a whole (synthesis) is standard scientific practice. The scientist goes through cycles of analysis and synthesis which may be simultaneous or consecutive, and of different scales of magnitude.

Quantitative epidemiological research has consisted mainly of analysis, with the necessary simplification and selection. Typical sequences of analysis are: the epidemic as a whole (stem rust epidemic), the recurrent infection cycle (uredocycle), a specific phase within that cycle (e.g., spore germination). The scale diminishes from the field over the individual seedling plant to the microscope slide. At that point, the work becomes irrelevant to epidemiology, because in nature rust does not occur on glass slides.

By synthesis, all the known details should be put together into one sweeping picture that explains the whole epidemic. A verbal, qualitative synthesis has been attempted by many epidemiologists, with more or less success.

Quantitative synthesis has become possible with the advent of fast computers and of suitable software.

A model is a simplified representation of reality. Details irrelevant to its purpose are omitted to obtain a stylized picture consisting only of the main features of the reality represented. Purposes and forms are many—qualitative or quantitative, material or abstract, descriptive or explanatory (11, 13, 31). In this context, quantitative models are models in which the pathogen or disease are quantified.

Material models.—(*a*) In a way, any field or growth chamber experiment is a material model of (a part of) an epidemic. (*b*) Recently, polycyclic studies on the development of an epidemic in a growth cabinet were successful (25). (*c*) Typical details of an epidemic process can be studied experimentally in the laboratory, where abiotic analogs may replace biotic materials. The deposition of spores on grass in the field was studied in a wind-tunnel, substituting fungal spores with labeled *Lycopodium* spores and grass with PVC strips (21).

Qualitative models.—(*a*) Simple, qualitative models are the life cycles of pathogens or disease cycle charts found in text-books. They are pictorial models, using stylized drawings, arrows, and as few words as possible (1, 2, 79). (*b*) Detailed description of processes occurring within a life cycle, based on extensive experimental work, can be condensed into a few statements, forming together a "biological model" (30, p. 62). This is a verbal model often used as a milestone on the way from observation to abstraction. (*c*) The biological model can sometimes be moulded into a "meteorological model" (30, p. 103), containing only a set of statements on weather requirements. For *P. infestans,* a meteorological model was designed similar to a binary decision system. (*d*) The meteorological model is "translated" into a "synoptic model" (30, p. 109) which uses synoptic weather maps to determine whether or not a warning should be issued. This synoptic model is a description of a few typical weather situations easily recognized on the synoptic weather chart (13).

Quantitative models.—(*a*) Aprioristic models are necessary for efficient sampling. If we want to know the severity of a disease in a field, we ask biological questions like "local lesion or systemic disease?", statistical questions like "normal or Poisson distribution?", and mixed questions like "clustering occurs?" meaning "foci or hot spots present?" A statistician should be consulted for the design of a mensurational model. (*b*) Typically descriptive mathematical models are Schrödter's (92, 93) equations for disease ecology. (*c*) A mathematical equation, which has explanatory value, is Van der Plank's (103) "mathematical model for resistance":

$$dx_t/dt = R_c \cdot (x_{t-p} - x_{t-p-i}) \cdot (1 - x_t) \qquad 3.$$

where x is disease severity ($0 \leq x \leq 1$), R_c is the corrected infection rate, *t* is time, *p* is latent period, and *i* is infectious period. (*d*) Complex situations are often described in a set of interrelated differential equations (5, 46, 47). Computer calculations can replace analytical solutions of these equations. (*e*) Characteristically, the epidemiologist thinks of models representing disease versus time. Models without the factor time often represent three-dimensional spatial distributions of the pathogen, or disease gradients (8, 50, 102). (*f*) In recording field work, where many independent and, at times, also interdependent variables, are monitored in the hope to explain a dependent variable, e.g. disease severity, the relations found are often reported in the form of a multiple regression equation, see below (32, 37, 88, 93).

Simulation models.—(*a*) The analog computer can integrate a set of differential equations over time. The product usually is a series of graphs. The term "simulation" is applied because the process in the computer is analogous to that in nature (80). (*b*) Simulation can also be performed by a digital computer, using an iterative process. The process cycles through a set of equations, here difference equations. The output of the last cycle serves as the input for the next one. The end result is a series of tables or graphs. The older simulation models (107, 123, 126) use FORTRAN, a computer language not so suitable for simulation. (*c*) New computer languages have been designed to simulate analog processes on digital computers (14, 22, 80). One, CSMP, is particularly suitable for the simulation of ecological models. It is customer-designed, and each statement has biological meaning. The analogy between simulation model and nature can approach isomorphism (18).

Systems analysis.—Systems analysis is, in part, a new terminology applied to old methods; it adds to the old method the view of disciplines in the area between mathematics and electronics, such as cybernetics and information theory (81, 112, 123). A system is "a portion of the world which at a given time can be characterized by a given state, together with a set of rules that permit the deduction of the state from partial information. The state of a system (in its hard sense) is a set of values of certain variable quantities at the moment of time in question" (84).

A system is represented by a model, ideally isomorphic. There are two distinct approaches. One defines the system, regards it as a mysterious black box, and restricts itself to the study of the responses at the output side to the various stimuli at the input side (4). The other, the mechanistic approach, tries to open the black box and study the interrelations of the component parts; the white box approach. In epidemiology typical black boxes and white boxes are rare, most are gray. Many multiple regression models are gray boxes, studying inputs and outputs only, but selecting meaningful inputs and outputs on the basis of existing knowledge about the mechanics of the system. This approach is appropriate in operational research, with the immediate task of disease control.

Forecasters, in a way, condense all available knowledge on the mechanics of the system into a black box: if this stimulus occurs, then spray. In the eyes of untrained farmers, like the potato growers of the Bolivian Andes, *P. infestans* epidemiology is a black box. If they treat their crop when a given rain gauge is full to the rim, their crop is safe (20).

The white box or mechanistic approach has already been discussed with respect to analysis. Synthesis, on a quantitative basis, is possible only by means of simulation techniques. The simulation model has to integrate existing knowledge on host, parasite, weather, and their interactions, at the lower integration level, and predict the course of the epidemic at the higher integration level. The available models are few, incomplete, but promising (107, 123, 126). They have great explanatory value, but their operational value, for actual forecasting, cannot yet be assessed. Present simulation models have one drawback: they cannot represent the development in time of three-dimensional spatial arrangements so typical in epidemiology (focus or hot spot development, disease gradients). A beginning has been made (6).

Deterministic or stochastic models.—Quantitative models can be stochastic or deterministic. Though these types are distinct, they are not contrasting (31). The movement of an individual gas molecule is unpredictable; the behavior of groups of molecules can be described in terms of probabilities, the stochastic method; the pressure of billions of molecules on the walls of a vessel is completely determined: e.g. PV = RT, the law of Boyle–Gay Lussac. The stochastic approach represents the microscopic, the deterministic approach the macroscopic view of the same process (5, 6). In botanical epidemiology, there is little reason to use purely stochastic models except, perhaps, when we regard the glasshouse as the smallest group of individuals in the case where an epidemic spreads from glasshouse to glasshouse. Plant pathologists go deterministic but err in doing so, because many of the variables used (latent period, percentage germination, etc.) are in fact stochastic variables. These stochastic aspects of the variables used in deterministic simulation models can and must be built into the simulation models.

Description and Explanation

The logistic equation $dx/dt = r.x. (1-x)$, where x is severity of disease ($0 \leq x \leq 1$) or concentration of a chemical, is an explanatory model when describing a second order chemical reaction, a descriptive model when used as a model for *P. striiformis* or *P. infestans* epidemics (83, 100, 116), and a heuristic model in Van der Plank's (100) theoretical treatise. The models based on regression analysis (32, 37, 88) are typically descriptive models. The equations describe the process up to perfection, the coefficients of correlation and determination are high, but the variables in the equation are not arranged in such a way that the process is explained. These models are little more than curve fitting, with the merit of condensing a large body of information into a few variables and parameters. " 'Quantify and clarify' has been the paradigm of much contemporary study, and the illusion of synecology as a 'hard' science has been provided by widespread use and misuse of statistical methods, which have enormous predictive appeal if little explanatory power" (81, p. 1). The regression model is a valid and often precise tool for disease or loss forecasting, but its validity is restricted to the circumstances, limited in space and time, under which it has been developed.

There are three main types of explanation: causal, mechanistic, and teleological. The causal explanation states simply that one antecedent is always followed by one and the same sequent. If the button is pushed, the bell rings. The mechanism of the relationship, covering several integration levels (hands pushing, electrons running), is not necessarily included in the causal explanation. The mechanistic explanation uncovers the whole chain of events, switching from one integration level to the other. A quantitative, mechanistic explanation is reached when all relevant relations at the lower integration level are known and when put together in a suitable model for synthesis, yield a correct quantitative description of the process at the next higher integration level. In epidemiology, the lower integration level could be the recurrent infection cycle with its various phases all quantitatively related to environmental factors and to each other, and the next higher integration level could be the epidemic as a whole consisting of a large number of overlapping infection cycles. In epidemiology, "explanation" in this mechanistic sense has not yet been completed, but certainly it has been approximated by the simulation models mentioned.

Teleologic explanation is completely different. In biology, it claims that all species aim at survival, and consequently, that all observable phenomena have some meaning in terms of survival value (106). This thesis helps to interrelate a great number of more or less unrelated phenomena. Good examples are the biological and meteorological models for *P. infestans,* and a tentative model for *P. graminis* (122). A typical example is as follows (57): the coprophilous *Sordaria fimicola* shoots its ascospores towards the light, away from the substrate, with some probability that the spores hit a grass blade that will be eaten by an animal, thereby completing the life cycle. These teleological explanations have heuristic value: even the size and shape of a spore have a function, so let's find it!

Phytopathology in general, and botanical epidemiology in particular, are service-rendering sciences. The aspect of service has left a deep imprint on the whole science. An epidemic must be stopped; the question, whether it will be understood, ranks second. Typically, control chemicals have usually been applied long before their mode of action was elucidated. The operational side of epidemiology has received more attention than the explanatory side (see 113). Thumbing through the journals one cannot escape the impression that the discrepancy was wider in America than in Europe. The discrepancy has, at times, been so big that epidemiology was hardly accepted as a science. I believe that future advance in the control of epidemics, especially control with a restricted application of chemicals, will depend more on explanatory research than ever before.

LITERATURE CITED

1. Agrios, G. N. 1969. *Plant Pathology*. New York: Academic Press. 629 pp.
2. Alexopoulos, C. J. 1962. *Introductory Mycology*. New York: Wiley. 613 pp.
3. Anderson, E. 1952. *Plants, Man and Life*. Boston: Little, Brown & Co. 245 pp.
4. Ashby, W. R. 1957. *An Introduction to Cybernetics*, 2nd Imp. New York: Wiley. 295 pp.
5. Bailey, N. T. J. 1957. *The Mathematical Theory of Epidemics*. London: Griffin & Co., 194 pp.
6. Bailey, N. T. J. 1967. The simulation of stochastic epidemics in two dimensions. *Proc. 5th Symp. Math. Statist. Probab.*, Berkeley. vol. IV: 237–57.
7. Baldwin, J. G. 1966. Dormancy and time of bud burst in the sultana vine. *Aust. J. Agr. Res.* 17:55–68.
8. Baker, R., McClintock, D. L. 1965. Populations of pathogens in soil. *Phytopathology* 55:495
9. Becker, G. J. F. 1955. Onderzoek naar de afrijpingsziekten van tarwe. *Tienjarenplan voor graanonderzoek*, Wageningen. 2:87–95
10. Benedict, W. G. 1971. Differential effect of light intensity on the infection of wheat by *Septoria tritici Desm.* under controlled environmental conditions. *Physiol. Plant Pathol.* 1:55–66
11. Bertels, K., Nauta, D. 1969. *Inleiding tot het model begrip*. Bussum: De Haan. 183 pp.
13. Bourke, P. M. Austin. 1970. Use of weather information in the prediction of plant disease epiphytotics. *Ann. Rev. Phytopathol.* 8:345–70
14. Brennan, R. D., De Wit, C. T., Williams, W. A., Quattrin, V. 1970. The utility of a digital simulation language for ecological modeling. *Oecologia* (Berlin) 4:113–32
15. Brook, P. J. 1969. Effect of light, temperature, and moisture on release of ascospores by *Venturia inaequalis* (Cke.) Wint. *N.Z. J. Agr. Res.* 12:214–27
16. Brook, P. J. 1969. Stimulation of ascospore release in *Venturia inaequalis* by far red light. *Nature*, London 222:390–92
17. Browning, J. A., Frey, K. J. 1969. Multiline cultivars as a means of disease control. *Ann. Rev. Phytopathol.* 7:355–82
18. Bunge, M. 1970. Analogy, simulation, representation. *Gen. Syst.* 15:27–34
19. Campbell, D. T., Stanley, J. C. 1969. *Experimental and quasi-experimental designs for research*. Chicago: Rand McNally & Co. 4th Print. 84 pp.
20. Castaño, J. J., Thurston, H. D. 1965. Aspersiones de Maneb a distintos intervalos y niveles de lluvia para control de Phytophthora infestans en la papa. *Agr. Trop.* 21:25–32
21. Chamberlain, A. C. 1967. Deposition of particles to natural surfaces. *Airborne microbes.* ed. P. H. Gregory, J. L. Monteith. Cambridge: Univ. Press. 138–64
22. Charlton, P. J. Computer languages for system simulation. See Ref. 29:53–70
23. Chester, K. S. 1943. The decisive influence of late winter weather on wheat leaf rust epiphytotics. *Plant Dis. Reptr., Suppl.* 143: 133–44
24. Chester, K. S. 1950. Plant disease losses: Their appraisal and interpretation. *Plant Dis. Reptr., Suppl.* 193:190–362
25. Cohen, Y., Rotem, J. 1971. Field and growth chamber approach to epidemiology of Pseudoperonospora cubensis on cucumbers. *Phytopathology* 61:736–37
26. Committee, A. P. S. 1967. *Source Book of Laboratory Exercises in Plant Pathology*. San Francisco: Freeman & Co. 388 pp.
27. De Groot, A. D. 1969. *Methodology, Foundations of inference and research in the behavioral sciences.* The Hague: Mouton. 490 pp.
29. Dent, J. B., Anderson, J. R. 1971. *Systems Analysis in Agricultural Management.* Sydney: Wiley Australasia. 394 pp.
30. De Weille, G. A. 1964. Forecasting crop infection by the potato

blight fungus. *Kon. Ned. Meteorol. Inst.* (Royal Neth. Meteorol. Inst.), *Meded. Verh.* No. 82:1–144. The Hague: Staatsdrukkerij.

31. De Wit, C. T. 1968. *Theorie en Model.* Wageningen: Veenman. 13 pp.
32. Dirks, V. A., Romig, R. W. 1970. Linear models applied to variation in numbers of cereal rust urediospores. *Phytopathology* 60:246–51
33. Emge, R. G. 1958. The influence of light and temperature on the formation of infection-type structures of Puccinia graminis var. tritici on artificial substrates. *Phytopathology* 48: 649–52
34. Eriksson, J. 1897. Vie latente et plasmatique de certaines Urédinées. *C.R. Acad. Sci.* Paris. 124:475–77
35. Eriksson, J. 1902. Sur l'origine et la propagation de la rouille des céréales par la semence (Suite). *Ann. Sci. Nat., 8me Sér. (Bot.)* 15:1–160
36. Eriksson, J., Henning, E. 1896. *Die Getreideroste, Ihre Geschichte und Natur sowie Massregeln gegen dieselben.* Stockholm: Norstedt & Söner. 463 pp.
37. Eversmeyer, M. G., Burleigh, J. R. 1970. A method of predicting epidemic development of wheat leaf rust. *Phytopathology* 60: 805–11
38. F.A.O. 1970. *Crop loss assessment methods. FAO manual on the evaluation and prevention of losses by pests, diseases and weeds.* Rome: F.A.O. loose leafed, ca. 200 unnumbered pages.
39. Finney, D. J. 1952. *Statistical method in biological assay.* London: Griffin & Co. 662 pp.
42. Gale, A. H. 1959. *Epidemic diseases.* Harmondsworth (Msex): Pelican Books A 456, Penguin Books. 159 pp.
43. Gäumann, E. 1951 *Pflanzliche Infektionslehre.* 2te Aufl. Basel: Birkhäuser. 681 pp.
44. Geyger, E. 1971. Green area indices of grassland communities and agricultural crops under different fertilizing conditions. *Ecol. Stud.* Berlin: Springer 1: 68–71
46. Goffman, W. 1966. Stability of epidemic processes. *Nature,* London 210:786–87
47. Goffman, W. 1966. Mathematical approach to the spread of scientific ideas—the history of mast cell research. *Nature,* London 212:449–52
48. Good, C. V., Scates, D. E. 1954. *Methods of Research, Educational, Psychological, Sociological.* New York: Appleton. 920 pp.
49. Grainger, J. 1950. Crops and diseases. I. A digest of results of the disease phenology plots maintained and recorded by the Department of Plant Pathology West of Scotland Agricultural College, Auchincruive, Ayr. 1945–49. *Res. Bull.* 9:1–55
50. Gregory, P. H. 1968. Interpreting plant disease dispersal gradients. *Ann Rev. Phytopathol.* 6:189–212
51. Hempel, C. G. 1965. *Aspects of Scientific Explanation.* New York: Free Press, 505 pp.
52. Hempel, C. G. 1966. *Philosophy of Natural Science.* Found. Philos. Ser., Englewood Cliffs, N.J.: Prentice-Hall. 116 pp.
53. Hirschhorn, N., Greenough III, W. B. 1971. Cholera. *Sci. Am.* 225:15–21
54. Hirst, J. M., Schein, R. D. 1965. Terminology of infection processes. *Phytopathology* 55:1157
55. Hogg, W. H., Hounam, C. E., Mallik, A. K., Zadoks, J. C. 1969. Meteorological factors affecting the epidemiology of wheat rusts. *World Meteorol. Organ., Tech. Note.* No. 99:1–143
56. Horsfall, J. G., Dimond, A. E. 1960. *Plant Pathology, Vol. III. The diseased population, epidemics and control.* New York: Academic 675 pp.
57. Ingold, C. T. 1953. *Dispersal in fungi.* Oxford: Clarendon Press, 206 pp.
58. James, W. C. 1971. An illustrated series of assessment keys for plant diseases, their preparation and usage. *Can. Plant Dis. Surv.* 51:39–65
59. Király, Z., Klement, Z., Solymosy, F., Vörös, J. 1970. *Methods in plant pathology.* Budapest: Publ. House Hung. Acad. Sci. 509 pp.

60. Klomp, H., Van Montfort, M. A. J., Tammes, P. M. L. 1964. Sexual reproduction and underpopulation. *Arch. Néerl. Zool.* 16:105–10
61. Koch, R. 1891. Ueber bakteriologische Forschung. *Verh. X Int. Medicinischen Congr., Berlin, 1890.* Berlin: Hirschwald Verlag. 35–47.
62. Kranz, J. 1970. Schätzklassen für Krankheitsbefall. *Phytopathol. Z.* 69:131–39
63. Large, E. C. 1945. Field trials of copper fungicides for the control of potato blight. I. Foliage protection and yield. *Ann. Appl. Biol.* 32:319–29
64. Large, E. C. 1966. Measuring plant disease. *Ann. Rev. Phytopathol.* 4:9–28
65. Large, E. C., Doling, D. A. 1962. The measurement of cereal mildew and its effect on yield. *Plant Pathol.* 11:47–57
66. Leach, C. M. 1971. Regulation of perithecium development and maturation in Pleospora herbarum by light and temperature. *Trans. Brit. Mycol. Soc.* 57: 295–315
67. Lehninger, A. L. 1965. *Bioenergetics.* New York: Benjamin. 258 pp.
68. Makkink, G. F. 1971. De theoretische bovenbouw van de ecologie. *Contactblad Oecologen,* Amsterdam. 7:14–36
69. Margalef, R. 1968. *Perspectives in Ecological Theory.* Chicago: Univ. Chicago Press. 111 pp.
70. Marshall, J. K. 1968. Methods for leaf area measurement of large and small leaf samples. *Photosynthetica* 2:41–47
71. McKay, R. 1957. *Cereal diseases in Ireland.* Dublin: Arthur Guinness. 161 pp.
72. Mehta, Y. R., Zadoks, J. C. 1970. Uredospore production and sporulation period of Puccinia recondita f. sp. triticina on primary leaves of wheat. *Neth. J. Plant Pathol.* 76:267–76
73. Milner, C., Hughes, R. E. 1968. *Methods for the Measurement of the Primary Production of Grassland.* I.B.P. Handbook No. 6. Oxford: Blackwell Sc. Publ. 70 pp.
74. Muench, H. 1959. *Catalytic Models in Epidemiology.* Cambridge (Mass.): Harvard Univ. Press. 110 pp.
75. Nagel-De Boois, H. M. 1972. Preliminary estimate of production of fungal mycelium in forest soil layers. *Proc. 4th Colloq. Soil Zool. Comm. I.S.S.* In press
76. Newbould, P. J. 1967. *Methods for Estimating the Primary Production of Forests.* I.B.P. Handbook No. 2. Oxford: Blackwell. 62 pp.
77. Nienhaus, F. 1969. *Phytopathologisches Praktikum.* Berlin: Paul Parey. 167 pp.
78. Odum, E. P. 1959. *Fundamentals of Ecology.* 2nd Ed. Philadelphia: Saunders Cy. 546 pp.
79. Parris, G. K. 1970. *Basic Plant Pathology.* State College, Miss.: Parris. 442 pp.
80. Patten, B. C. 1971. A primer for ecological modeling and simulation with analog and digital computers. See Ref. 81, 3–121
81. Patten, B. C. 1971. *Systems Analysis and Simulation in Ecology.* Vol. 1. New York: Academic 607 pp.
82. Plant Pathology Laboratory. 1971. *Guide for the assessment of cereal diseases.* Agr. Devel. and Advis. Serv.: Harpenden (Herts.). Loose leaf annual handouts.
83. Rapilly, F. 1968. Les techniques de mycologie en pathologie végétale. *Ann. Epiphyties* 19. Numéro hors série: 1–102
84. Rapoport, A. 1970. Modern systems theory—An outlook for coping with change. *Gen. Syst. Yearb.* 15:15–25
85. Rescher, N. 1964. *Introduction to Logic.* New York: St. Martin's Press. 360 pp.
86. Riker, A. J., Riker, R. S. 1936. *Introduction to Research on Plant Diseases.* St. Louis: J. S. Swift Co. 117 pp.
87. Roelfs, A. P., Rowell, J. B., Romig, R. W. 1970. Sampler for monitoring cereal rust uredospores in rain. *Phytopathology* 60:187–88
88. Romig, R. W., Dirks, V. A. 1966. Evaluation of generalized curves for number of cereal rust uredospores trapped on slides. *Phytopathology* 56:1376–80
89. Rotem, J., Palti, J., Lomas, J.

1970. Effects of sprinkler irrigation at various times of the day on development of potato late blight. *Phytopathology* 60:839–43

90. Rotem, J., Reichert, I. 1964. Dew—a principal moisture factor enabling early blight epidemics in a semiarid region of Israel. *Plant Dis. Reptr.* 48:211–15
91. Rowell, J. B., Romig, R. W. 1966. Detection of urediospores of wheat rusts in spring rains. *Phytopathology* 56:807–11
92. Schrödter, H. 1965. Methodisches zur Bearbeitung phytometeoropathologischer Untersuchungen, dargestellt am Beispiel der Temperaturrelation. *Phytopathol. Z.* 53:154–66
93. Schrödter, H., Fehrmann, H. 1971. Ökologische Untersuchungen zur Epidemiologie von Cercosporella herpotrichoides. III. Die relative Bedeutung der meteorologischen Parameter und die komplexe Wirkung ihrer Konstellationen auf den Infektionserfolg. *Phytopathol. Z.* 71:203–22
94. Sharp, E. L. 1965. Prepenetration and postpenetration environment and development of Puccinia striiformis on wheat. *Phytopathology* 55:198–203
95. Shearer, B. L., Zadoks, J. C. 1972. On the latent period of Septoria nodorum in wheat. 1. The effect of temperature and moisture treatments under controlled conditions. *Neth. J. Plant Pathol.* In press
97. Stakman, E. C., Christensen, C. M. 1946. Aerobiology in relation to plant disease. *Bot. Rev.* 12:205–53
98. Stakman, E. C., Harrar, J. G. 1957. *Principles of Plant Pathology.* New York: Ronald Press. 581 pp.
99. Topley, W. W. C., Wilson, G. S. 1929. *Principles of bacteriology and immunity.* Vol. II. 5th Ed. London: Arnold & Co. 589–1300

99a. Tu, J. C., Hendrix, J. W. 1970. The summer biology of Puccinia striiformis in Southeastern Washington. II. Natural infection during the summer. *Plant Dis. Reptr.* 54:384–86

100. Van der Plank, J. E. 1963. *Plant diseases: Epidemics and control.* New York: Academic. 349 pp.
101. Van der Plank, J. E. 1965. Dynamics of epidemics of plant disease. *Science* 147:120–24
102. Van der Plank, J. E. 1967. Spread of plant pathogens in space and time. *Airborne microbes,* ed. P. H. Gregory, J. L. Monteith. Cambridge: University Press. 227–46
103. Van der Plank, J. E. 1968. *Disease Resistance in Plants.* New York: Academic Press. 206 pp.
104. Van der Zaag, D. E. 1956. Overwintering en epidemiologie van Phytophthora infestans, tevens enige nieuwe bestrijdingsmogelijkheden. *Tijdschr. Pl. Ziekten* (Neth. J. Plant Path.) 62: 89–156
105. Van Everdingen, E. 1926. Het verband tusschen de weersgesteldheid en de aardappelziekte (Phytophthora infestans). *Tijdschr. Pl. Ziekten* (Neth. J. Plant Pathol.) 32:129–40
106. Voûte, A. D. 1968. Ecology as a teleological science. *Acta Biotheor.* 18:143–64
107. Waggoner, P. E., Horsfall, J. G. 1969. Epidem, a simulator of plant disease written for a computer. *Bull. Connecticut Agr. Exp. Sta.,* New Haven. No. 698:1–80
108. Ward, H. M. 1903. On the histology of Uredo dispersa, Erikss., and the "Mycoplasm" hypothesis. *Phil. Trans.* (B) 196: 29–46
111. Watson, D. J. 1958. The dependence of net-assimilation on leaf-area index. *Ann. Bot.* (N.S.) 22:37–54
112. Watt, K. E. F. 1966. *Systems Analysis in Ecology.* New York: Academic. 276 pp.
113. Wijvekate, M. L. 1969. *Methoden van onderzoek.* Utrecht: Spectrum. 351 pp.
114. Wolf, A. 1955. Scientific method. *Encycl. Brit.* London, Vol. 20: 125–31
115. Zadoks, J. C. 1959. Het gele-roestonderzoek in 1959. *Tienjarenplan voor graanonderzoek.* Wageningen 6:139–50
116. Zadoks, J. C. 1961. Yellow rust on wheat, studies in epidemiology and physiologic specialization. *Tijdschr. Pl. Ziekten*

(Neth. J. Plant Pathol.) 67:69–256

117. Zadoks, J. C. 1966. Field races of brown rust of wheat. *Proc. Cereal Rust Conf., Cambridge, 1964:* 92–93
118. Zadoks, J. C. 1966. Problems in race identification of wheat rusts. *Savremena poljopriveda,* Novi Sad: 299–305
119. Zadoks, J. C. 1967. International dispersal of fungi. *Neth. J. Plant Pathol.* 73. Suppl. 1:61–80
120. Zadoks, J. C. 1967. Bespiegelingen over resistentie. *Zaadbelangen* 21:162–68
121. Zadoks, J. C. 1967. Epidemieën als populatiebiologisch verschijnsel. In: *Populatiebiologie.* Wageningen: PUDOC: 57–79
122. Zadoks, J. C. 1968. Meteorological factors involved in the dispersal of cereal rusts. *Proc. W.M.O. Regional Training Seminar on Agrometeorology.* Wageningen: 179–94
123. Zadoks, J. C. 1971. Systems analysis and the dynamics of epidemics. *Phytopathology* 61:600–10
124. Zadoks, J. C. 1972. Reflexions on disease resistance in annual crops. *Biology of Rust Resistance in Forest Trees.* ed. Bingham, R. T., Hoff, R. J., McDonald, G. I. *U.S. Dep. Agr., Forest Serv., Misc. Publ.* 1221: 681 pp. In press
125. Zadoks, J. C., Groenewegen, L. J. M. 1967. On light-sensitivity in germinating uredospores of wheat brown rust. *Neth. J. Plant Pathol.* 73:83–102
126. Zadoks, J. C., Rijsdijk, F. H. 1972. A calculated guess of future rust development in cereal crops. *Proc. Symp. Epidemiol., Lucknow, 1971.* In press

GEOPHYTOPATHOLOGY

3552

Heinrich C. Weltzien

Institut für Pflanzenkrankheiten, Universität Bonn, Germany

Introduction

We are late. It was in 1807, that Humboldt & Bonpland (39) summarized their experiences from observations and measurements made between 1799 and 1803 in the new world's tropical belt under the title "Ideas for a Geography of Plants." They stated that this new branch of botany was to study the distribution of plants in various climates, and differentiated well between characteristics based on different geographical latitude or elevation above sea level. The main climatic factors studied were temperature, humidity, atmospheric pressure, and electric tension. But a far wider scope of geographical aspects is presented to the reader. The varying plant communities, the spread of plant species, and the necessity for development of botanical maps were equally stressed. Conclusions and applications were postulated, such as the support for the theory of land bridges between Africa and America, the impact of agriculture on climate and plant growth, the search for primitive plant species under the aspect of descendence and development, the significance for paleo-botany and paleontology as well as for anthropology, agriculture, and human nutrition. The "ideas" end with a vision of the capacity for forecasting of future events by the scientist from his data.

While cryptogamic organisms were expressly included in this scheme, almost nothing was known at this time about microorganisms as causes of diseases of men, animals, or plants. But diseases characterized by typical syndromes can be studied under the aspect of geographic distribution without much knowledge about their causes. Thus the literature on human diseases revealed a geographic approach as early as 1792 (Finke 27). More geographical studies of medical problems appeared during the 19th century (Hirsch 35, Jusatz 42) but only in 1942 Zeiss (103) published the first atlas to demonstrate the relationship between epidemic diseases and geographical features. In 1931 (104) he had already stressed the necessity of differentiating between "medical geography" and "geomedicine," a concept fully adopted in the World Atlas of Epidemic diseases by Rodenwald (65, 66) and Behnke (4). The first steps toward applying Humboldt's ideas to plant diseases were made by Reichert (60) in 1950. He understood plant pathology as a contribution to biogeography, and together with Palti later developed the concept of "patho-geography" in several articles (61–63). In 1967, Weltzien

(94) first published some contributions to a concept of "geoplantpathology" or "geophytopathology."

Definition and Aims

The two different terms used by Reichert (60) and Weltzien (94) originally were chosen independently to describe the same phenomenon. The first author mentioned the geographic distribution of plant diseases, their places of origin, and their migration routes, while the second author added the explanation of the causes to the study of the distribution phenomenon. But a closer look at both terms reveals differences, which have also led to distinguishing between medical geography and geomedicine (Zeiss 104). Geography by name and definition is descriptive and supplies basic facts for other sciences. Its typical contributions are of topographic character. The facts described can well be plant diseases or pathogens. Thus the mapping of the actual occurrence of a plant disease in any region, country, continent, or in the world can be called plant-pathological geography or patho-geography, according to Reichert (60). But plant pathology has a pronounced medical aspect. This means that the science of the diseased plant is aimed toward the art of disease prevention or cure (Horsfall 37). If seen under this applied aspect, geography becomes the assisting discipline to plant pathology and a far wider scope is opened to geographic phytopathology. We recognize that almost all aspects of plant pathology can also be studied from a geographic viewpoint, and a vast collection of data may fit into a new picture, if properly placed on maps. Improvement of disease control as well as prognosis of disease occurrence and its economic significance are possible results of the application of cartographic methods to our science. Thus the term geophytopathology is used throughout this article with the understanding that it also covers pathogeography as one of its fundamentals. So we can try to discuss its methods and possibilities by developing model systems for geopathological studies.

The Geographic Distribution of Plant Diseases

It is a rather obvious but important statement that all plant diseases do not occur everywhere. This justifies continued work on plant disease check lists in all parts of the world. They are mostly compiled on a national basis, but regional lists are sometimes developed. The Commonwealth Mycological Institute (CMI) has recently published all available check lists (15, 16–17, 18) and mentioned about 100 diseases by country. They are by no means complete or satisfactory and Central Europe especially appears highly underrepresented, as modern lists for France, Belgium, Germany a.o. are not available. It must be clearly understood that reliable and reasonably complete lists are of fundamental importance for geopathological work. Some rules must be set to make them useful tools for further studies. For practical reasons they will generally follow political or administrative boundaries. They are equally useful on the basis of communities, counties, or other limited districts, as of states, countries, or international unions. Actually a more thorough study of a limited area is easier to achieve and allows much more precision in description of the exact localization and the relevant environmental factors. Lists of larger areas should, if possible, be composed out of lists from smaller areas and not directly compiled from the literature, to avoid unrealistic generalizations. All records should include notes on the actual geographic location, the disease intensity, the affected area, and, if possible, the economic significance. Only if interested pathologists all over the world continue to invest some of their time in careful observation, can we hope to reach the point where the assembled data would have sufficient accuracy to be used for true location maps, showing the plant pathological topography in various scales. The map by Schnathorst et al (71) sets a good example showing the true locations of *Xanthomonas malvacearum* occurrence on cotton in California (Figure 1). Others are those by McCubbin (52), Brandenburg & Buhl (8), Fritsch (28), Crispin & Dongo (22), Roberts et al (64), Lehmann (48), and Cotton et al, (21) from various parts of the world and in various scales, dealing with different diseases. In general it seems that mapping of plant disease occurrence is still the most commonly used geographic approach to plant pathology. Out of a collection of 805 maps dealing with phytomedical aspects (Weltzien, unpublished) 365, or 45%, can be classified as localization maps.

If these localizations of diseases on maps are based on systematic field

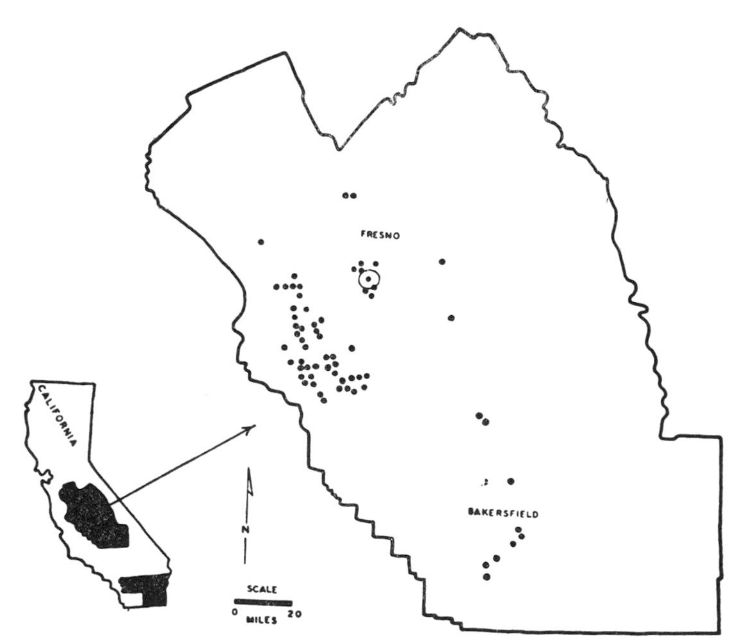

FIG. 1. Location map for angular leaf spot of cotton, *Xanthomonas malvacearum,* in California (Schnathorst 71).

studies, they allow direct conclusions on the disease distribution. But distribution maps should include the existing border problems and thus show us areas rather than spots of disease occurrence. The CMI distribution maps of plant diseases (14) represent the best known example. But as they are based on check lists or single disease records they fail to give a true estimate of the area covered by a pathogen. However, they allow differentiation between such general characteristics as disjunct or closed areas and cosmopolitan or endemic species (Weltzien 94).

Reading of disease distribution maps is hazardous, if the geographic distribution of the host is not known. Geopathology deals with disease occurrence more than with pathogen distribution, and thus includes the host problem. The host distribution follows its own environmental interdependences as shown in all geobotanical studies (Good, 30). In agricultural crops anthropogenous factors such as economics, history, or standards of technical developments complicate the picture. More detailed and improved maps on the distribution of agricultural crops are therefore necessary for improved disease mapping. For this purpose the cartographic method used by the U. S. Department of Agriculture (84) seems to be more suitable to geophytopathology than the one applied in the World Atlas of Agriculture (41). Good examples are also offered by Broekhuizen in his Atlas of the cereal growing areas in Europe (11).

Other problems arise from the host range phenomenon. Distribution map No. 96 (14) on *Cercospora beticola* e.g. includes all known hosts and therefore cannot explain the disease distribution on sugar beets. Also the apparent absence of the disease in some areas often only reflects the absence of the host. How much the simultaneous mapping of disease and host can improve the interpretation was shown by Byford & Hull (12) with *Peronospora farinosa* on sugar beets, which is much favored by neighboring seed beet crops, serving as overwintering sources (Figure 2). The maps by Hinds & Jones (34) on *Hypoxylon* canker of aspen, by Large (46, 47) on potato blight, by Atkins et al (1) on Hoja blanca on rice, or by Crowell (23) on *Gymnosporangium* spp. on various alternate hosts offer other solutions to the problem of host and disease maps. The need to recognize the limitations of different scales in mapping disease areas has been pointed out earlier (Weltzien 94). Other difficulties arise from the constant change of area sizes, due to changes in host cultivation and to environmental factors, which make it sometimes difficult to give definite borders for areas on small scale maps.

Areas of Different Disease Intensity

The uneven distribution of diseases within a given area needs special attention. If we consider the applied medical aspect as mentioned above, the area border must partly be defined on an economic basis. Biogeographically the scattered existence of diseased individuals may justify the inclusion of these locations into the area. However, geopathologically we have to register frequency of disease occurrence, disease intensity, and extent of damage.

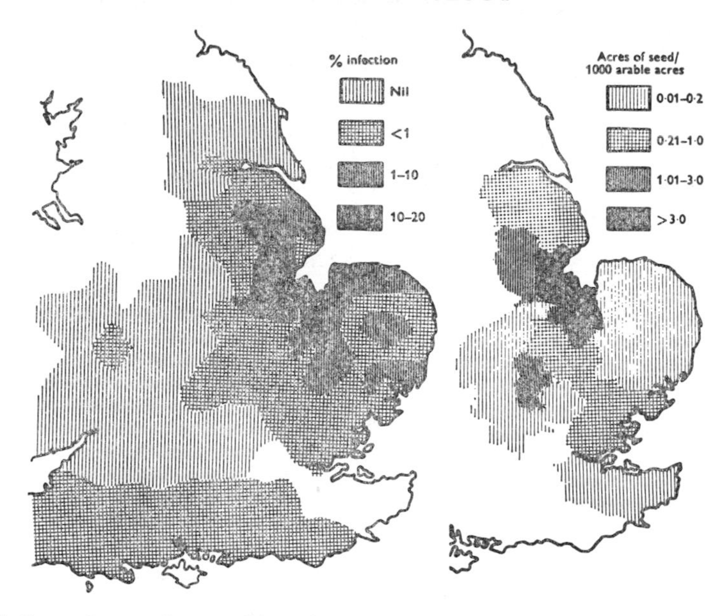

FIG. 2. Sugar beet and mangold seed crops in relation to downy mildew occurrence in sugar beets (*Peronospora farinosa*) in England (from Byford & Hull 12).

Consequently this leads to area zonation according to pathological judgment. Three zones were considered to be sufficient in differentiation of sugar beet leaf diseases (Weltzien 95) and may prove equally useful in other crops.

The *area of main damage* can be defined as one where regular epidemics occur whenever a susceptible crop is grown without protection. This includes yield losses or/and quality reductions of the harvested products and consequently means regular economic losses.

The next zone may be called *area of marginal damage*. It can be characterized by irregularly occurring epidemics with significant losses in some seasons only, while in others disease occurrence is insignificant.

All other zones or places where the disease is observed, belong to what may be called *areas of sporadic attack*. Here the disease does occur, but never, or only occasionally, shows epidemic development and crosses the threshold of economic damage. All three zones together represent the *area of disease distribution.*

This system can be applied to closed as well as to disjunct areas and may be used in all mapping scales. Weltzien (95), Drandarewski (24), and Bleiholder & Weltzien (5) have applied it to *Erysiphe betae* and *Cercospora beticola* (Figure 3) on sugar beets respectively. If sufficient observations and data of yield and control experiments are available, mapping of the three zones can be done without further experimental studies. Zone borders are of different accuracy and stability. The area of marginal damage will often be a

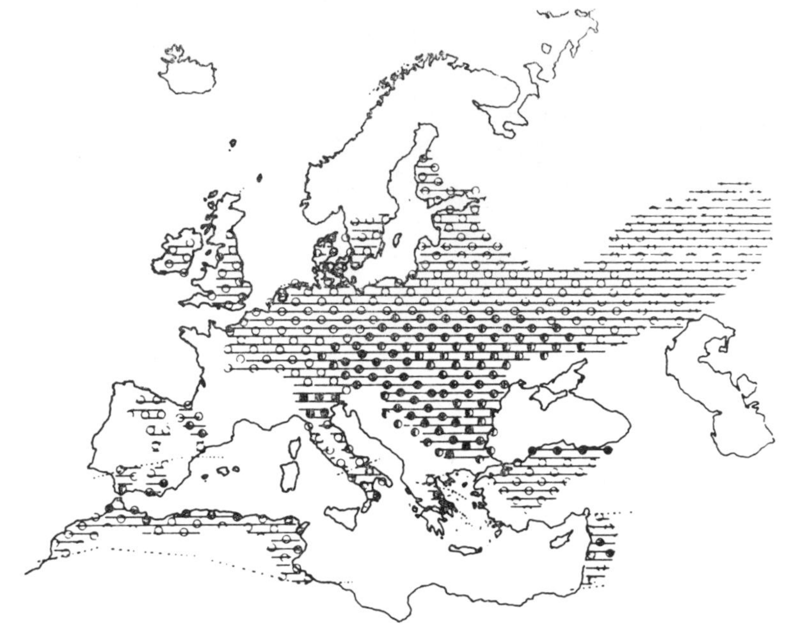

FIG. 3. Zones of different disease intensity for leaf spot of sugar beets, *Cercospora beticola,* in Europe and Mediterranean countries (Bleiholder & Weltzien 5).

≡ : area of sugar beet cultivation
● : area of main damage;
◐ : area of marginal damage
○ : area of sporadic attack, summer crops
◎ : area of sporadic attack, winter crops
..·˙: beets cultivated also in winter

transition zone, open to changes in size from both sides. On the other hand, these zones may well occur separated from each other with a comparatively high degree of stability.

Various attempts have already been made to map different intensities of attack. In some studies, we find intensity combined with location maps as with Steudel & Heiling (80) on sugar beet yellow virus (Figure 4), Schnathorst et al (70) on lettuce powdery mildew, or Elton (25) on dutch elm disease. Other authors have also documented various zones of disease intensity, thus arriving at maps showing the epidemiological situation of an area at a given time, as Zogg (105) on *Puccinia sorghi* on corn in Switzerland (Figure 5), Baldacci & Orsenigo (3) on chestnut blight in Italy, Hull (38) on sugar beet yellow virus in England, Warmbrunn (91) on dwarf bunt of wheat in Germany, and Scharif et al (69) on Ascochyta-blight of chickpeas in Iran. Stevens (79) used yield losses instead of disease intensity, and based his

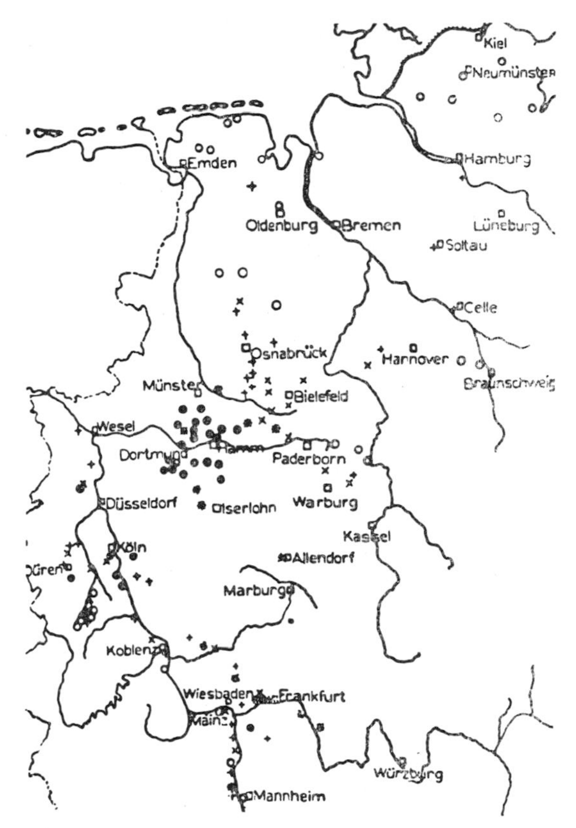

FIG. 4. Locations and disease intensity of sugar beet yellow virus in Germany (Steudel & Heiling 80).

0: 0–5%; +: 5–20%; ×: 20–40%; ✳: 40–60%; ●: 60–100%

maps on 10-year averages, thus picturing the long term threat to some crops such as peaches or sweet potatoes by fruit or storage rot fungi. But only if clear definitions as given by Bleiholder & Weltzien (5) are used for all zones of disease intensity, can a complete picture of the geopathological situation for a disease or the diseases of a special crop be drawn.

THE SPREAD OF EPIDEMICS

The mobility of pathogens makes the spread of epidemics possible. This dynamic aspect has attracted much attention to geographic studies in pathology. Here the connection to geomedicine as mentioned above is obvious, as epidemics of men, animals, and plants follow the same pattern of spread. Mapping of disease routes is almost the only possible documentation. Among the earliest examples are the spread of *Sphaerotheca morsuvae* on *Ribes* spp.

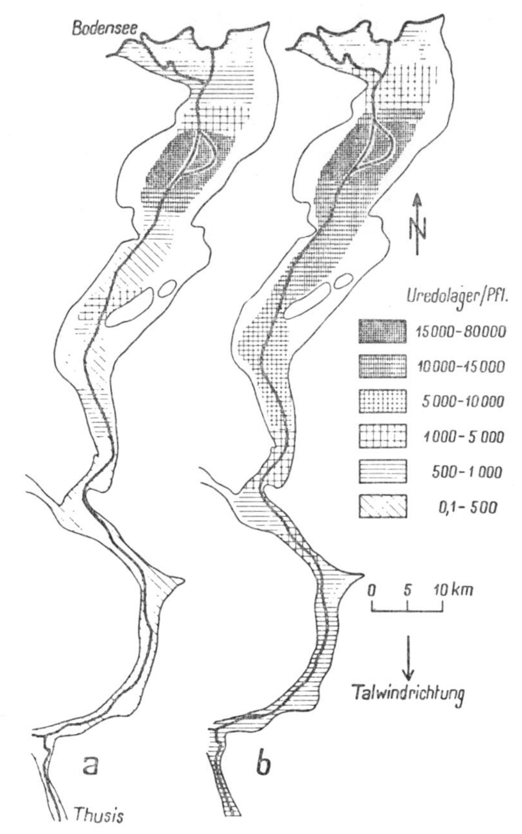

FIG. 5. Disease intensity and spread of corn rust, *Puccinia sorghi,* in the Rhine valley, Switzerland 1946 (Zogg 105).
a: Beginning of July;
b: Middle to end of July

in Europe since 1900, as documented by Eriksson (26) and Herter (33) (Figure 6) and the slowly moving epidemic of *Endothia* canker on *Castanea dentata* in north America (Rankin 59). For years the studies on the annual epidemics of cereal rusts in the U. S. have remained a classic in geographic epidemiology (Stakman & Harrar 76), (Figure 7). One may mention the results of Stakman et al (74, 75, 77), Wallace (88), and Peltier (56) on South-North dissemination of uredospores and distribution of physiologic races. A similar analysis for South America was made by Vallega (85), who arrived at four regions of race distribution for the subcontinent. The more diversified situation in Europe has been demonstrated by Zadoks (101, 102) for stem rust and stripe rust, who also delineated various epidemic centers within the region. A comprehensive review of the worldwide knowledge of

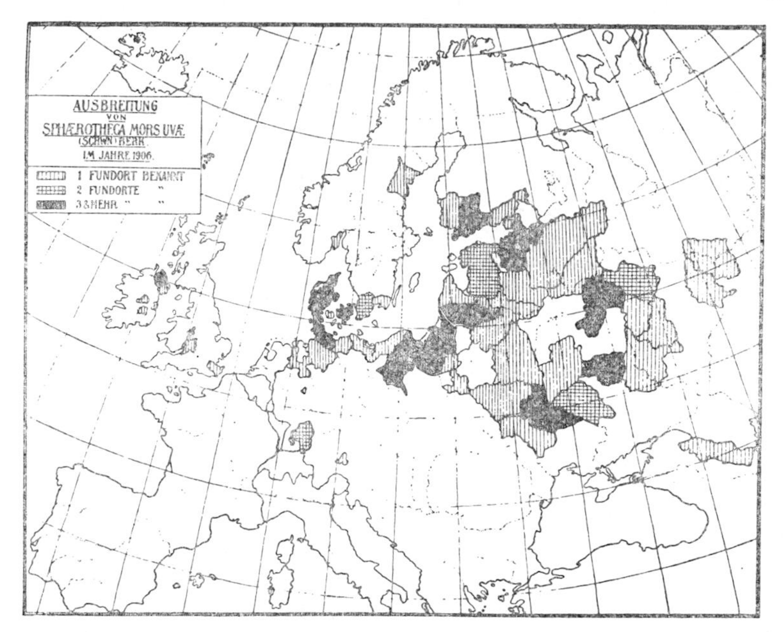

FIG. 6. The epidemiological situation of the American gooseberry powdery mildew, *Sphaerotheca mors uvae,* in Europe 1906 (Herter 33).

: 1 location; : 2 locations; : 3 & more locations

FIG. 7. The wheat stem rust epidemic, *Puccinia graminis,* in the U. S., 1923 (Stakman & Harrar 76).

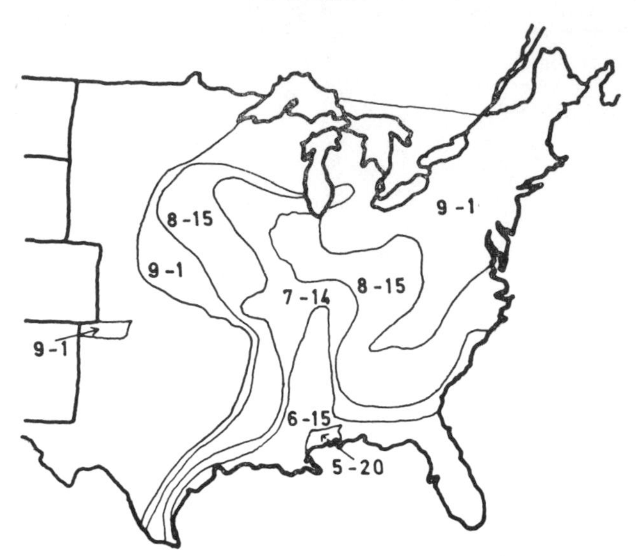

FIG. 8. The epidemic of Southern corn leaf blight, *Helminthosporium maydis,* in the U. S., between May 20 and Sept. 1, 1970 (from Moore 54).

wheat rust epidemics in relation to meteorological factors was published by Hogg et al (36).

The intercontinental spread of pathogens has also often been recorded with some accuracy. Blumer (7) summarized the Eurasian epidemic of *Uncinula necator* on grapes between 1845 and 1952. The continuing migration of the fire blight organism *Erwinia amylovora* since 1790 is shown on the maps by van der Zwet (86). But few of the world wide plant epidemics seem to have been followed up more thoroughly than the tobacco blue mold, caused by *Peronospora tabacina,* after its introduction to Europe in 1958 (Corbaz 19, 20; Kroeber 45; Klinkowski 44; Populer 57; Zachos 100), or the southern corn leaf blight caused by *Helminthosporium maydis* in its recent spread (Moore 54, Figure 8). There is almost no doubt that mapping of the spread of plant epidemics will continue to be one of the more frequent contributions to cartographic geophytopathology.

ECOLOGICAL INTERDEPENDENCIES

Geophytopathology also implies the causal explanation of the described phenomena and consequently needs ecological interpretation (Weltzien 96). The actual occurrence of diseases or their ecological existence is supplemented by studies of their ecological potential to their autecology (Tischler 81). Among the abiotic factors, the climate obviously has the greatest influence. Already Reichert & Palti (63) have related the distribution of eight different diseases to climatic factors. For *Synchytrium endobioticum* soil temperatures above 21–23°C limit disease spread. Thus, areas with soil temperatures above 30° were considered safe, but a map showing the areas concerned was not developed. *Phytomonas citri* cannot cause damage at temperatures above 20°C or with less than 150 mm of monthly rain at lower temperatures, and *Elsinoe citri* is never epidemic in citrus-growing areas with regular drought periods. The relation of cotton anthracnose, caused by *Glomerella gossypii,* to summer rainfall data in the U. S. has been demonstrated on a map by Miller (53) showing the area of regular attack limited by less than 10–13 inches of annual rainfall (Figure 9). For *Spongospora subterranea* Wuerzer (99) found a close relation between disease intensity and rainfall data, provided the average July temperature does not exceed 17°C. The disease does not occur in the Baden-Wuerttemberg area of South Germany, where the average precipitation in August drops below 80 mm. The same ecological relations were used by Wenzl (97) for development of a map of the disease distribution in Europe and for determination of the barriers preventing further spread toward the east in Austria. The correlation between severity of the soil-borne wheat spindle streak mosaic virus and soil temperatures was shown on the maps of Slykhuis (73). Prolonged periods of 4.4–12.8°C in fall and spring are necessary for heavy disease outbreaks. The dependence of apple scab, *Venturia inaequalis* on precipitation was used to explain the disease distribution in Washington by Blodgett & Semler (6).

Low rainfall areas, however, can favor other diseases, as shown by Atkinson & Grant (2) with wheat streak mosaic virus in Canada, and by Weltzien (93) and Drandarewski (24) with *Erysiphe betae* on sugar beets. The con-

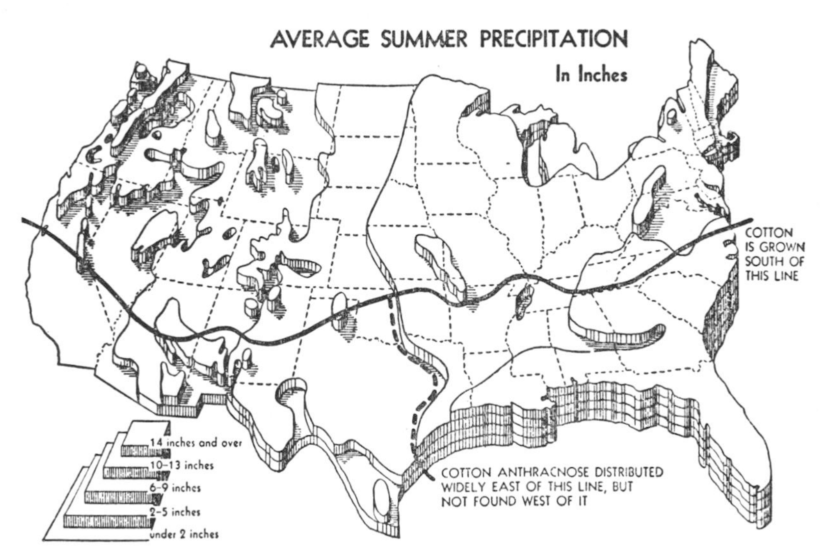

FIG. 9. Cotton anthracnose occurrence, *Glomerella gossypii* in the U. S. in relation to summer rainfall and host distribution (Miller 53).

tradictory effect of low temperatures and high rainfall on virus infection and *Phytophthora* attack of potatoes was mapped for the USSR by Roschdestwensky (67) including also disease intensity and economic loss zonation.

A systematic application of the ecological interrelationships for drawing of disease areas was made by Bleiholder & Weltzien (5) and Drandarewski (24) on *Cercospora beticola* and *Erysiphe betae* of sugar beets respectively. The area of main damage was characterized for *C. beticola* by climates of Type V (IV) and V (VI) according to Walter & Lieth (90) with June-September temperatures above 20°C and rainfall data above 100 mm monthly during most of the growth period, as represented by the plain of the river Po, Italy. On the contrary, the area of sporadic attack for the same disease is represented by climates with arid growth periods and temperatures above 15–20°C, e.g. most of the mediterranean climates or the coastal climate of Chile. These zones, however, are areas of main damage for *E. betae*. Similar definitions were found for areas of marginal damage of both diseases. Based on the climatic diagrams by Walter & Lieth (90), maps for beet growing zones of all continents were developed, using as much information as possible from world literature on disease intensity and economic losses. Palti (55) recently gave a comprehensive report on the pathogeography of *Leveillula taurica,* with special reference to temperature and humidity effects on distribution and disease severeness.

The close interrelation between elevation and disease occurrence is also based on changes of climatic factors. It has been well documented in studies of virus transmission and was used to delineate areas with low disease intensity for seed production of potatoes. Brod (10) proved for Baden-Württemberg, Germany that the best locations were above 700m and Rademacher (58) demonstrated the delay in occurrence of aphids between 340 and 800 m to be about one month in the same area. Savulescu & Ploaie (68) analyzed the geographic distribution of clover phyllody virus in Rumania and found host and transmitters confined to areas between 300–800 m elevation with yearly average temperatures between 6–10°C and an average rainfall in June between 89–123 m (Figure 10).

In analyzing the suitability of Cornwall, England for potato seed growing, Staniland (78) included exposure to winds in his maps, thus widening the recommended areas from the hills to the coast. In Nebraska, Weihing & Vidaver (92) were able to correlate a *Pseudonomas syringae* epidemic on corn to heavy winds and rain 6 days prior to disease outbreak.

Although soils do have pronounced influences on plant diseases, this phenomenon has been used only rarely for studies on geographical distribution. While Whitehead (98) related the discontinuous distribution of *Meloidogyne* spp. in East Africa on different soils and altitudes mainly to different soil temperatures, Seinhorst (72) found a close relation between clay content and the frequency of stem eelworm *Ditylenchus dipsaci* in onion cultures on the Dutch island of Goerre-Overflakkee (Figure 11). With clay content above 30% the disease intensity was very heavy, due to higher survival rates. On

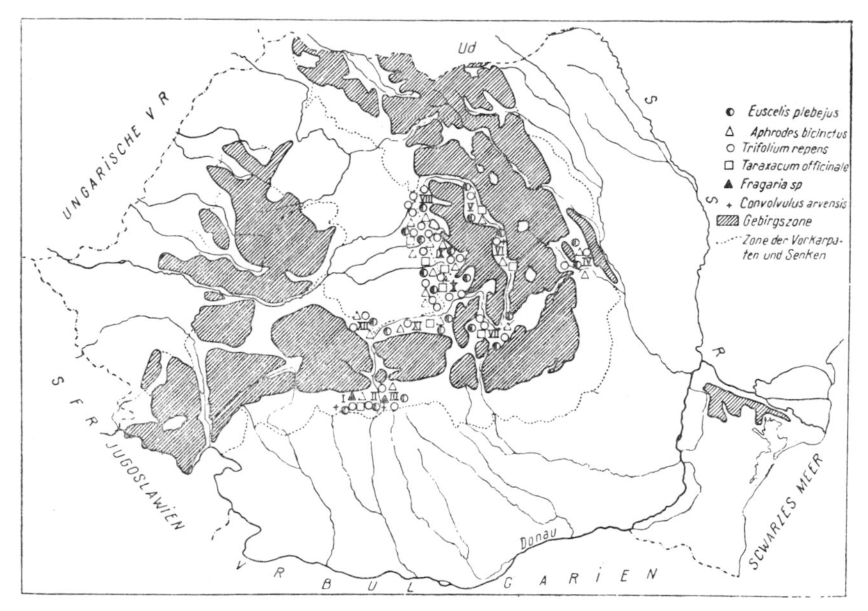

FIG. 10. Geographic distribution of green petal virus of clover host plants, and transmitting insects in Roumania in relation to mountain and foothill zones (Savulescu & Ploaie 68).

////: mountains ...: foothills

the contrary the general nematode population in sandy soils was 2–6 times higher than in clayey soils. Seinhorst's statement, that rotation is not the cause of eelworm attack but that on the contrary, rotation follows the degree of eelworm infestation and that soil is one of the major factors in their ecology, may prove valid also for other pathogens. Thus soil maps may in future prove to be valuable aids in geophytopathology.

While the importance of host maps for a geographic analysis has already been stressed, the significance of alternate hosts for disease distribution is an independent ecological factor. The importance of peach trees as winter hosts for the virus-transmitting aphid *Mycus persicae* east of the −11°C January minimum isotherm in Germany has been demonstrated by Heinze & Proffit (31). *Cronartium ribicola* has been thoroughly studied, and carefully conducted surveys as the one by McCubbin (52) were the basis for the synoptical map on blister rust epidemiology developed by Gaeumann (29). World distribution of *Puccinia graminis* in relation to aecial and telial hosts and its spread from the gene centers in South Asia to North and South America was demonstrated by Leppik (51). The importance of alternate uredial hosts for epidemics of *Puccinia striiformis* in California was shown by Tollenaar & Houston (83). Here wild *Hordeum, Elymus* and *Sitanion* spp. in the Sierra Nevada, susceptible to the same rust race as the one occurring on cul-

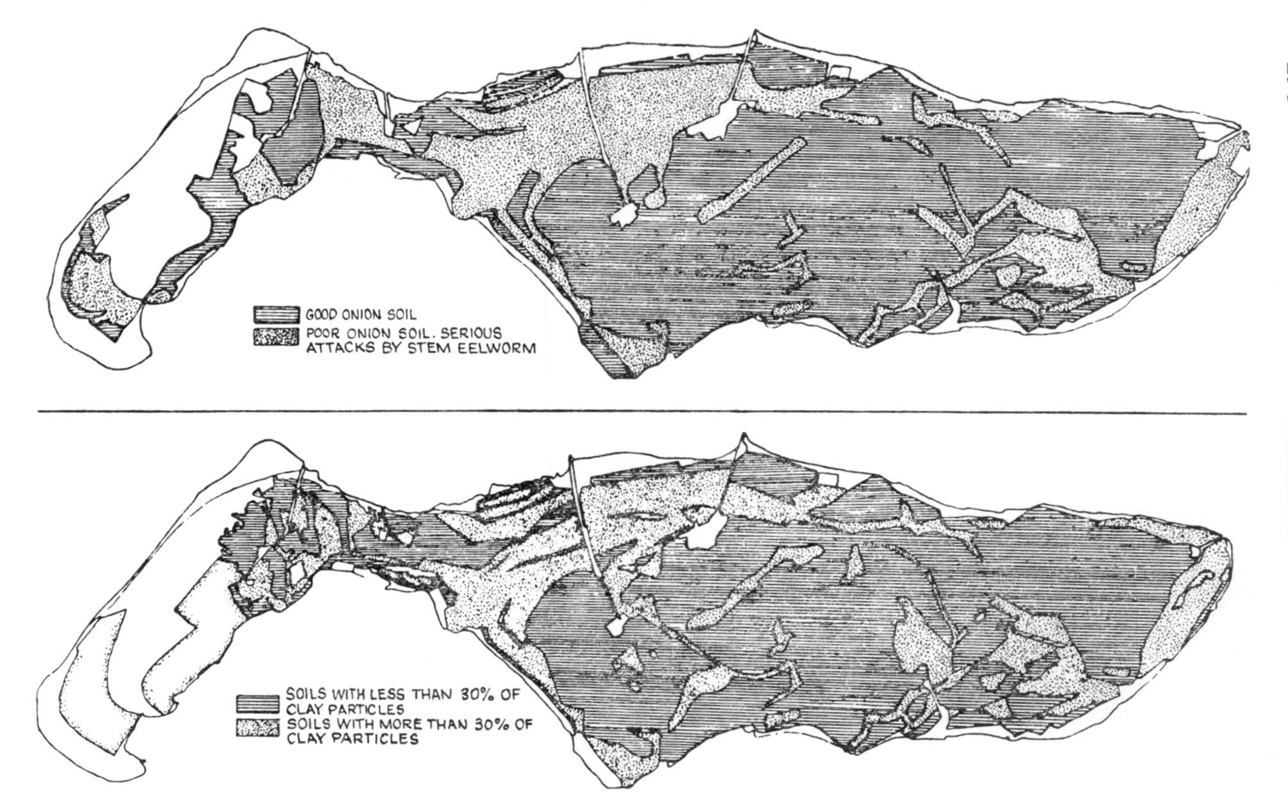

FIG. 11. Stem eelworm attack of onion, *Ditylenchus dipsaci* (above) in relation to clay content of soil (below) on the Dutch island of Goerre-Overflakkee (Steinhorst 72).

: Nematode attack unimportant; clay content less than 30%
: Nematode attack serious; clay content more than 30%

tivated wheat, serve as oversummering hosts during periods of host absence and after temperatures in lower areas rise too high for survival of the fungus (Figure 12). Mean temperatures of 22.3°C or a mean maximum of 32.4°C for 10 days are lethal.

Seed production and seed transmission offer other possibilities of geophytopathological analysis. Klemm (43) mapped the area of *Sclerotinia trifoliorum* attack in relation to red clover seed production, while Byford & Hull (12) demonstrated the close interdependence of sugar beet seed production and sugar beet downy mildew, *Peronospora farinosa*. The geographic distribution of *Tilletia* spp. is easily studied by seed sampling, as shown by Tisdale, Leighty, & Boener (82) for the U. S. Wallen & Sutton (89) based their map of *Xanthomonas phaseoli* occurrence in Ontario on the same technique and

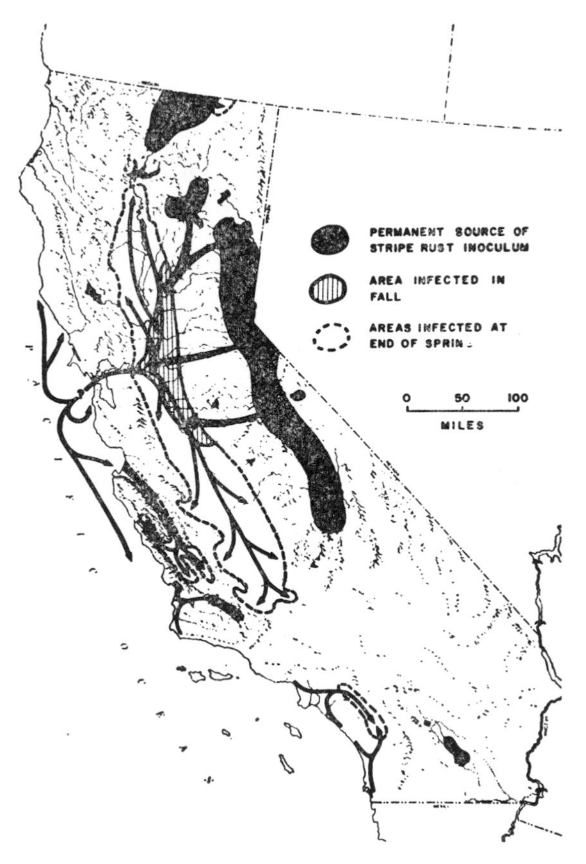

FIG. 12. The annual stripe rust epidemic of wheat, *Puccinia striiformis,* in California, with dissemination routes from oversummering areas to wheat growing areas with fall and spring infection (Tollenaar & Houston 83).

FIG. 13. Locations of sunflower downy mildew *Plasmopara halstedii,* detected from seed samples, and assumed gene centers of the fungus: 1, 2, 3 (Leppik 50).

Leppik (49, 50) even devised world distribution maps for *Plasmopora halstedii* (Figure 13) and *Alternaria sesami* by study of seed samples.

The clear interdependence of plant diseases with ecological factors thus makes true maps of disease distribution and severity possible for all parts of the world where sufficient meteorological data are available and the autecology of the causing organism is well known.

The Prognosis Problem

If the geographic distribution of a pathogen and its host is known and sufficient information on their ecological requirements is available, it would be possible to predict disease occurrence in previously uncontaminated areas, or in places where the host is introduced as a new crop. Thus, a true disease prognosis is one of the major applications of geophytopathology. Reichert & Palti (62) have stressed this point and used the terms "potential invader" for organisms moving from one climatic region into the other, and "intraregional invader" for those spreading within the same region. Again, development of maps seems to be the most suitable and comprehensive documentation technique. Unfortunately it has only been applied in a few cases. Most probably many plant protection organizations are at present much more concerned with short-term disease forecasting, than with long-term prognosis. Wagner (87) has tried mapping of the potential occurrence of dwarf bunt of wheat, *Tilletia contraversa* in Bavaria, Germany, based on the intensity of wheat cultivation. Also the map by Wuerzer (99) on *Spongospora subterranea* is basically a prognosis map, based on temperature and summer rainfall data, while Atkinson & Grant (2) used only summer rainfall data for delineation of the wheat area in Alberta, Canada, less threatened by streak mosaic virus. Seinhorst's (72) study of *Ditylenchus dipsaci* in relation to soil texture also allows a prognosis of the severity of stem eelworm attack, and Large (46) tried to develop zones for England and Wales, according to the dates of *Phytophthera infestans* outbreaks on potatoes. The climatic analysis of the ecological requirements was used by Drandarewski (24) (Figure 14) and Bleiholder & Weltzien (5) for world wide prognosis maps of *Erysiphe betae* and *Cercospora beticola* respectively on sugar beets. Using the principle of zoning the area of disease occurrence according to disease intensity and economic importance as mentioned above, a prognosis can be made for all sugar beet growing areas, as well as for any area in the world where sugar beet cultiva-

FIG. 14. Prognosis map for powdery mildew of sugar beets, *Erysiphe betae*, in sugar beet growing areas of South America (Drandarewski 24).

▤: area of sugar beet cultivation
●: probable area of main damage
◑: probable area of marginal damage
○: probable area of sporadic attack

tion may be started in the future. While here the prognosis was based on climatic diagrams developed by Walter & Lieth (90), other characteristics may be used for the prognosis of other diseases. However, there is little doubt that a worldwide prognosis for many plant diseases is now possible with the information available on the autecology of the pathogens. How much genetic changes of pathogens and/or host may affect the validity of the prognosis is a matter of question. Breeding for resistance and development of physiologic races do not seem to affect the climatic regions favorable for a certain disease, but rather serve as control techniques within the areas of main and marginal damage. However, if new ecological races of pathogens develop, the prognosis and zonation of disease areas may have to be changed.

Outlook

While documentation, analysis, and prognosis of plant epidemics seems to be an appropriate theme for maps as basic contributions to geophytopathology, the geographic approach can be applied to other subjects as well. There is a great need today for development and improvement of the organization of plant protection and disease control all over the world. Maps can be helpful to document the status quo and to suggest improvements. The regional plant protection organizations of the world have been mapped by Hes (32), and Chiarappa (13) showed the present intensity of technical assistance in plant pathology offered by FAO. The locations of international yellow rust trials in relation to host frequency were mapped by Zadoks (101), and even the analysis of the geographic distribution of publication intensity (Humphrey 40, Bremer 9), can be helpful in planning the advancement of our science.

A comprehensive view of the facts and problems of geographic distribution of plant diseases suggests the necessity of atlases dealing with these problems. Maybe the project of the European Cereal Atlas Foundation to develop an "Atlas of Cereal Diseases and Pests in Europe and the Mediterranean Countries" will set a first example, though limited in crop range and geographic area. An outlook to neighboring disciplines reveals that entomology has provided more geographic and ecological data in plant pests than pathology has on plant diseases. An Atlas covering pests and diseases of selected crops thus seems to be a possible project. It is therefore hoped that this article may encourage research and extension workers everywhere, to publish existing maps, to develop new ones, and to concentrate in the future more on the phenomenon of uneven geographic distribution of plant diseases and pests, thus helping to develop the concept of "geophytopathology" towards one that may be called "geophytomedicine." We may then find that the fascinating story of plant epidemics and their control offers a very striking example for what Humboldt (39) called a specific contribution to geographic botany: The demonstration that plants do affect the moral and political history of man.

LITERATURE CITED

1. Atkins, J. G., Newson, L. D., Sprink, W. T., Lindberg, G. D. 1960. Occurrence of hoja blanca and its insect vector, *Sogara ariziola* Muir, on rice in Louisiana. *Plant Dis. Reptr.* 44: 390–93
2. Atkinson, T. G., Grant, M. N. 1967. An evaluation of streak mosaic losses in winter wheat. *Phytopathology* 57:1188–92
3. Baldacci, E., Orsenigo, M. 1952. Chestnut blight in Italy. *Phytopathology* 42:38–39
4. Behnke, A. R. 1952. Foreword to: *World atlas of epidemic diseases.* Hamburg: Falk. Vol. I:7
5. Bleiholder, H., Weltzien, H. C. 1971. Beitrage zur Epidemiologie von *Cercospora beticola* Sacc. an Zuckerruben. *Phytopathol. Z.* In press
6. Blodgett, E. C., Semler, L. F. 1970. Apple scab in the Yakima valley of Washington. *Plant Dis. Reptr.* 54:63–65
7. Blumer, S. 1933. *Die Erysiphacceer Mitteleuropas mit besonderer Berücksichtigung der Schweiz.* Beiträge zur Kryptogamenflora der Schweiz 7:1–483
8. Brandenburg, T., Buhl, C. 1955. Uber das Vorkommen von Molybdänmangel bei Blumenkohl in Westdeutschland and seine Verhutung. *Z. Pflanzenkr.* 62:514–28
9. Bremer, H. 1963. Zur geographischen Verbreitung pflanzlicher Virosen. *Z. Pflanzenkr.* 70:1–3
10. Brod, G. 1966. Abgrenzung der fur die Pflanzkartoffelvermehrung geeigneten Anbaulagen in Baden-Wurttemberg in Beziehung zu Blattlausbefall und mittlerer Wertzahl. *Z. Pflanzenkr.* 73:577–97
11. Broekhuizen, S. 1969. *Agro-ecological atlas of cereal growing in Europe* Vol. II: Atlas of the cereal growing areas in Europe. Wageningen: Pudoc, 60 maps, 156 pp.
12. Byford, W. J., Hull, R. 1967. Some observations on the economic importance of sugar-beet downy mildew in England. *Ann. Appl. Biol.* 60:281–96
13. Chiarappa, L. 1969. International assistance in plant pathology in developing countries with particular reference to FAO programs. *FAO Plant Protec. Bull.* 17:1–8
14. Commonwealth Mycological Institute. 1962. *Distribution maps of plant diseases.* Kew: Commonw. Mycol. Inst.
15. Commonwealth Mycological Institute. 1968. Regional and territorial lists of plant diseases. *Plant Pathologist's Pocketbook.* Kew: Commonw. Mycol. Inst. 86–96
16. Commonwealth Mycological Institute. 1968. A bibliography of lists of plant diseases and fungi. I. Africa. *Rev. Appl. Mycol.* 42:553–58
17. Commonwealth Mycological Institute. 1970. A bibliography of lists of plant diseases and fungi. II. Asia. *Rev. Plant Pathol.* 49: 103–8
18. Commonwealth Mycological Institute. 1971. A bibliography of lists of plant diseases and fungi. III. America. *Rev. Plant Pathol.* 50:1–7
19. Corbaz, R. 1961. Considérations sur l'epidémie de mildious du tabac en Europe. *Phytopath. Z.* 42:39–44
20. Corbaz, R. 1964. Evolution de l'Epidémie de mildiou du tabac. *Phytopath. Z.* 51:191–92
21. Cotton, F. R., Presley, J. T., Darvish, F. 1969. Distribution of *Verticillium* wilt in cotton growing areas of the U. S. *Plant Dis. Rept.* 53:116–17
22. Crispin, A., Dongo, S. 1962. New physiologic races of bean rust, *Uromyces phaseoli typica,* from Mexico. *Plant Dis. Reptr.* 46: 411–13
23. Crowell, J. H. 1940. The geographic distributions of the genus *Gymnosporangium. Can. J. Bot.* 18:469–88
24. Drandarewski, C. 1969. Untersuchungen ueber den echten Ruebenmehltau *Erysiphe betae* (Vanha) Weltzien III. Geophytopathologische Untersuchungen. *Phytopathol. Z.* 65:201–18
25. Elton, C. S. 1966. *The ecology of*

invasion by animals and plants. London: Methuen. 181 pp.

26. Eriksson, J. 1906. Der Amerikanische Stachelbeermehltau in Europa, seine jetzige Verbreitung und der Kampf gegen ihn. *Z. Pflanzenkr.* 16:83–90
27. Finke, L. L. 1792. *Lehrbuch der Medizinischen Geographie.* nach Jusatz, H. J. in Zeiss, H. 1942. *Seuchen-Atlas*
28. Fritsch, K. 1958. Untersuchungen Ueber Vorkommen und Bedeutung des latenten Kupfermangels. *Z. Pflanzenkr.* 65:259–67
29. Gaeumann, E. 1951. *Pflanzliche Infektionslehre.* Basel, Birkheuser, 681 pp.
30. Good, R. 1964. *The geography of the flowering plants.* London: Longmans, 3rd. 518 pp.
31. Heinze, K., Profft, J. 1940. Ueber die an der Kartoffel lebenden Blattlausarten und ihren Massenwechsel im Zusammenhang mit dem Auftreten von Kartoffelviren. *Mitt. Biol. Reichsanst. Berlin,* Heft 60:164 pp.
32. Hes, I. W. 1966. Intra- und interkontinentale Pflanzenschutzorganisationen. *Ges. Pflanzen* 18:233–36
33. Herter, W. 1907. Weitere Fortschritte der Stachelbeerpest in Europa. *Zentralbl. Bakteriol.* 2. Abt. 18:825–30
34. Hinds, T. E., Jones, J. R. 1965. *Hypoxylon* canker of aspen in Arizona. *Plant Dis. Reptr.* 49: 480
35. Hirsch, 1881. *Handbook of historical and geographical pathology.* n. Behnke, A. R. in Rodenwaldt, E. 1952. *World atlas of epidemic diseases*
36. Hogg, W. H., Hounam, C. E., Mallik, A. K., Zadoks, J. C. 1969. *Meteorological factors affecting the epidemiology of wheat rusts.* World Meteorological Organization, Geneva, Tech. Note 99, 143 pp.
37. Horsfall, J. G. 1959. A look to the future—the status of plant pathology in biology and agriculture. *Plant Pathology, Problems and Progress. 1908–1958,* Madison: Univ. Wisc. Press 63–70
38. Hull, R. 1953. Assessment of losses in sugar beet due to virus yellows in Great Britain. 1942–52. *Plant Pathol.* 2:39–43
39. Humboldt, A. v., Bonpland, A. 1807. *Ideen zu einer Geographie der Pflanzen.* Tuebingen: Cotta. Neudruck Darmstadt: Wiss. Buchges. 1963. 182 pp.
40. Humphrey, H. B. 1932. Report of the twenty-third annual meeting of the American Phytopathological Society. Report of the Editor in Chief. *Phytopathology* 22:475–90
41. International Association of Agricultural Economists, 1969. *World Atlas of agriculture.* Novara: Instituto Geogr. De Agostini, 52 maps
42. Jusatz, H. J. 1942. Aufgaben und Methoden der medizinischen Kartographie. 7 pp. in: Zeiss, H., 1942, *Seuchen-Atlas*
43. Klemm, M. 1938. Schadgebiete des Kleekrebses (*Sclerotinia trifoliorum* Eriks.) in Deutschland, Kleesamenanbau und Witterung. *Z. Pflanzenkr.* 48:605–18
44. Klinkowski, M. 1961. Der Blauschimmel des Tabaks. *Deut. Landwirt.* 12:229–32, 237–39
45. Kroeber, H. 1961. Untersuchungen ueber die Blauschimmelkrankheit des Tabaks in Deutschland. *Nachrichtenbl. Deut. Pflanzenschutzdienst* (*Braunschweig*) 13:41–44
46. Large, E. C. 1956. Potato blight forecasting and survey work in England and Wales. *Plant Pathol.* 5:39–52
47. Large, E. C. 1958. Losses caused by potato blight in England and Wales. *Plant Pathol.* 7:39–48
48. Lehmann, H. 1965. Untersuchunger Ueber die Typhula-Faeule des getreides. *Phytopathol. Z.* 53:255–88
49. Leppik, E. E. 1964. *Alternaria sesami,* a serious seed-borne pathogen of world-wide distribution. *FAO Plant Protec. Bull.* 12:13–16
50. Leppik, E. E. 1964. Mapping the world distribution of seed-borne pathogens. *Proc. Int. Seed Test Assoc.* 29:473–77
51. Leppik, E. E. 1965. A pathologist's viewpoint in plant exploration

and introduction. FAO Plant Introduction News Letter No. 15:1–5

52. McCubbin, W. A. 1917. Does *Cronartium ribicola* winter on the currant? *Phytopathology* 7: 17–31
53. Miller, P. 1966. The effect of weather on prevalence of disease. *The American Biology Teacher* 28:469–72
54. Moore, W. I. 1970. Origin and spread of southern corn leaf blight in 1970. *Plant Dis. Reptr.* 54:1104–8
55. Palti, J. 1971. Biological characteristics, distribution and control of *Leveillula taurica* (Lev.) Arn. *Phytopathol. Mediter.* 10: 139–53
56. Peltier, G. L. 1933. Physiologic forms of wheat stem rusts in Kansas and Nebraska. *Phytopathology* 23:343–56
57. Populer, C. 1965. Le mildiou du tabac, *Peronospora tabacina* Adam. Chronologie de l'apparition annuelle de foyers en Europe. *Parasitica* 21:37–39
58. Rademacher, B. 1954. Regionale Pflanzenpathologie Suedwestdeutschlands. *Mitt. Biol. Bundesanst.* Berlin, Heft 80:34–50
59. Rankin, W. H. 1914. Field studies on the *Endothia* canker of chestnut in New York State. *Phytopathology* 4:233–60
60. Reichert, I. 1953. A biogeographical approach to phytopathology. *Proc. 12. Int. Bot. Congr.* Stockholm 1950:730–31
61. Reichert, I. 1958. Fungi and plant diseases in relation to biogeography. *Trans. N.Y. Acad. Sci. Ser. II.* 20:333–39
62. Reichert, I., Palti, J. 1966. On the pathogeography of plant diseases in the Mediterranean region. *Proc. I. Congr. Medit. Phytopathol. Union, Bari.* 273–80
63. Reichert, I., Palti, J. 1967. Prediction of plant disease occurrence; a patho-geographical approach. *Mycopathol. et Mycol. Appl.* 32:337–55
64. Roberts, E. T., Reading, N.A.A.S., Large, E. C. 1963. Surveys of verticillium wilt in Lucerne, England and Wales 1958–60. *Plant Pathol.* 12:47–58
65. Rodenwaldt, E. 1952, 1966, 1961. *World Atlas of Epidemic Diseases.* Hamburg, Falk. 3 Vols.
66. Rodenwaldt, E. 1952. Introduction to: *World atlas of epidemic diseases.* Hamburg: Falk, Vol. I: 11–12
67. Roschdestwensky, R. 1931. Kartoffelkrankheiten und -ernten in Russland. *Sorauer Handb. Pflanzenkr. Berlin; Parey 1933 I,* 1. Teil p. 454
68. Savulescu, A., Ploaie, P. G. 1967. Virogeographische Studien ueber das Kleeverlaubungsvirus und seine Vektoren. *Phytopathol. Z.* 58:315–22
69. Scharif, G., Niemann, E., Ghanea, M. 1967. Chickpeablight in Iran. *Mycosphaerella rabiei* Kov.-*Ascochyta rabiei* (Pass.) Lab. *Entomol. Phytopathol. Appl.* Teheran 25:10–15
70. Schnathorst, W. C., Crogan, R. G., Bardin, R. 1958. Distribution, host range and origin of lettuce powdery mildew. *Phytopathology* 48:538–43
71. Schnathorst, W. C., Halisky, P. M., Martin, R. D. 1960. History, distribution races and disease cycle of *Xanthomonas malvaceaum in* California. *Plant Dis. Reptr.* 44:603–8
72. Seinhorst, J. W. 1956. Population studies on stem eelworms (*Ditylenchus dipsaci*). *Nematologica* 1:159–64
73. Slykhuis, J. T. 1970. Factors determining the development of wheat spindle streak mosaic caused by a soil borne virus in Ontario. *Phytopathology* 60: 319–31
74. Stakman, E. C. 1947. International problems in plant disease control. *Proc. Am. Phil. Soc.* 91:95–111
75. Stakman, E. C. 1955. Progress and problems in plant pathology. *Ann. Appl. Biol.* 42:22–33
76. Stakman, E. C., Harrar, J. G. 1957. *Principles of plant pathology.* New York: Ronald 581 pp.
77. Stakman, E. C., Levine, M. N., Wallace, J. M. 1929. The value of physiologic-form surveys in the study of the epidemiology

of stem rust. *Phytopathology* 19:951–59

78. Staniland, L. N. 1943. A survey of potato aphides in the southwestern agricultural advisory province. *Ann. Appl. Biol.* 30:33–42
79. Stevens, N. E. 1933. Some significant estimates of losses from plant diseases in the United States. *Phytopathology* 23:975–84
80. Steudel, W., Heiling, A. 1949. Ueber die Verbreitung der Vergilbungskrankheit und des Mosaiks der *Beta*-Rueben in Westdeutschland. *Z. Pflanzenkr.* 56: 380–85
81. Tischler, W. 1965. *Agraroekologie* Jena: Fischer. 499 pp.
82. Tisdale, W. H., Leight, C. E., Boerner, E. G. 1927. A study of the distribution of *Tilletia tritici* and *T. laevis* in 1926. *Phytopathology* 17:167–74
83. Tollenaar, H., Houston, B. R. 1966. A study on the epidemiology of stripe rust, *Puccinia striiformis* West., in California. *Can. J. Bot.* 45:291–307
84. U. S. Dept. Agr. 1948. *Agricultural geography of Europe and the Near East.* Misc. Publ. No. 665, U.S. govt: Washington D.C.
85. Vallega, J. 1955. Wheat rust races in South America. *Phytopathology* 45:242–46
86. van der Zwet, T. 1968. Recent spread and present disease distribution of fire blight in the world. *Plant Dis. Reptr.* 52: 698–702
87. Wagner, F. 1956. Warum tritt der Zwergbrand des Weizens nur im Suedlichen Bayern auf? *Z. Pflanzenbau Pflanzenschutz* 51: 28–32
88. Wallace, J. 1932. Physiologic specialization as a factor in the epidemiology of *Puccinia graminis tritici*. *Phytopathology* 22:105–42
89. Wallen, V. R., Sutton, M. D. 1965. *Xanthomonas phaseoli* var. *fuscans* (Burkh.) Starr and Burkh. on field beans in Ontario. *Can. J. Bot.* 43:437–46
90. Walter, H., Lieth, H. 1966. *Klimadiagramm-Weltatlas.* Jena: Fischer
91. Warmbrunn, K. 1952. Untersuchungen ueber den Zwergsteinbrand. *Phytopathol. Z.* 19:441–82
92. Weihing, J. L., Vidaver, A. K. 1967. Report of *Holcus* leaf spot (*Pseudomonas syringae*) epidemic on corn. *Plant Dis. Reptr.* 51:396–97
93. Weltzien, H. C. 1965. Der echte Mehltau der Rueben. Mittlg. Biol. Bundesanst. Berlin, Helft 115:188–91
94. Weltzien, H. C. 1967. Geopathologie der Pflanzen. *Z. Pflanzenkr.* 74:176–89
95. Weltzien, H. C. 1970. Krankheitsverhuetung durch Standortwahl bie Zuckerrueben. *Arch. Pflanzenschutz* 6:217–24
96. Weltzien, H. C. 1971. Weltweiter Pflanzenschutz als oekologisches Problem. *Meded. Rijksfac. Landbouwwetensch. Gent.* 36: 13–19
97. Wenzl, H. 1962. Beitraege zur Oekologie des Kartoffelschorfes. *Pflanzenschutzberichte Wien* 29: 33–64
98. Whitehead, A. G. 1969. The distribution of root knot nematodes (*Meloidogyne spp.*) in tropical Africa. *Nematology* 15:315–33
99. Wuerzer, B. 1964. *Ergaenzende Untersuchungen ueber den Pulver-schorf der Kartoffel und dessen Erreger Spongospora subterranea* (Wallr.) Lagerh. Diss. Landw. Hochschule Hohenheim 104 pp.
100. Zachos, D. G. 1963. Evolution de l'epidémie du mildiou du tabac en Grece durant l'Année 1962. *Ann. Phytopathol. Benaki* 5: 191–203
101. Zadoks, J. C. 1961. Yellow rust on wheat, studies in epidemiology and physiologic specialization. *Tijdskr. Plantenziekten* 67:69–256
102. Zadoks, J. C. 1965. Epidemiology of wheat rusts in Europe. *FAO Plant Prot. Bull.* 13:97–108
103. Zeiss, H. 1942. *Seuchen-Atlas.* Gotha: Perthes, 2 Vols.
104. Zeiss, H. 1942. Medizinische Kartographie und Seuchenbekaempfung. in Zeis: *Seuchen-Atlas:* 3 pp.
105. Zogg, H. 1949. Untersuchunger ueber die Epidemiologie des Maistrostes *Puccinia sorghi* Schw. *Phytopathol. Z.* 15:143–90

THE ROLE OF PHYTOPHTHORA CINNAMOMI IN AUSTRALIAN AND NEW ZEALAND FORESTS

3553

F. J. Newhook and F. D. Podger

Botany Department, University of Auckland, New Zealand, and Forest Research Institute, Department of National Development, Canberra, Australia

In 1922, while R. D. Rands reflected on the "rather narrowly limited" host relationships of the cinnamon *Phytophthora* he had just discovered in the mountains of tropical Western Sumatra (124), his fungus was already established some 3,000 miles to the south-east on a wide range of native plants in temperate Western Australia. Had he been able then to visit the native forests near Perth, Rands could have seen in its infancy an epidemic disease that causes wreckage in more than 100 species of the native flora and converts tall forests to virtual barrens. Yet almost half a century passed before *Phytophthora cinnamomi* Rands was recognized as the cause of this disease known as jarrah dieback (Figure 1). Today Rands would find his fungus unquestionably the most destructive plant pathogen ever recorded in native vegetation of this and possibly any region. It is unmatched in the variety of plants and the range of communities it affects. In marked contrast with the chestnut blight, dutch elm disease, and white pine blister rust pathogens, each epidemic on only one or two genera, *P. cinnamomi* kills plants in 48 families. Already it has devastated complex forest, woodland, and heath communities on more than 100,000 hectares in Western Australia (W.A.) and Victoria (Vic.) and is the cause of grave fears for the future of much of the indigenous vegetation of southern Australia. By contrast, in New Zealand (N.Z.) and in eastern Australia the role of the fungus in native vegetation is less clearly understood.

Although *P. cinnamomi* also causes serious loss in nursery crops, in conifer shelterbelts, and in avocado and pineapple plantations its importance in indigenous communities dictates that the emphasis of this review should be on its role in forestry and conservation. Because this importance has been recognized only recently and research findings are yet largely unpublished we have drawn extensively on personal communication from generous colleagues who share our interest in the present distribution of *P. cinnamomi,* its variability, its role in a number of diseases of uncertain cause, the factors that determine the variety of its effects in the flora of our region, its indigenous or exotic origin, its potential to cause further damage, and the prospects for its control.

FIGURE 1. Jarrah dieback disease in Western Australia. The large dead trees are jarrah (*Eucalyptus marginata*); note healthy forest with dense understory in background.

Historical review and current importance of P. cinnamomi in our region.—*P. cinnamomi* was first recorded in the region by Ashby (123) who in 1930 identified the *Phytophthora* sp. that Simmonds (137) reported to be the cause of pineapple wilt and top rot, diseases described by Tryon (150) as epidemic in Queensland (Q.) during 1887 and 1892. Today *P. cinnamomi* is probably the most serious pathogen of avocado and pineapple in subtropical New South Wales (N.S.W.) and Q. (45, 102).

First isolations for other parts of the region are New South Wales, Fraser 1948 (44); New Zealand, Smith 1950 (140); Western Australia, Zentmyer 1964 (109); Victoria, Jenkins & Wauchope 1964 (68); Tasmania (Tas.), Zentmyer 1965 (47); and South Australia (S. Aust.), Davison 1969 (36).

Until 5 years ago *P. cinnamomi* was considered the most important pathogen in commercial nurseries in North Island, N.Z. Without control by soil sterilization and nursery hygiene, conversion to container-grown and summer irrigated stock would not have been economically feasible (71). In Victoria and New South Wales *P. cinnamomi* is a serious pathogen in nurseries; but surprisingly its importance in W.A. nurseries is not known (39), and there are few records in Queensland (138). The fungus causes serious losses in many Australian and South African species of the Proteaceae and Leguminosae as well as *Prunus, Camellia, Erica, Calluna, Rhododendron, Boronia, Eriostemon, Jacaranda, Eucalyptus, Thryptomene, Juglans, Olea, Passiflora,* and many conifer species (15, 38, 45, 67, 68, 71). In N.Z. forest nurseries, control measures are rarely warranted (10), although occasionally losses up to 70% in *Pinus radiata* (11) and 34% in *Pseudotsuga mensiesii* (63) have occurred. In Q. *Pinus* nurseries control treatments are regularly applied (22).

The association of *P. cinnamomi* with epidemic losses in conifer shelterbelts was reported (90) in 1959 and later (14, 74, 151) in Australia. In a N.Z. survey in 1956 about 50% of 1,970 *Pinus radiata* and 863 *Cupressus macrocarpa* shelterbelts were seriously damaged (145). On the Swan Coastal Plain, W.A. conifer shelterbelts on irrigated and poorly drained soil have been virtually eliminated (14).

Littleleaf disease of *Pinus* was first reported in N.Z. by Hepting & Newhook (62). The role of the fungus in plantations of *Pinus radiata* has been investigated by Newhook in N.Z. (94) and by Hartigan (54) and Davison & Bumbieris (155) in Australia. Oxenham & Winks have discussed its effects in nurseries (96, 97) and in plantations (96) of southern pines in Queensland.

P. cinnamomi was first recognized as a pathogen of Australian native plants by Fraser (44) in 1948 when she associated the fungus with occasional deaths of ground flora species in eucalypt woodland near Sydney, N.S.W. In N.Z. Newhook (91) first isolated *P. cinnamomi* from native vegetation in 1957, but like Fraser found no evidence of serious effects in the wild. The fungus has made its greatest impact in the dry sclerophyll forests of jarrah (*Eucalyptus marginata*) in southwestern Australia (Figs. 1, 2) where 80,000 hectares or 5% of the commercially productive forest have been devastated and serious damage done to scenic and conservation values. The disease, first noted in 1921 when it was already well established (154), was not recognized as epidemic until 1950 (156) or associated with *P. cinnamomi* until 1964 (109). Later Podger (105, 106) defined the ecological occurrence of the disease, and proved pathogenicity under field conditions. He concluded that *P. cinnamomi* was introduced in W.A., and recommended changes in forest management to reduce spread of inoculum (104) which have been adopted (153).

The severity of the epidemic was recognized in 1965 by Cromer, then Director, Forest Research Institute, Department of National Development, and by Harris, then Conservator of Forests, W.A. This led to funding of research in two universities, commissioning of specialist reports (93, 166) and formal establishment of a national research working group for forest diseases which recommended urgent investigation on *P. cinnamomi* in eastern Australia. The number of scientists working on the fungus in indigenous plant communities and exotic forests in Australia has grown from the equivalent of one full time in 1966 to the equivalent of fifteen in 1971 (25 individuals work full or part time in 3 commonwealth, 5 state, and 4 university laboratories).

The Pathogen and the Diseases Associated With It

There are two distinct interpretations of the epidemic. In one, *P. cinnamomi* is considered indigenous, already occupying most habitats potentially available to it, and displaying definite variation. Those who hold this view attribute disease occurrences to disturbance of the normal balance between host and pathogen especially where man's activities have altered the environment. In the other view, *P. cinnamomi* is considered a relatively uniform and virulent pathogen, moving into a great diversity of new habitats in two countries that have been separated since the late Cretaceous and in whose territories, spanning almost 50° of latitude, distinctive floras have evolved in the absence of the fungus. The great variety of plant response to the invasion is explained by the influence of environmental factors on interaction between host and pathogen.

Distribution of P. cinnamomi.—Much of the information on the distribution of *P. cinnamomi* has been determined by the use of selective agar based on Eckert & Tsao (41) and lupin baiting techniques based on Chee & Newhook (27). Some limitations to the lupin baiting method have been overcome by presoaking (90), aeration (84), addition of peptone (30), repeated baiting (99, 105), and the use of excised radicles (84). The method fails to detect some other pathogens of woody plants that sometimes co-occur with *P. cinnamomi,* e.g. *P. cambivora* and *P. cactorum* (26). These pathogens may be detected by concurrent use of other baits, e.g. pine needles (35). In some cases *P. cinnamomi* has been determined by examination of sporangia on baits in lieu of plating, risking confusion with other non-papillate species (vide 157).

Our information on the distribution of *P. cinnamomi* is derived from host records for which localities are given (10, 38, 39, 83, 103, 107, 111, 114–116, 119, 127, 160) from local surveys in south-western Australia (105), in the Australian Capital Territory (A.C.T.) (66), and in the Brisbane Ranges, Vic. (160), and from an extensive survey in eastern Australia (118). In southwestern Australia and the Brisbane Ranges, Vic., *P. cinnamomi* is increasing its distribution from numerous widespread centers of infection marked by severe effects on the flora. In these areas and in east Gippsland, Vic. the fungus has not been isolated from adjacent healthy forest (83, 105, 160). In most of southern Australia the boundaries of *P. cinnamomi* infestation are so sharply marked by changes in the health of understory and ground flora that they can be mapped from aerial photographs (Figure 2). In an aerial-photographic survey of 462,000 ha of jarrah forest it was found that the area infected had increased to 23,000 ha at the rate of 4% per annum, in the period 1943–65 (153).

The fungus is reported to be widespread in forested areas in N.S.W., A.C.T., Tas., and Q. but there are few reports of damage to native vegetation in these areas. In N.Z. the fungus occurs in native vegetation in Auckland Province, Nelson, and Westland but is not ubiquitous (10, 78, 111).

In W.A. and Vic. the spread of the disease is irregular in both space and time and is correlated with spread of the fungus (105, 160). Weste & Taylor

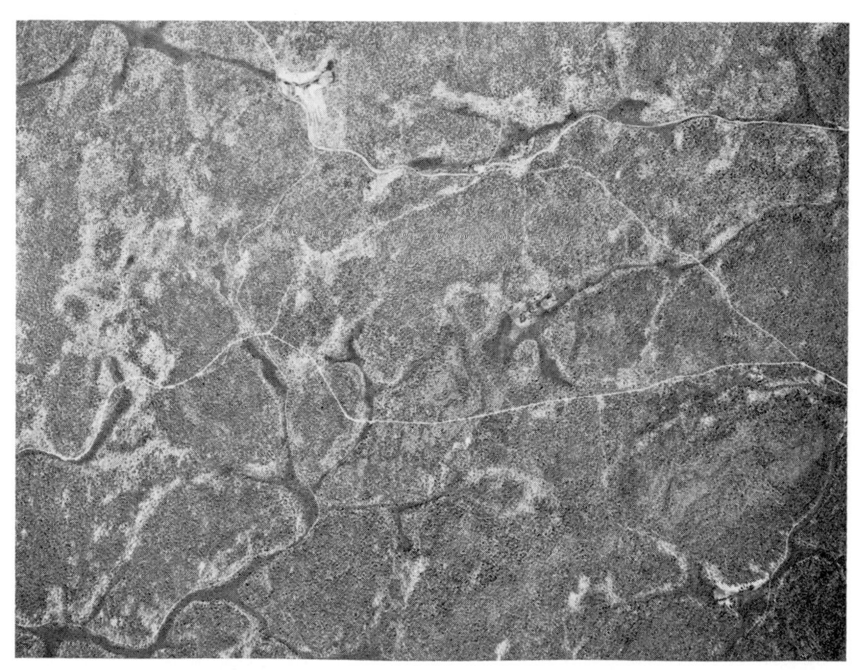

FIGURE 2. Aerial view from 6,700 m above jarrah forest in Western Australia. In the light-toned areas most of the vegetation is dead or dying.

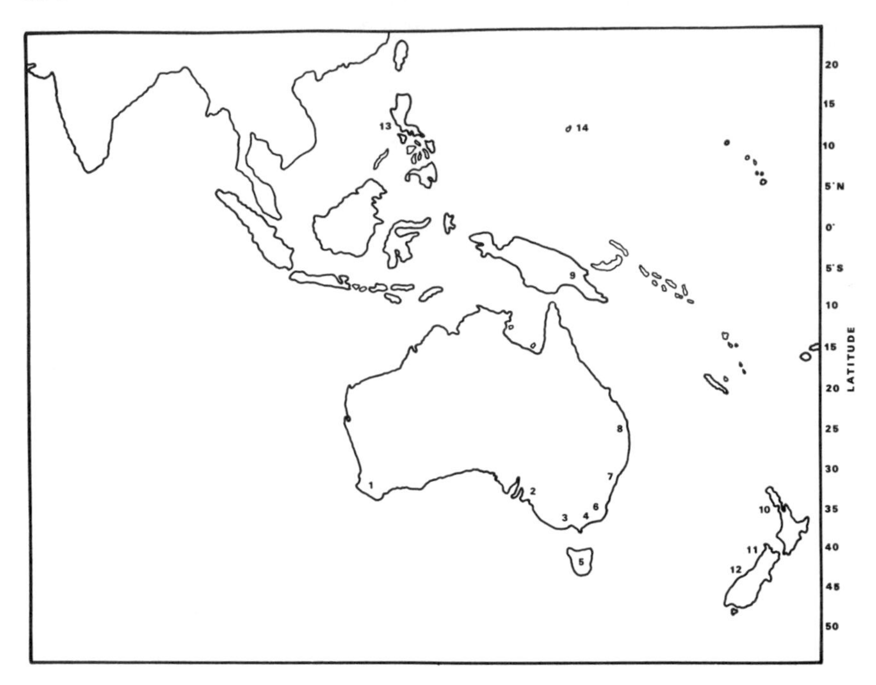

FIGURE 3. Outline map indicating main localities referred to in text: 1, jarrah forests of southwestern Australia; 2, Adelaide Hills, South Australia; 3, Brisbane Ranges, Victoria; 4, East Gippsland, Vic.; 5, Tasmania; 6, Canberra, A.C.T.; 7, Sydney, New South Wales; 8, "wallum" heaths, southwest Queensland; 9, New Guinea; 10, Auckland Province (including Northland) N.Z.; 11, Nelson Province, N.Z.; 12, Westland, N.Z.; 13, Manila; 14, Guam.

(160) have shown that spread of the fungus over a 5-month period varied from nil to 10m on the contour, up to 29m downslope, and was less than 1.5m upslope.

In W.A., the Brisbane Ranges, and east Gippsland, there is strong correlation of disease with logging, roading, and vehicle movement (13, 83, 105, 107). Survival of inoculum has been demonstrated in small pieces of infected root (105) and quantities of field soil (105) much smaller than are carried and shed by vehicles (12) or shifted with gravel for road making. In W.A. the widespread use of sandy gravels from the surface horizon of lateritic soils has been a major factor in the development of the epidemic (106), where gravels were taken from dying forest areas until jarrah dieback was shown to be caused by a soil-borne organism (153). This factor has been less important in much of eastern Australia and in N.Z. where road surfaces are constructed from in situ materials, quarried stone, and river gravels.

Other important means of dispersal are infected nursery stock (36, 39, 96, 102, 143), vegetative propagating material (76), and bulk supplies of garden loam (143). Field evidence in W.A. suggests that dispersal by boots and animals is unimportant (105), but this might not be so with adhesive soils (vide 168). Passive dispersal has also been demonstrated in surface runoff (99). The fungus has been isolated from the banks of the Murrumbidgee River, N.S.W. (143) and presumably is being dispersed to irrigated areas of the Murray Valley. Zoospores may be important in these instances as well as in localized autonomous dispersal. Palzer reports (99) extension of motility as a result of polyplanetism, increase in motility with increasing zoospore concentration, and increase in the extent and rate of root rot with increasing zoospore dosage. Palzer (99) has shown rapid growth of *P. cinnamomi* in large roots of several hosts, a mechanism whereby the pathogen can move uphill and during dry periods.

Variation in P. cinnamomi.—The extent of variability within and between the various Australasian populations of *P. cinnamomi* is not yet clear. Shepherd et al (134) report differences in the geographic distribution of 4 thermoauxotypes. Other workers (28, 45, 79, 84, 148) have noted variation between isolates in growth rate, morphology, and sporulation but none sufficient to indicate distinct races. While natural occurrence of oospores has not been reported in the region, where only the A^2 mating strain is known to occur, in vitro demonstration of interspecific mating (29) and the recent demonstration that microbial metabolites stimulate oospore formation in single A^2 isolates (21) suggests that recombination may occur.

It might be expected, in such a widely distributed pathogen with a broad host range, that adaptation to particular hosts would have produced distinct strains. However, from all reports the picture is one of remarkably uniform pathogenicity among isolates from a wide range of hosts and geographic sources. Nonetheless there have been reports of differences in pathogenicity, mainly from individual tests on a limited number of hosts (84, 105). In some of these tests the differences are due in part to use of genetically variable host material, and in part to unexplained variation in the conditions of test. Thus Bertus (15) found that, despite use of the same conifer cultivars and the same four isolates, he obtained different results in successive tests.

Effects of P. cinnamomi on natural vegetation in the region.—From 35 published and unpublished sources we find 404 Australian and 40 N.Z. native species recorded as hosts or probable hosts. In most of these it is clear that *P. cinnamomi* is the cause of severe chlorosis and stunting and often death, though a few records may represent chance saprophytic colonization (167) of roots damaged by other agents. The taxonomic spectrum is remarkable. It includes a lycopod, a cycad, podocarps, and a wide range of angiosperms. In the total of 444 species in 131 genera of 48 families, 75% are in 4 families: Proteaceae (13 genera, 48 species), Leguminosae (13 genera, 39 species), Epacridaceae (14 genera, 37 species), Myrtaceae (25 genera, 202 species). Many species in the Proteaceae are particularly susceptible to extensive invasion of large roots (105) in which starch-laden parenchyma is extensively developed (122). Marked differences have been demonstrated in the susceptibility of subgenera in the important genus *Eucalyptus* (108). Members of the subgenus *Monocalyptus* (121), which includes the Renanthera of Blakely's classification (17), are highly susceptible, whereas members of *Symphyomyrtus* are generally resistant. The physiological basis for this difference is not known. However there is increasing evidence that *Symphyomyrtus* and *Monocalyptus* are respectively tolerant and intolerant of waterlogging. It is possible that susceptibility to *P. cinnamomi* and sensitivity to waterlogging are related in some way. Information on within-species variation in resistance to *P. cinnamomi* is limited. All provenances of *E. marginata* have been found to be highly susceptible (99, 105) and all *E. diversicolor* provenances tolerant (98). Although Palzer (99) has demonstrated differences between individual *E. marginata* wildlings in response to wound inoculation, such variation seems to have little significance for survival in the field epidemic.

P. cinnamomi occurs in a wide range of forest, woodland, and heath communities, but only in the jarrah forest of W.A. has its role been clearly established. There distribution of the fungus is marked by severe damage in the understory and shrub layers and later in the overstory; despite frequent attempts *P. cinnamomi* has not been isolated from adjacent healthy forest (105). Pathogenicity has been demonstrated under field conditions; typical symptoms of jarrah dieback followed pure-culture inoculation of previously healthy mature forest at three places. Similar symptoms were produced by inoculation of healthy forest with soil taken from a naturally infested area, but not with soil from a healthy area (105). The outbreaks in the Brisbane Ranges, Vic. (160) and in east Gippsland, Vic. (83) are almost identical insofar as the effects on the community and the distribution of the fungus are concerned, but proof by field inoculation has yet to be obtained.

In each of these outbreaks two important criteria for recognizing *P. cinnamomi* infestation are apparent. First is the high susceptibility of almost all species of three floristically important families of the understory and shrub layers Proteaceae, Epacridaceae, and Dilleniaceae. Second is the high susceptibility of species of *Monocalyptus* and field tolerance of the eucalypt subgenera *Symphyomyrtus* and *Corymbia*. Pot inoculation studies confirm these observations particularly for the Proteaceae (149) and for *Eucalyptus* (108).

In the few small areas of native vegetation in N.Z. where severe damage has been attributed to *P. cinnamomi* (111) indigenous species in the Proteaceae and Epacridaceae are among those that have been killed.

There are several other disease syndromes in eucalypt forests in eastern Australia in which the role of *P. cinnamomi* is not yet clear (42, 55, 66, 108). Other factors considered important include *Armillaria mellea* (20) and psyllid attack (55), as well as waterlogging, drought-stress and nutrient deficiencies (66, 142). *A. mellea* is often a secondary invader (48) and could follow infection by *P. cinnamomi* (vide 34, 129, 130).

P. cinnamomi has been also isolated from forests in which there are no symptoms above ground e.g. under inherently susceptible *Agathis australis* in North Island, N.Z. (111) and *Nothofagus* in Westland, N.Z. (10). In Q. inherently susceptible species of *Eucalyptus* (108) and of *Banksia* (149) are largely unaffected in forests and heathlands immediately adjacent to severely affected pineapple plantations. Obviously the epidemic cannot be explained in terms of interactions between susceptible hosts and pathogen alone.

Importance of Environment in the Interactions Between P. cinnamomi and its Hosts

The modifying influence that environment exerts on interaction of host and pathogen has been variously emphasized. Some consider environment to be of overriding importance and *P. cinnamomi* a secondary factor (49, 125, 142, 147). Others (117) accept that the fungus is important but believe that disturbance of the environment is a prerequisite for disease. Often undue emphasis has been given single environmental factors. For example, recognition of the tendency for disease to be more frequent on soils with impeded drainage has often been followed by the suggestion that disease is unlikely on well drained soils. However severe losses have been reported on well drained soils in all parts of our region (45, 68, 102, 105, 115, 145) as well as the U.S.A. (24, table 7). In the epidemics in conifer shelterbelts (90) and the patch death of *Agathis australis* forest in N.Z. (111), prolonged rainfall has been reported to be a critical factor in disease development. Yet *P. cinnamomi* has extended its area and caused new damage in the jarrah forest in years of relatively low rainfall (105), and forests of inherently susceptible *Nothofagus* (10) in Westland, N.Z. are apparently free of damage despite the fungus and regularly heavy summer rainfall. Obviously no single environmental factor is of overriding importance. The great diversity of factors involved and the complexity of their interactions are illustrated by four examples from our region.

P. cinnamomi and diseases of exotic conifers.—In the diseases of exotic conifers environment is dominant in determining the outcome of host-pathogen-environment interactions. They illustrate the fallibility of attempts to generalize from other situations. In the southeast U.S.A. vigorous trees of shortleaf pine (*Pinus echinata*) with profuse ectotrophic mycorrhiza, growing on well drained, moderately fertile soils are much less susceptible to littleleaf disease than are unthrifty trees, with poorly developed mycorrhiza, growing on eroded soils prone to waterlogging. It might also be expected that concentration of host plants in plantations would favor epidemic development more than would single row shelterbelt plantings. For conifers in N.Z. however, the reverse is often the case. Newhook's accounts (90, 94) of shelterbelt epidemics and of "littleleaf" disease in *Pinus radiata* show how the outcome of infection is determined by shifting balances between shoot-root ratios, changing susceptibility with age, the timing of infection in relation to seasonal patterns of shoot growth, rootlet regeneration and evaporative stress, and the overriding importance of excessive rainfall. These influences are of greater significance for relatively tolerant hosts such as *Pinus radiata* in which infection is mainly restricted to fine roots, than they are for hosts such as *Chamaecyparis* in which the fungus invades major roots (90).

In *Pinus* spp. susceptibility is high in the seedling stage but, with occasional exceptions (96), field susceptibility is low between 3–20 years, after which it rapidly increases (14, 24, 74, 90, 151). The basis of resistance in the intermediate stage has not been investigated, although it has been suggested (90) that root regeneration capacity is then high but declines with age. The major change in resistance around 20 years of age is probably associated with a general change in host physiology, as it is also around the age that *P. radiata* becomes resistant to *Dothistroma pini* (50), heartwood formation begins, and branching habit changes. The relative tolerance of *P. radiata* is due largely to its high rootlet regeneration capacity and to rapid reduction of leaf area by abscission (90), a characteristic less strongly developed in some more highly susceptible hosts. In most years these adjustments maintain shoot-root balance and serious plant water deficits are avoided. Years of shelterbelt epidemics are characterized by abnormally long periods of rainfall when soil temperatures favor infection, followed by wet winters when rootlet regeneration is poor. In the shelterbelts, vigorous deep-crowned trees commence spring growth with root systems unfit to meet increasing water demands, then wilt rapidly, while trees with littleleaf symptoms resulting from earlier infection may survive because of lower water demand. Likewise, littleleaf trees in plantations on eroded, phosphate-deficient, clay soils in N.Z. (7, 62, 94) survive without change in apparent health. Nor is there any spectacular disease response during such seasons in vigorous plantations on infested soils; it has been suggested (94) that the water demand of short-crowned trees in closed-canopy forest is less than that of exposed and deep-crowned trees in shelterbelts. The importance of transpiration and interception is also illustrated by accelerated deterioration of plantations with littleleaf following removal of understory shrubs (94).

Other complex interactions are involved. Although littleleaf disease in N.Z. is essentially a reflection of nutrient deficiencies, since it is corrected by single applications of fertilizers, nutrient uptake is reduced by root-rot, adverse soil conditions, and lack of mycorrhiza. The symptoms are similar in all places (24, 62, 94), but the critical element is not always the same, being phosphorus in N.Z. (8, 161, 163) and nitrogen in U.S.A. (24). At Waipoua, N.Z., heavy applications of nitrogen accentuated, while phosphorus corrected, littleleaf in *P. echinata*. Phosphate application initiates a sequence of interactions resulting in improved mycorrhizal development and crown density which leads to increased interception and transpiration, and thus to improved soil moisture relations. Since populations of *P. cinnamomi* are unaltered Newhook (94) suggests that there is indirect control of the pathogen through mycorrhizal protection, reduced opportunity for infection by zoospores, and improved conditions for host recovery. Bowen's demonstration (18) that P-deficient plants of *P. radiata* exude more amino acids than normal plants, suggests that P applications may reduce chemotaxis of zoospores. Phosphate application does not always provide protection, as pastures under diseased shelterbelts have regularly received the nutrient. The complexity of the interactions is further illustrated by Roy's demonstration that dung and urine contain a sporangial stimulation factor (131, 132), and that applications of dung to *P. radiata* seedlings resulted in increased root regeneration and increased root-rot, but reduced mycorrhizal development (131). It is possible that elution of root and microbial exudates from the rhizosphere by prolonged and heavy rain may be important. In South Island, N.Z., Prior (120) suggested that manure stacks, stock camping, and drought before and after flooding, all contributed to shelterbelt deaths. Rewetting of Northland clays, which normally develop nonwetting properties during summer, is facilitated by deep cracking and fracturing during protracted drought (90). In Australia shelterbelt epidemics are associated with irrigation or poorly drained soils (14) and heavy summer and autumn rainfall (74, 151). Other *Phytophthora* spp. sometimes involved are *P. citricola, P. cryptogea,* and *P. syringae* in N.Z. shelterbelts (90), a *Phytophthora* similar to *P. drechsleri* in *Pinus* forest in S.Aust. (155), and *P. boehmeriae* in *Pinus patula* plantations in Q. (96).

Jarrah dieback and similar diseases in native communities in southern Australia.—This group of diseases in W.A. and Vic. illustrates how vulnerability and susceptibility to an introduced pathogen leads to great destruction, despite the existence of environmental conditions usually considered unfavorable for *Phytophthora* activity. In these areas most rain falls in winter when soil temperatures are too low for infection. Opportunities for infection occur during short periods of rainfall in the warmer months. In contrast to the clays of Northland, N.Z., the sandy and gravelly soils of these areas are rapidly rewetted and relatively light falls of rain provide sufficient moisture for completion of infection cycles. Infection of lupin radicles has been obtained within 24 hours of wetting summer dry soils (106) and Palzer (99) has shown that the lack of sporangial stimulation by summer soil extracts reported from N.Z. (30) does not obtain in Western Australia. Podger (106) suggested that these soils may also have a low level of microbial antagonism. High inoculum levels develop by mycelial growth in the rich food base provided by the large roots of many susceptible species which grow in intimate and almost continuous contact. Summer stress on plants whose growth rhythms are out of phase with rainfall distribution (141) is an important factor in the death of infected plants (106).

The range of field responses, from slow decline to sudden death, almost certainly reflects differences in interaction between inoculum density, timing of infection, crown vigor, and post-infection stress, because all provenances of *Eucalyptus marginata* tested are susceptible to a wide range of W.A. isolates of *P. cinnamomi* (99, 105). In the maritime climate near the south coast of W.A. the fungus spreads rapidly but jarrah dies slowly, whereas in the Mediterranean climate farther north spread is slower but jarrah dies more rapidly. The field tolerance of other eucalypts may be due to their greater root regeneration capacity, e.g. *E. diversicolor* (98, 99), or to superior ability to reduce water loss during dry periods, e.g. *E. calophylla* (51). The inherent tolerance of hosts such as *P. pinaster,* which thrives as a replant on jarrah dieback soils, may be overcome on more severe sites, e.g. on poorly drained soils on the Swan Coastal Plain, W.A.

Despite the great reduction of food base which follows the initial wave of infection the fungus remains active for many years (105) even after the forest cover has been destroyed and soil temperatures may reach 40°C. In spring and autumn soil temperature and soil moisture levels may be more favorable for infection in those open areas. Both natural regeneration and replants of susceptible species continue to be infected and the sites slowly convert to open woodland dominated by *E. calophylla* with a sparse ground cover of tolerant species.

Although Heather et al (58) have shown in laboratory experiments that $CaCl_2$ amendment of jarrah forest soil can reduce infection of lupin seedlings, host nutrition is apparently of little significance in the field (154), and field applications of fertilizers have had no remedial effect (56).

The east Gippsland, Vic. dieback is sporadic and associated with unusually heavy summer rainfall (5). Sapling regrowth tends to be more persistent than in W.A. despite severely rotted and shallow root systems. The differences from jarrah dieback are possibly due to better distribution of rainfall and lower drought stress.

Occasionally other *Phytophthora* spp. have been found under dying native communities in W.A. Of these *P. citricola* and *P. nicotianae* var. *parasitica* were pathogenic on *E. marginata* whereas *P. megasperma* var. *sojae* was not.

P. cinnamomi in indigenous vegetation in N.Z.—This is a situation in which climatic factors normally mask the effects of interactions between a virulent pathogen and susceptible hosts. Susceptibility to *P. cinnamomi* has been demonstrated in 3 species of *Nothofagus* (10), *Agathis australis,* and 16 other native species (111, 127). Although the fungus has been isolated from soil beneath many communities and from roots of several species the outward health of the community is usually unaffected. This may be due to high root regeneration capacity, lack of environmental stress (in a mild climate with well distributed rainfall), and the action of soil factors on reproduction of the fungus (111). It is also likely that the dense vegetation maintains soil temperatures below optimum for *P. cinnamomi,* especially at higher altitudes and latitudes. Nevertheless, decline and death have been observed in a few cases in *Agathis australis* regrowth on soils of impeded drainage. This decline was triggered by unusually prolonged and heavy rainfall in the summer of 1956, and affected 90–100 year old trees. At this age *Agathis* loses its conical form and develops a broadly branching habit similar to that in *P. radiata* shelterbelt trees as they change from the resistant to the susceptible stage. Once littleleaf symptoms had developed in the forest canopy, interception and transpiration were reduced and the sites tended to remain waterlogged. In all strata the root systems of plants that persisted were shallow, and rootlet death and replacement were observed.

The behavior of *Eucalyptus* in exotic plantations in N.Z. also supports the hypothesis that soil and climatic factors limit the effects of *P. cinnamomi* infection in this area. Although the fungus is probably a limiting factor in the establishment of seedling *E. delegatensis* on the Mamaku Plateau (10), established trees of eight highly susceptible species (including *E. marginata*) are healthy on *P. cinnamomi* infested clay soils in high rainfall districts in Northland (136). Isolates from six N.Z. native hosts were as pathogenic to *E. marginata* as 48 W.A. isolates (105).

Other *Phytophthora* species isolated from N.Z. native forests include *P. heveae, P. syringae, P. megasperma* var. *megasperma* from plants, and *P. nicotianae* var. *nicotianae* from soil.

P. cinnamomi and diseases in native vegetation of subtropical Australia. —Until recently, there had been a puzzling absence of disease reports[1] for native vegetation despite a history of serious loss in nearby pineapple (76, 77, 95, 100, 137, 150), in avocado (101), and in *Pinus* nurseries (22, 96, 97) and the presence (33) of susceptible species of *Xanthorrhoea, Banksia* (149), and *Eucalyptus* (108).

Little is known of the distribution of *P. cinnamomi* in the subtropics, or of factors controlling its dispersal and interaction with its potential hosts in native communities. It was only in 1971 following exceptionally heavy summer rain that deaths of native plants and isolation of *P. cinnamomi* were first noted[2] in the "wallum" heath, Q. (102). Although the warm and moist soil conditions of summer may favor biological control by antagonists and competitors, the normal absence of symptoms might be due to lack of stress on plants whose growth rhythms are "in-phase" with rainfall distribution. The coincidence of high rainfall and high temperatures would seem to favor infection, but soil temperatures are above the upper limits for activity of *P. cinnamomi* (28) for a considerable part of the time and might reach levels which, in wet soils, are lethal for the fungus. Soil cooling by rainfall and the duration of such cooling might be important determinants of opportunity for

[1] Recent observation indicates that *P. cinnamomi*-associated disease of native vegetation in Queensland is more widespread than previously realized (102).

[2] Photographs taken in 1971 by Pegg (102) show severe damage in *Banksia* woodland communities in the "wallum." These and 1951 photographs taken by Coaldrake (102) indicate that there has been similar damage in the past.

infection. It is also possible that where nonsusceptible grasses dominate the ground cover the effectiveness of *P. cinnamomi* inoculum is reduced by lower accessibility of host roots and by antagonistic effects.

The great floristic diversity of this zone where tropical and temperate floras overlap (23), together with the detailed documentation of the ecology and land-use of southeastern Q. (158) and northeastern N.S.W., make it particularly attractive for study of host-pathogen-environment complexes.

In the above four examples we see how environment so modifies interactions between *P. cinnamomi* and its hosts that under some conditions the reaction of inherently susceptible communities to invasion by a virulent pathogen might come to resemble that between indigenous organisms mutually adapted. Let us consider evidence that the pathogen is in fact an invader.

Indigenous or Introduced

The origin of *P. cinnamomi* is an important and much debated question. The debate stems, at least in part, from different interpretations of terms. Take *endemic* for example. The essence of authoritative definitions (1, 2, 59) is restriction of host or pathogen to one place or country. In pathology the connotation "constant presence" is added to distinguish endemic disease from epidemic and sporadic disease (Webster). Walker's (152) definition of endemic disease which includes the connotation but not the essential meaning is sometimes (e.g. 118) inappropriately applied to the pathogen. Like Robinson (128) we are of the opinion that "endemic" has dubious value if applied to plant disease i.e. to the interaction between host and pathogen.

To avoid further confusion we explain our usage of some important terms; *indigenous* (or native)—an organism natural to a place and in dynamic balance with the environment including those other organisms natural to the place; *endemic*—natural to a place, and occurring only at that place; *ubiquitous*—occurring at all possible sites in a defined area; *introduced*—having moved to a place, either by migration or by transfer; the organism may still be extending its area or may already occupy all potential habitats, but it will still be adjusting toward a balance with its environment.

A regional perspective—New Zealand.—In 1960 Newhook (91) suggested that the fungus might have been introduced by European settlers, but commented "if the fungus was introduced it certainly became remarkably well dispersed and well established." Later, following recovery of *P. cinnamomi* from healthy virgin forest on the rugged northern face of Little Barrier Island, a remote wildlife sanctuary in the Hauraki Gulf, he concluded, "*Phytophthora* spp. are so widespread and common in native forests that they should be regarded as indigenous" (92). In 1968 he reconsidered in the light of archaeological evidence on Little Barrier and of the rates of autonomous spread of *P. cinnamomi* demonstrated elsewhere (105), and concluded (94) that even occurrence on northern slopes might be accounted for by autonomous dispersal from early Maori settlements.

Robertson (127), reporting pathogenicity tests with 23 indigenous species, concluded that the limited effects of root infection on foliar symptoms and growth in many of the species, together with extensive root regeneration of inoculated plants, indicate evolution of "a certain measure of tolerance toward *P. cinnamomi* and lend support to the theory that the fungus is indigenous rather than an introduction of more recent times." If this were so, one might also expect evidence of selection within the fungus. However, comparison of N.Z. and overseas isolates provided no evidence of local strain formation (28). Podger & Newhook (111) have suggested that distribution of the fungus and its limited effects on the native flora of New Zealand can be accounted for whether *P. cinnamomi* is indigenous or introduced. The fungus might have been introduced along with kumara (sweet potato, *Ipomoea batatas*) during Maori settlement, or later by European man, and then disseminated even into remote areas by feral pigs, goats, and deer. On local evidence the problem of indigenous or exotic origin seems not to be readily soluble.

Importance of the question for Australia and a history of opinion.—Resolution of the question is urgent; whether the fungus is indigenous and widely distributed or recently introduced and still increasing its area, influences approaches to control, the need for intra-regional quarantine, and the directions which research should take.

The earliest relevant information is in Tryon's 1904 (150) account of pineapple wilt disease. There were several introductions of pineapple from as early as 1837 and doubtless opportunities for *P. cinnamomi* to enter and become widely dispersed on vegetative propagating material. If *P. cinnamomi* was indigenous and already widely distributed it is surprising that wilt was not reported to be epidemic before 1887 (150). Although the disease history to 1904 might be explained by either an introduced or an indigenous pathogen which became epidemic only on unsuitable soils and during unusually wet and cool (for Queensland) years, Lewcock (76), in 1935, reported that new-land plantings were generally free of wilt and that the fungus was disseminated in propagating material. Later Oxenham & Winks (96) warned that infestation of *Pinus* nurseries led to the spread of a potentially dangerous pathogen into virgin forest areas.

Because it affected the choice of control methods, the origins of *P. cinnamomi* were questioned as soon as the fungus was implicated in the mass dying of jarrah. Podger, Doepel & Zentmyer (109) suggested that the fungus was introduced because of (*a*) the absence of marked resistance in individual jarrah trees, (*b*) discrete boundaries between healthy and diseased forest, and (*c*) continuing spread of diseased patches. However, Bjorkman (16) reporting on a brief visit in 1966, suggested that the fungus might be native and widespread but that it seemed to require waterlogged conditions for activity and therefore caused disease symptoms only on wet sites. He suggested this might explain the patch-like occurrence and indicated that the disease might be controlled by improving soil aeration. It seems that his tour could not have included occurrences on well-drained sites. In 1968 Podger (105) advanced evidence that *P. cinnamomi* was an introduced fungus in southwestern Australia. However, Titze (148) has suggested that the fungus might have spread from niches where it was indigenous and to which until recently it had remained restricted. This seems unlikely since the flora of these places should be resistant, and no such niche has been found (106). Indeed it is difficult to envisage where they might have occurred without their drainage waters carrying inoculum into susceptible communities downslope.

In eastern Australia the situation is less clear. In some areas where the evidence is similar to that in the jarrah forest e.g. the Brisbane Ranges, Vic. (107, 160) Wilsons Promontory, Vic. (159) and east Gippsland, Vic. (73, 83) it seems almost certain that the fungus is introduced. In S. Australia Warcup & Davison (155) attribute the limited occurrences of *P. cinnamomi* in the Adelaide Hills, to introduction of infected plants from Vic. and W.A. No definite conclusions are yet possible for much of coastal N.S.W. and Q. where *P. cinnamomi* is reported to have a widespread occurrence in native forest (118) but seems to cause little damage. Nor is it possible to judge from accounts for the A.C.T. (66) or Tas. (42) whether the fungus is introduced or indigenous. In the A.C.T. the fungus is present in both diseased and healthy communities.

The introduced vs indigenous debate and the "indigenous-disturbance hypothesis."—Among Australian commentators there is a remarkable polarization of opinion on whether *P. cinnamomi* is introduced or indigenous. Those working with seriously affected communities almost unanimously consider the fungus introduced, at least to their own areas. Others consider it to be indigenous, at least in N.S.W. (45), throughout southern and eastern Australia (118), or in all its Australian occurrence (143). They argue that the native communities normally tolerate the pathogen until the balance is upset by high levels of infection in abnormally wet years, stress in the host during drought, or disturbance of the environment by man. Titze (147) and Stahl & Jehne (142) consider that physiological stress in hosts in disturbed or naturally unfavorable environments is the primary cause of death and that the role of *P. cinnamomi* is seçondary. Pratt et al (117) consider that disturbance of native communities by logging, roading, or burning is a major factor in diseases associated with *P. cinnamomi* and note that soil-water and light regimes are affected. This view of a potential pathogen existing as an innocuous soil organism together with susceptible hosts in a normally buffering environment, until the balance is disturbed by climatic shift or by man's activities, has also been expressed for *P. cinnamomi* by Woods (164) in southeastern U.S.A. Such an explanation, which might be called the "indigenous-disturbance hypothesis," has been discussed in relation to jarrah dieback (106) and rejected because it is inconsistent with the local patterns of disease development and occurrence of the fungus, and because of the strong historical and palynological evidence that the native communities are well adapted to disturbance. While it is indisputable that the kinds of disturbance listed by Pratt et al alter soil temperature and moisture regimes in the immediate environs and down the drainage lines it is difficult to explain how such effects could extend considerable distances upslope from the focus or to accept that such levels of disturbance have not been equalled by natural events in the history of the communities. Protagonists of the indigenous-disturbance hypothesis have not yet explained why so much of the disease is so recent despite the long history of severe disturbance of native communities. Nor is the postulated sensitivity of native communities to disturbance consistent with what is known of their reactions to fire and partial clearing.

Fire has long been a factor in the evolution of the flora (32, 46) and in the development of local vegetation patterns (70, 88). It may even be necessary for maintenance of general health in some communities (40, 88). Many native communities, some with dominants highly susceptible to *P. cinnamomi,* are able to re-establish rapidly following wildfires of great intensity e.g. tall forests of *E. obliqua* and *E. regnans* in Vic. in 1939, and jarrah forest in W.A. in 1961. A heath in Vic. has been described (52) where the entire biomass above ground was consumed by fire; 25 years later the stand of 25,000 kg/ha included 9 species which have been shown elsewhere (107, 160) to be highly susceptible to *P. cinnamomi.*

Over the last century vast areas of Australia's native vegetation have been disturbed by partial clearing and logging. Yet in many places healthy plants of *P. cinnamomi*-susceptible species have re-established on heath and forest lands which had been bulldozed, burned, and ploughed for agriculture or forest plantations. Despite this long history of disturbance the diseases associated with *P. cinnamomi* are relatively new events in most areas. In some places jarrah dieback has developed only recently in stands last disturbed more than 40 years ago (105). In our view it is the increased opportunity for introduction and spread of inoculum during clearing, logging and roading that is important rather than disturbance per se.

In examining the possible origins of *P. cinnamomi* and the manner of its dispersal within our region let us consider the evidence for local origin, autonomous migration, introduction by prehistoric man, and recent introduction.

Origin in or autonomous migration to the area.—Acceptable hypotheses of the origin of *P. cinnamomi* must be consistent with the histories of disease development wherever the fungus occurs. They must account for the world distribution of a relatively uniform organism unadapted for aerial dispersal.

On Australian evidence it might be argued that *P. cinnamomi* is indigenous at least in northeastern Australia. However, similar claims could be made for N.Z. and already have been made for southeast U.S.A. (164), Asia (34), and England (37). We need not consider theories of independent multiple origin as they call for too much coincidence. The last opportunity for plants to have migrated autonomously either into or out of the region across a land bridge to Asia occurred sometime in the early Tertiary. If for Australia we propose an ancient origin or entry we need to explain why *P. cinnamomi* has only recently exerted selection pressure on a complex and highly susceptible flora, "all the main elements of which were present in the Tertiary assemblages" (23).

Any hypothesis that requires co-evolution of the pathogen and highly susceptible floras is at variance with Vavilow's rule: "where both host and parasite have long been associated in their centers of origin every new and more virulent race of parasite must necessarily have eliminated most of the susceptible individuals in the local population" (75). The high proportions of susceptible species in the Brisbane Ranges and W.A., together with evidence of their evolution since the Cretaceous (23), excludes the possibility that these areas were either centers of origin of *P. cinnamomi* or within the range of autonomous dispersal from such centers within the period under discussion.

Opportunities for introduction by prehistoric man.—Colonization of Australasia by aboriginal man began some time before 30,000 B.P. (19). Since this man was a gatherer and a hunter, not a cultivator (162), he is unlikely to have introduced or spread *P. cinnamomi.* However the fungus might have spread into Polynesia from an Asiatic or intermediate center of origin in the later, complex series of human migrations (9, 25, 65, 126) which carried tuberous crops both eastward and westward in the Pacific. Although New Guinea was involved in much of this migration, *P. cinnamomi* has not been detected there (112, 133). The fungus must either have evolved in Australia after 5,000–10,000 B.P. when the last opportunity for movement to New Guinea across the Pleistocene land bridges occurred (69) or have been introduced since then by mariners. Human contacts between Asia, Central America, and the South Pacific Islands (25) could have carried *P. cinnamomi* to Polynesia as early as 500 A.D. and then to N.Z. during Maori colonization. If this were so there might have been time for wide dispersal in N.Z. and some selection toward tolerance in its flora. Unfortunately we have no information on the occurrence of *P. cinnamomi* in undisturbed native vegetation in the Pacific Islands that might help resolve this question. There is no evidence of Polynesian contact with Australia. The apparent absence of *P. cinnamomi* from New Guinea is of such importance that Shaw's survey (133) should be extended before chance introduction occurs.

Evidence for introduction in historic times.—Crandall & Gravatt (34) suggest that the Australasian occurrences are secondary. They deduce that *P.*

cinnamomi probably reached Africa in the 15th or 16th century and Europe before 1772, and entered southeastern U.S.A. prior to 1780. This last estimate is based on old accounts of losses of *Castanea* and *Gordonia* in an epidemic spreading inland one thousand miles from the probable ports of entry, Mobile, Ala. and Savannah, Ga. If *P. cinnamomi* had originated in Australia, Crandall & Gravatt's African dates are untenable. The fungus must then have become established in Europe and U.S.A. within a remarkably short time of the arrival in 1769 of Capt. James Cook, R.N. on the Australian coast. Had Cook been the vector and Westminister known the potential of this scourge, delivery would surely have been made at Boston, Mass., rather than at Savannah, Ga.

Prior to Cook there appear to have been no European contacts in which plants and soil could have been removed from Australia. It has been suggested that Spanish mariners probably played an important role in the distribution of *P. cinnamomi* (34). It is known that they introduced many plants to Manila and Guam (144), but there seems to be no record of Spanish contact with Australia or N.Z. If *P. cinnamomi* was introduced before settlement by the British it must have entered during the few chance contacts by Dutch and English mariners or during Macassan contact. Although Macassans visited the northern coastline of Australia from at least the early 16th until the 19th century they had little direct impact beyond the mangrove fringe and introduced few plants (89).

Possible centers of origin of P. cinnamomi.—If the fungus is not recently evolved, Vavilov's rule excludes as centers of origin areas in which highly susceptible species are indigenous, e.g. Europe, Australia, South Africa, Central America, Chile, and both southeastern and northwestern U.S.A. We might also exclude N.Z. because the advent of *P. cinnamomi* in the U.S.A. probably pre-dates the first contact between the two countries. In some respects the nomination of southeast Asia (34) offers the best explanation. However if the fungus is of ancient occurrence there it is surprising that in the 5,000 years or more of crop culture in Asia it did not become sufficiently widespread to be transferred to New Guinea during human migrations. It is tempting to postulate a more limited natural occurrence in Asia and isolation from the opportunity of dispersal until historic times. If so, a useful approach in determining its origin might be to apply Vavilov's rule to such pan-Asiatic genera as *Vaccinium* and *Rhododendron,* which are rich in endemic species, extend into New Guinea (139), and include tolerant and susceptible species. If on the other hand *P. cinnamomi* evolved recently, and its relative uniformity indicates that this might be so, it will be difficult to trace its origin. It seems that the only hypothesis which (*a*) accounts for the distribution and activity of *P. cinnamomi,* (*b*) is not at variance with Vavilov's rule, and (*c*) is consistent with the historical record elsewhere, is that the fungus has been introduced, probably no earlier than the late eighteenth century. For Australia the indigenous disturbance hypothesis fails each of these tests. For N.Z. it fails the historical test and must be rejected.

This invasion of our region by *P. cinnamomi* affords perhaps the first opportunity for plant pathologists to observe from its early stages the development of a major epidemic in natural communities which have not been affected by shifting cultivation. It also presents a challenge to develop methods to control the epidemic and to forecast its outcome.

Control

Although there are intra-regional quarantines in our region which have been invoked for other pathogens, e.g., *Dothistroma pini* in N.Z., no action seems yet to have been taken for *P. cinnamomi*. Attempts at chemical eradication in small infected areas of forest have been unsuccessful (6, 159). The use of fire and mechanical methods to reduce inoculum by removing alternative hosts is being investigated (72). Containment by ditching (72, 159) and by clearing, cultivation, and replacement with resistant species (64) is being evaluated. More extensively affected areas in W.A. (72) and Vic. (84, 160) are being replanted with resistant species.

It has been suggested that the varied effects of *P. cinnamomi* in our region may be associated with differences in natural levels of microbial antagonism. In preliminary investigations (87, 99, 113) such factors have been demonstrated in vitro. On this basis it has been suggested (113), for Australian native forests, that "the possibility of controlling *P. cinnamomi* by natural or induced antibiosis is attractive." While this approach might provide effective protection in nurseries the establishment and maintenance of suitable associations on a forest scale is likely to be difficult. Many basidiomycetes compete for mycorrhizal sites in *Eucalyptus* (31) and *Pinus* (81, 165). More than one mycorrhiza may occur on a single root (82) and they may be differentially affected by temperature (86, 146), soil aeration (53), and stage of root development (82). Fungal symbionts are not all equally effective against *P. cinnamomi* even for different isolates of the one species, e.g. *Boletus* (*Suillus*) *luteus* (vide 85, 113). Breakdown of mycorrhizal protection may occur during prolonged rainfall (11, 80, 90), by bacterial lysis (43), and by rupture of the mantle during root elongation (82). In hosts in which *P. cinnamomi* invades progressively larger roots even occasional failure of mycorrhizal protection may have serious consequences. A great deal of research, or good fortune, may be needed before long-lasting control of *P. cinnamomi* on a field scale can be achieved by manipulation of biological balances. If useful antagonists are found they will need to be even more successful at invading new territory than *P. cinnamomi*.

It seems to us that except in nurseries the epidemic has long passed the point where *P. cinnamomi* can be eradicated or even contained within its present boundaries. The rate at which the problem will increase depends largely upon how widely the need for intraregional quarantine is recognized. Those who hold to the "indigenous-disturbance hypothesis" must inevitably underestimate the importance of inoculum transfer and are therefore unlikely to implement the necessary control measures. They are also likely to advocate that site-disturbance must be avoided in native communities, many of which depend for their health and renewal upon periodic disturbance such as burning.

Prognosis

There can be no doubt that the epidemic in native vegetation in Australia will cause even greater destruction than we now see. While quarantine could delay the entry of *P. cinnamomi* to New Guinea and reduce its spread in Australia it seems inevitable that the entire region is open to the chance of infestation. The chances of the fungus surviving and the damage that it might cause in the wide range of environments available are difficult to forecast. Attempts to define disease hazards are being made in W.A. by Havel (57) who incorporates the approaches of both Hepting (61) and Campbell & Copeland (24). It is to be expected that predictions may be upset by climatic change (60), by unusual meterological events, or by irrigation.

We expect that *P. cinnamomi* will become more important in horticulture, forest nurseries, and shelterbelts. In plantations *Pinus* may suffer little loss, except perhaps on poorly drained sites where mounding is often necessary. Establishment losses in *Pseudotsuga* and *Larix* may be so high in some areas as to limit the usefulness of these species. In N.Z., *Agathis* regrowth that developed after extensive milling last century may be reaching an age of greatly increased susceptibility. Although it causes little damage in extensive undisturbed forests of *Nothofagus* in N.Z., it is possible that as logging intensifies, *P. cinnamomi* may affect regeneration particularly on warmer aspects. In Australia the fungus will seriously reduce productivity in *Monocalyptus* which provides more than 40% of the log supply (3) for Australia's forest industry; jarrah, Australia's second most important timber species faces commercial extinction.

The epidemic poses a serious problem for conservation in dry sclerophyll heath, woodland, and forest communities, not only for the flora including the wild flowers (110) for which these communities are famous, but also for many dependent species in the fauna. Many of Australia's national parks and reserves are in jeopardy; it is a matter of the utmost urgency that they be protected from *P. cinnamomi* infestation for as long as possible. Without protection all the highly susceptible elements of vulnerable communities will be eliminated. Ultimately a new state of fluctuating biological balance will develop in each habitat and become self-adjusting for relatively slow evolutionary and climatic changes. At that distant time pathologists, who for some reason had no access to the history of the epidemic and who applied the same tests that we did, would find that a much improverished flora and a more variable *P. cinnamomi* were so well adapted that the fungus might then seem to be an indigenous pathogen.

Acknowledgement

Our grateful thanks are due to many colleagues who have permitted us to refer to manuscripts and data yet unpublished and who have answered a questionnaire. We are particularly grateful to R. N. Allen and S. R. Pennycook for helpful criticism during preparation of the manuscript.

LITERATURE CITED

1. Abercrombie, M., Hickman, C. J., Johnson, M. L. 1962. *A Dictionary of Biology*. Chicago: Aldine Publ. Co. 254 pp.
2. Ainsworth, G. C. 1961. *Ainsworth & Bisby's Dictionary of the Fungi*. 5th Ed., London: Commonw. Mycol. Inst. 517 pp.
3. Anonymous unpublished records. Commonwealth Forestry & Timber Bureau Canberra, Australia
4. Anonymous unpublished records. Queensland Dep. Forestry, Brisbane, Australia
5. Anonymous unpublished records. Victorian Forests Commission, Melbourne, Australia
6. Ashton, D. H. Personal communication
7. Atkinson, I. A. E. 1959. Soils and the growth of *Pinus radiata* at Cornwallis, Auckland. *N.Z. J. Sci.* 2:443–72
8. Ballard, R. 1970. The phosphate status of the soils of Riverhead Forest in relation to growth of radiata pine, *N.Z. J. Forest.* 15: 88–99
9. Barrau, J. 1963. Introduction. In *Plants and the Migrations of Pacific Peoples,* ed. J. Barrau, 1–6 Honolulu: Bishop Museum Press. 136 pp.
10. Bassett, C. Personal communication
11. Bassett, C., Will, G. M. 1964. Soil sterilisation trials in two forest nurseries. *N.Z. J. Forest.* 9:50–58
12. Batini, F. Personal communication
13. Batini, F. 1970. *Preliminary report on the MIADS programme. Perth: Forest. Dep. W. Aust.,* typed 9 pp.
14. Batini, F., Podger, F. D. 1969. Shelterbelt mortalities on the Swan Coastal Plain. *Aust. Forest. Res.* 3:39–45
15. Bertus, A. L. 1968. *Phytophthora cinnamomi* Rands on conifers in New South Wales. *Agric. Gaz. N.S.W.* 79:751–54
16. Bjorkman, E. 1966. *On the jarrah* (Eucalyptus marginata) *dieback in Western Australia—its cause and control.* Canberra: *Commonw. For. Timb. Bur.,* typed 3 pp.
17. Blakely, W. F. 1965. *A key to the Eucalypts.* 3rd Ed. Canberra Comm. Aust. Forest. Timb. Bur. 359 pp.
18. Bowen, G. D. 1969. See Harley 1969 p. 313
19. Bowler, J. M., Jones, R., Allen, H., Thorne, A. G. 1970. Pleistocene human remains from Australia: a living site and human cremation from Lake Mungo, western New South Wales. *World Archaeol.* 2:39–60
20. Bowling, P. J., McLeod, D. E. 1967. A note on the presence of *Armillaria* in second growth eucalypt stands in southern Tasmania. *Aust. Forest. Res.* 3:38–40
21. Brasier, C. M. 1971. Induction of sexual reproduction in single A^2 isolates of *Phytophthora* species by *Trichoderma viride. Nature New Biol.* 231:283
22. Brown, B. N. Personal communication
23. Burbidge, N. T. 1960. The phytogeography of the Australian region. *Aust. J. Bot.* 8:75–211
24. Campbell, W. A., Copeland, O. L. 1954. *Littleleaf disease of shortleaf and loblolly pines, U.S. Dept. Ag. Circ. 940.* 41 pp.
25. Carter, G. F. 1963. *Movement of people and ideas across the Pacific. Plants and the Migrations of Pacific Peoples,* ed. J. Barrau, 7–22. Honolulu: Bishop Museum Press. 136 pp.
26. Chee, K. H. 1964. *Experimental studies on* Phytophthora cinnamomi *Rands.* M.Sc. thesis. Univ. Auckland, N.Z. 73 pp.
27. Chee, K. H., Newhook, F. J. 1965a. Improved methods for use in studies on *Phytophthora cinnamomi* Rands and other *Phytophthora* species. *N.Z. J. Agr. Res.* 8:88–95
28. Chee, K. H., Newhook, F. J. 1965b. Variability in *Phytophthora cinnamomi* Rands. *N.Z. J. Agr. Res.* 8:95–103
29. Chee, K. H., Newhook, F. J. 1965c. Variability in sexual re-

production of *Phytophthora cinnamomi* Rands. *N.Z. J. Agr. Res.* 8:947–50

30. Chee, K. H., Newhook, F. J. 1966. Relationship of microorganisms to sporulation of *Phytophthora cinnamomi* Rands. *N.Z. J. Agr. Res.* 9:32–43
31. Chilvers, G. A. 1968. Some distinctive types of eucalypt mycorrhiza. *Aust. J. Bot.* 16:49–70
32. Churchill, D. M. 1968. The distribution and prehistory of *Eucalyptus diversicolor* F. Muell., *E. marginata* Donn. ex Sm. and *E. calophylla* R. Br. in relation to rainfall. *Aust. J. Bot.* 16:125–51
33. Coaldrake, J. E. 1961. *The ecosystem of the coastal lowlands ("wallum") of southern Queensland. C.S.I.R.O. Aust. Bull. No. 283.* 138 pp.
34. Crandall, B. S., Gravatt, G. F. 1967. The distribution of *Phytophthora cinnamomi. Ceiba.* 13:43–53, 57–78
35. Dance, H. M., Newhook, F. J., Cole, J. S. A new method for the recovery of *Phytophthora* spp. from the soil. *N.Z. J. Bot.* In press
36. Davison, E. 1970. *Phytophthora cinnamomi* in South Australia. *Aust. Plant Dis. Recorder* 22:18
37. Day, W. R. 1939. Root rot of sweet chestnut and beech caused by species of *Phytophthora* II. Inoculation experiments and methods of control. *Forestry* 13:36–58
38. Dingley, J. M. 1969. *Records of plant diseases in New Zealand. N.Z. D.S.I.R. Bull. 192.* 298 pp.
39. Doepel, R. F. Personal communication
40. Ellis, R. C. 1964. Dieback of alpine ash in north-eastern Tasmania. *Aust. Forest.* 28:75–90
41. Eckert, J. W., Tsao, P. H. 1962. A selective antibiotic medium for isolation of *Phytophthora* and *Pythium* from plant roots. *Phytopathology* 52:771–77
42. Felton, K. C. 1970. *Eucalypt diebacks in Tasmania. Rep. Res. Working Gp. 7. Hobart: Tas. Forest. Comm.* mimeo. 8 pp.
43. Foster, R. C., Marks, G. C. 1967. Observations on the mycorrhizas of forest trees II. The rhizosphere of *Pinus radiata* D. Don. *Aust. J. Biol. Sci.* 20:915–26
44. Fraser, L. R. 1956. *Phytophthora cinnamomi* attacking native plants. *Aust. Plant Dis. Recorder.* 8:30
45. Fraser, Lilian R., Walker, J., Bertus, A. L., Broadbent, Patricia. Personal communication
46. Gardner, C. A. 1957. The fire factor in relation to the vegetation of Western Australia. *West. Aust. Naturalist* 5:166–73
47. Geard, I. D. Personal communication
48. Gilmour, J. W. 1954. *Armillaria mellea* (Vahl) Sacc. in New Zealand forests. *Forest. Res. Notes* 1:1–40
49. Gilmour, J. W. 1960. The importance of climatic factors in forest mycology *N.Z. J. Forest.* 8:250–60
50. Gilmour, J. W. 1967. Distribution and significance of the needle blight of pines caused by *Dothistroma pini* in New Zealand. *Plant Dis. Reptr.* 51:727–30
51. Grieve, B. J. 1956. Studies in the water relations of plants. I. Transpiration of Western Australian (Swan Plain) sclerophylls. *J. Roy. Soc. West. Aust.* 40:15–30
52. Groves, R. H., Specht, R. L. 1965. Growth of heath vegetation I. Annual growth curves of two heath ecosystems in Australia. *Aust. J. Bot.* 13:261–80
53. Harley, J. H. 1969. *The Biology of Mycorrhiza.* 2nd Ed. London: Leonard Hill. 334 pp.
54. Hartigan, D. T. 1964. Some observations on the effect of *Phytophthora* in Monterey pine. *Forest & Timber* 2:2 pp.
55. Hartigan, D. T. Personal communication
56. Hatch, A. B. Personal communication
57. Havel, J. Personal communication
58. Heather, W. A., Bellany, G., Pratt, B. H. 1971. *The effect of mineral nutrition of soil on the infection of roots of* Lupinus angustifolius *by* Phytophthora cinnamomi. Presented at Aust. Pl. Path. Conf., Hobart
59. Henderson, I. F., Henderson, W. D. 1960. *A Dictionary of Scientific terms* 7th Ed., ed. J. H.

Kenneth, London: Oliver & Boyd. 595 pp.
60. Hepting, G. H. 1963. Climate and forest diseases. *Ann. Rev. Phytopathol.* 1:31–50
61. Hepting, G. H. 1964. *Appraisal and prediction of international forest disease hazards. Symposium on internationally dangerous forest diseases and insects,* Oxford, 20–29 July, 1964. Documents—Vol. 1., 14 pp. Rome: F.A.O./I.U.F.R.O.
62. Hepting, G. H., Newhook, F. J. 1962. A pine disease in New Zealand resembling littleleaf. *Plant Dis. Reptr.* 46:570–71
63. Hood, I. A. Personal communication
64. Hopkins, E. R. Personal communication
65. Howard, A. 1967. Polynesian origins and migrations, a review of two centuries of speculation and theory. In *Polynesian Culture History*. eds. G. A. Highland et al, p. 85. Honolulu: Bishop Museum Press. 594 pp.
66. Jehne, W. 1971. The occurrence of *Phytophthora cinnamomi* and tree dieback in the A.C.T. *Aust. Forest. Res.* 5:47–52
67. Jenkins, P. T. 1970. Root rot of azalea and rhododendron. *Aust. Plant Dis. Recorder*. 22:34
68. Jenkins, P. T., Wauchope, D. G. Personal communication
69. Jennings, J. N. 1971. Sea level changes and land links. In *Aboriginal Man and Environment in Australia,* eds. D. J. Mulvaney, J. Golson, 1–13. Canberra: Aust. Nat. Univ. Press. 389 pp.
70. Jones, R. 1968. The geographical background to the arrival of man in Australia and Tasmania. *Archaeol. & Phys. Anthrop. Oceania*. 3:186–215
71. Jordan, R., Heseltine, A., Rainey, G. Personal communication
72. Kimber, P. C. Personal communication
73. Lee, H. M., 1962a. Death of Eucalyptus spp. in East Gippsland. *Forest. Comm. Vic. Tech. Paper* 8:14–18
74. Lee, H. M. 1962b. A disorder of *Pinus radiata* Don. in shelterbelts in southern Victoria. *Forest. Comm. Vic. Tech. Paper* 8: 18–25
75. Leppik, E. E. 1970. Gene centers of plants as sources of disease resistance. *Ann. Rev. Phytopathol.* 8:323–44
76. Lewcock, H. K. 1935a. Pineapple wilt disease and its control. *Queensl. Agr. J.* 43:9–17
77. Lewcock, H. K. 1935b. Top rot of pineapples and its control. *Queensl. Agr. J.* 43:145–49
78. McAlonan, M. J. 1970. *An undescribed* Phytophthora *sp. recovered from beneath stands of* Pinus radiata. M.Sc. thesis Univ. Auckland, N.Z. 63 pp.
79. McGechan, J. K. 1966. *Phytophthora cinnamomi* responsible for a root rot of grape vines. *Aust. J. Sci.* 28:354
80. Marks, G. C. 1965a. The pathological histology of root rot associated with late damping off in *Pinus lambertiana. Aust. Forest.* 29:30–38
81. Marks, G. C. 1965b. The classification and distribution of the mycorrhizas of *Pinus radiata. Aust. Forest.* 29:238–51
82. Marks, G. C., Foster, R. C. 1967. Succession of mycorrhizal associations on individual roots of radiata pine. *Aust. Forest.* 31: 193–201
83. Marks, G. C., Kassaby, F. Y. 1971. *Dieback in the coastal hardwoods in eastern Victoria associated with soil infection by* Phytophthora cinnamomi. Presented at Aust. Pl. Path. Conf., Hobart
84. Marks, G. C., Kassaby, F. Y. Personal communication
85. Marx, D. H. 1969. The influence of ectotrophic mycorrhizal fungi on the resistance of pine roots to pathogenic infections I. Antagonism of mycorrhizal fungi to root pathogenic fungi and soil bacteria. *Phytopathology* 59:153–63
86. Marx, D. H., Bryan, W. C., Davey, C. B. 1970. Influence of temperature on aseptic synthesis of ectomycorrhizae by *Thelephora terrestris* and *Pisolithus tinctorius* on loblolly pine. *Forest. Sci.* 16:424–31
87. Mills, J. F. 1970. *Basidiomycete antibiosis to* Phytophthora cinnamomi *Rands.* B.Sc. For. (Hons.) thesis. Aust. Nat. Univ. 45 pp.
88. Mount, A. B. 1969. *Eucalypt ecol-*

ogy as related to fire. Proc. 1969 Tall Timbers Fire Ecol. Conf. 75–108 Talahassee, Florida

89. Mulvaney, D. J. 1969. *The Prehistory of Australia.* London: Thames & Hudson, 276 pp.
90. Newhook, F. J. 1959. The association of *Phytophthora* spp. with mortality of *Pinus radiata* and other conifers I. Symptoms and epidemiology in shelterbelts. *N.Z. J. Agr. Res.* 2:808–43
91. Newhook, F. J. 1960. Climate and soil type in relation to *Phytophthora* attack on pine trees. *Proc. N.Z. Ecol. Soc.* 7:14–15
92. Newhook, F. J. 1964: *Forest disease situation, Australasia. Symposium on Internationally Dangerous Forest Diseases and Insects, Oxford, 20–29 July 1964.* Documents, Vol. 1, 15 pp. Rome: F.A.O./I.U.F.R.O.
93. Newhook, F. J. 1968. *Report on visit to Western Australian Jarrah Dieback area 28 April-4 May 1968. Canberra: Commonw. For. Timb. Bur.* mimeo. 3 pp.
94. Newhook, F. J. 1970. *Phytophthora cinnamomi* in New Zealand. In *Root Diseases and Soil-borne Pathogens.* eds. T. A. Toussoun, R. V. Bega, P. E. Nelson, 173–76 Berkeley: Univ. Calif. Press. 252 pp.
95. Oxenham, B. L. 1957. Diseases of the pineapple. *Queensl. Agr. J.* 83:13–25
96. Oxenham, B. L., Winks, B. L. 1963a. *Phytophthora* root rot of *Pinus* in Queensland. *Queensl. J. Agr. Sci.* 20:355–66
97. Oxenham, B. L., Winks, B. L. 1963b. *Pinus* damping-off investigations in southern Queensland. *Queensl. J. Agr. Sci.* 20:455–61
98. Palzer, C. R. 1968. *Is* Phytophthora cinnamomi *a threat to karri forests?* Presented at 5th Conf. Inst. For. Perth, Aust.
99. Palzer, C. R. Personal communication
100. Pegg, K. G. 1969. Pineapple top rot control. *Queensl. Agr. J.* 95: 458–59
101. Pegg, K. G. 1970. Root rot of the Avocado. *Queensl. Agr. J.* 96: 412–14
102. Pegg, K. G. Personal communication
103. Pegg, K. G., Alcorn, J. L. Personal communication
104. Podger, F. D. 1965. *Report on jarrah dieback. Dieback Seminar, Harvey. Canberra: Commonw. Forest. Timb. Bur.* mimeo. 20 pp.
105. Podger, F. D. 1968. *Aetiology of jarrah dieback.* M.Sc. Forest. thesis, Univ. Melbourne, Aust. 292 pp.
106. Podger, F. D. *Phytophthora cinnamomi* a cause of lethal disease in indigenous plant communities in Western Australia. *Phytopathology.* In press
107. Podger, F. D., Ashton, D. H. 1970. *Phytophthora cinnamomi* in dying vegetation on the Brisbane Ranges, Victoria. *Aust. Forest. Res.* 4:33
108. Podger, F. D., Batini, F. 1971. Susceptibility to *Phytophthora cinnamomi* root rot of thirty-six species of *Eucalyptus. Aust. Forest. Res.* 5:9–20
109. Podger, F. D., Doepel, R. F., Zentmyer, G. A. 1965. Association of *Phytophthora cinnamomi* with a disease of *Eucalyptus marginata* forest in Western Australia. *Plant Dis. Reptr.* 49: 943–47
110. Podger, F. D., Palzer, C. R., Batini, F. E. 1967. *Phytophthora cinnamomi* in the jarrah forests of Western Australia. *Commonw. Phytopathol. News* (4): 1–2
111. Podger, F. D., Newhook, F. J. 1971. *Phytophthora cinnamomi* in indigenous plant communities in New Zealand. *N.Z. J. Bot.* 9: 625–38
112. Powell, J. N. 1971. The history of agriculture in the New Guinea highlands. *Search* 1:199–200
113. Pratt, B. H. 1971. Isolation of basidiomycetes from Australian eucalypt forest and assessment of their antagonism to *Phytophthora cinnamomi. Trans. Brit. Mycol. Soc.* 56:243–51
114. Pratt, B. H., Heather, W. A., Bolland, L., Brown, B. *Phytophthora cinnamomi* detected in natural vegetation in Queensland. *Aust. Plant Dis. Recorder.* In press
115. Pratt, B. H., Heather, W. A., Sedgely, J. H., Shepherd, C. J. 1971. *Phytophthora cinnamomi*

in natural vegetation in the A.C.T. *Aust. Plant Dis. Recorder*. In press

116. Pratt, B. H., Heather, W. A., Shepherd, C. J. 1971a. *Recovery and identification of* Phytophthora cinnamomi *from soil by lupin baiting*. Presented at Aust. Plant Pathol. Conf., Hobart.
117. Pratt, B. H., Heather, W. A., Shepherd, C. J. 1971b. *Association of environmental factors with* Phytophthora cinnamomi *disease in native forest*. Presented at Aust. Plant Pathol. Conf., Hobart
118. Pratt, B. H., Heather, W. A., Shepherd, C. J. 1971c. *Distribution of* Phytophthora cinnamomi *in Australia, with particular reference to forest plants*. Presented at Aust. Plant Pathol. Conf., Hobart
119. Pratt, B. H., Wrigley, J. W. 1970. New host records for *Phytophthora cinnamomi* Rands. *Aust. Plant Dis. Recorder*. 22: 34–36
120. Prior, K. W. 1961. Shelterbelt mortality in South Canterbury, Otago and Southland. *N.Z. J. Forest*. 8:498–507
121. Pryor, L. D., Johnson, L. A. S. 1971. *A classification of the Eucalypts*. Canberra: Aust. Nat. Univ. Press. 102 pp.
122. Purnell, H. M. 1960. Studies of the family Proteaceae I. Anatomy and morphology of the roots of some Victorian species. *Aust. J. Bot*. 8:38–50
123. Purss, G. S. Personal communication
124. Rands, R. D. Streepkanker van kaneel, veroozaarkt door *Phytophthora cinnamomi* n.sp. *Meded. Inst. voor Plantenziekten Dept. Lanb. Nijv. en Handel*. 54:1–53 (author's English version deposited Univ. Calif., & Commw. Mycol. Inst.)
125. Rawlings, G. B. 1957. The pathology of *Pinus radiata* as an exotic. *N.Z. For. Ser. Tech. Pap. 20*. 16 pp.
126. Robbins, R. G. 1963. *Correlations of plant patterns and population migration into the Australian New Guinea highlands. Plants and the Migrations of Pacific Peoples*. ed. J. Barrau, 45–59. Honolulu: Bishop Museum Press. 136 pp.
127. Robertson, G. I. 1970. Susceptibility of exotic and indigenous trees and shrubs to *Phytophthora cinnamomi* Rands. *N.Z. J. Agr. Res*. 13:297–307
128. Robinson, R. A. 1969. Disease resistance terminology. *Rev. Appl. Mycol*. 48:593–606
129. Ross, E. W. 1970. Sand pine root rot—pathogen: *Clitocybe tabescens*. *J. Forest*. 68:156–58
130. Ross, E. W., Marx, D. H. Personal communication
131. Roy, S. C. 1968. *The influence of nitrogenous compounds and animal excreta on infection by* Phytophthora cinnamomi *Rands*. M.Sc. thesis, Univ. Auckland, N.Z. 89 pp.
132. Roy, S. C., Newhook, F. J. 1970. The influence of cattle excreta on sporulation of *Phytophthora cinnamomi* Rands. *N.Z. J. Agr. Res*. 13:308–14
133. Shaw, Dorothy, E. Personal communication
134. Shepherd, C. J., Pratt, B. H., Smith, A. 1971. *Temperature relations of* Phytophthora cinnamomi *Rands*. Presented at Aust. Pl. Path. Conf., Hobart
135. Shepherd, C. J. Personal communication
136. Shirley, J. W. 1968. *A survey of* Eucalyptus *species in Northland, N.Z., with respect to the incidence of* Phytophthora cinnamomi. *B.Sc. project rep. Univ. Auckland, N.Z.* 11 pp.
137. Simmonds, J. H. 1929. Diseases of Pineapples. *Queensl. Agr. J*. 32: 398–405
138. Simmonds, J. H. 1966. *Host index of plant diseases in Queensland*. Brisbane: *Queensl*. Dep. Primary Ind. 111 pp.
139. Sleumer, H. 1966. *Ericaceae. Flora Malesiana series 1, 6* ed. C. G. G. J. van Steenis. 649–914 Groningen: Noordhoff
140. Smith, H. C. Personal communication
141. Specht, R. L., Rayson, P. 1957. Dark Island heath (Ninety-mile Plain, South Australia) I. Definition of the ecosystem. *Aust. J. Bot*. 5:52–85
142. Stahl, W., Jehne, W. 1971. *Ecolog-*

ical factors in the occurrence of tree dieback in the Australian Capital Territory and Tasmania. Presented at Aust. Plant Pathol. Conf., Hobart

143. Stahl, W., Jehne, W. Personal communication
144. Stone, B. C. 1971. America's asiatic flora: the plants of Guam. *Am. Sci.* 59:308–19
145. Sutherland, C. F., Newhook, F. J., Levy, J. 1959. The association of *Phytophthora* spp. with mortality of *Pinus radiata* and other conifers. II. Influence of soil drainage on disease. *N.Z. J. Agr. Res.* 2:844–58
146. Theodorou, C., Bowen, G. D. 1971. Influence of temperature on the mycorrhizal associations of *Pinus radiata* D. Don. *Aust. J. Bot.* 19:13–20
147. Titze, J. F. 1971. Phytophthora cinnamomi—*a disease or a complex of stresses?* Presented at Aust. Plant Pathol. Conf., Hobart
148. Titze, J. F. Personal communication
149. Titzc, J. F., Palzer, C. R. 1969. Host list of *Phytophthora cinnamomi* Rands with special reference to Western Australia. *Dep. Nat. Dev. Forest Timb. Bur. Tech. Note 1.* 58 pp. + *1970 Addenda.* 7 pp.
150. Tryon, H. The pineapple disease. *Queensl. Agr. J.* 15:477–84
151. Wade, G. C. 1962. Deaths of *Cupressus* and *Pinus* trees on King Island. *Inst. Forest. Aust., Newsletter* 3:7–9
152. Walker, J. C. 1969. *Plant Pathology.* New York: McGraw Hill. 819 pp.
153. Wallace, W. R. 1969. *Progress report on jarrah dieback research in Western Australia. Perth: Forest. Dep. W. Aust.* mimeo. 153 pp.
154. Wallace, W. R., Hatch, A. B. 1953. *Crown deterioration in the northern jarrah forests.* Perth: For. Dep. W. Aust., typed 46 pp.
155. Warcup, J. H., Davison, Elaine. Personal communication
156. Waring, H. D. 1950. *Report on a brief investigation into the death of jarrah* (Eucalyptus marginata) *in the Dwellingup Division, Western Australia. Canberra: Commonw. Forest. Timb. Bur.*, typed 35 pp.
157. Waterhouse, G. 1970. Taxonomy in Phytophthora. *Phytopathology* 60:1141–43
158. Webb, L. J., Tracey, J. G., Williams, W. T., Lance, G. N. 1971. Prediction of agricultural potential from intact forest vegetation. *J. Appl. Ecol.* 8:99–121
159. Weste, Gretna, M. Personal communication
160. Weste, G. M., Taylor, P. 1971. The invasion of native forest by *Phytophthora cinnamomi.* I. Brisbane Ranges, Victoria. *Aust. J. Bot.* 19:281–94
161. Weston, G. C. 1956. Fertiliser trials in unthrifty pine plantations at Riverhead Forest. *N.Z. J. Forest.* 7:35–46
162. White, J. P. 1971. New Guinea and Australian Prehistory. In *The Neolithic Problem. Aboriginal Man and Environment in Australia,* eds. D. J. Mulvaney, J. Golson, 182–195. Canberra: Aust. Nat. Univ. Press. 389 pp.
163. Will, G. M. 1965. Increased phosphorus uptake by radiata pine in Riverhead Forest following superphosphate applications. *N.Z. J. Forest.* 10:33–42
164. Woods, F. W. 1963. Disease as a factor in the evolution of forest composition. *J. Forest.* 51:871–73
165. Zak, B., Marx, D. H. 1964. Isolation of mycorrhizal fungi from roots of individual slash pines. *Forest. Sci.* 10:214–22
166. Zentmyer, G. A. 1968. *Report on jarrah dieback problem in Western Australia. Canberra: Commonw. Forest. Timb. Bur.,* mimeo. 8 pp.
167. Zentmyer, G. A., Mircetich, S. M. 1966. Saprophytism and persistence in soil by *Phytophthora cinnamomi. Phytopathology* 56: 710–12
168. Zentmyer, G. A., Paulus, A. O., Burns, R. M. 1967. Avocado root rot. *Calif. Agr. Exp. Sta. Circ. 511* 16 pp.

SOIL FUNGISTASIS—A REAPPRAISAL 3554

Andrew G. Watson and Eugene J. Ford
Resource Management & Research, Berkeley, California, and Compania Agricola de Rio Tinta, La Lima, Honduras, Central America

INTRODUCTION

Soil fungistasis was reviewed by Lockwood in this journal in 1964 (70). Our reappraisal is not intended as a bibliography of the many studies dealing directly or indirectly with the phenomenon, but as an integrating interpretation of the corpus of information, including the major research published since 1964. Bibliographies of fungistasis can be found in the reviews of Dobbs et al (35), Lockwood (70), Jackson (57, 58), and Weltzien (115).

We hope our interpretation will reconcile the different hypotheses that have been used to explain the phenomenon. We think these differences have arisen from too specialized experimental designs and interpretations aimed at consolidating a particular hypothesis; whereas research aimed at integrating the diverse observations made by different workers has been lacking. The brevity of the cited literature will enable us to present with greater clarity the basic concepts and our interpretations of the evidence regarding fungistasis without obscuring this view with references to every supporting or refuting experimental paper.

Definition

The term soil fungistasis describes the phenomena whereby: (*a*) viable fungal propagules not under the influence of endogenous or constitutive dormancy (107) do not germinate in soil in conditions of temperature and moisture favorable for germination, or (*b*) growth of fungal hyphae is retarded or terminated by conditions of the soil environment other than temperature and moisture. Fungistasis is a dynamic phenomenon and can be further classified into three stages: induction and maintenance of, and release from fungistasis. Soil spore fungistasis has been studied in greater detail than soil hyphal fungistasis. Though the term "fungistasis" was originally proposed by Dobbs & Hinson (34), Dobbs and his co-workers (29) have subsequently used the term "mycostasis" on the grounds that the latter term is etymologically purebred, whereas "fungistasis" is a Latin-Greek half-breed. The English language has mixed origins, and as both "fungus" and "stasis" are English words, we can see no valid reason for divorcing the two elements of the English word "fungistasis," especially as it was thus originally used. It has been widely accepted, and has an obviously acceptable relative in the term "fungicide."

THEORY

The ability of research workers in the physical sciences to direct their efforts by theory and deductive reasoning has resulted in impressive advances in knowledge in these areas. Biologists have rarely used this approach because life does not generally lend itself to an exact analysis or the establishment of comprehensive formulae. We will, however, attempt to use evolutionary and ecological theory to deduce the characteristics of fungistasis, and then examine the evidence. These deductions are being made with the advantages of hindsight, true, but that is a criticism only of our powers of creative imagination in that we did not deduce these predictions from theory before experimental evidence was available; it is not a valid criticism of the objective use of deductive reasoning in this or any other case.

All environments of all organisms contain, in a general sense, stimulators and inhibitors of those organisms.—These stimulators and inhibitors of the metabolic processes delimit an organism's life. Many environmental factors (e.g., temperature, water content, ionic concentration, soluble substrate concentration, and heavy metal concentration) may be stimulatory over parts of their concentration or energy ranges and inhibitory over other parts of their ranges. Moreover, quantitatively identical environmental factors may be stimulatory to one process in an organism's life cycle, while they may be inhibitory to another process.

The evolutionary success of any organism depends on its ability to strike a balance between the stimulatory and inhibitory factors of its environment.—The exogenous control of an organism's metabolism is the evolutionary result of an organism's ability to balance the stimulatory and inhibitory factors of its environment to its selective advantage. The "balance of nature" is a general and unconscious expression of this concept. We consider that response to fungistasis is a vital process in the life cycle of any soil fungus. Evolutionary value can justifiably be ascribed to any population characteristic of an organism, but the determination of the mechanism of selection of that characteristic is more speculative. The soil environment is one of intense competition for temporally and spatially discontinuous substrates. The vast majority of soil fungi have to survive inter-substrate periods; most often this survival is in the form of dormant propagules. During this period, spontaneous germination of propagules or continued hyphal growth would be highly disadvantageous to survival; hence, the selection for fungistasis. Some dormancy may be endogenous, but exogenously imposed fungistasis must be controlled by an environmental balance of stimulatory and inhibitory factors. As selection of the phenomenon of fungistasis has taken place in an environment containing both stimulators and inhibitors, fungistasis could not possibly have evolved to be controlled solely by stimulators or inhibitors. Stimulatory factors stimulate vegetative growth of a fungus and thus release from fungistasis; inhibitory factors retard vegetative growth and thus induce or maintain fungistasis.

As a result of the severe competition for substrates, soil fungi must have evolved to occupy diverse nutrient environments. The broadest classification of these has been described by Garrett (43), and one ramification has been the evolution of parasitism. Closer analysis would reveal even more specificity of substrate habit. The many papers dealing with degradative fungal successions in soil (23, 24, 108) and other habitats (21, 100) indicate that a fungus species is able to succeed only in a particular place in the degradative succession of a particular substrate within a certain range of other environmental conditions. This specificity implies that the exogenous control of fungistasis must also be specific and involve a balance of different concentrations of either many stimulators or many inhibitors or both. Indeed, stimulators for some fungal species in respect to fungistasis may be inhibitory for other species.

Stimulators

The factors that release fungistasis must be mobile and ephemeral in soil and closely associated with the microenvironment wherein the given fungus has a good chance for completing a successful life cycle. We can deduce their mobile and ephemeral nature because any material, in order to have been selected as a stimulator, must be soluble or volatile and must not accumulate in soil. Release from fungistasis at a greater distance from the favorable substrate than the fungus can traverse by hyphal growth or zoospore taxis, will have been selected against—implying that these mobile and ephemeral stimulators must be constituents of microhabitats surrounding islands of substrate in the soil. It is not possible to deduce whether or not stimulators will themselves be specific in action; we can only deduce that the balance of stimulators and inhibitors should be specific. It would, however, be surprising if all the specificity of response is imparted by the inhibitor component of the balance and none by the stimulator.

Inhibitors

The deduced characteristics of inhibitors are slightly different from those of stimulators because they must provide a uniform balancing background for the same organic substrates, and hence similar concentrations of the same stimulators, in many different soil types. If the effective inhibitor concentration for a particular fungus varies greatly from one soil type to another, it would allow spontaneous germination in the absence of a suitable substrate in some soils and prevent release of fungistasis in other soils, even when a suitable substrate is present in the microhabitat. Obviously, a material varying in concentration from one soil to the next will not be selected for by a fungus species inhabiting a wide range of soils (which is the situation with most soil fungi) as its inhibitor. The inhibitors should be mobile and could conceivably be of biotic or abiotic origin. Certain mineral compounds of iron, aluminum, and silica are common to all soils, and a soluble fraction of these could have been selected as an inhibitor. It seems doubtful, however, that this would be the case because (*a*) the equilibrium concentrations of soluble minerals vary considerably with the soil pH and other factors; (*b*) the specificity of fungistatic release would have to be completely controlled by the stimulatory side of the balance. Inhibitors of biological origin would appear to be more logical candidates for a role in the control of fungistasis. Nearly all soils exhibit some degree of biological activity and there would be ample opportunity for selection of specificity among the infinite array of metabolic products produced in soil (73). Furthermore, this specificity could be directly related to the activity of a particular organism or association of organisms, indicating that an advantageous time in a succession has been reached. Like the stimulators, these inhibitors must be mobile and ephemeral. Biological production that results in accumulation in any soil would render such materials unacceptable for selection as inhibitors in the fungistatic balance, since opportunities for exploitation of a suitable substrate would be lost in such environments. At the same time, the concentration of inhibitors should not fall to a point where spontaneous germination is allowed in the absence of a suitable substrate. One way to ensure uniform concentrations of inhibitors is to have overproduction with respect to the effective inhibitor concentrations—even in soils with relatively low absolute metabolic activity—and a rapid loss of the inhibitors through physical or biological degradation. The law of mass action, operating in the degradative processes, will tend to ensure more uniform concentrations of inhibitors in different soils in that the higher the concentration, the greater will be the rate of degradation. Biological degradation may offer an additional advantage regarding the maintenance of uniform concentrations in different soils. In situations where there is a high rate of production of inhibitors, high overall soil biological activity and a high rate of biological degradation might also be expected. The characteristics of the inhibitors deduced here imply that these materials are acting at the microhabitat level, and are probably produced very close to the fungus structures under their influence.

Summary of Theory

From a theoretical point of view, we predict that the dynamic phenomenon of soil fungistasis—including the stages of induction, maintenance, and release of fungistasis—should be controlled by balances of stimulatory and inhibitory compounds. We would expect a degree of specificity of these balances for different species of fungi and for the three dynamic stages of fungistasis for the same fungus. The stimulatory materials should be mobile in soil, ephemeral, and closely associated in time and space with particular substrate microhabitats. They will probably be specific for different fungi. The inhibitory compounds should be mobile, ephemeral, and probably of biological origin, although abiotic materials may be involved in some situations. They should be active at the microhabitat level, associated with substrate sources, the affected fungus, or both, and will very probably be specific in action. The phenomenon of soil fungistasis should be general, but it is likely that there are a multitude of complex mechanisms which will be elucidated only after more exacting soil microbial research.

EVIDENCE—THE PHENOMENON

Occurrence

Soil fungistasis was first defined by Dobbs & Hinson in 1953 (24) and its widespread nature has been thoroughly attested to by the numerous observations cited in the reviews of Dobbs et al (35), Lockwood (70), and Jackson (58). Since the last review, there have been many more observations of the phenomenon, and many papers dealing with various aspects of the work.

The phenomenon has been described in soils from tropical (47, 48, 56, 106), sub-tropical (49), temperate (13, 33, 35, 59, 114), and sub-arctic (18) areas. While a wide range of soil types exhibit the phenomenon (35, 67), various workers report that the level of fungistasis differs among soil types. Payen (85) found a heath soil to be highly fungistatic, and a peat soil not fungistatic towards *F. culmorum* macroconidia on buried slides, whereas Chinn (13) found a peat soil to be the most fungistatic of those tested toward spores of *Cochliobolus sativus* on buried slides. These differences are what we would expect to find if the specificity of the fungistatic phenomenon, which we have predicted, is indeed one of its characteristics. A unitary explanation of fungistasis would not be supported by these different results. Although Schüepp & Frei found no correlation between fungistasis and soil depth (99), there seems to be general agreement among other investigators that fungistasis is more pronounced in surface than in sub-soil layers of any one profile (12, 25, 47, 56). Soil fungistasis towards *Mucor ramannianus* was, however, more pronounced in the humus and mineral layers of a beech-oak forest soil than in the litter layer (32).

Fungal Structures Affected and Stages of Fungistasis

Most work has been done on the maintenance of and release from fungistasis in asexual spores, conidia or sporangia, directly in soil or in conditions supposedly reproducing the fungistatic effect of soil. More recent work has been carried out on (*a*) the induction of fungistasis as seen in the formation of chlamydospores from macroconidia of *Fusarium solani* (40–42) (*b*) the retardation of hyphal growth of *Pythium* and *Thanatephorus* (113, 114) (*c*) the response of zoospores of *Phytophthora parasitica* (74), and chlamydospores of *Thielaviopsis basicola* (50) and *Phytophthora cinnamomi* (77) to soil fungistasis, and (*d*) the maintenance and release of fungistasis in sclerotia of *Sclerotium cepivorum* (4, 14–16, 59), *Sclerotium gladioli* (14), and *Verticillium albo-atrum* (38, 97). Many papers dealing with the rhizosphere or spermosphere phenomenon also cover the release from fungistasis, although direct reference to this release is seldom made. In this respect, release from fungistasis of *Pythium* oospores (6), *Fusarium* chlamydospores (17), *Phytophthora* chlamydospores (77, 78), and propagules of many other fungi has been demonstrated (95, 98).

Species Affected

At least 116 species of fungi representing 53 genera were recorded in 1964 as being affected by soil fungistasis (70). Undoubtedly others have been added to the list since then. Many reports are of fungi that are not soil inhabitants, and for which the selection of propagules susceptible to soil fungistasis and able to survive in soil would have no evolutionary value. Other reports are of structures of soil-inhabiting fungi other than inter-substrate survival stages of fungi that again, would not derive evolutionary advantage from response to soil fungistasis. Evidence gained from studies of the behavior of these fungi and propagules in soil cannot be used to determine the characteristics of biologically valid soil fungistasis. Their behavior in an alien environment is incidental to their biology. By analogy, a flightless bird projected through the air will have aerodynamic properties, but information gained from studies of this are unlikely to throw much light upon mechanics of highly evolved bird flight.

We appeal to future workers in the area of soil fungistasis to work with structures of fungi that represent biologically effective inter-substrate soil survival stages.

Origin

The term "microbial mycostasis" was introduced in 1965 by Dobbs & Gash (31) to describe fungistasis of biological origin. This becomes "microbial fungistasis" under the terminology used here. Most studies have been on this type of fungistasis. The characteristics of microbial fungistasis that lead us to conclude that it is of microbial origin have been thoroughly described (70); simply stated, factors that reduce soil biological activity reduce the level of microbial fungistasis. Dobbs (27) distinguished between microbial fungistasis that he described as "heat-and-sugar-sensitive," and residual fungistasis that he described as "heat-and-sugar-resistant." Residual fungistasis is of nonbiological origin, and was first described by Dobbs & Bywater (30). Since then, its presence has been confirmed in a number of soils (27, 31, 47). The different types of fungistasis will be discussed under "Mechanisms."

Seasonal Variation

The fungistatic effect in 12 forest soils, assayed by the cellophane method using *Mucor ramannianus,* showed marked seasonal variation, reaching a maximum in summer and a minimum in winter (30, 32, 35). The concentrations of reducing sugars in the same soils were also found to follow a seasonal variation reaching a maximum in winter and a minimum in summer (25, 33). A similar seasonal variation in inhibitory levels of extracts from nursery soils toward *Pythium* and *Thanatephorus,* with maximum inhibition in the late summer, has been recorded (114).

EVIDENCE—THE MECHANISMS

Direct and Indirect Methods Using Soil

The methods used to study fungistasis have generally been satisfactory for demonstrating the existence of the phenomenon, but quite inadequate for elucidating the controlling mechanisms. Direct methods used include: (*a*) the recovery of fungal propagules from soil surfaces using glass slides, cellulose fiber, agar discs, and plastic films, (*b*) recovery of fungal propagules from within the soil mass using glass fibers, fiberglass tape, nylon mesh, flotation, or smear techniques. Indirect methods used include the observation of spores on transparent water-permeable materials such as cellulose film, cellulose acetate membranes, or various agar media placed on, in, or above soil.

References to most of the methods used can be found in Lockwood's review (70). Methods developed since then include those dealing with assays for volatile inhibitory compounds released from soil. A soil-emanation technique in which fungistasis was demonstrated in agar discs held above live soil has been described (52, 53). Fungistasis in water-saturated soil blocks enclosed in tubes over water or solutions of heavy metals has also been reported (5, 63). A direct observational technique has been described by Ko (60), and further refinements of this method should yield valid information. Fluorescent brighteners have been used as an aid for direct observation of fungistasis in *Phytophthora* (109) and nematophagous fungi (39).

Criticisms of Methods Used

Objections to the use of nonsoil fungi in the fungistasis assay have been stated above. In addition, the methods of culture and preparation of the inoculum may cause fungal propagules to behave abnormally with respect to fungistasis.

Most studies aimed at elucidating the mechanisms of soil fungistasis have used distilled water in preparing the fungal inoculum and as the nonnutrient control solution in determining the nutritional requirements for germination. We have reservations regarding the possible effects of distilled water on fungal propagules.

Distilled water is a solution not normally encountered by soil fungi; it totally lacks dissolved compounds of carbon nutrition, minerals, and osmotic factors. Its use as a wash prior to determination of nutrient requirements for germination apparently gained impetus from the work of Lin (65). She found that nutrients were contaminants on the surface of spores washed from tube cultures, affecting subsequent nutritional studies on spore germination. Spores washed with distilled water or collected by a suction process would not germinate in distilled water but would germinate with the addition of carbon nutrition. Unfortunately, it was not possible to determine from these experiments whether the effect of carbon nutrition was to stimulate germination by satisfying a nutritional requirement, or by modifying the physical properties of distilled water.

As early as 1901, Duggar (37), studying the germination of fungal spores, discovered a possible "distilled water inhibition" of germination for some of the fungi studied. Spores of *Aspergillus flavus, Sterigmatocystis nigra,* and *Penicillium glaucum* did not germinate in distilled water, but did germinate in bean decoction or a solution of mineral salts plus sucrose. In a mineral salt solution, or a sucrose solution alone, there was 25% germination. The reduced germination of spores observed in sucrose alone, compared with sucrose plus mineral salts, indicates that the failure of spores to germinate in distilled water was not solely due to lack of carbon nutrition. More recent work has shown that the addition of distilled water to normally growing hyphae of several fungi caused inhibition, distortion, and in severe cases, bursting of the extreme tips of the hyphae (84, 89–91). Additional evidence gained from studies of *Candida utilis* suggested that immersion of cells in distilled water removed amino acids rapidly from the "expendable pool" of these materials, and greatly affected their physiology (19).

These results show that distilled water greatly affects the physiology of fungal cells, and indicate that caution should be used in interpreting results where distilled water is used to prepare inoculum, and to serve as the nonnutrient control. Certainly distilled water environments are not comparable to those of the soil where free water is held in micropores and contains a variety of dissolved materials imparting osmotic potential and other properties to the solution.

Fungal propagules produced in axenic culture on artificial media may have different properties from those produced in their natural environment. Spores of parasitic fungi grown on their natural host have been shown to germinate better than spores produced on artificial media (36). Differences in pathogenicity of *Fusarium roseum* have been recorded for macroconidia produced on different media (86). Furthermore, there is evidence to suggest that naturally produced soil-fungus propagules may be associated with certain microbial populations (20). A fungal propagule produced in the soil, or a hypha growing through the soil, will exhibit behavior that will depend partly upon its past micro-environment (through its effect on the internal constitution of fungus structure) and partly upon the past micro-environmental modifications brought about by the fungus. The degree to which the behavior of naturally produced fungus structures in the soil differs from artificially produced structures introduced into the soil is largely unknown.

All methods, both direct and indirect, have involved removing soil from the field and placing it in various container systems in the laboratory. Burges (9), pointed out that when soil is broken up, it seems to lose its fungistatic activity. It is not known if this effect is through redistribution and release of nutrients through loss of effective concentrations of inhibitors (especially volatile compounds), or both, or some other mechanism. The observations by Yarwood (118) on the effect of soil tillage on plant growth suggest far-reaching effects of physical disturbance on soil ecology. A period of equilibration in containers may restore the natural in situ properties of soil. Definitive work on the time period, volume of soil, and type of container required for this restoration is lacking.

The introduction of cellophane, cellulose acetate, nylon mesh, agar, or glass into soil as carrying agents for the fungus being assayed for fungistasis obviously creates artificial environments. The relatively large, uniform surfaces of materials possessing different properties with respect to permeability, thermal characteristics, and surface chemistry from the rest of the soil, will have surrounding micro-environments different from those of natural soil. In addition, the nutritive qualities of some of these materials allow increased microbial activity at their surfaces (67). Lingappa & Lockwood (67) correctly questioned whether fungistasis seen in indirect methods was generated on the surface of the assay medium.

Extraction of Biotic Inhibitors from Soil

Waksman (112) stated that we can realistically expect to find any naturally occurring organic compound in soil and the list of compounds isolated from and identified in soil (73) bears out this statement. Some are potential inhibitors of fungi (55, 101, 104). The extraction of different inhibitory compounds from soil, and their separation and identification, are somewhat academic with respect to determination of the fungistatic mechanism. The significant question to be asked is, "What is the concentration of the effective inhibitors in critical microhabitats in the soil, in this case in the immediate microhabitat of the soil fungus subject to fungistasis?"

Bulk extractions of a soil mass by aqueous solutions will remove certain fractions of all the water-soluble components from the microhabitats, some fungistatic and others nonfungistatic, represented in that soil. The final equilibrium concentrations of these components will be further affected by their stability in the water extract and the degree to which they are absorbed or otherwise removed during any preparative procedures. The final concentrations of inhibitors in the extract, unfortunately, tells us nothing about distribution, concentration, or biological effect within different microhabitats in the soil, except that the concentration in the extract is certainly much lower than that in certain microhabitats in the soil.

Because we expect the fungistatic balance to be somewhat specific, it is not surprising that Lockwood (70) reported very limited evidence to support the existence of inhibitory materials in extracts from soil masses. By analogy, the existence and identity of antibiotics produced by soil organisms in culture was known for many years before production in natural soil habitats was detected (105, 116, 117). In the case of biotic inhibitors involved in fungistasis, we do not yet know the identity of the compounds involved, so that positive detection is even more difficult.

Nevertheless, Lockwood (70) concluded in his review that convincing evidence of inhibition in sterile extracts was presented by Dobbs et al (25, 35). Since then, other examples of sterile-soil extracts containing inhibitors of biotic origin have been cited. Leachates collected from tension plate lysimeters placed 12 inches below the surface of a sandy nursery soil were found to affect hyphal growth of *Pythium ultimum* and *Thanatephorus praticolus* causing strong inhibition or mycolysis (114). Unlike the inhibitory extracts of Dobbs et al (25, 35), the inhibitors were heat stable and could be concentrated, but nonetheless were of biological origin. King & Coley-Smith (59) expressed soil moisture at 2 tons per square inch and detected a thermolabile fungistatic factor which suppressed germination of sclerotia of *Sclerotium cepivorum*. The unstable nature of inhibitory fungistatic factors has been recorded in many papers (26, 29, 31, 35), and Burges (9) has suggested that volatile compounds might be involved. A fungistatic condition was induced in soil held over water and enclosed in tubes (15, 63). This fungistatic condition was prevented if the soil was held over silver nitrate or mercuric perchlorate solutions, leading the authors to conclude that fungistasis in this situation was being induced by volatile hydrocarbons. Subsequently, Hora & Baker (52) and Hora et al (53) induced a fungistatic condition in agar discs, which were held on sterile glass slides in covered petri dishes with 50 g of field soil without direct contact between the agar discs and soil.

Additional evidence of inhibitory fungistatic compounds present in soils has been gained from studies of chlamydospore-inducing factors. Compounds that induce chlamydospore formation by *Fusarium solani* f. sp. *phaseoli* have been extracted from several soils (3, 41, 42, 64). Although chemical characterizations have not been completed, it is clear that more than one chemical entity is involved, and that specificity of response for different clones of the fungus exists (40). The chlamydospore-inducing chemicals are heat stable and can be isolated from field soil. Cultures of common soil-inhabiting bacteria belonging to the genera *Arthrobacter, Bacillus,* and *Chromobacter* produced compounds of similar biological activities (41, 88). Chlamydospore formation in the presence of partially purified inducing compounds from soil was dependent on a balance between these compounds and the level of carbon nutrition in the system (42). Chlamydospore induction in soil thus seems to be controlled by a balance of inhibitory and stimulatory factors, showing some specificity, very similar to the mechanism proposed for other stages of fungistasis.

Extraction of Abiotic Factors from Soil

McFarlane (72) demonstrated that a high pH in soil solutions prevented germination of *Plasmodiophora brassicae* and was responsible for the failure of clubroot to become a serious problem in soils with a pH of greater than 7.4. Other fungi have similarly shown greater susceptibility to fungistasis at higher pH values (99), while *Trichoderma koningi* (99) and *Mucor ramannianus* (66), on the other hand, were more affected by fungistasis at lower pH values (44). It was not determined whether the pH was directly involved in causing fungistasis in these cases.

Residual fungistasis, first reported by Dobbs & Bywater (30), confirmed by Dobbs et al (35) and Dobbs & Gash (31), and now known to be of widespread occurrence (28), is of abiotic origin. Washing with dilute hydrochloric acid removed the inhibition and both calcium carbonate and iron compounds were found to contribute to residual fungistasis (31).

It seems clear that inhibitors of both biotic and abiotic origin, including pH, contribute to inducing and maintaining the fungistatic condition in soil. The diversity of responses and types of compounds reported as being involved supports our contention and that of others (26, 31) that many probably complex mechanisms are responsible for the generally observed phenomenon of fungistasis.

Extraction of Stimulators from Soil

The release from many types of fungistasis by stimulators or nutrients was well documented in Lockwood's review (70). He commented on the conflicting results obtained by different methods, nutrients, and fungi. We would expect to find this pattern of response if the stimulator component of the fungistatic balance imparts some of the specificity required for evolutionary success of the release from fungistasis. There are as many potential candidates for roles as stimulators in the fungistatic balance in soil as there are potential inhibitors. Like inhibitors, these compounds are distributed in varying concentrations in many diverse microhabitats throughout the soil mass, and the significant question to be asked is, "What is the concentration of the effective stimulators in the critical microhabitats where fungistasis is being released?"

If we consider just one such microhabitat, that represented by the plant roots and rhizospheres, we find that amino acids, soluble sugars, organic acids, nucleotides, flavanones, enzymes, vitamins, and other compounds have been identified (95, 96). Although the release from fungistasis by the rhizosphere effect has been well documented (98), determination of the concentrations of specific compounds that actually effect this release is largely lacking. Many substances have been used to duplicate this release, including plant residues, extracts, and exudates, specific sugars, amino acids, and mixtures, and mineral salts and inorganic nitrogen compounds (70). These substances have often been added to the soil in arbitary concentrations, and the subsequent release from fungistasis interpreted as being effected by the same mechanism that occurs in nature.

We consider this approach to be invalid. The application to soil of an arbitrary concentration of an inhibitor, known to be present in soil, to simulate the induction or maintenance of fungistasis would be an analogous situation.

It seems to us that much attention has been given to the isolation and characterization of inhibitors involved in the fungistatic balance, while little has been given to the specific effective stimulator (or nutrient) concentrations in the micro-environment. At the same time, important interpretations have been based on the success or failure to isolate inhibitors satisfying arbitrary requirements (61, 103), while stimulators have not been subjected to the same search or requirements.

Fortunately, some observations of specific stimulators of fungus propagules under fungistasis have recently been made. The timely work of Coley-Smith and his associates (14–16, 59) has shown that the specific release from fungistasis of sclerotia of *Sclerotium cepivorum* in live soil is by volatile stimulators associated with *Allium* roots. These materials were identified as volatile alkyl sulphides which were produced when alkyl cysteine sulphoxides, originating from *Allium* roots, were metabolized by soil bacteria (14). A similar specificity of release from fungistasis has been observed for sclerotia of *Sclerotium gladioli,* stimulated to germinate by members of the *Iridaceae* (14), and for microsclerotia of *Verticillium albo-atrum* (38). Menzies & Gilbert (76) and Owens et al (80) investigated the volatile compounds given off by plant residues and found several low-molecular-weight alcohols and aldehydes that markedly stimulated germination of soil fungi. Materials stimulating germination of *Phytophthora parasitica* chlamydospores when added to field soil were shown to have no effect in in vitro axenic situations, and vice versa (110). Hawthorne (50) and Tsao & Hawthorne (111) were unable to stimulate germination of the nutritionally independent *Thielaviopsis basicola* by addition of certain sugars, including glucose, to natural soil, whereas complex materials such as carrot juice or alfalfa meal stimulated germination.

These demonstrations of a degree of specificity in respect to the release from fungistasis in natural soil hopefully indicate a trend towards a critical analysis of the stimulatory components of microenvironments and their roles in maintaining and releasing from fungistasis. Successful investigations of either the inhibitors or the stimulators cannot be carried out without due consideration of both.

Model Systems

To gain evidence in support of the "nutrient sink" hypothesis for the mechanism of fungistasis, Ko & Lockwood (61) exposed spores on Millipore filters to continuous leaching with glass-distilled water or 0.01 M phosphate buffer at rates of 10–30 ml/hr. This system sets up a steep gradient for all materials (except phosphates) diffusing from spores to an equilibrium concentration of zero. In soil we would expect that gradients are far shallower, having different values for different materials, and not equilibrating to such low levels. The claim is unfounded that the validity of the leaching system as a model for soil fungistasis is indicated by the correlation between the pattern of spore germination occurring in the leaching system and on natural soil. The production of similar results by different methods does not necessarily indicate a duplication of mechanisms involved. However, this method does add further evidence that the nutrient status of spores, both endogenous and exogenous, affects their germination in vitro and their response to soil fungistasis. A similar intense leaching process was used with chlamydospores of *Thielaviopsis basicola* and it was found that when chlamydospores were removed from this system, or from prolonged incubation on natural soil, they germinated in deionized water (51, 111).

Though not proposed as model systems for soil fungistasis, staling and self-inhibition of fungus cultures (11, 83, 84, 93, 94), and mutual inhibition of germinating spores in vitro (45, 66, 71, 81), have some of the characteristics of soil fungistasis (83). The degree of inhibition is affected by the nutrient level of the system (46, 66). Robinson & Park (93) showed that volatile compounds were involved in inhibition of spore germination in culture and subsequently some of these sporostatic compounds were identified, including acetaldehyde, propionaldehyde, n-butyraldehyde, n-propanol, isobutanol, ethylacetate, iso-butyl acetate, and acetone (92). Acetaldehyde was identified as a volatile product of *Trichoderma* spp. (22). We consider that it is possible that self or mutual inhibition may be involved in some cases of soil fungistasis, and that the identification of some of the materials involved in this mechanism in cultures will assist in the verification of the operation of this mechanism in soil.

Summary of Methods

The study of microbiological mechanisms in soil environments is at present necessarily accomplished by imperfect methods and model systems that mostly provide correlative evidence. Our critical analysis of the methods used is not intended to negate evidence gained from methods other than perfect, rather it is intended to indicate the limitations of the evidence thus far available, and to motivate the use of methods that yield more valid and direct evidence. We suggest that the ideal methods would involve the remote sensing of qualitative and quantitative factors of the micro-environment of a soil fungus during the different dynamic stages of fungistasis, together with remote manipulation of selected factors. "A man on the moon by 1970!"—"A man in the soil sporosphere by 1980?"

INTERPRETATIONS AND HYPOTHESES PROPOSED TO EXPLAIN FUNGISTASIS

Prior to Dobbs & Hinson's work in 1953 (34) it was generally assumed by most soil biologists that fungal spores did not germinate in soil owing to lack of nutrients, oxygen, or perhaps excess of CO_2. After 1953, though the original paper clearly stated that the fungistatic factors were counterbalanced by glucose (34), the emphasis in the interpretation was placed on the inhibitors. The hypothesis proposed by Dobbs & Hinson was that there exists in the soil a widespread fungistatic factor of microbial origin. As the general nature of the phenomenon became apparent, it was often assumed that only one mechanism was involved; the search for an inhibitory material as widespread in soil as the phenomenon was quite naturally unsuccessful (66, 70).

Park (82, 83) suggested that inhibitors produced in staled cultures might play some role in soil fungistasis. Recent work (92, 93) shows that volatile inhibitors are involved in the phenomenon of staling, and these results continue to parallel those obtained from soil (5, 52, 53, 63) with respect to fungistatic factors.

Lockwood originally thought that inhibitors present in the soil were involved in fungistasis (69), but later, through consistent failures to prepare sterile inhibitory extracts from soils, he shifted to the hypothesis that fungistasis is a localized result of microbial activation in the sporosphere, probably caused by antibiotic production (66, 67). Spores of several fungi were shown to release nutrients when added to soil, demonstrating that they could act as micro-substrates (66). It was also shown that bacteria and *Streptomyces* were able to use these nutrients in axenic systems, and that their activity inhibited spore germination of five test fungi (66). In addition, the same workers demonstrated that when peptone was added to the top layer of double-agar-layer plates in contact with live soil, the bottom layer became inhibitory in 3 days; without peptone 7–8 days were required. The presence of a diffusible inhibitor of spore germination produced at the surface of the assay medium in contact with soil was demonstrated (67). This hypothesis was later abandoned, largely because of the difficulty of explaining prolonged fungistasis, since spore exudates would eventually be exhausted. The possibility that some metabolic change occurs in the spore as a result of the localized microbial activity has been suggested (70). Further evidence on this point was provided by work in which exposure of *Thielaviopsis basicola* chlamydospores to soil inhibited the initiation of the germination process, possibly by changing their permeability (51, 111). There are many attractive features of Lockwood's first hypothesis; it was accepted that stimulators (nutrients) released fungistasis, inhibitory compounds would be confined to the microhabitat of the structure under the influence of fungistasis, and this mechanism would readily allow specificity of effect for different fungi. We believe that some cases of soil fungistasis, controlled by activity of microorganisms in the sporosphere may yet be found.

Lockwood and his co-workers proposed a second hypothesis in 1964—that fungistasis is caused solely by the depletion of nutrients in the micro-environment of the fungus as the result of microbial activity (54, 61, 62, 70, 102, 103). A corollary was attached to the simple nutrient-deficiency hypothesis to explain the failure of nutritionally independent spores to germinate in soil. This is the "nutritional sink" idea (61), which states that the intense nutrient diffusion gradient away from the spore set up by microbial activity in soil deprives the spore of nutrients essential for germination. Lockwood's two hypotheses are to some extent contradictory: in the first hypothesis, nutrients in the assay materials placed in soil stimulated antibiotic production, while, in the second hypothesis, these materials were so nutrient-free that they deprived the spores of essential nutrients for germination (26).

The reasons for proposing the nutrient-deficiency-and-sink hypothesis were the failure to find inhibitory soil extracts and the desire to "more simply explain many puzzling aspects of fungistasis" with a unitary explanation (61). Neither reason seems to us sufficiently compelling to justify the establishment of an hypothesis that denies the existence of fungal inhibitors in soil micro-environments.

The evidence cited (54, 61, 62, 70, 102, 103) to support this hypothesis is, we think, the result of too simple an analysis of a balance situation in terms of only one side of that balance. Consider a beam balance with two sides, *A* and *B*. If we examine only side *A* and ignore *B*, the beam will descend or ascend on side *A* according to the amount of weight on *A*. We could explain all movement of the beam in terms of the weight on *A*. If we examine only side *B* of the same balance we would conclude that all movement of the beam was due to the weight on *B*. Both explanations are correct in part, but the complete explanation is that the beam descends at *A* when the weight on *A* overcomes the weight on *B*, or ascends at *A* when the weight on *B* overcomes the weight on *A*. In any environment there are stimulators and inhibitors, and the soil is no exception. One side of the balance is represented by the concentrations of stimulators, and the other side by the concentration of inhibitors. Germination of spores in soil occurs when the concentration of stimulators overcomes the concentration of inhibitors (8). Furthermore, the concentrations of inhibitors and stimulators (nutrients) in soils are probably positively correlated, and the greater the biological activity and nutrient depletion rate, the greater the rate of production of inhibitors. Hence, it is easy to understand how the solely nutrient deficiency explanation for fungistasis arose through consideration of only one side of the balance.

The consideration of a balance leads us to the final hypothesis which we can attribute originally to Dobbs and his co-workers, but which has since been added to by other workers (2, 5, 14, 26, 27, 31, 42, 57, 79, 113, 114). We state this hypothesis: that the general but complex phenomenon of soil fungistasis is caused by the presence in soil microenvironments of complex inhibitors of biotic or abiotic origin effective at the low concentrations of stimulators (mostly nutrients) present in the soil. The control of fungistasis is effected by specific balances of inhibitor and stimulator concentrations.

Recent detection and isolation of volatile inhibitors (5, 14, 15, 27, 52, 53, 59) and inhibiting factors of abiotic origin (26, 27, 31) have strengthened the evidence for the presence of inhibitors in soil microhabitats, removing one of the chief objections, though a negative one, to the balance hypothesis. The work of Coley-Smith and his co-workers in isolating and identifying a specific nonnutritive stimulator for *Sclerotium cepivorum* is strong evidence that specificity of the fungistatic balance exists in nature (14).

BIOLOGICAL AND ECONOMIC SIGNIFICANCE

The biological significance of the phenomenon has been covered in the section entitled "Theory." Briefly, any soil fungus (indeed any soil organism) has to survive inter-substrate periods, usually in the form of dormant propagules, to complete a successful life cycle; much of that dormancy is under exogenous control and is termed "fungistasis" if conditions of moisture and temperature would otherwise permit germination.

The economic significance is potentially enormous. Induction and maintenance of and release from fungistasis are three dynamic processes a soil-borne plant-pathogenic fungus must successfully go through to maintain or increase its inoculum density between encounters with susceptible hosts. Any procedure, chemical or cultural, that prevents the induction and maintenance, stimulates inopportune release, or prevents opportune release of soil fungistasis with respect to a particular plant pathogen, will tend to reduce the inoculum density of that pathogen in soil (1, 7, 10, 45, 72, 75, 87, 119). The converse is also true.

CONCLUSIONS

Soil fungistasis has been shown to be general in natural soils, and we have differentiated three dynamic stages: induction, maintenance, and release. A theoretical consideration of the phenomenon enables us to deduce certain characteristics regarding both the phenomenon and the mechanism, which are largely borne out by the available evidence. Soil fungistasis is, we believe, controlled by a complex balance of stimulators and inhibitors in soil microenvironments. Both exogenous and endogenous factors, acting concurrently, consecutively, or both, may be involved in causing fungistasis. The stimulators are mostly of biotic origin, and may act as nutrients, while inhibitors are of both biotic and abiotic origin. The characterized inhibitors are water soluble and may be volatile. Both the stimulators and the biotic inhibitors are mobile and ephemeral, and are difficult to isolate and identify. The proposed abiotic inhibitory factors include calcium carbonate, iron, and pH. Nothing is known of the effective concentrations of any of these materials in soil microhabitats. There is some evidence for a specificity of the balance affecting the dynamic stages of fungistasis. We encourage future research to be directed towards elucidating the effective concentrations of specific stimulators and inhibitors of the fungistatic balance in soil micro-environments.

LITERATURE CITED

1. Adams, P. B., Papavizas, G. C. 1969. Survival of root-infecting fungi in soil. X. Sensitivity of propagules of *Thielaviopsis basicola* to soil fungistasis in natural alfalfa-amended soil. *Phytopathology* 59:135–38.
2. Agnihotri, V. P., Vaartaja, O. 1967. Effects of amendments, soil moisture contents, and temperatures on germination of *Pythium* sporangia under the influence of soil mycostasis. *Phytopathology* 57:1116–20
3. Alexander, J. V., Bourret, J. A., Gold, A. H., Snyder, W. C. 1966. Induction of chlamydospore formation by *Fusarium solani* in sterile soil extracts. *Phytopathology* 56:353–54
4. Allen, J. D., Young, J. M. 1968. Soil fungistasis and *Sclerotium cepivorum* Berk. *Plant Soil* 29: 479–80
5. Balis, C., Kouyeas, V. 1968. Volatile inhibitors involved in soil mycostasis. *Ann. Inst. Phytopathol. Benaki,* Athens 8 n.s.: 145–49
6. Barton, R. 1957. Germination of oospores of *Pythium mamillatum* in response to exudates from living seedlings. *Nature* 180:613–14
7. Barton, R. 1958. Occurrence and establishment of *Pythium* in soils. *Trans. Brit. Mycol. Soc.* 41:207–22
8. Brown, W. 1922. On the germination and growth of various fungi at various temperatures and in various concentrations of oxygen and of carbon dioxide. *Ann. Bot.* 36:257–83
9. Burges, A. 1960. Discussion. In *The Ecology of Soil Fungi,* ed. D. Parkinson, J. S. Waid, 180. Liverpool: Liverpool Univ. Press 324 pp.
10. Burke, D. W. 1965. *Fusarium* root rot of beans and behavior of the pathogen in different soils. *Phytopathology* 55:1122–26
11. Carlile, M. J., Sellin, M. A. 1963. An endogenous inhibition of spore germination in fungi. *Trans. Brit. Mycol. Soc.* 46:15–18
12. Chacko, C. I., Lockwood, J. L. 1966. A quantitative method for assaying soil fungistasis. *Phytopathology* 56:576–77
13. Chinn, S. H. F. 1967. Differences in fungistasis in some Saskatchewan soils with special reference to *Cochiliobolus sativus. Phytopathology* 57:224–26
14. Coley-Smith, J. R., Cooke, R. C. 1971. Survival and germination of fungal sclerotia. *Ann. Rev. Phytopathol.* 9:65–92
15. Coley-Smith, J. R., Hickman, C. S. 1957. Stimulation of sclerotium germination in *Sclerotium cepivorum. Nature* 180:445
16. Coley-Smith, J. R., Holt, R. W. 1966. The effect of species of *Allium* on germination in soil of sclerotia of *Sclerotium cepivorum* Berk. *Ann. Appl. Biol.* 58:273–78
17. Cook, R. J., Snyder, W. C. 1965. Influence of host exudates on growth and survival of germlings of *Fusarium solani* f. *phaseoli* in soil. *Phytopathology* 55:1021–25
18. Cooke, R. C. 1967. Mycostasis in soils recently exposed by a retreating ice cap. *Nature* 213: 295–96
19. Cowie, D. B., McClure, F. T. 1959. Metabolic pools and the synthesis of macromolecules. *Biochim. Biophys. Acta.* 31: 236–45
20. Curl, E. A., Hansen, J. D. 1964. The microflora of natural sclerotia of *Sclerotium rolfsii* and some effects upon the pathogen. *Plant Dis. Reptr.* 48:446–50
21. D'Aeth, H. R. X. 1939. A survey of interactions between fungi. *Biol. Rev.* 14:105–31
22. Dennis, C., Webster, J. 1971. Antagonistic properties of species-groups of *Trichoderma.* II. Production of volatile antibiotics. *Trans. Brit. Mycol. Soc.* 57:41–48
23. Dix, N. J. 1964. Colonization and decay of bean roots. *Trans. Brit. Mycol. Soc.* 47:285–92
24. Dix, N. J. 1967. Mycostasis and root exudation: factors influencing the colonization of bean

roots by fungi. *Trans. Brit. Mycol. Soc.* 50:23–31
25. Dobbs, C. G. 1963. Factors in soil mycostasis. In *Recent Progress in Microbiology,* pp. 235–243 Toronto: Univ. Toronto Press, 718 pp.
26. Dobbs, C. G. 1966. Soil-spores of the fungi. *Proc. Welsh Soils Discussion Group,* No. 7:4–13
27. Dobbs, C. G. 1971. Soil mycostasis—The self-soil test. *Soil Biol.* 13:31–33
28. Dobbs, C. G. 1971. Residual soil mycostasis. Resume of paper read at *1st Int. Mycol. Congr.* Exeter, Engl. Sept. 1971
29. Dobbs, C. G., Bywater, J. 1957. Studies in soil mycology. I. *G. Brit. Forest. Comm., Rept. Forest Res.,* 1957: 92–94
30. Dobbs, C. G., Bywater, J. 1959. Studies in soil mycology. II. *G. Brit. Forest. Comm., Rept. Forest Res.,* 1958: 98–104
31. Dobbs, C. G., Gash, M. J. 1965. Microbial and residual mycostasis in soils. *Nature* 207:1354–56
32. Dobbs, C. G., Griffiths, D. A. 1961. Studies in soil mycology. IV. *G. Brit. Forest. Comm., Rept. Forest Res.,* 1960: 87–92
33. Dobbs, C. G., Griffiths, D. A. 1962. Studies in soil mycology, V. Mycostasis in soils. *G. Brit. Forest. Comm. Rept. Forest Res.,* 1961: 95–100
34. Dobbs, C. G., Hinson, W. H. 1953. A widespread fungistasis in soils. *Nature* 172:197–99
35. Dobbs, C. G., Hinson, W. H., Bywater, J. 1960. Inhibition of fungal growth in soils. In: *The Ecology of Soil Fungi.* ed. D. Parkinson, J. S. Waid, 130–47. Liverpool: Liverpool Univ. Press. 324 pp.
36. Doran, W. L. 1922. Effect of external and internal factors on the germination of fungus spores. *Bull. Torrey Bot. Club* 49:313–40
37. Duggar, B. M. 1901. Physiological studies with reference to the germination of certain fungus spores. *Bot. Gaz.* 31:38–66
38. Emmatty, D. A., Green, R. J. 1969. Fungistasis and the behavior of the microsclerotia of *Verticillium albo-atrum* in soil. *Phytopathology* 59:1590–95
39. Eren, J., Pramer, D. 1968. Use of a fluorescent brightener as aid to studies of fungistasis and nematophagous fungi in soil. *Phytopathology* 58:644–46
40. Ford, E. J., Gold, A. H., Snyder, W. C. 1970. Soil substances inducing chlamydospore formation by *Fusarium. Phytopathology* 60:124–28
41. Ford, E. J., Gold, A. H., Snyder, W. C. 1970. Induction of chlamydospore formation in *Fusarium solani* by soil bacteria. *Phytopathology* 60:479–84
42. Ford, E. J., Gold, A. H., Snyder, W. C. 1970. Interaction of carbon nutrition and soil substances in chlamydospore formation by *Fusarium. Phytopathology* 60:1732–37
43. Garrett, S. D. 1951. Ecological groups of soil fungi—a survey of substrate relationships. *New Phytologist* 50:149–66
44. Garrett, S. D. 1970. *Pathogenic Root-Infecting Fungi.* London: Cambridge Univ. Press. 294 pp.
45. Gilbert, R. G., Griebel, G. E. 1969. The influence of volatile substances from alfalfa on growth and survival of *Verticillium dahliae* in soil. *Phytopathology* 59:1400–02
46. Griffin, G. J., 1970. Exogenous carbon and nitrogen requirements for chlamydospore germination by *Fusarium solani:* dependence on spore density. *Can. J. Microbiol.* 16:1366–68
47. Griffiths, D. A. 1966. Vertical distribution of mycostasis in Malayan soils. *Can. J. Microbiol.* 12:149–63
48. Griffiths, D. A. 1966. Sensitivity of Malayan isolates of *Fusarium* to soil fungistasis. *Plant Soil.* 24:269–78
49. Hashmi, M. H., Mallik, M. A. B. 1967. Mycostasis in semi-arid soil of West Pakistan. *Pak. J. Sci. Ind. Res.* 10:254–58
50. Hawthorne, B. T. 1969. *Germination of chlamydospores of Thielaviopsis basicola: physiology, ecology in soil and nature of soil fungistasis.* Ph.D. Thesis. Univ. Calif., Riverside. 129 pp.

51. Hawthorne, B. T., Tsao, P. H. 1969. Inadequacy of the nutrient hypothesis to explain soil fungistasis in relation to chlamydospores of *Thielaviopsis basicola. Phytopathology* 59: 1030
52. Hora, T. S., Baker, R. 1970. Volatile factor in soil fungistasis. *Nature* 225:1071–72
53. Hora, T. S., Romine, M., Baker, R. 1971. Water solubility of a volatile factor inducing soil fungistasis. *Phytopathology* 61: 895
54. Hsu, S. C., Lockwood, J. L. 1971. Responses of fungal hyphae to soil fungistasis. *Phytopathology* 61:1355–62
55. Hurst, H. M., Burges, N. A. 1968. Lignin and humic acids. In: *Soil Biochemistry* ed. A. D. McLaren, G. H. Peterson, 260–86. N.Y.: Marcell Dekker, Inc. 509 pp.
56. Jackson, R. M. 1958. An investigation of fungistasis in Nigerian soils. *J. Gen. Microbiol.* 18: 248–55
57. Jackson, R. M. 1960. Soil fungistasis and the rhizosphere In *The Ecology of Soil Fungi,* 168–176. ed. D. Parkinson, J. S. Waid. Liverpool Univ. Press, 324 pp.
58. Jackson, R. M. 1965. Antibiosis and fungistasis of soil microorganisms. In *Ecology of Soil-borne Plant Pathogens,* 363–69. ed. K. F. Baker, W. C. Snyder. Berkeley, Univ. Calif. Press. 571 pp.
59. King, J. E., Coley-Smith, J. R. 1969. Suppression of sclerotial germination in *Sclerotium cepivorum* Berk. by water expressed from four soils. *Soil Biol. Biochem.* 1:83–87
60. Ko, W. H. 1971. Direct observation of fungal activities on soil. *Phytopathology* 61:437–38
61. Ko, W. H., Lockwood, J. L. 1967. Soil fungistasis: Relation to fungal spore nutrition. *Phytopathology* 57:894–901
62. Ko, W. H., Lockwood, J. L. 1970. Mechanism of lysis of fungal mycelia in soil. *Phytopathology* 60:148–54
63. Kouyeas, V., Balis, C. 1968. Influence of moisture on the restoration of mycostasis in air-dried soils. *Ann. Inst. Phytopathol. Benaki,* Athens 8 n.s.:123–44
64. Krikum, J., Wilkinson, R. F. 1963. Production of chlamydospores by *Fusarium solani* f. sp. *phaseoli* insoil extracts sterilized by passage through different types of filters. *Phytopathology* 53:880
65. Lin, C. K. 1945. Nutrient requirements in the germination of the conidia of *Glomerella cingulata. Am. J. Bot.* 32:296–98
66. Lingappa, B. T., Lockwood, J. L. 1964. Activation of soil microflora by fungus spores in relation to soil fungistasis. *J. Gen. Microbiol.* 35:215–27
67. Lingappa, B. T., Lockwood, J. L. 1961. The nature of the widespread soil fungistasis. *J. Gen. Microbiol.* 26:473–85
69. Lockwood, J. L. 1959. *Streptomyces* spp. as a cause of natural fungi-toxicity in soils. *Phytopathology* 49:327–31
70. Lockwood, J. L. 1964. Soil fungistasis. *Ann. Rev. Phytopathol.* 2: 341–62
71. Macko, V., Staples, R. C., Renwick, J. A. A. 1971. Germination self-inhibitor of sunflower and snapdragon rust uredospores. *Phytopathology* 61:902
72. McFarlane, I. 1952. Factors affecting the survival of *Plasmodiophora brassicae* Wor. in the soil and its assessment by a host test. *Ann. Appl. Biol.* 39:239–56
73. McLaren, A. D., Peterson, G. H. 1967. Introduction to the biochemistry of terrestrial soils. In *Soil Biochemistry,* 1–15, ed. A. D. McLaren, G. H. Peterson, New York: Marcel Dekker, Inc., 509 pp.
74. Menyonga, J. M. 1966. *Nature of response of zoospores of* Phytophthora parasitica *to soil fungistasis.* Ph.D. Thesis. Univ. California, Riverside. 106 pp.
75. Menzies, J. D., Griebel, G. E. 1967. Survival and saprophytic growth of *Verticillium dahliae* in uncropped soil. *Phytopathology* 57:703–09
76. Menzies, J. D., Gilbert, R. G. 1967. Responses of the soil microflora to volatile compounds in plant residues. *Soil Sci. Soc. Am. Proc.* 31:495–96

77. Mircetich, S. M. 1966. *Saprophytic behavior and survival of* Phytophthora cinnamomi *in* soil. Ph.D. Thesis. Univ. Calif., Riverside, 159 pp.
78. Mircetich, S. M., Zentmyer, G. A. 1970. Germination of chlamydospores of *Phytophthora*. In *Root Diseases and Soil-borne Pathogens* 112–15 ed. T. A. Toussoun, R. V. Bega, P. E. Nelson, Berkeley, Univ. Calif. Press. 252 pp.
79. Old, K. M. 1965. Fungistatic effects of soil bacteria on root-rotting fungi with particular reference to *Helminthosporium sativum*. *Phytopathology* 55: 901–05
80. Owens, L. D., Gilbert, R. G., Griebel, G. E., Menzies, J. D. 1969. Identification of plant volatiles that stimulate microbial respiration and growth in soil. *Phytopathology* 59:1468–72
81. Page, W. J., Stock, J. J. 1971. Regulation and self-inhibition of *Microsporum gypseum* macroconidia germination. *J. Bacteriol.* 108:276–81
82. Park, D. 1960. Antagonism—the background to soil fungi. In *The Ecology of Soil Fungi*. 148–59 ed. D. Parkinson, J. S. Waid, Liverpool: Liverpool Univ. Press, 324 pp.
83. Park, D. 1961. Morphogenesis, fungistasis and cultural staling in *Fusarium oxysporum* Snyder & Hansen. *Trans. Brit. Mycol. Soc.* 44:377–90
84. Park, D., Robinson, P. M. 1967. Internal pressure of hyphal tips of fungi and its significance in morphogenesis. *Ann. Bot.* 31: 735–38
85. Payen, J. 1962. Recherches sur le comportement de champignons dans les sols. I. Experiences préliminaires, *Bull. Ecole Nat. Super. Agron. Nancy* 4:52–70
86. Phillips, D. J. 1965. Ecology of plant pathogens in soil. IV. Pathogenicity of macroconidia of *Fusarium roseum* f. sp. *cerealis* produced on media of high or low nutrient content. *Phytopathology* 55:328–29
87. Popov, V. I., Zdrozhevskaya, S. D. 1969. Regulation of soil fungistasis, a method for controlling soil pathogens. *Mikol. Fitopatol.* 3:48–52
88. Ram, C. S. V. 1952. Soil bacteria and chlamydospore formation in *Fusarium solani*. *Nature* 170:889
89. Robertson, N. F. 1958. Observations on the effect of water on the hyphal apices of *Fusarium oxysporum*. *Ann. Bot. Lond.* 22:159–73
90. Robertson, N. F. 1965. The fungal hypha. *Trans. Brit. Mycol. Soc.* 48:1–8
91. Robertson, N. F., Rizvi, S. R. H. Some observations on the water relations of the hyphae of *Neurospora crassae*. *Ann. Bot.* 32: 279–91
92. Robinson, P. M., Garrett, M. K. 1969. Identification of volatile sporostatic factors from cultures of *Fusarium oxysporum*. *Trans. Brit. Mycol. Soc.* 52:293–99
93. Robinson, P. M., Park, D. 1966. Volatile inhibitors of spore germination produced by fungi. *Trans. Brit. Mycol. Soc.* 49: 639–49
94. Robinson, P. M., Park, D., Garrett, M. K. 1968. Sporostatic products of fungi. *Trans. Brit. Mycol. Soc.* 51:113–24
95. Rovira, A. D. 1965. Plant root exudates and their influence upon soil microorganisms. In *Ecology of Soil-borne Plant Pathogens*. 179–84, ed. K. F. Baker, W. C. Snyder, Berkeley, Univ. Calif. Press. 571 pp
96. Rovira, A. D., McDougall, B. M. 1967. Microbiological and biochemical aspects of the rhizosphere. In *Soil Biochemistry*. 417–63, ed. A. D. McLaren, G. H. Peterson, New York: Marcel Dekker. 509 pp
97. Schreiber, L. R., Green, R. J. 1963. Effect of root exudates on germination of conidia and microsclerotia of *Verticillium albo-atrum* inhibited by the soil fungistatic principle. *Phytopathology* 53:260–64
98. Schroth, M. N., Hildebrand, D. C. 1964. Influence of plant exudates on root-infecting fungi. *Ann. Rev. Phytopathol.* 2:101–32
99. Schüepp, H., Frei, E. 1969. Soil fungistasis with respect to pH

and profile. *Can. J. Microbiol.* 15:1273–79

100. Shigo, A. L. 1967. Successions of organisms in discoloration and decay of wood. *Int. Rev. Forest. Res.* 2:237–99
101. Steelink, C., Tollin, G. 1968. Free radicals in soil. In *Soil Biochemistry,* ed. A. D. McLaren, G. H. Peterson, 147–72. N. Y.: Marcel Dekker, Inc., 509 pp.
102. Steiner, G. W., Lockwood, J. L. 1969. Soil fungistasis: Sensitivity of spores in relation to germination time and size. *Phytopathology* 59:1084–92
103. Steiner, G. W., Lockwood, J. L. 1970. Soil fungistasis: Mechanism in sterilized reinoculated soil. *Phytopathology* 60:89–91
104. Stevenson, F. J. 1968. Organic acids in soil. In *Soil Biochemistry.* 119–46. Ed. A. D. McLaren, G. H. Peterson, New York: Marcel Dekker, Inc. 509 pp.
105. Stevenson, I. L. 1956. Antibiotic activity of actinomycetes in soil as demonstrated by direct observation techniques. *J. Gen. Microbiol.* 15:372–80
106. Stover, R. H. 1955. Flood-fallowing for eradication of *F. oxysporum* f. *cubense.* III. Effect of oxygen on fungus survival. *Soil Sci.* 80:397–412
107. Sussman, A. S., Halvorson, H. O. 1966. *Spores: Their Dormancy and Germination.* N.Y.: Harper & Row 354 pp.
108. Tribe, H. T. 1961. Microbiology of cellulose decomposition in soil. *Soil Sci.* 92:61–71
109. Tsao, P. H. 1970. Applications of the vital fluorescent labeling technique with brighteners to studies of saprophytic behavior of *Phytophtora* in soil. *Soil Biol. Biochem.* 2:247–56
110. Tsao, P. H., Bricker, J. L. 1968. Germination of chlamydospores of *Phytophthora parasitica* in soil. *Phytopathology* 58:1070
111. Tsao, P. H., Hawthorne, B. T. 1970. Soil fungistasis soil amendments, lysis, and biological control of *Thielaviopsis basicola. Proc. 7th Int. Congr. Plant Prot.* Paris. 534–35
112. Waksman, S. A. 1938. *Humus,* 2nd ed. Baltimore: Williams & Wilkins 526 pp.
113. Vaartaja, O. Agnihotri, V. P. 1966. An unusually stable inhibitor of *Pythium* and *Thanatephorus* (*Rhizoctonia*) in a nursery soil. *Phytopathology* 56:905
114. Vaartaja, O., Agnihotri, V. P. 1967. Inhibition of *Pythium* and *Thanatephorus (Rhizoctonia)* by leachates from a nursery soil. *Phytopathol. Z.* 60:63–72
115. Weltzien, H. C. 1963. Untersuchungen über die Ursachen der Keimhemmung von Pilzsporen im Boden. *Zentralbl. Bakt. Par. Infekti. Abt.* II, 116:131–70
116. Wright, J. M. 1956. The production of antibiotics in soil. III. Production of gliotoxin in wheatstraw buried in soil. *Ann. Appl. Biol.* 44:461–66
117. Wright, J. M. 1956. The production of antibiotics in soil. IV. Production of antibiotics in coats of seeds sown in soil. *Ann. Appl. Biol.* 44:561–66
118. Yarwood, C. E. 1968. Tillage and plant diseases. *Bioscience* 18: 27–30
119. Zentmyer, G. A., Mircetich, S. M. 1966. Saprophytism and persistence in soil by *Phytophthora cinnamomi. Phytopathology* 56: 710–12

INFLUENCE OF WATER POTENTIAL OF SOILS AND PLANTS ON ROOT DISEASE

R. J. Cook and R. I. Papendick

Research Plant Pathologist, Plant Science Research Division; and Soil Scientist, Soil & Water Conservation Research Division, U. S. Department of Agriculture, Pullman, Washington

Introduction

The favorable and unfavorable effects of water on root diseases have long been known, and reviews related to the subject are available (24, 32, 48, 59, 61, 118). Most of these reviews emphasize the effects of soil water on the pathogen and say little about the host or the competing soil microflora. In this review we examine the effects of water on the host-pathogen-soil microorganism system and how soil water relates to root disease development and control.

Concepts and Terminology

Research on plant-soil-water relationships has been made needlessly complicated by the use (and abuse) of a large number of terms such as suction force, water-holding capacity, diffusion pressure deficit, etc. Water, like heat, flows from regions of high to regions of low energy. Thus, it is best to think about the movement of water in terms of changes in its potential energy. The simple concept of "water potential" permits this kind of thinking because it provides a single unit of measurement for all forces that influence movement of water from soil, through plants, and into the atmosphere. The reviews of Griffin (59, 61) and Cook & Papendick (32) discuss the various terms used to describe soil water potential and their significance in plant pathology and soil microbiology. Reviews on techniques for the measurement of water potential are also available (13, 32, 97, 139).

Water potential is a measure of the capacity of the water at some point in a system to do work, as compared with the work capacity of pure, free water at the same temperature. Water potential is equal to the "chemical potential" (as used in solution chemistry) divided by the partial molar volume of water. Thus, it is dimensionally equivalent to pressure and allows the use of bars or atmospheres as the unit of measurement. Pure free water at atmospheric pressure is arbitrarily assigned a value of zero potential energy. In soil at less than saturation, or where salts are present, the potential is less than zero (negative); work must be expended to return this water to the reference state. As the soil dries, the water potential decreases.

Total water potential (3) may be divided into: (*a*) osmotic potential due to solutes (always negative); (*b*) gravitational potential caused by changes in elevation (positive or negative); (*c*) matric or capillary potential, which includes both adsorption and capillary effects of the solid phase such as mineral or organic materials (always negative); and (*d*) pressure potential, which results from external pressure applied to the soil water (positive or negative).

In soil, the liquid (film and pore water) and the vapor phase (drained pores) equilibrate rapidly and thus both have the same water potential. The water potential can be inferred from the equilibrium relative humidity (RH) in the equation

$$\psi = \frac{RT}{V} \log \text{RH}$$

where Ψ is the water potential, R is the ideal gas constant, T is the absolute temperature, V is the volume of a mole of water, and log is the natural logarithm. For example, water having an equilibrium RH of 98% at 25C would have a potential of −27.7 bars (1 bar = 0.987 atmospheres). The equation emphasizes the marked sensitivity of water potential to change in RH—14 bars = approximately 1% change in RH.

Total water potential of a living cell is the sum of the turgor (positive), matric (negative), and osmotic (negative), potential (109). In turgid cells, the matric component is insignificant compared to the osmotic component (46).

For microorganisms buried in soil or other media, the total water potential of the microbial cell tends to equilibrate with that of the external environment, including the substrate. This may be expressed by the relationship: $\Psi_{\text{external environment}} = \Psi_{\text{cell turgor}} + \Psi_{\text{cell osmotic}}$, neglecting the matric component. As the soil dries, the cell tends to lose water to the outside, or, if the water potential of the cell decreases below that of the external environment, water tends to flow from the higher energy status outside the cell into the cell so that equilibrium is maintained.

Higher plants differ from most microorganisms in that they occupy two environments at the same time—roots in soil and tops in the atmosphere. Water uptake by roots, movement in the vascular system, and transpiration occur in response to a water potential gradient (109). Transpiration (the flow of water vapor from a higher energy status within the leaf to a lower energy status in the atmosphere) is limited by the energy available for evaporation, and by resistance to water movement through the soil-plant system. Water flow into the root is from higher energy status in the soil to lower energy status in the root cortex. The root in turn is generally at a higher water potential than the leaf. The water potential of the leaf and root is dominated by osmotic forces whereas that of the connecting xylem is primarily matric (capillary). The internal water deficit is determined by the soil water potential, which sets the base level of the plant water potential, and by the extent that the flow within the plant lags behind evaporative demand. The leaf water potential at which stomata close to decrease transpiration is not clearly defined; it probably varies with plant species. The water potential at which permanent wilting occurs also varies with plant species and may not always be −15 bars as often reported (108).

Examples of Soil Water Effects on Root Diseases

The literature contains much valuable information on the importance of soil water to plant diseases despite many early deficiencies of concept and technique. Field observations made repeatedly and reported by earlier plant pathologists are as reliable as those made today. Greenhouse demonstrations of the influence of soil water are less easily interpreted, but even here valuable clues are revealed.

Only a few root diseases are favored by dry soils (Table 1); most are favored by wet soils (Table 2). Those caused by species of *Phythium, Phytophthora, Aphanomyces, Snychytrium,* and other aquatic phycomycetes (48) and those caused by nematodes (135) are favored by near-saturated soils and need no enumeration.

Fusarium wilts and possibly verticillium wilt are apparently favored by wet soils but illustrate some special problems in interpretation. Fusarium wilt of peas (84), tomato (25, 121), cotton (125), and celery (99) were shown in greenhouse tests to be more severe in wet than in dry soil. These reports apparently contradict the earlier ones of Gilman (53) and Tisdale (128) for cabbage yellows, and of Humbert (66) for fusarium wilt of tomato, who observed severe disease following hot dry weather and little or none during cool moist weather. Another pattern, demonstrated best for banana wilt (120),

Table 1. Diseases favored by dry soil.

Disease	Pathogen	Reference
Seedling blight of cereals	Fusarium roseum f sp cerealis 'Graminearum'	37
Seedling blight of cereals	*F roseum* f. sp. *cerealis* 'Culmorum'	26, 87, 106
Wheat root and foot rot	*F. roseum* f. sp. *cerealis* 'Culmorum'	28
Seedling-blight of clover	*F roseum*	58
Stem rot of sweet potato	*F. solani* f. sp. *batatas*	65
Root and stem rot of peas	*F. solani* f. sp. *pisi*	72
Pox of sweet potato	*Streptomyces iponeae*	95
Common scab of potato	*Streptomyces scabies*	56, 74, 77, 100
Charcoal rot of sorghum	*Macrophomina phaseoli*	42
Charcoal rot of cotton	*Macrophomina phaseoli*	52
Seed decay of wheat	*Penicillium* and *Aspergillus* sp.	60

TABLE 2. Diseases favored by wet soil

Disease	Pathogen	Reference
Cotton root rot	*Phymatotrichum omnivorum*	123
Take-all of wheat	*Ophiobolus graminis*	34, 44, 47
Cephalosporium stripe of wheat	*Cephalosporium gramineum*	14
Bare-patch of wheat	*Rhizoctonia solani*	67
Black root rot of tobacco	*Thielaviopsis basicola*	7
Armillaria root rot	*Armillaria melea*	48
Southern blight	*Sclerotium rolfsii*	23, 132
Sclerotinia disease of vegetables and field crops	*Sclerotinia sclerotiorum* and related species	23, 132

but implicated for other fusarium wilts as well (112), is that sandy soils apparently are more conducive than loams or clays to spread of fusarium wilts (112). Sandy soils are better drained, hold less water, dry faster than clays, and thus favor multiplication and survival of fusaria (119). How is it possible that the disease is (*a*) favored more by sandy than clay soils, (*b*) more severe with high than low water potentials in greenhouse pot trials, and yet (*c*) most severe in the field following hot, dry weather? Similar apparent contradictions probably exist in the literature on other diseases.

Different experimental techniques emphasize the effects of soil water on different stages in the disease cycle. In fusarium wilt of banana, for example, the differential influence of sandy versus clay soils may be apparent only after several years of banana culture (120) and probably reflects an effect of the soil on the ecology of the pathogen, perhaps through enhanced multiplication and survival (119). The greenhouse studies of fusarium wilts (25, 84, 99, 121, 125), generally involve wound-inoculated plants in steamed soil. This circumvents possible long-term effects of soil water on pathogen ecology. Field observations (53, 66, 128) on the importance of hot, dry weather may be related to postinfection phenomena, e.g., inability of transpiration to meet evaporative demand because of vascular plugging.

Another complication in the fusarium wilts can arise when root-knot nematodes are present. Resistance to fusarium and verticillium wilt is decreased when the plants are infected with nematodes, particularly *Meloidogyne incognita* (4, 96). Drying of the soil to reduce nematode infections should also reduce the severity of wilt.

DIRECT LIMITATION OF THE PATHOGEN BY LOW SOIL WATER POTENTIALS

Every microorganism has optimal and minimal water potentials for growth (59, 104). Plant pathogens are no exception (Table 3). Water potentials for optimal growth may differ with temperature (15, 62) and some patho-

TABLE 3. Water potentials (osmotic, matric, or from air relative humidity) at which growth of select and soil-borne pathogens is optimal, reduced by half, or prevented

Pathogen	Water potential control	Growth/-bars water potential			
		Optimal	Reduced by ½	Prevented	Reference
Rhizoctonia solani	osmotic	>5	25–30	<50–56	39
	matric	>3	15	<32	39
	in air	0	16	<45–50	103
Verticillium albo-atrum	osmotic	10–15	50–60	100–120	88
	matric	>1	40–50	80–100	88
	in air	0	24	<100	103
V. albo-atrum, spore germination	osmotic	>5	20–25	<30–40	90
Phytophthora cinnamomi	osmotic	5–15	20–25	40–50	2, 113
	matric	>5	10–20	20–35	2
Thielaviopsis basicola	in air	0	22	<80	103
Ophiobolus graminis	osmotic	>1	20	45–50	33
	matric	>1	20	45–50	33
	in air	0	18	<45	103
Fusarium nivale	osmotic	2–5	17		15
Typhula idahoensis	osmotic	2–5	17–18		15
Cephalosporium gramineum	osmotic	2–6	21–22	100–110	16
Fusarium roseum f. sp. *cerealis* 'Culmorum'	osmotic	8–10	45–50	80–85	33
	matric	0–5	40–45	70–75	33
	in air	0–3	40–45	<120	103
'Avenaceum'	in air	0–13	45	130	103
F. solani	in air	0–32	80–85	<125	103
F. solani 'Coeruleum'	in air	0–3	40	<140	103
F. oxysporum f. sp. *lycopersici*	in air	0–3	60	<140	103
f. sp. *lini*	in air	0–3	44	<120	103
f. sp. *vasinfectum*	osmotic	0–30	50–60	100–120	88
	matric	10–15	50–60	90–100	88

gens respond differently to osmotic than to matric water potentials (2, 33).

The simplest explanation of why most root disease occurs in wet soil is that water potential is too low for growth of the pathogen in drier soils. This is true particularly with diseases in which the pathogen is active at or very near the soil surface. Examples include *Cercosporella herpotrichoides* on wheat (54), *Sclerotium rolfsii* on various hosts (11) and *Rosellinia arcuata* on plantation crops in the tropics (48).

Sclerotia of *S. rolfsii* do not germinate readily at relative humidities below

99% (= − 14 bars) (1). Since the minimum RH for growth and sporulation of *Rosellinia* and *Sclerotium* spp. is probably greater than 95–96%, and RH a few mm below the soil surface probably falls below this level on most warm, sunny days, it is little wonder that these pathogens are destructive only during wet, overcast periods or where shading is heavy.

Water potentials that prevent growth of most root-infecting fungi (−30 to −50 bars or lower) are well below those that exist beneath the soil surface except under prolonged drought conditions. Since most pathogens can grow at water potentials well below the minimum for growth of most higher plants (0 to −15 bars), Griffin (61) suggests that water potential per se probably does not directly limit pathogens of growing plants. He cites the case of decay of seeds in soil too dry to permit their germination (60) as the only known example of "a disease in which (a low water potential) is known to be important through its direct effect on the fungus."

It must be significant, however, that fungi with the highest water potential requirements for growth are those that cause severe disease in wet soils, and

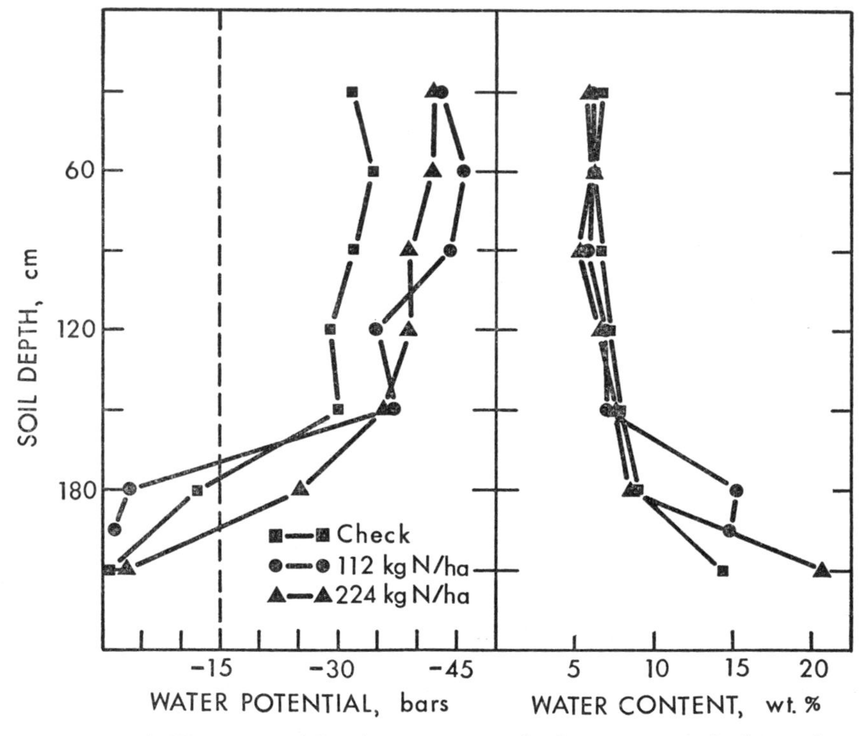

FIGURE 1. Water potential and water content in the root zone of winter wheat near the end of the 1970 growing season as affected by three rates of nitrogen application. After Papendick et al (92).

conversely, those with the lowest water potential requirements for growth are those that cause severe disease in dry soils (Table 3). *Verticillium albo-atrum* and the wilt fusaria may be exceptional in that they are capable of growth in dry soil, but cause the most severe disease in wet soil (see pages 21 and 22).

The inoculum of most root parasites exists primarily in the surface 24 cm of soil. The evaporative demand is high during the growing season in arid, semi-arid, and often humid regions as well, and the dry soil boundary can move several cm deep. The pathogens may thus be exposed to a restrictive water potential while the host continues to absorb water from deeper in the soil.

This is well illustrated in the case of dryland wheat in the northwestern U.S.A. where *Fusarium roseum* f. sp. *cerealis* 'Culmorum' but not *Ophiobolus graminis* is destructive (33). Water potentials in the surface 10 cm of soil (where 90% of the 'Culmorum' population resides) are commonly below −100 bars at seeding time, and for a few weeks thereafter, until fall rains begin. The seed is planted in moist soil at a depth of 12–15 cm. The shoot then emerges through the "dust mulch" above this depth. Winter rains maintain high soil water potentials, but then temperatures are generally too low for activity of these pathogens. During the spring and summer, water potentials become more favorable for Culmorum, and less so for *O. graminis*. In June, the surface 30 cm may again be drier than −100 bars, because of evapotranspiration (92). But then, the water potential below this very dry layer is commonly −30 to −45 bars and this may extend to 150 cm or deeper (92) (Figure 1). Root absorption is responsible for these low water potentials, which can only mean that the root itself, or at least the cortex, is at −30 to −45 bars water potential—otherwise it could not have drawn the water potential of the surrounding soil down by such an amount; leaf osmotic

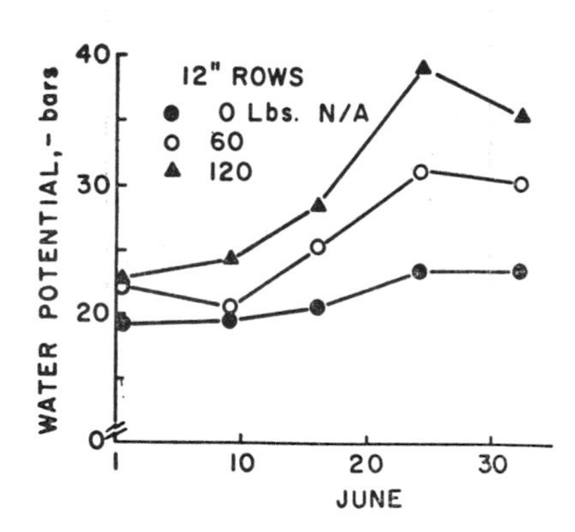

FIGURE 2. Leaf osmotic water potentials during the 1971 growing season for winter wheat planted in rows 12 inches apart and fertilized with 0, 60, and 120 lbs nitrogen/acre (Papendick & Cook, unpublished).

potential is of the same magnitude (Figure 2). From this we can infer that all parts of the plant have a water potential of −30 to −45 bars or lower. These water potentials inside as well as outside the host are clearly too low for *O. graminis* but are ideal for Culmorum (Table 3).

Root infecting fungi can also be restricted by low water potentials in furrow- or sub-irrigated fields. This effect is best revealed by describing water and salt movement from a water table in soils of different textures.

At low evaporative demand, capillary rise is faster in coarse- than in fine-textured soils but eventually becomes higher in fine-textured soils. With increased evaporative demand, the wet-dry boundary may persist deeper into a fine- than a coarse-textured soil, because the flow of water is slower. Osmotic forces can become very important where the water evaporates and salts accumulate (Figure 3). Salts lower the vapor pressure of water and hence slow transport upward in the vapor phase. This increases the tendency for soil above the wet-dry boundary to dry down and thus equilibrate with the above-ground atmosphere. Where salt concentrations are very high, osmotic water potentials may be less than −40 to −50 bars, even if the soil is wet (−⅓ bar matric). As a rule (130), the osmotic potentials are reduced by one-half (e.g., to −80 to −90 for very high salt concentrations) when the matric potential is lowered from −⅓ to −15 bars. Total soil water potentials well

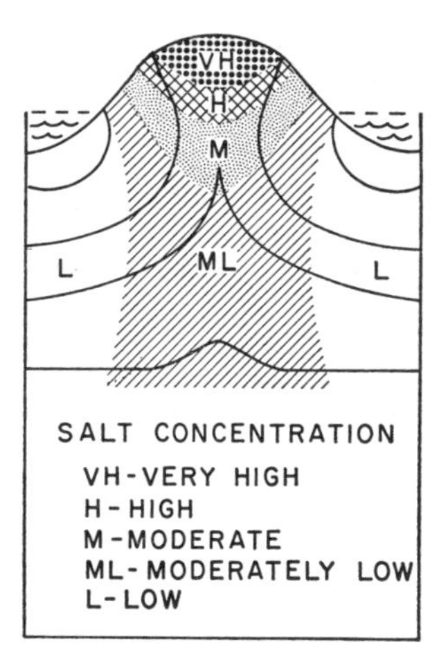

FIGURE 3. Schematic of salt concentrations that develop with furrow irrigation of a saline soil. Converting the salt values given by Wadleigh & Fireman (131) to osmotic potentials in bars for the different regions at a water content corresponding to −⅓ bar matric potential gives VH, > −45; H, −7 to −45; M, −4 to −7; ML, −3 to −4; and L, the soil below the furrow center, < −1. At −15 bars matric potential, the osmotic potentials would be less than ½ the values given for ⅓ bar.

below −100 bars are probably common in the surface layers of a saline soil if matric potentials are below −15 bars.

A low water potential may develop in the rhizosphere because of water absorption by the root. At matric potentials of −1 bar or less, silt or clay soils often have higher conductivities than do sands, and thus the gradients will be steeper in sandy than in loam or clay soils. Precise or even approximate values for water potential in the rhizosphere are not available from experimental data, but presumably approach those of the root cortex. Conceivably, they are lowest during periods of high transpiration and high rates of photosynthesis, and return to those of the surrounding moist soil when water uptake is slowed.

Although water usually flows from roots to stems to leaves, the reverse may occur if the water potential gradient is reversed. This has been shown for bean hypocotyls surrounded by dry soil when the roots were in wet soil (102). Water presumably flowed out through stomata on the hypocotyl. This has also been shown for tomato roots in dry soil when above-ground parts were subjected to a water mist (12). Barley seedlings with roots in wet soil guttated water through hydathodes at the shoot apex during emergence through a layer of dry soil (Cook & Griffin, unpublished).

Probably little water (liquid or vapor) flows into soil from older, suberized roots, or where natural openings such as hydathodes or stomata are lacking. In dry-land wheat in June, for example, water is conducted from the root tips 150–180 cm below ground to the leaves 50–60 cm above the ground. If the loss of water from roots over this long distance was appreciable, the water could never reach the tops. Moreover, the profile gradient would be gradual over the entire vertical distance rather than uniform the entire distance downward to the moist zone (Figure 1). Pathogens probably gain little if any benefit from water vapor or liquid water moving into dry soil around the plant from organs that have no natural openings, or are mature and suberized.

At less than optimal water potentials, pathogens do not display their full pathogenic potential. In essence, a low water potential decreases inoculum potential (*sensu* Garrett, 49) by decreasing the capacity of the pathogen to compete with other organisms and to overcome host barriers.

Limitation of the Pathogen by the Size of Soil Pores

Nematodes, motile bacteria, and aquatic phycomycetes may be limited in movement if the water-filled pores or pore necks in the soil are too small to permit their passage (63, 117, 134). Larvae of *Heterodera schachtii* moved at maximum speed through columns of sand when water-filled pore diameters were 30–60μ (134). *Pseudomonas aeruginosa* apparently requires water-filled pores 1–1.5μ or larger in diameter (63). Stolzy et al (117) suggested that zoospores of *Phytophthora* species require water-filled pores at least 40–60μ in diameter. The maximum pore diameter (d), in microns, that remains filled with water as matric potential (Ψ_m) is lowered from saturation can be determined by the simplified capillary rise equation, $\Psi_m d = 2.94$.

Pore size distribution is an equally important limitation on movement because it determines the frequency or number of pores that are large enough to accommodate the pathogen (63, 134). This is a function of particle size and thus of soil type (Figure 4) (63, 134). Maximum movement of a given organism becomes possible at that matric potential that allows the greatest frequency of water-filled pores large enough to accommodate the organism, and thus will vary with soil type. Wallace (134) showed that movement of larvae of *Heterodera schachtii* was optimal at matric potentials of 25–40 millibars where the particle size was 150–250μ and pore diameters were 40–60μ.

High matric potentials are essential for maximum rates and distances of diffusion of exudates from the host (32, 61). This has been indicated for exudation of pea seeds as it influences activity of *Fusarium solani* f. sp. *pisi* (30) and *Pythium ultimum* (45, 69), and bean seeds as it influences activity of *F. solani* f. sp. *phaseoli* and *Pythium ultimum* (115). Pore-size distribution (Figure 4) may explain why a given amount of pea seed exudation occurred at progressively smaller matric potentials if the soil contained a progressively greater percentage of fine particles and hence small pores (69).

Water in soil pores may limit diffusion of oxygen to pathogens. This influence of the liquid water phase has been shown for *Armillaria elegans* (closely related to *A. mellea*). Rhizomorphs grew rapidly when oxygen was supplied to the growing tip at high rates of diffusion via the central canal of

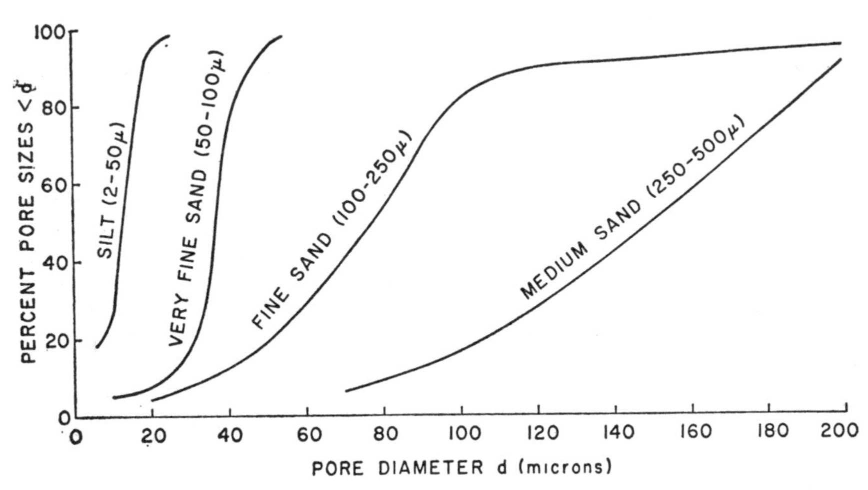

FIGURE 4. Percentage of pore sizes having diameters of less than d microns for a silt and for three different particle sizes of sand. Data drawn from Leamer & Lutz (80).

the rhizomorph (111). By contrast, oxygen in excess of 0.04 atm in contact with the outside of the rhizomorph growing point was inhibitory to rhizomorph growth apparently because it stimulated activity of p-diphenol oxidase which led to formation of a brown pigment within the intercellular spaces of the rind. Water films surrounding the tips of the rhizomorphs acted as a barrier to oxygen diffusion to the surface. At high water potentials, the films were inadequate to provide such protection. Optimal growth was observed with progressively lower matric potentials if the soil contained progressively more silt and clay—lower matric potentials are required to drain the predominant pore class with fine textured soils. It is thus easier to retain a water film over the rhizomorph apex in fine- than in coarse-textured soils (111). An increase in bulk density, because it reduces pore size, would also facilitate film maintenance on the rhizomorph tip.

Propagule Desiccation

The sclerotia of *Phymatotrichum omnivorum* may be killed within a few days if in soil allowed to air dry (70, 98, 124). Populations of *Pseudomonas solanacearum* can be reduced in Central America, in fields destined for banana plantings, by repeated disking of the soil to 25 cm during the dry season, and by allowing the field to lie fallow for 9 months (105). Disking enhanced the drying effect during fallow as compared with no tillage.

Tolerance to dehydration may depend on conditioning received by the propagule prior to exposure to desiccation. Hyphae of *Rhizoctonia solani* from an actively growing colony on agar (> −2 bars osmotic potential) died within a few days after abrupt exposure in soil at −100 to −200 bars water potential (39). In contrast, hyphae grown at −30 bars water potential were tolerant of the abrupt desiccation treatment, as were those grown on ordinary medium, preconditioned for a few days at −30 bars or drier, and then exposed to desiccation.

Chlamydospores of *Thielaviopsis basicola* (83) and *F. solani* f. sp. *phaseoli* (27) germinated poorly following air-dry storage for several months in soil, but recovered their full germination potential if remoistened in soil for a few days. Papavizas & Lewis (91) were unable to confirm the poor germinability of chlamydospores of *T. basicola* following an air-dry treatment, but probably their "air-dry" soil was in equilibrium with air at an RH considerably greater (more typical of eastern U. S.) than generally occurs in Berkeley, California where the work of Linderman & Toussoun was done (83).

The harmful effects of low water potentials on propagules of plant pathogens may relate, in part, to effects on bound water in macromolecules of the cell, including proteins, enzymes, RNA, and DNA. In DNA, water molecules are bound between the helices and apparently stabilize the duplex helical configuration (43, 73). According to Falk et al (43) at 80% RH or higher, all sites in the DNA molecule are hydrated. When equilibrated with 55–75% RH, 4–5 water molecules per nucleotide pair are lost and structural integrity of the DNA molecule is consequently lost. This is a reversible process, but below 55% RH the DNA molecule becomes irreversibly unstructured.

Webb et al (136) demonstrated greatest loss of water from bacterial cells (*Serratia marcescens*) beginning when the RH dropped below a threshhold of 75 to 80%; they proposed that dehydration of DNA, RNA, proteins, and possibly other macromolecules were responsible for weakening and death of the cells. Dispersed bacterial cells very likely would equilibrate internally with the external RH, and thus macromolecules conceivably would react to the RH as they do in vitro (43). On the other hand, fungus spores, nematode cysts, bacterial spores, and other survival structures of microorganisms with thick outer protective walls can perhaps maintain their internal water potential above that of a surrounding dry environment, or at least slow the rate of internal-external equilibration. This may be why unconditioned hyphae of *Rhizoctonia solani* were sensitive to desiccation while preconditioned ones were not (39); unconditioned hyphae may have been less able to prevent or delay their equilibration with the external environment. This may also explain why growth, even of xerophytic fungi such as *Pencillium* and *Aspergillus* species, is prevented or greatly slowed at 75–80% RH. At this RH and drier, the internal water potential of the cell must be above, not below, the external environment to prevent temporary or permanent damage to macromolecules. The cells of *Pseudomonas solanacearum* (105) in banana fields of Central America, and the sclerotia of *Phymatotrichum omnivorum* (70, 98, 124) in southwestern U.S.A. may be poorly equipped to protect against internal equilibration with harmfully-low external water potentials. Chlamydospores of *F. solani* f. sp. *phaseoli* and *Thielaviopsis basicola* may exemplify cells that tend to equilibrate internally with RH between 55 and 75%, whereupon reversible unstructuring of DNA occurs. Probably no cell wall can protect completely and forever against desiccation, but rather, only slows the process with some propagules being more resistant than others.

Conidia of *Cochliobolus sativus* apparently are highly resistant to desiccation (81). Indeed, they survived longer (up to 52 months) at 10 and 50% RH than at 90 or 100% RH (viability declined after 7 months at 25°C and 12 months at 15°C). At higher RH or temperatures, endogenous respiration probably proceeds at a faster rate, thereby lowering food reserves and shortening life of the spore. Desiccation may also prolong survival of *Cephalosporium gramineum* (17) and *Gibberella zeae* (19) (= *Fusarium roseum f.* sp. *cerealis* 'Graminearum') in wheat straw, by reducing metabolism provided they are not replaced in the straw by competitors during periods when their own metabolism is reduced. At 15°C, *C. gramineum* survived in straw in atmospheres drier than 82% RH. Between 86 and 90%, however, it was largely replaced by *Penicillium* species. Above 95% RH, metabolism of *C. gramineum* was higher, and the fungus retained possession of the straw through antibiotic production. Antibiotic production is apparently maximal at near zero water potential and progressively less down to −60 to −70 bars

(16). Survival of *Giberella zeae* at 25°C was good at about 33 and 100% RH and poor at 75 and 85% RH where *Penicillium* and *Aspergillus* sp., but not *G. zeae,* were active (19). At 10°C, survival was good at all water potentials tested and at 35°C, was progressively poorer as the RH was increased to 100%.

A few hours drying followed by rewetting stimulates germination of sclerotia of *Sclerotium rolfsii* (110). In addition, the drying causes cracks to develop in the rind which, upon rewetting of the soil, results in leakage of nutritive substances, colonization of the sclerotium by soil organisms, and sclerotial death. Smith (110) has suggested a biological control system for this fungus whereby the tillage layer is allowed to dry, then kept wet by irrigation for two weeks prior to planting, to permit sclerotial germination and decay. Staten & Cole (116) suggested a pre-irrigation treatment to reduce infection of cotton seedlings by *Rhizoctonia solani.* This might now be attributed to injury of the sclerotia during the dry period followed by decay during a wet period.

Limitation of the Pathogen Through Antagonistic Microorganisms

A drop in water potential sufficient to reduce growth of a pathogen by a certain proportion would probably have minimal influence on that pathogen if growth of all surrounding and potentially-antagonistic microorganisms were affected by the same amount. This probably never happens. Instead, the change in water potential places the pathogen at a greater or lesser competitive advantage, relative to the other organisms. Baker & Cook (4) refer to this as an effect on the "Relative Competitive Advantage" of the organism. The concept is patterned after that of Leach (79) for seedling blight at different temperatures as influenced by the relative effect of temperature on rate of seedling emergence versus rate of pathogen growth. A water potential unfavorable to the antagonist, but even more unfavorable to the pathogen, will enhance biological control; a water potential unfavorable to the pathogen, but even more unfavorable to the antagonist, will favor the pathogen.

The influence of water potential on a pathogen versus its competitors in soil is illustrated with germlings of *Fusarium roseum* f. sp. *cerealis* 'Culmorum.' In sterile soil, hyphal growth is maximal at water potentials near zero, and progressively slower at lower water potentials down to about −80 bars (33). In nonsterile soil, continuing hyphal growth occurs only if the soil is at water potentials drier than −8 to −10 bars (31). Above this range, initial chlamydospore germination is not suppressed, but the germ tubes quickly lyse or convert back into dormant chlamydospores. Bacterial activity slows as the soil dries, and virtually ceases at −10 to −15 bars (32), which coincides with the highest water potential at which Culmorum germlings can grow in natural soil without lysing (31). Streptomycin and neomycin delayed germ tube lysis in wet soil (31). By its ability to grow at low water potentials Culmorum escapes the antagonism of organisms (mainly bacteria in this case) unable to grow at such low water potentials.

Saprophytic colonization of wheat straw in soil by Culmorum was maximal at −80 to −90 bars (29), even though at this water potential chlamydospore germination and germling growth of the fungus both were nearly nonexistant (31, 33). The little growth possible at −80 to −90 bars is apparently of great advantage to the fungus, because of the relatively greater reduction of antagonism.

Whether or not decreased antagonism accounts for the greater incidence of fusarium seedling blight and foot rot of cereals in dry soils is less certain; the host may become more susceptible to *Fusarium* as the soil dries (37, 93). According to Colhoun & Park (26), *Fusarium roseum* f. sp. *cerealis* 'Culmorum', 'Graminearum', and 'Avenaceum' each caused more wheat seedling blight in dry than in wet soil when placed as conidia on the seeds prior to planting. Reduced seedling resistance to the fusaria might account for the greater death rate in dry compared to wet soil, but cannot readily explain the increased numbers of infection sites (lesions) which they found in the drier soils. In other words, resistance in the seedlings at the higher water potential might not be sufficient to prevent initial establishment of the fungus, but might slow development of an established lesion. The number of lesions probably relates to the number of fungus propagules that survived the effects of antagonists outside the host long enough to infect. Decreased antagonism rather than host resistance probably accounts for the increased activity of *F. solani* f. sp. *pisi* on pea epicotyls in dry soil (30).

With common scab of potatoes caused by *Streptomyces scabies,* infection occurs during brief periods of drying and is prevented if soil near the surface of the tuber is kept wet (75). Infection is through lenticels on the tuber surface and must take place during the period of lenticel development, before suberization occurs (74, 76). Lewis (82) suggested that an increased population of antagonistic bacteria may limit lenticel infection by *S. scabies* in wet soil. In aseptic tissue culture, infection of tubers by *S. scabies* was greatest at sites where free water collected, suggesting that high water potential per se is not limiting to this organism (6). Take-all of wheat caused by *Ophiobolus graminis* is severe in eastern Washington fields newly reclaimed from native vegetation after two or more consecutive irrigated wheat crops (34). The disease has never been reported from those fields of the state with a long history of dryland wheat alternated with fallow; it occurs only rarely (114) in the subhumid Palouse, in fields cropped annually to dryland wheat or barley. The fungus is present in practically all soil samples, including dryland field soils; thus there is no lack of inoculum (107). This suggests that the fungus is maintained in the dryland area through limited parasitism on the wheat. Low water potentials cannot account entirely for the suppression of disease, however, because it does not develop, or does so only in limited areas, even when dryland wheat fields are converted to irrigated wheat and cropped annually (107).

An as yet unidentified, but transferable, biological factor antagonistic to *O. graminis* has been demonstrated in soils with a long history of wheat culture (107). This factor does not occur in, or is present at ineffective concentrations in virgin soils. The antagonism apparently is similar to that associated with take-all decline reported in England (55), Holland (51), and Switzerland (140). Wheat fields presumably acquire a microflora more specific than exists naturally in the virgin areas. The new microbiota may include saprophytic organisms which are selected by exudates from the wheat plant and may be antagonistic to root pathogens such as *O. graminis*. When irrigation makes the environment favorable for *O. graminis,* the flora established during years of dryland wheat culture buffers against the pathogen.

Menzies (89) reported for the Columbia Basin that potato scab (*Streptomyces scabies*) would develop in soils newly reclaimed from native vegetation, but was greatly suppressed in "old soils" subjected to years of irrigated agriculture. Weinhold et al (137) grew potatoes in monoculture under irrigation in Southern California; scab increased for eight years, then declined. In irrigated fields of potatoes in Nebraska, Goss & Afanasiev (57) observed after several years that scab was less severe with continuous potatoes than with potatoes grown in 2- or 3-year rotations with sugar beets, barley, or corn. In contrast, a study with dryland potatoes in Nebraska revealed that scab did not decline but remained severe in continuous potatoes (138). This exception is highly significant because it supports the observations of Lapwood & Herring (75, 76) and Lewis (82) that *S. scabies* is limited by antagonism in wet but not in dry soils. This indicates that the buildup of antagonism to *S. scabies,* and thus scab decline, are achieved mainly, perhaps exclusively, where soils are kept sufficiently wet. The response of *S. scabies* to low water potentials is unknown, but generally actinomycetes are most active in dry soils (18, 71). This pathogen may likewise be capable of relatively good growth and thus have a high competitive advantage in soils of low water potential.

Influence of Water Potential on Host Resistance

If some factor of the environment is unfavorable to the host, various physiological processes (35, 50) including disease resistance (4) are commonly impaired. The predisposing effect of stress generally is not on initial establishment of the pathogen in the host, but rather on development of established infections.

Macrophomina phaseoli, cause of charcoal rot of many plant species, is apparently dependent for its pathogenicity on plant water stress, at least in sorghum and cotton (42, 52). In sorghum, production of typical severe disease in the greenhouse was possible only if plants grown in heated beds (35–40°C) were subjected to water stress (soil dried from 80–25% of "available soil moisture"). Inoculation was accomplished by direct insertion of sclerotia into the stem by toothpick puncture, thus lack of disease in moist soil cannot be attributed to inability of the pathogen to invade the plant. Charcoal rot of sorghum is most rapid and severe when a vigorous crop is subjected to a week or more of hot, dry weather between bloom and maturity (42). The combined stresses of blooming and lack of soil water predispose the plant to disease. The pathogen apparently infects early, but remains latent until the host is stressed, at which time the disease progresses rapidly (42).

Similarly in cotton, typical field symptoms of root rot were produced only on seedlings subjected to water stress prior to inoculation of the roots with *M. phaseoli* (52). The addition of Czapeks solution as a source of nutrition for the pathogen in moist soil did not increase infection, suggesting that disease of water-stressed plants did not result from increased exudation from the roots as sometimes happens with roots in dried, remoistened soil (68).

In the semi-arid portions of the northwestern U.S.A., infection of winter wheat by *Fusarium roseum* f. sp. *cerealis* 'Culmorum' occurs during fall (28), but subsequent decay at the base of the plants does not become rapid until late June-early July. The sudden onset of disease is favored by low soil moisture coupled with hot weather. This is very similar for *Macrophomina phaseoli* on sorghum (42). If the spring is wet, and particularly if precipitation is above normal in June, disease is suppressed.

We have studied water stress in wheat to determine the effects of different varieties, rates of nitrogen fertilization, and row spacings. Osmotic water potentials were determined in leaves by thermocouple psychometry (22). A limited number of water potential measurements (osmotic plus turgor) were made on intact leaves on plants in situ in the field. Beginning in early June (Figure 3), high rates of nitrogen fertilization and the close spacing between rows caused the lowest leaf osmotic potential and the greatest incidence of foot rot by mid-July. Percentage infection was the same at all levels of nitrogen and row spacings, but disease development was greater with large amounts of nitrogen, close spacing, or both.

Nitrogen fertilization increased rooting depth, root density, and plant leaf area, thus increasing the transpirational capacity of the plant. In addition, nitrogen increased the chlorophyl content of leaves, which presumably increased photosynthesis per unit leaf area and lowered the osmotic water potential through increased manufacture of sugars. We consistently obtained lower osmotic water potentials for deep green leaves than for yellow ones, even if taken from the same plant. Total water potential was generally similar on all leaves of an individual plant indicating that differences in osmotic potentials were balanced by differences in turgor pressure. Thus, a plant can develop a low osmotic potential through high N fertilization even if soil water is ample. The pathogen is favored by low osmotic potentials.

Mycelium of Culmorum within a host is presumably at about the same water potential as the host cells. Growth of Culmorum is stimulated by osmotic potentials down to −8 to −10 bars; it grows as well at −35 as at −1 bars (Table 3). Wider spacing can alleviate stress caused by limited soil water, but does not prevent development of low osmotic potentials within the host caused by high nitrogen supply or high evaporative demand.

Fomes annosus (129) and possibly *Fusarium solani* f. sp. phaseoli (21) are also favored by host water stress. *F. annosus* progressed more rapidly in loblolly pine trees subjected to water stress than in comparable trees well watered throughout the course of a field experiment (129). Pine seedlings in the greenhouse similarly showed more root rot if subjected to frequent but temporary wilting following inoculation than did comparable seedlings not stressed. *Fusarium solani* f. sp. *phaseoli* causes more disease of beans with narrow than with wide (20) row spacing; this disease can be reduced in irrigated fields of central Washington by subsoil tillage to break a hard pan at the plow depth and thus improve root penetration (21).

Vascular parasites have a close dependence on internal water relations of the plant since they exist within the xylem vessels and are carried as microconidia in the transpirational stream (38, 122).

In fusarium wilt of banana (10) and other hosts (8), vascular occlusions form within the vessel elements of both susceptible and resistant varieties in response to both parasitic and saprophytic microorganisms (9). Wilting results from the increased resistance to water flow in the xylem (40, 41, 86, 101), thereby preventing water flow from keeping pace with evaporative demand. Duniway (40, 41) points out, that resistance to water flow in stem xylem is normally so low that the increased resistance due to plugging cannot account for wilting. Leaves wilted only when resistance to waterflow was high in the petioles, presumably because of vascular occlusions formed there.

Because rapid upward spread of vascular parasites is accomplished largely by spores carried in the transpirational stream, it seems likely that treatments that reduce transpirational flow will slow upward spread, and thus complement the occlusion reaction of the host. Conversely, treatments that increase transpirational flow may, in effect, hasten upward spread of the parasite and hence onset of wilt.

Increased transpiration may explain why earlier workers (25, 84, 99, 121, 125) consistently observed more rapid development of wilt in wound-inoculated plants in the greenhouse with wet than with dry soils. Wet soil would not necessarily increase transpiration, but dry soil would decrease it, and thus reduce water flow and upward movement of the pathogen in the stem. Water flow through the plant generally cuts off sharply at some critical water potential (109). This could also explain the observation that rapidly growing susceptible tomato plants became "resistant" to wilt when in soil that was allowed to dry and susceptible again when the soil was made moist, thereby inducing rapid succulent growth of the host (25). Many conditions that favor growth of tomato plants also favor wilt (133). This may be just another way of saying that certain conditions unfavorable to tomato plant growth may reduce transpiration and hence pathogen spread. The frequent observation that wilt is most severe following hot, dry weather in the field may also relate to transpirational rate. Provided the soil is sufficiently moist to keep pace with the evaporative demand, an increase in transpiration rate will increase the rate of distribution of the pathogen throughout the plant.

The roots of winter wheat plants may be damaged during winter months in water-saturated soils subjected to frost heaving (14). This breaks roots and opens portals of entry for *Cephalosporium gramineum*. Soil saturated with water reduces oxygen diffusion to citrus roots thus increasing their susceptibility to *Phytophthara citrophthora* (117). Severe drying of soil has been shown to increase leakage of exudates from roots when the soil is subsequently watered. Such exudates may stimulate pathogens (68).

Disease Control by Management of Soil Water Potential

Control of soil water potential could provide a means of disease control, either alone or in combination with other controls. Too often water management is considered solely in terms of satisfying the crop water requirement and not in terms of its effect on disease.

With irrigation, the frequency and method of water application combined with certain tillage practices provides the most direct means for controlling soil water status. Under full irrigation, or during periods of frequent rainfall, high soil water potentials (often well above -1 bar) are maintained in much of the root zone with only limited drying between water applications. Although the wetter conditions minimize plant water stress, they may be conducive to diseases favored by wet soil (Table 3).

Plants may also be more or less susceptible to adverse effects of water stress depending on the stage of growth. For example, the most critical period for moisture in the growth of spring wheat planted in December in Arizona was the jointing stage (36). Stress at either the flowering or dough stage was less detrimental to yield. Where *O. graminis* is a hazard, an irrigation program designed to alleviate stress during critical growth states while allowing drier interim periods, might offer a measure of disease control. Water stress apparently is most critical to potato production during tuber formation (M. Campbell, unpublished). Heavy irrigation during this critical period, with lighter irrigations before and afterward might control common scab (78), but on the other hand, might increase damage by rootknot nematodes (127). The critical stress period in grain sorghum occurs shortly after bloom (42). Water applied during this period may control charcoal rot. For disease control, irrigation timing cannot be haphazard.

With furrow irrigation, the water spreads out in a cylindrical pattern below each furrow (64). When the wetting front contacts moist soil below, flow downward increases and lateral water movement decreases.This allows the surface layers on the raised beds where plants are grown to remain relatively dry while the roots receive ample water from lower depths. The height of water rise in the raised bed adjacent to the furrow can be increased by increasing the depth of water in the furrows, making lower beds, or decreasing the distance between furrows. The ideal combination will depend on soil texture and the capacity of the underlying soil to conduct water (126).

A more recent irrigation method employs pipes underground in the root zone where water is made available slowly and continually. Sold commercially under the trade name "Drip-feed"[1] the method makes possible a restricted soil zone of very high water potential (~ 0 bar) with the water application rate being determined by the evaporative demand. Surface soil and zones between rows can be kept relatively dry. Other than through direct effects on water availability to plants, the implications of this method to disease control are unknown, but may hold promise for control of pathogens dependent on high water potentials at or near the soil surface.

Cultivation can conserve water in the profile if applied so as to disrupt continuity of capillaries from the deeper layers to the surface where evaporation must take place. The practice hastens formation of a dry surface layer which resists both liquid and vapor flow to the atmosphere. A bare, undisturbed surface will often lose more water to the atmosphere than one that is cultivated, and much of this loss will occur from the shallow parts of the rooting zone. Subsoil tillage constitutes another tillage practice to improve soil water infiltration or plant water use for the reduction of plant disease (21). By conserving water in the profile, pathogens dependent for pathogenicity on plant water stress can be suppressed.

Development of crop varieties with different rooting patterns or water-use and transpirational characteristics has great potential but has been little attempted. Waukena White cotton plants, for example, form a few major lateral roots in the upper 45 cm of soil whereas Acala SJ-1 forms many lateral roots in this zone (94). Waukena White has field resistance to verticillium wilt whereas Acala is susceptible. In this case, field resistance may merely be escape since the inoculum of the pathogen is mostly in the surface layers of the soil. Such plants will also encounter more uniform and generally higher soil water potentials than those with roots in the surface layers where water potential fluctuations are great. In the northwestern U.S.A., some soft white winter wheat varieties appear to stress more slowly than others; they also resist *Fusarium roseum* f. sp. *cerealis* 'Culmorum' the longest (93). Resistance to this pathogen is now being sought, not only by conventional tests based on disease reactions of various varieties, but also by tests that select wheats for

[1] Trade names are mentioned here solely to provide specific information, and do not constitute a warranty of the product by the U. S. Department of Agriculture, or an endorsement by them over other products not mentioned.

their tendency to use soil water conservatively and thus stress slowly.

In conclusion, much can be done through breeding and management to affect water relations of the soil-plant-microbial system and hence disease severity. Some of these practices have already proven effective, but many must still be tried or applied to other diseases. Nevertheless, the possibilities are there, and largely await man's ingenuity to make them work.

LITERATURE CITED

1. Abeygunawardena, D. V. W., Wood, R. K. S. 1957. Factors affecting germination of sclerotia of *Sclerotium rolfsii*. *Trans. Brit. Mycol. Soc.* 40:221–31
2. Adebayo, A. A., Harris, R. F. 1971. Fungal growth responses to osmotic as compared to matric water potentials. *Soil Sci. Soc. Am. Proc.* 35:465–69
3. Aslyng, H. C. 1963. Soil physics terminology. *Int. Soc. Soil. Sci., Bull.* 23:1–4
4. Baker, K. F., Cook, R. J. 1973. *Biological control of plant pathogens*. W. H. Freeman, San Francisco. In press
5. Bakshi, B. K. 1957. Wilt disease of Shisham (*Dalbergia sissoo* Roxb.). IV. The effect of soil moisture on the growth and survival of *Fusarium solani* in the laboratory. *Indian For.* 83: 505–11. *Rev. Appl. Mycol.* 37: 117
6. Barker, W. G., Page, O. T. 1954. The induction of scab lesions on aseptic potato tubers cultured *in vitro*. *Science* 119: 286–87
7. Bateman, D. F. 1961. The effect of soil moisture upon the development of poinsettia root rots. *Phytopathology* 51:445–51
8. Beckman, C. H. 1968. An evaluation of possible resistance mechanisms in broccoli, cotton, and tomato to vascular infection by *Fusarium oxysporum*. *Phytopathology* 58:429–33
9. Beckman, C. H., Halmos, S. 1962. Relation of vascular occluding reactions in banana roots to pathogenicity of root-invading fungi. *Phytopathology* 52:893–97
10. Beckman, C. H., Halmos, S., Mace, M. E. 1962. The interaction of host, pathogen, and soil temperature in relation to susceptibility to Fusarium wilt of bananas. *Phytopathology* 52: 134–40
11. Boyle, Lytton W. 1961. The Ecology of *Sclerotium rolfsii* with emphasis on the role of saprophytic media. *Phytopathology* 51:117–19
12. Breazeole, E. L., McGeorge, W. T. 1953. Exudation pressure in roots of tomato plants under humid conditions. *Soil Sci.* 75: 293–98
13. Brown, R. W. 1970. Measurement of water potential with thermocouple psychrometers: construction and application. *U. S. Dept. Agr. Forest Serv. Res.* Paper INT80, 27 pp
14. Bruehl, G. W. 1968. Ecology of Cephalosporium stripe disease of winter wheat in Washington. *Plant Dis. Reptr.* 52:590–94
15. Bruehl, G. W., Cunfer, B. 1971. Physiologic and environmental factors that affect the severity of snow mold of wheat. *Phytopathology* 61:792–99
16. Bruehl, G. W., Cunfer, B., Toiviainen, M. 1972. Influence of water potential on growth, antibiotic production, and survival of Cephalosporium gramineum. *Can. J. Plant Sci.* 52: In press
17. Bruehl, G. W., Lai, P. 1968. Influence of soil pH and humidity on survival of Cephalosporium gramineum in infested wheat straw. *Can. J. Plant Sci.* 48: 245–52
18. Bumbieris, M., Lloyd, A. B. 1966. Influence of soil fertility and moisture on lysis of fungal hyphae. *Aust. J. Biol. Sci.* 20: 103–12
19. Burgess, L. W., Griffin, D. M. 1968. The recovery of Gibberella zeae from wheat straws. *Aust. J. Exp. Agr. Animal Husb.* 8:364–70
20. Burke, D. W. 1965. Plant spacings and Fusarium root rot of beans. *Phytopathology* 55:757–59
21. Burke, D. W., Miller, D. E., Holmes, L. D., Barker, A. W. 1972. Counteracting bean root rot by loosening the soil. *Phytopathology* 62:306–09
22. Campbell, G. S., Zollinger, W. D., Taylor, S. A. 1966. Sample changer for thermocouple psychrometers: Construction and some applications. *Agron. J.* 58:315–18
23. Chupp, C. 1925. *Manual of vegetable-garden diseases*. The Macmillan Co. New York. 647 pp.

24. Chupp, C. 1946. Soil temperature, moisture, aeration, and pH as factors in disease incidence. *Soil Sci.* 61:31–36
25. Clayton, E. E. 1923. The relation of soil moisture to the Fusarium wilt of tomato. *Am. J. Bot.* 10:133–47
26. Colhoun, J., Park, D. 1964. Fusarium diseases of cereals I. Infection of wheat plants, with particular reference to the effects of soil moisture and temperature on seedling infection. *Trans. Brit. Mycol. Soc.* 47: 559–72
27. Cook, R. J. 1964. *Influence of the nutritional and biotic environments of soil on the bean root rot Fusarium.* Ph.D. Thesis. 81 pp. Univ. Calif., Berkeley
28. Cook, R. J. 1968. Fusarium root and foot rot of cereals in the Pacific Northwest. Phytopathology 58:127–31
29. Cook, R. J. 1970. Factors affecting saprophytic colonization of wheat straw of *Fusarium roseum* f. sp. *cerealis* 'Culmorum'. *Phytopathology* 60:1672–76
30. Cook, R. J., Flentje, N. T. 1967. Chlamydospore germination and germling survival of *Fusarium solani* f. sp. *pisi* in soil as affected by soil water and pea seed exudation. *Phytopathology* 57:178–82
31. Cook, R. J., Papendick, R. I. 1970. Soil water potential as a factor in the ecology of *Fusarium roseum* f. sp. *cerealis* 'Culmorum'. *Plant Soil* 32:131–45
32. Cook, R. J., Papendick, R. I. 1971. Effect of soil water on microbial growth, antagonism and nutrient availability in relation to soil-borne fungal diseases of plants. p. 81–88. In *Root Diseases and Soil-borne Pathogens.* Ed. T. A. Toussoun, R. V. Bega, P. E. Nelson. Univ. Calif. Press. Berkeley
33. Cook, R. J., Papendick, R. I., Griffin, D. M. 1972. Growth of two root-rot fungi as affected by osmotic and matric water potentials. *Soil Sci. Soc. Am. Proc.* 36:78–82
34. Cook, R. J., Huber, D., Powelson, R. L., Bruehl, G. W. 1968. Occurrence of take-all in wheat in the Pacific Northwest. *Plant Dis. Reptr.* 52:716–18
35. Crafts, A. S. 1968. Water deficits and physiological processes. pp. 85–133. In *Water Deficits and Plant* Growth, Vol. II. ed. T. T. Kozlowski. Academic, New York & London
36. Day, A. D., Intalap, S. 1969. Don't stress your wheat for water! *Progr. Agr. Ariz.* 21:8–10
37. Dickson, J. G. 1923. Influence of soil temperature and moisture on the development of the seedling-blight of wheat and corn caused by *Gibberella saubinetii. J. Agr. Res.* 23:837–69
38. Dimond, A. E. 1970. Biophysics and biochemistry of the vascular wilt syndrome. *Ann. Rev. Phytopathol.* 8:301–32
39. Dube, A. J. 1971. *Studies on the growth and survival of* Rhizoctonia solani. Ph.D. thesis. Univ. Adelaide, Adelaide, Aust. 144 pp.
40. Duniway, J. M. 1971. Resistance to water movement in tomato plants infected with Fusarium. *Nature* 230:252–53
41. Duniway, J. M. 1971. Water relations of Fusarium wilt in tomato. *Physiol. Plant Pathol.* 1: 537–46
42. Edmunds, L. K. 1964. Combined relation of plant maturity, temperature, and soil moisture to charcoal stalk rot development in grain sorghum. *Phytopathology* 54:514–17
43. Falk, M., Hartman, K. A., Jr., Lord, R. C. 1963. Hydration of deoxyribonucleic acid. III. A spectroscopic study of the effect of hydration on the structure of deoxyribonucleic acid. *J. Am. Chem. Soc.* 85:391–94
44. Fellows, H., Ficke, C. H. 1939. Soil infestation by *Ophiobolus graminis* and its spread. *J. Agr. Res.* 58:505–19
45. Flentje, N. T., Saksena, H. K. 1964. Pre-emergence rotting of peas in south Australia. III. Host-pathogen interaction. *Aust. J. Biol. Sci.* 17:665–75
46. Gardner, W. R., Ehlig, C. F. 1965. Physical aspects of the internal water relations of plant tissues. *Plant Physiol.* 40:705–10
47. Garrett, S. D. 1934. Factors af-

fecting the severity of take-all. III. The climatic factors. *South Aust. Dept. Agr. J.* 37:976–83
48. Garrett, S. D. 1944. *Root disease fungi.* Chronica Botanica Co. Waltham, Mass. 177 pp.
49. Garrett, S. D. 1956. *Biology of root-infecting fungi.* Cambridge Univ. Press, Cambridge, 293 pp.
50. Gates, C. T. 1964. The effect of water stress on plant growth. *J. Aust. Inst. Agr. Sci.* 30:3–22
51. Gerlagh, M. 1968. Introduction of *Ophiobolus graminis* into new polders and its decline. *Neth. J. Plant Pathol.* 74: Suppl. No. 2, 97 pp.
52. Ghaffar, A., Erwin, D. C. 1969. Effect of soil water stress on root rot of cotton caused by *Macrophomina phaseoli. Phytopathology* 59:795–97
53. Gilman, J. C. 1916. Cabbage yellows and the relation of temperature to its occurrence. *Ann. Missouri Bot. Garden.* 3:25–81
54. Glynne, Mary D. 1953. Production of spores by *Cercosporella herpotrichoides. Trans. Brit. Mycol. Soc.* 36:46–51
55. Glynne, Mary D. 1965. Crop sequence in relation to soil-borne pathogens. p. 423–435. In *Ecology of soil-borne plant pathogens.* Ed. K. F. Baker, W. C. Snyder. Univ. Calif. Press, Berkeley
56. Goss, R. W. 1937. The influence of various soil factors upon potato scab caused by *Actinomyces scabies. Nebr. Agr. Exp. Sta. Res. Bull.* 93
57. Goss, R. W., Afanasiev, M. M. 1938. Influence of crop rotations under irrigation on potato scab, Rhizoctonia, and Fusarium wilt. *Nebr. Agric. Exp. Sta. Bull.* 317
58. Graham, J. H., Sprague, V. G., Robinson, R. R. 1957. Damping-off of Ladino clover and Lespedeza as affected by soil moisture and temperature. *Phytopathology* 47:182–85
59. Griffin, D. M. 1963. Soil moisture and the ecology of soil fungi. *Biol. Rev.* 38:141–66
60. Griffin, D. M. 1966. Fungi attacking seeds in dry seed-beds. *Proc. Linnean Soc. N.S.W.* 91:84–89
61. Griffin, D. M. 1969. Soil water in the ecology of fungi. *Ann. Rev. Phytopathol.* 7:289–310
62. Griffin, D. M. 1970. Effect of soil moisture and aeration on fungal activity: an introduction. p. 77–80. In *Root diseases and soil-borne pathogens.* Ed. T. A. Toussoun, R. V. Bega, P. E. Nelson. Univ. Calif. Press, Berkeley
63. Griffin, D. M., Quail, G. 1968. Movement of bacteria in moist, particulate systems. *Aust. J. Biol. Sci.* 21:579–82
64. Haise, Howard R. 1948. Flow pattern studies in irrigated coarse-textured soils. *Soil Sci. Am. Proc.* 13:83–89
65. Harter, L. B., Whitney, W. A. 1927. The relation of soil temperature and soil moisture to the infection of sweet potatoes by the stem rot organism. *J. Agr. Res.* 34:435–41
66. Humbert, J. G. 1918. Tomato diseases in Ohio. *Ohio Agr. Exp. Sta. Bull.* 321
67. Hynes, H. J. 1937. Studies on Rhizoctonia root-rot of wheat and oats. *Sci. Bull.* 58:42 pp. Dept. Agr. N.S.W. Sydney
68. Katznelson, H., Rouatt, J. W., Payne, T. M. B. 1955. The liberation of amino acids and reducing compounds by plant roots. *Plant Soil* 7:35–48
69. Kerr, A. 1964. The influence of soil moisture on infection of peas by *Pythium ultimum. Aust. J. Biol. Sci.* 17:676–85
70. King, C. J., Eaton, E. D. 1934. Influence of soil moisture on longevity of cotton root-rot sclerotia. *J. Agr. Res.* 49:793–98
71. Kouyeas, V. 1964. An approach to the study of moisture relations of soil fungi. *Plant Soil* 20:351–63
72. Kraft, J. M., Roberts, D. D. 1969. Influence of soil water and temperature on the pea root rot complex caused by *Pythium ultimum* and *Fusarium solani* f. sp. *pisi. Phytopathology* 59:149–52
73. Langridge, R., Seeds, W. E., Wilson, H. R., Hooper, C. W., Wilkins, M. H. F., Hamilton, L. D. 1957. Molecular structure of deoxyribonucleic acid (DNA). *J. Biophys. Biochem. Cytol.* 3:767–78

74. Lapwood, D. H. 1966. The effects of soil moisture at the time potato tubers are forming on the incidence of common scab (*Streptomyces scabies*). *Ann. Appl. Biol.* 58:447–54
75. Lapwood, D. H., Herring, T. F. 1968. Infection of potato tubers by common scab (*Streptomyces scabies*) during brief periods when soil is drying *Eur. Potato J.* 11:177–87
76. Lapwood, D. H., Herring, T. F. 1970. Soil moisture and the infection of young potato tubers by *Streptomyces scabies* (common scab). *Potato Res.* 13: 296–304
77. Lapwood, D. H., Lewis, B. G. 1967. Observations on timing of irrigation and the incidence of potato common scab (*Streptomyces scabies*). *Plant Pathol.* 16:131–35
78. Lapwood, D. H., Wellings, L. W., Rosser, W. R. 1970. The control of common scab of potatoes by irrigation. *Ann. Appl. Biol.* 66:397–405
79. Leach, L. D. 1947. Growth rates of host and pathogen as factors determining the severity of preemergence damping-off. *J. Agr. Res.* 75:161–79
80. Leamer, R. W., Lutz, J. F. 1940. Determination of pore-size distribution in soils. *Soil Sci.* 49: 347–60
81. Ledingham, R. J. 1970. Survival of *Cochliobolus sativus* conidia in pure culture and in natural soil at different relative humidities. *Can. J. Bot.* 48:1893–96
82. Lewis, B. G. 1970. Effects of water potential on the infection of potato tubers by *Streptomyces scabies* in soil. *Ann. Appl. Biol.* 66:83–88
83. Linderman, R. G., Toussoun, T. A. 1967. Behavior of chlamydospores and endoconidia of *Thielaviopsis basicola* in nonsterilized soil. *Phytopathology* 57:729–31
84. Linford, M. B. 1928. A *Fusarium* wilt of peas in Wisconsin. *Wisc. Agr. Exp. Sta. Res. Bull.* 85
85. Linford, M. B. 1931. Transpirational history as a key to the nature of wilting in the *Fusarium* wilt of pea. *Phytopathology* 21:791–96
86. Ludwig, R. A. 1952. Studies on the physiology of hadromycotic wilting in the tomato plant. *MacDonald Col. Tech. Bull.* No. 20, McGill Univ., Montreal
87. Malalasekera, R. A. P., Colhoun, J. 1968. *Fusarium* diseases of cereals III. Water relations and infection of wheat seedlings by *Fusarium culmorum*. *Trans. Brit. Mycol. Soc.* 51:711–20
88. Manandhar, J. B. 1971. *Water relations of* Fusarium *and* Verticillium *wilt fungi in vitro*. M.S. Thesis. Dept. Plant Pathol., Washington State Univ., Pullman
89. Menzies, J. D. 1959. Occurrence and transfer of a biological factor in soil that suppresses potato scab. *Phytopathology* 49:648–52
90. Munzumder, B. K. G., Caroselli, N. E., Albert, L. S. 1970. Influence of water activity, temperature, and their interaction on germination of *Verticillium albo-atrum* conidia. *Plant Physiol.* 46:437–39
91. Papavizas, G. C., Lewis, J. A. 1971. Survival of endoconidia and chlamydospores of *Thielaviopsis basicola* as affected by soil environmental factors. *Phytopathology* 61:108–13
92. Papendick, R. I., Cochran, V. L., Woody, W. M. 1971. Soil water potential and water content profiles with wheat under low spring and summer rainfall. *Agron. J.* 63:731–34
93. Papendick, R. I., Cook, R. J., Shipton, P. J. Plant water stress and the development of Fusarium foot rot in wheat. *Phytopathology* 61:905
94. Phillips, D. J., Wilhelm, S. 1971. Root distribution as a factor influencing symptom expression of Verticillium wilt of cotton. *Phytopathology* 61:1312–13
95. Poole, R. F. 1925. The relation of soil moisture to the pox or ground rot disease of sweet potatoes. *Phytopathology* 15:287–93
96. Powell, N. T. 1971. Interactions between nematodes and fungi in disease complexes. *Ann. Rev. Phytopathol.* 9:253–74
97. Rawlins, Stephen L. 1971. Some new methods for measuring the

components of water potential. *Soil Sci.* 112:8–16
98. Rogers, C. H. 1939. The relation of moisture and temperature to growth of the cotton root-rot fungus. *J. Agr. Res.* 58:701–09
99. Ryker, T. C. 1935. Fusarium yellows of celery. *Phytopathology* 25:578–600
100. Sandford, G. B. 1923. The relation of soil moisture to the development of common scab of potato. *Phytopathology* 13: 231–36
101. Scheffer, R. P., Walker, J. C. 1953. The physiology of Fusarium wilt of tomato. *Phytopathology* 43:116–25
102. Schippers, B., Schroth, M. N., Hildebrand, D. C. 1967. Emanation of water from underground plant parts. *Plant Soil* 27:81–91
103. Schneider, R. 1954. Untersuchungen uber feuchtigkeitsanspruche parasitischer Pilze. *Phytopathol. Z.* 21:61–78
104. Scott, W. J. 1957. Water relations of food spoilage microorganisms. *Advan. Food Res.* 7:83–127
105. Sequeira, L. 1958. Bacterial wilt of bananas: Dissemination of the pathogen and control of the disease. *Phytopathology* 48:64–69
106. Shen, C. I. 1940. Soil conditions and the *Fusarium culmorum* seedling blight of wheat. *Ann. Appl. Biol.* 27:323–29
107. Shipton, P. J., Cook, R. J., Sitton, J. W. 1973. Occurrence and transfer of a biological factor in soil that suppresses take-all of wheat in eastern Washington. *Phytopathology* 63:In press
108. Slatyer, R. O. 1957. The influence of progressive increase in total soil moisture stress, on transpiration, growth, and internal water relationships of plants. *Aust. J. Biol. Sci.* 10: 320–36
109. Slatyer, R. O. 1967. *Plant Water Relationships.* Academic Press, Inc., New York, N. Y. 366 pp.
110. Smith, A. M. 1972. Drying and wetting sclerotia promotes biological control of *Sclerotium rolfsii. Soil Biol. Biochem.* 4: In press
111. Smith, A. M., Griffin, D. M. 1970. Oxygen and the ecology of *Armillariella elegans.* Heim. *Aust. J. Biol. Sci.* 24:231–62
112. Smith, Shirley N., Snyder, W. C. 1971. Relationship of inoculum density and soil types to severity of Fusarium wilt of sweet potato. *Phytopathology* 61: 1049–51
113. Sommers, L. E., Harris, R. F., Dalton, F. N., Gardner, W. R. 1970. Water relations in three root-infecting *Phytophthora* species. *Phytopathology* 60: 932–34
114. Sprague, R. 1948. Cereal disease situation in eastern Washington this season. *Plant Dis. Reptr.* 32:392–94
115. Stanghellini, M. E., Hancock, J. G. 1971. Radial extent of the bean spermosphere and its relation to the behavior of *Pythium ultimum. Phytopathology* 61:165–68
116. Staten, G., Cole, J. F. Jr. 1948. The effect of pre-planting irrigation on pathogenicity of *Rhizoctonia solani* in seedling cotton. *Phytopathology* 38:661–64
117. Stolzy, L. H., Letey, J., Klotz, L. J., Labanauskas, C. K. 1965. Water and aeration as factors in root decay of *Citrus sinensis. Phytopathology* 55:270–75
118. Stolzy, L. H., Van Gundy, S. D. 1968. The soil as an environment for microflora and microfauna. (Symposium on Microenvironment and Biology of Soil) *Phytopathology* 58:889–99
119. Stover, R. H. 1953. The effect of soil moisture on Fusarium species. *Can. J. Bot.* 31:693–97
120. Stover, R. H. 1962. Fusarial wilt (Panama disease) of bananas and other Musa species. *Commonw. Mycol. Inst. Phytopathol. Paper* No. 4, 117 pp.
121. Strong, M. C. 1946. The effects of soil moisture and temperature on Fusarium wilt of tomato. *Phytopathology* 36:218–25
122. Talboys, P. W. 1968. Water deficits on vascular disease. In *Water deficits and plant growth,* Vol. II. 255–311. Ed. Kozlowski, T. T. New York, London: Academic
123. Taubenhaus, J. J., Dana, B. F. 1928. The influence of moisture and temperature on cotton root

rot. *Tex. Agr. Exp. Sta. Bull.* 386, 23 pp.

124. Taubenhaus, J. J., Ezekiel, W. N. 1936. Longevity of sclerotia of *Phymatotrichum omnivorum* in moist soil in the laboratory. *Am. J. Bot.* 23:10–12
125. Tharp, W. H., Young, V. H. 1939. Relation of soil moisture to Fusarium wilt of cotton. *J. Agr. Res.* 58:47–61
126. Thorne, D. W., Peterson, H. B. 1954. *Irrigated Soils* (2nd ed.). New York, Toronto: Blakiston Co., Inc.
127. Thorne, G. 1942. Distribution of the root-knot nematode in high ridge plantings of potatoes and tomatoes. *Phytopathology* 32: 650
128. Tisdale, W. B. 1923. Influence of soil temperature and soil moisture upon the Fusarium disease in cabbage seedlings. *J. Agr. Res.* 24:55–86
129. Towers, B., Stambough, W. J. 1968. The influence of induced soil moisture stress upon *Fomes annosus* root rot of loblolly pine. *Phytopathology* 58:127–268
130. U. S. Salinity Laboratory Staff. 1954. Diagnosis and improvement of saline and alkali soils. *Agr. Handbook No. 60,* US Dep. Agr.
131. Wadleigh, C. H., Fireman, M. 1948. Salt distribution under furrow and basin irrigated cotton and its effect on water removal. *Soil Sci. Soc. Am. Proc.* 13:527–30
132. Walker, J. C. 1952. *Diseases of vegetable crops.* New York: McGraw-Hill. 529 pp.
133. Walker, J. C. 1971. Fusarium wilt of tomato. Monograph No. 6, 56 pp. *Am. Phytopathol.* Soc., St. Paul
134. Wallace, H. R. 1958. Movement of eelworms. I. The influence of pore size and moisture content of the soil on the migration of larvae of the beet eelworm, *Heterodera schactii* Schmidt. *Ann. Appl. Biol.* 46:74–85
135. Wallace, H. R. 1963. *The biology of plant parasitic nematodes.* New York: St. Martins Press. 280 pp.
136. Webb, S. J., Cormack, D. V., Morrison, H. G. 1964. Relative humidity, inosital and the effect of radiations on air-dried microorganisms. *Nature* 201:1103–05
137. Weinhold, A. R., Oswald, J. W., Bowman, T., Bishop, J., Wright, D. 1964. Influence of green manures and crop rotation on common scab of potato. *Am. Potato J.* 41:265–73
138. Werner, H. O., Kiesselbach, T. A., Goss, R. W. 1944. Dry-land crop rotation experiments with potatoes in northwestern Nebraska. *Nebr. Agr. Exp. Sta. Bull.* 363. 43 pp.
139. Wiebe, H. H., Campbell, G. S., Gardner, W. H., Rawlins, S. L., Cary, J. W., Brown, R. W. 1971. Measurement of plant and soil water status. *Utah Agr. Exp. Sta. Bull.* 484. 71 pp
140. Zogg, H. 1951. Studien über die pathogenitäl von Erregernischen bei Getreldefusskrankheiten. *Phytopathol. Z.* 18:1–54

FACTORS AFFECTING THE EFFICACY OF FUNGICIDES IN SOIL[1]

3556

DONALD E. MUNNECKE
Department of Plant Pathology, University of California, Riverside

INTRODUCTION

In preparing this article I have been impressed with the extent of knowledge on the fate of pesticides in soil. Although perhaps 20 years ago little was known about soil fungicides, a great deal is now known, as attested by an article of 93 pages and approximately 600 references (33), and an annotated bibliography (98). Much of the information concerns insecticides and herbicides, but is readily applicable to fungicides. While some fundamental advances follow empirical discoveries, experiments based upon fundamental information are more apt to lead to success. For this reason a brief review of some of the fundamental knowledge of soil pesticides is presented here.

A list of chemicals discussed, and their names or abbreviations used in the text will be found in Table 1.

FUNDAMENTAL FACTORS THAT AFFECT THE FIELD PERFORMANCE OF FUNGICIDES

PHYSICAL ASPECTS

The soil is an extremely complex physical-chemical-biological system of interactants (5, 29, 30), yet chemicals applied to soil are governed by specific physical laws and principles derived primarily from systems less complex than those existing in soil. Thus, the laws for diffusion for both liquids and gases apply in soil. Henry's constant (the ratio of concentration of a pesticide in water to its concentration in air) may be used partially to predict activity of a pesticide in soil. Other physical properties such as water and fat solubility, molecular weight, and vapor pressure are known for most compounds, and are useful in determining the fate of a chemical in the soil.

Leistra (48, 49) showed how the application of physical laws can be used to follow fate of a pesticide in soil. He devised a FORTRAN computer program for the diffusion of 1,3-dichloropropene from a plane source in soil. He claimed that sufficient information may be obtained in this way to use the model to predict the efficacy of the fumigant and to determine the effect of

[1] This review covers the literature from 1964 to 1971. For brevity some pertinent papers were omitted; others may have been inadvertently overlooked. I apologize for any such omissions.

TABLE 1. Some pesticide chemicals named in the text, and the abbreviations used

Name or abbreviation used	Active chemical
Fungicides	
benomyl	methyl 1-(butylcarbamoyl)-2-benzimidazolcarbamate
captan	*N*-trichloromethylmercapto-4-cyclohexene-1,2-dicarboximide
Ceresan M	ethylmercury *p*-toluene sulfonanilide
dazomet (mylone)	3,5-dimethyltetrahydro-1,3,5-2*H*-thiadiazine-2-thione
dexon	*p*-(dimethylamino)benzenediazo sodium sulfonate
ferbam	ferric dimethyldithiocarbamate
MMDD (Panogen)	methylmercury dicyandiamide
nabam	disodium ethylene*bis*dithiocarbamate
PCNB	pentachloronitrobenzene
Plantvax (oxycarboxin)	2,3-dihydro-5-carboxanilido-6-methyl-1,4-oxathiin-4,4-dioxide
PMA	phenylmercury acetate
Semesan	2-chloro-4-(hyroxymercuri) phenol
thiram	tetramethylthiuram disulfide
vapam	sodium *N*-methyldithiocarbamate dihydrate
Vitavax (carboxin)	2,3-dihydro-5-carboxanilido-6-methyl-1,4-oxathiin
ziram	zinc dimethyldithiocarbamate
Fumigants	
chloropicrin	trichloronitromethane
CS_2	carbon disulfide
D-D	1,3-dichloropropene and 1,2-dichloropropane mixture
ethylene oxide	ethylene oxide
methyl bromide	methyl bromide
MIT	methyl isothiocyanate
propylene oxide	propylene oxide
Telone (1,3-D)	1,3-dichloropropene
Insecticides	
DDT	1,1,1-trichloro-2,2-*bis*(*p*-chlorophenyl)ethane
Dieldrin	1,2,3,4,10,10-hexachloro-6,7-epoxy-1,4,4a,5,6,7,8,8a octahydro-1,4-*endo*,*exo*-5,8-dimethanonaphthalene
Lindane	α-1,2,3,4,5,6-hexachlorocyclohexane

various conditions on diffusion of the gas. Studies of this nature are leading to a better understanding of the complex interactions in the soil. Goring (30) has stressed that we need to know, in addition to the physical-chemical data generally available, data on the organic matter-to-water ratios, the rate of decomposition of a compound in soil and the concentration/time (CT) products necessary to give a certain level of kill of the more important pathogens. For the most part these data are missing.

Munnecke et al (74) determined the CT necessary to give an LD_{95} dose of methyl bromide against *Armillaria mellea*. Pieces of citrus roots artificially infested with *A. mellea* were exposed to concentrations of methyl bromide in air controlled by the method of Kolbezen et al (45). After treating, the pieces containing the weakened *A. mellea* were buried in nonsterile orchard soil for 21 days exposed to natural biological antagonism. Pieces from the roots were plated on agar to determine fungus viability. A dosage response curve for 95% kill was plotted and from it a CT value of approximately 5000 ppm/h was obtained. More data of this type are needed for other pathogens.

The role of sorption in soil is of prime importance. Hartley (32) stated that sorption of pesticides by the clay fractions of soil had been overstressed, and that sorption by the organic matter is more important. Munnecke et al (70) followed the release of MIT from various clays mixed with sand. Although clays initially stimulated release of MIT, no evidence of sorption was found comparable to silica sand alone. Humic acid added as ammonium humate to a mineral soil also resulted in increased release of MIT, presumably by affecting changes in pH. The authors pointed out that other soil humus fractions such as the polysaccharides fraction might be more sorptive. An aspect of sorption that has been frequently overlooked, according to Marshall (54), is the physiochemical effect of solid-liquid interfaces on microbiol behavior. An apparent anomaly exists: most soil organisms occur in the sorbed state, and toxicants must be sorbed by organisms to be effective; yet sorption by soil decreases toxicity. Why? Marshall believes this is because relatively few organisms are present in soil compared to the almost infinite sorptive sites of soil. Chance alone dictates that most of the toxicant molecules cannot reach vital sites in the organism until large concentrations of the pesticide are attained. There may be differential rates of sorption between the different types of organic matter, for example, as contrasted to sorption by a bacterial cell wall. Soil microbiologists have collected evidence with the scanning electron microscope to support Marshall's views that bacteria and fungi are not nearly so numerous as formerly thought. However, Webley & Jones (100) claim that, if electron microscope techniques are accurate, the numbers of soil bacteria could be 4-10 times greater than the measure by direct count.

Another activity of sorption is the catalytic activity of soils. Usually compounds such as nabam, vapam, or dazomet are more rapidly broken down in soil than in dilute aqueous solution (70). Mortland (67) stressed that soil clays may activate breakdown more than has been realized, and perhaps less breakdown is due to biological activity than was thought.

Soil moisture is another vital factor in pesticide activity. Pesticides generally are more active and mobile in wet than in dry soil, whereas fumigants are less mobile in wet than dry soils. Spencer et al (92) noted that increasing soil moisture displaces pesticides from absorbed surfaces, increasing its concentration in soil solution and soil air, making the pesticide available at its site of activity or more available for volatilization. The authors calculated that values of the vapor pressure of lindane and dieldrin were several times higher than the published vapor pressures. They cautioned that vapor pressure should be rechecked in future studies.

Chemical Aspects

Much is known about the chemical breakdown of many individual pesticides in soil, and with increased interest, and the advent of new analytical techniques, information is accumulating rapidly (68). Crosby & Li (19) stressed that practically all herbicides are capable of photodecomposition, a point not generally appreciated since the reaction products are so similar to those produced by biological or chemical decay. The photodecomposition of dexon is probably the best known fungicide example. Photodecomposition of fungicides should be more carefully evaluated.

One compound may act upon another, resulting in alteration of the fungicidal action of the second compound. Miller & Lukens (60) reported that D-D reacted with vapam in soil, resulting in a more rapid release of MIT from the vapam. Although this may not be a common occurrence in soil, it deserves consideration in studies in the field.

Microbial Aspects

Excellent reviews are available concerning microbial breakdown of pesticides (4, 33, 56, 103). Helling et al (33) pointed out a number of factors sometimes overlooked. Laboratory conditions are not the same as in the field: pure cultures of microorganisms behave differently from naturally occurring mixed cultures; chemicals are not homogeneously distributed in the field; solvents and carriers may be harmful to the biotic environment; commercial formulations may contain contaminants that are responsible for the side-effects noted by the investigator.

Concept of "microbial infallibility."—Attitudes of researchers toward pesticides and soil microorganisms have changed. Alexander (2) was perhaps first to challenge the then-present notion of "microbial infallibility" (i.e., all organic molecules would be subject to attack by some microbe in the soil), citing among others, the persistence of DDT in the biological food chain. Before this time the persistence in soil of substances like DDT, dieldrin, lindane, and PCNB was looked upon as an exception to the rule of "microbial infallibility," and numerous attempts were made to develop other persistent compounds. As the hazards of persistent compounds became apparent, research efforts switched to developing compounds with less persistent activity. Coincidentally, researchers found that even DDT may be microbially degraded (actually it is degraded quite rapidly under anaerobic conditions). Also, soil microflora may be "trained" more readily to alter a pesticide when successively higher applications of the compound are applied (69). It appears that "microbial infallibility" may remain as a reasonable concept that applies to most situations in soil, although a number of compounds obviously resist microbial decomposition under most conditions.

Examples of reactions involved in the biological breakdown of pesticides.—Chacko et al (14) used an enrichment technique to develop and isolate organisms capable of degrading chlorinated hydrocarbons such as DDT, dieldrin, and PCNB. Eight species of fungi (*Aspergillus niger, Fusarium solani,* f. sp. *phaseoli, Glomerella cingulata, Helminthosporium victoriae, Mucor ramannianus, Myrothecium verrucaria, Penicillium frequentans, and Trichoderma viride*) degraded PCNB. Also, 8 isolates of actinomycetes reduced PCNB to pentachloroaniline (PCA). Degradation occurred only during active growth phases of the organisms. They proposed that growth in soil is not very active; hence the chlorinated hydrocarbons persist for a long time in natural soils. Kaufman (39) enlarged on this work, showing that in addition to PCA, methylthiopentachlorophenol was formed from PCNB.

An extensive literature has developed concerning biological breakdown of herbicides in soil. The data of Bollag & Alexander (11) should be cited as an excellent example of the status of such research. In addition to isolating organisms responsible for breakdown of compounds in soil, they isolated enzyme preparations from the bacteria capable of detoxifying the chlorinated aliphatic acid herbicides.

A series of careful thorough papers by Castro, Bartnicki & Belser (8, 9, 15–18) followed the chemistry of biological and nonbiological degradation of halogenated nematocides and their breakdown products. *Cis* and *trans* 1, 3,-dichloropropene (I) hydrolyze in moist soil to the corresponding 3-chlorlyl alcohols (II) which are biocidal. In laboratory experiments, cultures of a species of *Pseudomonas* isolated from soil previously enriched with (I) eventually converted the residues in steps to Cl^- and CO_2. The steps follow: (II) was converted into formylacetic acid (III) with the release of Cl^-. Rapidly, after the previous slow step, (III) is decarboxylated into CO_2. Their work is notable in that the data are confirmed by using materials-balance calculations, rate-curve analyses, stoichiometry, and stereochemistry.

Kaars Sijpesteijn & Vonk (38) summarized the microbial conversions of some dithiocarbamate fungicides. Thiram is reduced to dimethyldithiocarbamate by microbes or reducing agents. When soil pH is below pH 7, dimethyl dithiocarbamic acid is formed. Eventually this resolves to dimethyl amine and CS_2. The dimethyl amine may volatilize or be metabolized by soil bacteria. The ethylenebisdithiocarbamates, in contrast, have not been shown to be degraded microbiologically. However, they are unstable in aqueous solutions, and may rapidly form ethylene diamine and CS_2, which are both volatile. In aerated soil however, nabam forms ethylenethirammonosulfide (ETM) and ethylenethiourea (ETU) or ethylene diamine and CS_2. ETU appears to be the only decomposition product of bisdithiocarbamate fungicides that persists in soil. The authors neglected a report by Moje et al (64), who showed that with an acid soil treated with nabam, carbonyl sulfide (COS) and H_2S were gaseous products in air above the soil. While H_2S was quite innocuous to test fungi, vapors of COS were quite toxic to *Pythium irregulare* in vitro. These reactions with nabam illustrate how complex the breakdown of fungicides in soils may be.

The breakdown of mercury-containing compounds illustrates the need for factual information in assessing the impact of pesticides on the environment. As a result of several well-documented cases of poisoning of people by food containing methylmercury, the fungicides have been withdrawn from use. Several Japanese groups (25, 36, 58) have investigated pathways of mercury breakdown. Matsumura et al (58) reported that phenylmercuric acetate (PMA) was quickly metabolized by soil and aquatic organisms, and that one of the major products was diphenylmercury; methylmercury was not found, although it had been suspected. Although they pointed out that PMA does not yield the highly toxic methylmercury, the toxicological implications of diphenylmercury remain obscure and need investigating. In this respect, an early report by Spanis et al (91) demonstrated that mercury-containing fungicides (MMDD, Semesan) were rapidly detoxified (as determined by a bioassay using *Myrothecium verrucaria*) by bacteria and fungi, but neither the breakdown products nor the chemistry of these reactions was identified. Furukawa & Tonomura (25), in contrast, used an enzyme preparation from *Pseudomonas* sp. which decomposed phenylmercuric acetate (PMA) to metal mercury. They demonstrated that the inducible enzyme, a reduced NAD(P)-generating system, glucose dehydrogenase or arabinose dehydrogenase, and cytochrome C-I were required for decomposition of PMA.

Effect on soil microorganisms.—Soil fungicides, in common with other pesticides, are apt to have serious disruptive effects upon soil microorganisms. The more effective a chemical is as a fungicide, the more disruptive it is to soil microorganisms. This applies especially to fumigants such as methyl bromide and chloropicrin. In contrast, materials such as PCNB, captan, and dexon have only slight lasting effects on soil microflora (20, 68). Alconero & Hagedorn (1) followed the persistence of dexon in soil and the fungus population in the treated soil. Numbers of *Pythium* and *Aphanomyces* decreased, but other fungi in soil did not decrease, confirming earlier reports of the specificity of dexon toward Pythiaceae. Another relatively specific type of compound, PCNB, is commonly used to control *Rhizoctonia* and other plant pathogens. Chacko et al (14) first reported that PCNB was degraded to pentachloroaniline (PCA) by a total of 8 common soil fungi and much more so by 8 isolates of actinomycetes only during active growth of the organisms. Ko & Farley (43) followed this by determining whether the parent compound, PCNB, or its breakdown product, PCA, was the active fungicide in the soil. Both PCNB and PCA were inhibitory to soil actinomycetes and fungi in nutrient agar, but had no effect on bacteria. PCNB was generally more toxic, however, and the authors considered that the formation of PCA is a step towards detoxification against soil microorganisms. In another paper, Ko & Lockwood (44) reported that *R. solani* accumulated levels of PCNB or other chlorinated hydrocarbons above the ambient concentration in soil. Although the amount accumulated was small, this is another means of detoxifying pesticides.

Effect on plants.—Because they are phytotoxic, fumigants usually are not applied around living plants. Generally speaking, fungicides applied properly have had few untoward effects on crops subsequently grown in the soil. Thorn & Richardson (94, 95) carefully measured the effects of 9 fungicides applied as soil drenches on amino acid transport in tomato plants. They used nabam (I), MMDD (II), dexon (III), benomyl (IV), Vitavax (V), Plantvax (VI), and two proprietary compounds. After 24 hours the plants were topped, and the exudate collected from the stumps for the succeeding 6 hours. Compounds (I), (II), and (III) markedly reduced the volume of the exudate, but the others did not. Amino acid content was not changed appreciably, except that (I) decreased amino acids to low levels, and (II) increased the alanine content. In contrast, when cycloheximide was used, transport was markedly reduced, aminobutyric acid and sodium increased, and glutamine, α-ketogluterate, and potassium decreased. The authors suggested that cycloheximide may have other effects on plants than the inhibition of protein synthesis.

Fungicides may have genetic effects on plants. Prasad & Pramer (79) reported that ferbam (240–1000 ppm) induced mutations in *Aspergillus niger* and caused unusually high chromosome aberration in *Allium cepa* root tips. George et al (26) soaked barley seeds for 5–10 days in fungicides and determined the effects on chromosomes as well as upon growth. Ziram stunted plants and decreased germination of seeds; most importantly, it induced chromosome aberrations. Treatment with Ceresan, PCNB, and thiram also induced chromosome aberrations.

The effects of ziram on barley are reported by Pilinskaya (78) who also observed chromosome aberrations in leucocytes of 9 persons who worked in a factory producing it. He concluded that ziram presents a genetic danger for somatic human cells.

These reports of potential genetic dangers to man and plants illustrate the need for more research on this aspect of fungicides in soil. It should be noted, however, that the concentrations used were unusual and the systems were not typical.

The effects of fungicides on soil algae have rarely been reported but Moore (66) found that nabam and vapam were more toxic and inhibited photosynthesis and growth of *Euglena gracilis* strain Z more than did parathion or malathion.

Effect on disease control.—The direct effect of fungicides in controlling diseases is well known. Occasionally a fungicide does not control a pathogen by direct fungicidal action. One well-known example is the use of carbon disulfide to control *Armillaria mellea.* Low concentrations of the toxicant apparently result in an increased activity of soil microorganisms, *Trichoderma viride* being most commonly implicated. The result is that *A. mellea* is killed in field soil, even though the concentration of the gas reaching the fungus may be sub-lethal. Probably similar relations exist with other fungicide-pathogen systems.

Richardson (81) reported that the resultant protection for seedlings in soil treated with thiram against *Pythium ultimum* persisted longer than did the fungicide. Munnecke & Michail (71) used higher concentrations of thiram (250, 500, or 1000 ppm), and found that damping-off control was directly related to concentration of thiram remaining in the soil at planting time. They used a bioassay that was not specific for thiram. The authors concluded that damping-off control was more directly related to the presence of the fungicide than to a stimulation by thiram of a biotic environment antagonistic to *Pythium ultimum.* A similar situation has been reported using dexon in control of pythiacious damping-off. Using a potato-disc bioassay procedure, Khan & Baker (42) claimed that dexon concentration declined almost linearly for 28 days, but effectiveness in control of damping-off did not decline. Mitchell & Hagedorn (62), in more extensive work, concluded that it is necessary to provide and maintain relatively high residue levels (ca. 2 μg dexon/g of dry soil) to control pea root rot due to *Aphanomyces euteiches.* They thought that dexon is fungistatic, as usually claimed, but also is fungicidal. They suggested that the persistence of the effect of treatment into a second planting season after treatment probably reflects an initial reduction in active propagules and a delayed buildup of inoculum in subsequent plantings.

Although most disease control by soil fungicides probably is by direct action of the fungicide on the pathogen, there are enough cases reported to indicate that secondary control mechanisms may be involved to warrant more research.

Examples of chemical pathways involved in the biological breakdown of pesticides.—There have been recent attempts to increase or decrease persistence of pesticides in soil by altering chemical structures. Alexander & Lustigman (3) and Kaufman (39) pointed out in 1966 that there was surprisingly little information about the influence of chemical structure upon microbial degradation in soil. They reported that mono- and disubstituted benzenes, chloro, sulfonate, and nitro groups retarded degradation, but carboxyl and phenolic hydroxyl groups favored degradation. Helling et al (33) reported that introduction of polar groups such as OH, NH_2, >N-C(O), COO-, and NO_2 in chlorinated hydrocarbons often afford sites of attack.

Kearney & Plimmer (41) doubted that pesticide molecules could be designed to degrade rapidly in the environment, and still selectively control target organisms. In contrast, Kaufman et al (40) believed that it is feasible to control pesticide biodegradation and residue formation in soil, and continue to control the undesirable organisms.

METHODS USED TO INCREASE EFFICACY OF SOIL FUNGICIDES

Using microorganisms to augment fungicide activity.—The use of soil fungicides should aim to control the pathogen with a minimum upset of the natural environment. One problem is that some workers attempt to eradicate the pathogen from the soil. It is theoretically, as well as practically, almost impossible to eradicate a pathogen in the field (in contrast to soil held in containers). If one attempts to use an eradicative dose of a fungicide, it leads to numerous problems. The trend in recent years appears to be away from such aims, toward attempts to alter the natural conditions as little as possible. The reduction of inoculum density of the pathogen to economically controllable levels appears to be the key to use of soil fungicides. Wilhelm (101) discussed chemical treatments and inoculum density in soil. Ludwig (51) pointed out that a pathogen is more easily inhibited when its population density is low, and that less chemical is required to control a disease under such conditions. R. Baker (7) amplified these views in relation to the mathematics involved in the relationship of inoculum density and control by soil fungicides.

The role of crop rotation has evolved from "resting" the soil to stimulating the soil to "rejuvenate" itself. That judicious crop rotation leads to reduction in inoculum density as well as to excellent disease control is well known. However, there does not seem to have been research on whether a fungicide should preceed or follow nonsusceptible plants in a crop rotation. It is possible that a fungicide applied after a susceptible host has been cropped might be more efficacious than the same fungicide applied before the susceptible host is planted.

Much use has been made of soil amendments to control plant pathogens, but application to it of knowledge gained from fungicide research is quite recent. Lewis & Papavizas (50) and Papavizas & Lewis (77) demonstrated that decomposition products of cabbage, kale, mustard, turnips, and Brussels sprouts reduced root rot of peas caused by *Aphanomyces euteiches,* and that vapors from the decomposition contained a number of potentially fungicidal compounds such as mercaptans, sulfides, and isothiocyanates. They found that substances like vapam and dazomet, which release methylisothiocyanate in soil, controlled the disease when applied to the soil as a drench at concentrations of 50–200 ppm. The treatments were enhanced by enclosing the soil with a plastic cover. A striking similarity between biological and fungicidal control was thus shown. The study of products of host breakdown deserves more consideration, with the aim of stimulating formation of fungitoxicants in the soil. It is possible that disease control could be greatly enhanced by covering the soil immediately after plowing under the crop refuse to confine any fungitoxic vapors that possibly might be produced.

Using microorganisms to decrease fungicide activity.—There are times when any pesticide or its breakdown products may contaminate the soil or water. In such cases microorganisms may assist in detoxifying the environment. Suzuki et al (93) and Tonomura et al (96, 97) developed a novel detoxification process for industrial waste waters. They developed a strain of *Pseudomonas* (K62 strain) capable of sorbing mercurials and stimulating their vaporization. They incubated the organism for 6 hours in a culture medium containing one of 9 mercury compounds, mostly inorganic, but including PMA. After uptake the cells were collected by centrifuging, transferred to another medium containing casamino acid, and gently shaken for 6 hours. Most of the "bound" mercury vaporized, and was adsorbed on charcoal. The cells could be reused 3 times without loss of ability to remove mercury. The system was inhibited by the presence of sodium chloride, which severely limits its practical use. The bacterium was most efficient in removing PMA, wherein 80% of the mercury was removed from solutions of 100 ppm PMA. One g of dry cells removed 15 mg of mercury from 2.5 l of industrial waste water, and mercury content in the waste decreased to approximately 10% of the initial concentration. No explanation was given for the mechanisms by which mercurials were taken up and vaporized. While these processes are not immediately practical for use with soil fungicides, the idea of using bacterial cells to "selectively filter" a specific toxicant is intriguing.

Specific Methods Used to Enhance Efficacy

McNew (53) pointed out that control of plant disease is aimed at (*a*) escape of infection, (*b*) suppression of inoculum potential, (*c*) improving host resistance, or (*d*) improving recovery from infection. Whether by design or accident, most methods devised to enhance soil fungicide action have been aimed at reducing inoculum density of the pathogen in soil. Some of the novel ways to use fungicides or to handle soil to control plant diseases are: simultaneous fumigating and planting of seeds (37, 102); use of anhydrous ammonia as a fungicide as well as fertilizer (88); use of a fungicide in freshly steamed soil to prevent recontamination losses (21). An alteration or adaptation of old methods may make possible a more efficient use of a soil fungicide.

Increasing the concentration of the fungicide.—Munnecke et al (72) showed that increasing the dose of methyl bromide from 1 to 4 pounds per 100 sq ft of surface beneath an impervious (mylar) cover gave a pronounced enhancement of penetration of a sandy loam soil. The effect was more pronounced as soil depth increased. At the 1-foot depth the concentration of methyl bromide using 4 lb was approximately 2.5 times higher than the treatment using 1 lb. The maximum concentration attained at 6 ft however, was 12 times higher. Although there are well-known limits to the concentration that may be used, one of the easiest ways to obtain greater penetration of soil and increased kill by a soil fungicide is to increase the concentration applied.

Placement of the chemical in soil.—The strategy of enhancement of activity of a fungicide depends upon the habitat of the pathogen, the growth pattern of the host, and physical-chemical-biological factors in the soil. Goring (28) discussed the geometry of diffusion patterns of fumigants applied to the soil, pointing out the differences obtained when a fumigant is applied as a point, line, or plane source in the soil. The pattern of the placement of the fumigant greatly affects the behavior of the fumigant in soil.

Most fumigations are designed to treat the upper 18 inches of soil where most pathogens exist. All root pieces infected with pathogens such as *Armillaria mellea* are capable of becoming infected, and remaining in the soil as potential sources of inoculum. With such pathogens deep vertical distribution of a fumigant is of utmost importance. Hence, a fundamental difference in strategy is involved when treating soil to be planted to strawberry to control root-rotting fungi or *Verticillium,* compared to treating land infested with *Armillaria mellea* before planting to a woody perennial. Fumigants such as methyl bromide-chloropicrin mixtures are often applied as liquids in continuous-flow applications through tubes attached to chisels drawn by tractors. If the stratum to be treated is mostly the upper 2 feet of soil, the chisels are set 8–10 inches apart and 6–8 inches deep. If the upper levels of the soil are not so important but it is necessary to fumigate the soil as deep as possible, the chisels are set 3–5 feet apart and as deep as the terrain and equipment permits. Presently, attempts are being made to inject fumigants 3 ft deep for control of *A. mellea,* but this requires extremely heavy equipment not commonly available.

In areas where machine applications are not feasible, hand-application of fumigants is the only recourse. Holes may be bored or punched into the soil 3–5 ft deep and fungicides applied to the bottom and quickly covered with soil. A safe way to do this with methyl bromide is to chill 1-lb cans in a freezer or on dry ice. When cans are punctured and dropped in the hole, no gas escapes before the hole is filled with soil and tamped (46). The hand-applied charges might be useful in treating "problem" soils that contain deep plow soles, hardpans, or clay strata that prevent the downward diffusion of fumigants. In such cases holes may be bored through the obstruction layers and charges of liquid methyl bromide or carbon disulfide placed beneath the layers. It may be necessary to use shallow injections to fumigate the areas above the obstructing layers after the hole is filled, since the gases will diffuse horizontally and vertically beneath the cap. Still unsolved is the problem of how to obtain toxic concentrations of the gases in the clay layer itself.

Raski et al (80) have shown that split applications of 1,3-D are beneficial in fumigating soil infested with *Xiphinema* and other nematodes. A very high concentration of 1,3-D is applied (200 gal/A) 18–36 inches deep, followed by a lower concentration (50 gal/A) applied at 8–10 inches, to control nematodes to depths of 8 ft.

The careful placement of nonvolatile fungicides can enhance their activity also. It is common in floricultural practice to broadcast granular material such as PCNB, captan, or dexon over the surface and incorporate it into the top 4–6 inches by rototillers. With in-furrow applications, positioning PCNB in bands immediately above the seed increases effectiveness. The principle involved with most nonvolatile fungicides is that the fungicide must be in close proximity to the pathogen to be effective, since diffusion through water in soil is so slow.

Confinement of fungicides in soil following treatment.—Perhaps the greatest impetus to successful field use of highly volatile soil fumigants such as methyl bromide, was the development of relatively cheap polyethylene sheeting and machines capable of applying fumigants and covering large fields in one operation (101). It is unfortunate that polyethylene is permeable to methyl bromide and other gases, as compared to other types of films. Waack et al (99) showed the permeability of films [P = cc gas (STP)/sec/cm^2 for 1 mm thickness] to methyl bromide gas as follows: polyethylene @ 20C, 12.5×10^8P; Saran @ 30C, 0.13×10^8P; and Mylar @ 30C, 0.0022×10^8P. A substitute should be found for polyethylene, but other films to date are prohibitively expensive or unsuited for use in the field.

Kolbezen et al (46) have shown that when the relatively impermeable Mylar or Saran covers are used in lieu of polyethylene, the dose of methyl bromide may be greatly reduced in the field, retaining the same concentrations of gas deep in the soil. If impervious tarps were available commercially, deep penetration of soil from relatively shallow applications (12–18 inches) would be possible.

The possibility of increasing effectiveness of soil fumigations by increasing the thickness of the polyethylene covers has been investigated. Grimm & Alexander (31) compared the effects of confining methyl bromide-chloropicrin mixtures in soil using polyethylene covers 1 or 4 mil thick. They buried soil infested with *Phytophthora* in the plots, treated them for 4 days with the fumigants, removed the covers, and retrieved the samples 12 days later. They used doses of 0.25, 0.5, 0.75, and 1.0 lb/100 sq ft applied to the surface beneath the covers. They got 100% kill of *Phytophthora* at all depths to 4 feet with the 1 pound dose under the 4 mil tarp. In contrast, only approximately 60% of the samples were killed at the same depth under the 1 mil tarp. Munnecke et al (76) studied the effect of polyethylene covers 1, 4, or 6 mil thick on penetration of methyl bromide (66%)-chloropicrin (33%) mixtures in a commercial application of strawberry land. The concentration of methyl bromide attained at 6 inches depth after 60 hours for the 300 lb/A application was as follows: 1 mil, 200 ppm; 4 mil, 900 ppm; and 6 mil, 2000 ppm. The concentration observed with the 150 lb/A treatment under a 6 mil cover was 1300 ppm. Thus, increasing the thickness from 4 to 6 mil resulted in more than doubling the concentration in the soil.

In a previous review (68) I stated, "It is useless to use methyl bromide unless the soil is covered". Fortunately, this statement can be modified for deep-soil fumigations. Kolbezen et al (46) treated a uniform sandy loam soil dried to less than 15% moisture (by wt) with 4 lb methyl bromide applied as a point source 5 feet deep. The gas diffused laterally and vertically very rapidly so that even though some escaped to the air at the surface, the concentration×time (CT) products were lethal to *A. mellea* in a volume over 12 feet in diameter and 9 feet deep. In another experiment at the same location the dry soil was irrigated so that the upper 2 feet were wet and 4 lb methyl bromide was applied as a point source 4 feet deep. The wet soil layer at the top greatly slowed the diffusion of the gas to the surface and much higher concentrations of gas were obtained laterally and downward than in the unwetted soil. The CT values at the upper 6–12 inches were lethal for *A. mellea*. These experiments were made under ideal field conditions, but they illustrate that if soil is managed properly, it is possible to obtain excellent deep fumigations without using a cover. However, vertical gas escape is so rapid that it is unwise to discontinue use of the tarps, and much better fumigations are obtained by confining the gas with a plastic cover.

Formulation of fungicides.—Pesticides have often been altered to increase effectiveness. Several authors (10, 12) have claimed that surfactants have enhanced fungicidal activity in soil. Biehn & Dimond (10) reported that soil injected with benomyl plus Surfactant F or Tween 2 reduced Dutch elm disease 79% or 97% on trees treated prior to inoculation. The role of the surfactants was not determinable, since treatments without surfactants were not reported. Hock & Schreiber (34) tested 23 adjuvants without indication that they increased uptake of benomyl from planting medium by elm seedlings. It appears that adjuvants may confer some benefits to uptake, but the effects are not striking.

Occasionally a synergistic effect is claimed for mixtures of fungicides, as in the case of use of mixtures of 66% methyl bromide and 33 % chloropicrin for control of *Verticillium*. No evidence has been presented to justify these assumptions.

A potentially useful development with methyl bromide has been formulation of the gas in a thixotropic gel preparation. The preparation solidifies upon release in the soil and the gas volatilizes less rapidly than if liquid methyl bromide were released. We (76) have found that it may be possible to use gel preparations without covering the soil with tarps.

The controlled release of fertilizers in pelleted form provided an example for using soil fungicides in a similar fashion. Furmidge et al (24) discussed some of the physical-chemical problems involved with pelleted pesticides. Mills & Schreiber (61) have worked extensively on the use of latex-coated pellets for control of root rot of wheat. The preparations did not control root rot, nor did they accurately release the fungicides. The authors stated that if time-release pellets are to be used effectually, a mechanism of release independent of moisture needs to be devised.

Alteration of soil before or after treatment with fungicides.—The manipulation of field-soil moisture may be one of the most efficient means of increasing fungicide efficacy. Moisture greatly influences the susceptibility to fungicides of pathogens in soil. Munnecke et al (75) used a carefully controlled continuous flow of methyl bromide through columns of soil infested with *Pythium ultimum* or *Rhizoctonia solani* to determine the dosage response as affected by soil moisture. The effect of soil moisture on control of damping-off was pronounced in soil infested with *P. ultimum*, but not so pronounced in soil infested with *R. solani.* Most effective control of *P. ultimum* was obtained in moderately moist soil, next in very wet soil, and least in very dry soil. With *R. solani,* best control of damping-off was obtained in moderately moist soil. In contrast to results with *P. ultimum,* control of damping-off due to *R. solani* in very wet soil and in very dry soil was similar and only slightly poorer than that obtained in moderately moist soil.

Monro et al (65) experimented with ethylene oxide, propylene oxide, and methyl bromide fumigations against *Synchytrium endobioticum* and found, as is usually the case, less kill in dry than in wet soil. However, they attributed the decreased kill in dry soil to an increased adsorbance of the toxicants. The data are not adequate for a proper criticism of their conclusion, but I think there is additional evidence in other reports to indicate that resistance of organisms in dry soil is inherent in the organism, rather than in sorptive factors of the soil itself.

Soil moisture also affects chemical reaction, and soil distribution in liquid as well as in gaseous phases. Moisture content is probably the chief limiting factor in deep penetration of soils by fumigants. Some of the deeper fine-textured soils in our experiments (76) never dry sufficiently to allow adequate diffusion of gases such as methyl bromide or carbon disulfide to be fungitoxic to *Armillaria mellea.* This problem can be corrected partially by growing a cover crop of sudan grass or safflower. The upper 3–4 feet may be dried by withholding irrigation water during the hot dry season in California, but plants are needed to withdraw water from the deeper levels. Since cover crops most efficiently withdraw water late in the growing season, it is important to delay fumigating as long as possible, but before the fall rains occur.

Development of new compounds.—We are nearing a new era in the use of soil fungicides that become systemic in plants. Erwin (22) has summarized this work and pointed out the duality existing between systemic and soil fungicides. Benomyl has been tried in so many ways and by so many researchers that it is a subject for a review paper in itself. Among the numerous diseases against which it has been partially successful when applied to soil is Dutch elm disease. Smalley (87) reported that very high rates (308–382 kg/hectare) gave significant protection of elms to subsequent inoculation with the pathogen. However, Schreiber et al (86) reported that significant concentrations of benomyl were found in bark of elm seedlings, but not in the wood or leaves. When applied to soil or injected into the trunk, symptoms of trees were not reduced. The increased activity of benomyl in the acidified form may facilitate its action (13).

The remaining big gap in the use of systemic fungicides applied to soil is that no one has shown downward translocation of fungicides into the roots in concentrations sufficiently high to protect the host. This is important for control of the root-rotting and wilt-inducing fungi.

No new soil fumigants have been found that appear to be better than those 10 years ago.

PROGRESS MADE ON QUESTIONS RAISED 7 YEARS AGO

Approximately 7 years ago (68) I raised some questions concerning soil fungicides. This article presents an opportunity to re-examine some of them.

Why do Fungicides Fail?

The role of sorption and lack of gas penetration in soil is better known today. Many organisms have been reported that are capable of detoxifying pesticides in soil. It may be significant that *Pseudomonas* spp. frequently have been implicated as detoxifying organisms. Whether this is because *Pseudomonas* spp. are primarily responsible for detoxification, or because the genus is so ubiquitous and easy to isolate, is not known. Perhaps the group has unique enzymatic systems that deserve further study.

Host plants may detoxify soil pesticides. Richardson (82) reported that toxicity of thiram may be negated by exudates from the seed and roots of soybean. While this was demonstrated in vitro it is possible that similar events might occur in the field to lessen the effectiveness of fungicides.

It is well known that sometimes after PCNB is added to soil a significant increase in seedling damage by pathogenic fungi not sensitive to PCNB may occur. Farley & Lockwood (23) have presented evidence that reduced nutrient competition by soil microorganisms may be responsible for such PCNB-induced disease accentuation. The increased activity of PCNB-insensitive pathogens had been suggested by other authors to be due to suppression by PCNB of specific fungal antagonists of the pathogen. Farley & Lockwood proposed instead that fungi insensitive to PCNB may benefit because of a broad reduced microbial competition for nutrients. Whether the phenomenon is due to nutrient competition or to suppression of specific fungal antagonists remains undecided. The important point is that fungicides may fail, not because they do not control the target pathogen, but because of secondary effects involving other pathogens.

How do Fungicides Kill Fungi in the Soil Environment?

This is still a moot question. The direct fungitoxic effects of chemicals in soil are well known, however the role of organisms, antagonistic to pathogens, that gain ascendency in soil following treatment with fungicides is less known. The *Armillaria-Trichoderma* system is an interesting example.

Weindling (100a, 100b) originally described the parasitic nature of *Trichoderma* in relation to *Rhizoctonia* and other soil fungi. His discovery of the phenomenon, as well as the antibiotic from *Trichoderma* (subsequently named gliotoxin) led to early suppositions that the antagonism of *Trichoderma* to other fungi in soil was most directly related to antibiotic substances produced by *Trichoderma*. Bliss (10a) used the antagonism of *Trichoderma* to explain the killing of *Armillaria mellea* in infected tree roots by fumigations with CS_2 as due to an indirect effect, not to direct fungicidal action. He noted that *Trichoderma* invariably was isolated from *Armillari*-infested roots after treatment with CS_2, even though the dose was not lethal to *A. mellea* initially. After a delay of approximately a month however, *A. mellea* was killed and only *Trichoderma* was isolated. He thought that *Trichoderma,* being tolerant of CS_2, greatly increased following fumigation. The increased population of *Trichoderma* thus was capable of breaking down the protective marginal layers of mycelium in roots (pseudosclerotium) and killing *Armillaria* mycelium. Darley and Wilbur (19a) who continued Bliss' work, reported that CS_2 had a direct fungicidal effect on *Armillaria* at high doses, but indirect effects at low doses. This led to the concept of the Riverside group that *Armillaria* must be "weakened" in some manner before *Trichoderma* is able to exert its antagonistic secondary action following fumigation. Garrett elaborated on the results of Bliss and Darley & Wilbur. He confirmed their findings, adding that at least a portion of the *Armillaria* mycelium must be damaged by the fumigant to facilitate invasion by *Trichoderma* (25a), and also that the lethal parasitic activity of *Trichoderma* was related directly and quantitatively to its inoculum potential in soil (25b).

Research continued at Riverside has clarified the relationships of the two fungi and the toxicants. Ohr (76b) has found that a natural weak antibiosis occurs between *Armillaria* and *Trichoderma,* presumably allowing both fungi to exist together. When *Armillaria* is subjected to sub-lethal concentrations of methyl bromide, the antibiosis exhibited by *Armillaria* may be lost and the fungus is overrun by *Trichoderma.* Although the mechanism is still unknown, the phenomenon helps to explain how *Armillaria* may lie dormant in roots in soil for long periods, and to explain what the "weakening" effect of sub-lethal fungicide treatments is. Also, there are differences in antagonistic reactions of isolates of the two fungi, a fact which has been partially described by Mughogho (63). Finally, we (76a) have found that there is a long delay before growth of *Armillaria* resumes following fumigation with CS_2 or methyl bromide or treatment with heat, but that *Trichoderma* grows almost normally under the same conditions.

All these data indicate that one of the prime effects of treating soil with fungicides may be that antagonists are stimulated to rapid growth and sporulation, and that the pathogen may be "weakened" (i.e., lose growth potential, ability to form antibiotics, or capacity to form protective layers of tissue), and thus a secondary control of the pathogen be obtained.

How can Satisfactory Control be Obtained Without Leaving Harmful Residues in the Soil or in Plant Products?

Considerable progress has been made along these lines, in part due to public concern over environmental pollution. This has led to more efficacious use of old fungicides as well as to introduction of newer, presumably safer, compounds.

An auxilliary of the question above is "How can the *decreased* growth response (DGR), sometimes observed following soil fumigation, be avoided?" There are some exciting developments here.

The DGR following fumigations may be due to inadequate control of the target pathogen, to reintroduction of a pathogen, or to the differential stimulation of indigenous populations of pathogens. However, commercial appliers of soil fumigants occasionally observe a DGR of the type described by Martin et al (55). Sometimes soil treatment with steam, propylene oxide, CS_2, chloropicrin, D-D, or methyl bromide was followed by exceedingly poor growth in the treated soil. Citrus (55) or avocado (57) showed symptoms of nutrient deficiency, and levels of phosphate in leaves were abnormally low even though phosphate was present in soil in adequate amounts. The symptoms were also related to relatively high manganese levels in the leaves. The authors postulated that a toxin(s) was produced in soil by the action of some unspecified organism that upset phosphate uptake. The toxicity of the soils was offset by adding large quantities of phosphate fertilizers, and this practice is routine in the field where the problem has occurred. Usually the trouble has occurred sporadically and seemingly without pattern. Martin & Ervin (57) noted recent reports on the role of mycorrhiza and plant growth and suggested that lack of such a system in a fumigated soil might be the reason for poor growth after soil fumigation or steaming. However, they did not abandon the role of microbial toxins that they had first proposed as being responsible for the phenomenon. Gerdemann (27) noted that the vesicular-arbuscular mycorrhiza (VAM) are so ubiquitous in crop plants that the Cruciferae and Chenopodiaceae are the only families of importance in which the phenomenon is not known. His thesis is that one need not postulate the presence of toxins in order to explain DGR. *Endogone* spp. are the most common of the obligate or near-obligate parasites. The vital point is that *Endogone* increases growth of the host plants by enhancing the absorptive ability of the roots. When small amounts of nonsterile soil are added to fumigated or steamed soil, growth of plants increases, provided they become infected with VAM. Also, there were numerous cases where uptake of phosphorus was greatly enhanced by VAM in the roots. Rovira & Bowen (85) reported on the detoxification of heat-sterilized soil, and showed that the DGR resulting in a steamed (120C) soil could be offset by suspensions from soil, by cultures of various genera of bacteria, or by 5 species of fungi. *Endogone* was not mentioned. In contrast to the report by Gerdemann, when Rovira & Bowen eluted steamed soil with sterile water, the eluate was toxic to subterranean clover and tomato seedlings. They concluded that the toxicity of the soil was due to the formation of toxic organic substances. How it was detoxified by various organisms was not stated. More recently, Ross & Harper (83) and Ross (84) more closely allied absence of *Endogone* with the DGR phenomenon. Ross positively correlated infection of soybeans by chlamydospores of *Endogone* with uptake of phosphate and recovery of post-treatment stunting. While the evidence is not incontestable, the sporadic occurrence of toxicity in soils treated with various fumigants (DD, CS_2, propylene oxide, ethylene oxide, chloropicrin, or methyl bromide) or steam probably sometimes is due to disruption of the mycorrhizal systems of the roots. The toxin formation should not be discounted, however, and it is likely that both mechanisms may be responsible for the phenomenon.

How May Fungicides be Applied Around Living Plants?

Relatively few fumigants can be used to treat root diseases of living plants. A report by Zentmyer (104) on efforts to control *Phytophthora* on avocado with applications of dexon is encouraging. He treated large infected avocado trees in the field by continued applications (8–10) per year for up to 7 years) of dexon as a drench. Although results were variable, indications were that in a high-value crop such as avocado, under optimum conditions for disease control and with good prices for fruit, usage may be feasible.

An interesting strategy was used by Lyda & Burnett (52) for control of *Phymatotrichum omnivorum* on cotton, using benzimidazole fungicide. They noted that *P. omnivorum* grows superficially and surrounds the tap root. At early stages of flowering or later the mycelium ascends the root peripherally, forming a mantle like an ectomycorrhiza. When the fungus reaches the soil line and immediately above the area of lateral roots, the internal portions of the root are invaded and the plant dies. Their strategy is to get fungicide to this region of the root to deny access by the fungus. Benomyl controlled the pathogen in greenhouse tests, presumably in this manner. This is an example of how careful study of the host and pathogen may lead to more intelligent application of control procedures around living plants.

Can Fungicides of Increasing Selectivity be Developed?

Probably more selective fungicides may be developed, but is this a desirable aim in research?

Kreutzer (47) has argued persuasively for development of specific selective fungicides. K. Baker (6) also has pointed out that the use of chemicals toxic to a wide spectrum of organisms may be dangerous to subsequent plantings because of the indiscriminate destruction of beneficial soil microorganisms. Also, the over-kill sometimes occurring with broad-spectrum toxicants, such as methyl bromide, may reduce the mycorrhizal fungi naturally occurring in agricultural soils, and lead to DGR as discussed above.

On the other hand, use of selective fungicides may be harmful. The indiscriminate use of benomyl could easily lead to problems in this regard. Most Basidiomycetes and Phycomycetes are highly tolerant of benomyl. This specificity is useful, for example, in control of mushroom diseases (90). It could be harmful, as Smith et al (89) reported, when it stimulated occurrence of an unidentified Basidiomycete after applications on turf. Also, it may be easier for fungi to develop tolerance to fungicides having a narrow host range than to those having a broad host range.

Probably the broad-spectrum fungicides will remain the favorites, but as more is learned, the selective fungicides will have increased use. Horsfall & Lukens (35) have reviewed the details on the mechanisms involved in selectivity of fungicides.

Is it Feasible to Develop Standardized Assay Techniques for Comparing Fungicides?

Not much has been done along these lines. Since soils are so heterogeneous and the natural systems so complex, it is virtually impossible for one investigator to duplicate another's experiments. There have been developments in analytical techniques so that it is possible to measure gases accurately in the soil atmosphere in the field and in the laboratory (45). When investigators combine bioassays with chemical measurements, dosage response curves may be obtained (74), but much more data of this sort are needed for various organisms and chemicals. There is need to obtain standard inoculum and inoculation techniques. Munnecke & Moore (73) found that artificially infested UC-mix could be maintained at −18 C for relatively long periods. The incidence of damping-off of seedlings of pea or aster remained relatively constant for 2 years in soil infested with *Pythium ultimum* or *Fusarium oxysporum* f. sp. *callistephi,* but rapidly decreased thereafter. *Rhizoctonia solani* could not be stored satisfactorily. They pointed out that the cold-storage method could be useful for standardizing inoculum for periods of less than 2 years.

It seems obvious that more efforts are needed to provide more standardized systems with soil fungicide research.

LITERATURE CITED

1. Alconero, R., Hagedorn, D. J. 1968. The persistence of Dexon in soil and its effects on soil mycoflora. *Phytopathology* 58: 34–40
2. Alexander, M. 1965. Persistence and biological reactions of pesticides in soils. *Soil Sci. Soc. Am. Proc.* 29:1–7
3. Alexander, M., Lustigman, B. K. 1966. Effect of chemical structure on microbial degradation of substituted benzenes. *J. Agr. Food Chem.* 14:410–13
4. Audus, L. J. 1964. Herbicide behaviour in the soil. II. Interactions with soil micro-organisms. In *The Physiology and Biochemistry of Herbicides,* ed. L. J. Audus, 163–206. New York: Academic 555 pp.
5. Bailey, G. W., White, J. L. 1970. Factors influencing the adsorption, desorption and movement of pesticides in soil. *Res. Rev.* 32:29–92
6. Baker, K. F. 1970. Selective killing of soil microorganisms by aerated steam. In *Root Diseases and Soil-Borne Pathogens,* eds. T. A. Toussoun, R. V. Bega, P. E. Nelson, 234–239. Berkeley: Univ. Calif. Press. 252 pp.
7. Baker, R. 1971. Analyses involving inoculum density of soil-borne plant pathogens in epidemiology. *Phytopathology* 61: 1280–92
8. Bartnicki, E. W., Castro, C. E. 1969. Biodehalogenation. The pathway for transhalogenation and the stereochemistry of epoxide formation from halohydrins. *Biochemistry* 8:4677–80
9. Belser, N. O., Castro, C. E. 1971. Biodehalogenation—the metabolism of the nematocides *cis*– and *trans*–3–chlorallyl alcohol by a bacterium isolated from soil. *J. Agr. Food Chem.* 19:23–26
10. Biehn, W. L., Dimond, A. E. 1971. Prophylactic action of benomyl against Dutch elm disease. *Plant Dis. Reptr.* 55:179–82

10a. Bliss, D. E. 1951. The destruction of *Armillaria mellea* in citrus soils. *Phytopathology* 41:665–683

11. Bollag, J. M., Alexander, M. 1971. Bacterial dehalogenation

of chlorinated aliphatic acids. *Soil Biol. Biochem.* 3:91–96

12. Booth, J. A., Rawlins, T. E. 1970. A comparison of various surfactants as adjuvants for the fungicidal action of benomyl on *Verticillium. Plant Dis. Reptr.* 54:741–44
13. Buchenauer, H., Erwin, D. C. 1971. Control of Verticillium wilt of cotton by foliar sprays with acidic solutions of benomyl and thiabendazole. *Phytopathology* 61:1320.
14. Chacko, C. I., Lockwood, J. L., Zabik, M. 1966. Chlorinated hydrocarbon pesticides: degradation by microbes. *Science* 154:893–95
15. Castro, C. E., Bartnicki, E. W. 1965. Biological cleavage of carbon-halogen bonds metabolism of 3-bromopropanol by *Pseudomonas* sp. *Biochim. Biophys. Acta* 100:384–92
16. Castro, C. E., Belser, N. O. 1966. Hydrolysis of *cis*– and *trans*–1,3–dichloropropene in wet soil. *J. Agr. Food Chem.* 14: 69–70
17. Castro, C. E., Bartnicki, E. W. 1968. Biodehalogenation. Epoxidation of halohydrins, epoxide opening and transhalogenation by a *Flavobacterium* sp. *Biochemistry* 7:3213–18
18. Castro, C. E., Belser, N. O. 1968. Biodehalogenation. Reductive dehalogenation of the biocides ethylene dibromide, 1,2–dibromo–3–chloropropane, and 2,3–dibromobutane in soil. *Env. Sci. Tech.* 2:779–83
19. Crosby, D. J., Li, Ming-yu. 1969. Herbicide photodecomposition. In *Degradation of Herbicides,* ed. P. C. Kearney, D. D. Kaufman, 321–63. New York: Dekker. 394 pp.

19a. Darley, E. F., Wilbur, W. D. 1954. Some relationships of carbon disulfide and *Trichoderma viride* in the control of *Armillaria mellea. Phytopathology* 44:485

20. Domsch, K. H. 1965. Der Einfluss von Captan auf den Abbau von Glukose, Aesculin, Chitin, und Tannin im Boden. *Phytopath. Z.* 52:1–18
21. Engelhard, A. W., Miller, H. N., DeNeve, R. T. 1971. Etiology and chemotherapy of Pythium root rot on chrysanthemums. *Plant Dis. Reptr.* 55:851–55
22. Erwin, D. C. 1970. Progress in the development of systemic fungitoxic chemicals for control of plant diseases. *FAO Plant Prot. Bull.* 18:73–82
23. Farley, J. D., Lockwood, J. L. 1969. Reduced nutrient competition by soil microorganisms as a possible mechanism for pentachloronitrobenzene-induced disease accentuation. *Phytopathology* 59:718–24
24. Furmidge, C. G. L., Hill, A. C., Osgerby, J. M. 1968. Physicochemical aspects of the availability of pesticides in soil. II. Controlled release of pesticides from granular formulations. *J. Sci. Food Agr.* 19:91–95
25. Furukawa, K., Tonomura, K. 1971. Enzyme system involved in the decomposition of phenyl mercuric acetate by mercury-resistant *Pseudomonas. Agr. Biol. Chem.* 35:604–10

25a. Garrett, S. D. 1957. Effect of a soil microflora selected by carbon disulphide fumigation on survival of *Armillaria mellea* in woody host tissues. *Can. J. Microbiol.* 3:135–149

25b. Garrett, S. D. 1958. Inoculum potential as a factor limiting lethal action by *Trichoderma viride* Fr. on *Armillaria mellea* (Fr.) Quel. *Trans. Brit. Mycol. Soc.* 41:157–164

26. George, M. K., Aulakh, K. S., Dhesi, J. S. 1970. Morphological and cytological changes induced in barley (*Hordeum vulgare*) seedlings following seed treatment with fungicides. *Can. J. Genet. Cytol.* 12:415–19
27. Gerdemann, J. W. 1968. Vesicular-arbuscular mycorrhiza and plant growth. *Ann. Rev. Phytopathol.* 6:397–418
28. Goring, C. A. I. 1962. Theory and principles of soil fumigation. In *Advances Pest Control Res.* 5: 47–84
29. Goring, C. A. I. 1967. Physical aspects of soil in relation to the action of soil fungicides. *Ann. Rev. Phytopathol.* 5:285–318
30. Goring, C. A. I. 1970. Physical soil factors and soil fumigant action. In *Root Diseases and*

Soil-Borne Pathogens, ed. T. A. Toussoun, R. V. Bega, P. E. Nelson, 229–33, Berkeley: Univ. Calif. Press. 252 pp.

31. Grimm, G. R., Alexander, A. F. 1971. Fumigation of *Phytophthora* in sandy soil by surface application of methyl bromide and methyl bromide-chloropicrin. *Plant Dis. Reptr.* 55:929–31
32. Hartley, G. S. 1964. Herbicide behaviour in the soil. I. Physical factors and action through the soil. In *The Physiology and Biochemistry of Herbicides,* ed. L. J. Audus, 111–61. New York: Academic. 555 pp.
33. Helling, C. S., Kearney, P. C., Alexander, M. 1971. Behavior of pesticides in soils. *Advan. Agron.* 23:147–240
34. Hock, W. K., Schreiber, L. R. 1971. Effect of adjuvants on the uptake of benomyl from planting media by American elm seedlings. *Plant Dis. Reptr.* 55: 971–74
35. Horsfall, J. G., Lukens, R. J. 1966. Selectivity of fungicides. *Conn. Agr. Exp. Sta. Bull.* 676. 23 pp.
36. Isobe, K., Takeda, M., Tanabe, H. 1971. Metabolic fate of organomercuric compounds. II. Effect of thiol compounds on decomposition of organomercuric compounds. *Chem. Abstr.* 75: 110563S. 1971 (original not seen)
37. Johnson, Ronald C. 1969. Simultaneous fumigating and planting of sugar beets. *J. Am. Soc. Sugar Beet Tech.* 15:379–83
38. Kaars Sijpesteijn, A., Vonk, J. W. 1970. Microbial conversions of dithiocarbamate fungicides, *Meded. Fac. Landbouw. Wetenschappen Ghent* 35:799–804
39. Kaufman, D. D. 1970. Pesticide metabolism. In *Pesticides in the Soil: Ecology, Degradation and Movement,* 73–86. Int. Symp. on Pesticides in Soil. Mich. State Univ. 144 pp.
40. Kaufman, D. D., Blake, J., Miller, D. E. 1971. Methylcarbamates affect acylanilide herbicide residues in soil. *J. Agr. Food Chem.* 19:204–06
41. Kearney, P. C., Plimmer, J. R. 1970. Relation of structure to pesticide decomposition. In *Pesticides in the Soil: Ecology, Degradation and Movement,* 65–72. Int. Symp. on Pesticides in Soil. Mich. State Univ. 144 pp.
42. Khan, S., Baker, R. 1968. Residual activity of dexon. *Phytopathology* 58:1693–96
43. Ko, W. H., Farley, J. D. 1969. Conversion of pentachloronitrobenzene to pentachloroanaline in soil and the effect of these compounds on soil microorganisms. *Phytopathology* 59: 64–67
44. Ko, W. H., Lockwood, J. L. 1968. Accumulation and concentration of chlorinated hydrocarbon pesticides by microorganisms in soil. *Can. J. Microbi*ol. 14: 1075–78
45. Kolbezen, M. J., Abu-El-Haj, F. 1972. Fumigation with methyl bromide. I. Apparatus for controlled concentrations and continuous flow laboratory procedures. II. Equipment and methods for sampling and analyzing deep field soil atmospheres. *Pesticide Sci.* In press
46. Kolbezen, M. J., Munnecke, D. E., Stolzy, L. H. (unpublished results)
47. Kreutzer, W. A. 1970. The reinfestation of treated soil. In *Ecology of Soil-Borne Plant Pathogens,* ed. K. F. Baker, W. C. Snyder, 495–508. Berkeley: Univ. Calif. Press. 571 pp.
48. Leistra, M. 1970. Distribution of 1,3-dichloropropene over the phases of soil. *J. Agr. Food Chem.* 18:1124–26
49. Leistra, M. 1971. Diffusion of 1,3-dichloropropene from a plane source in soil. *Pesticide Sci.* 2: 75–79
50. Lewis, J. A., Papavizas, G. C. 1971. Effect of sulfur-containing volatile compounds and vapors from cabbage decomposition on *Aphanomyces euteiches. Phytopathology* 61:208–14
51. Ludwig, R. A. 1970. The role of chemicals in the biological control of soil-borne plant pathogens. In *Ecology of Soil-Borne Plant Pathogens,* ed. K. F. Baker, W. C. Snyder, 471–78. Berkeley: Univ. Calif. Press. 571 pp.

52. Lyda, S. D., Burnett, E. 1970. Influence of benzimidazole fungicides on *Phymatotrichum omnivorum* and Phymatotrichum root rot of cotton. *Phytopathology* 60:726–28
53. McNew, G. L. 1966. Progress in the battle against plant disease. *Scientific aspects of pest control.* No. 1402, 74–101
54. Marshall, K. C. 1971. Sorptive interactions between soil particles and microorganisms. In *Soil Biochemistry,* ed. A. D. McLaren, J. Skujins, 2:409–45. New York: Marcel Dekker. 527 pp.
55. Martin, J. P., Baines, R. C., Page, A. L. 1963. Observations on the occasional temporary growth inhibition of citrus seedlings following heat or fumigation treatment of soil. *Soil Sci.* 95: 175–85
56. Martin, J. P. 1972. Side effects of organic chemicals on soil properties and plant growth. In *Chemicals in the Soil Environment,* ed. C. A. I. Goring. (in manuscript)
57. Martin, J. P., Ervin, J. O. (Personal communication)
58. Matsumura, F., Gotoh, Y., Boush, G. M. 1971. Phenylmercuric acetate: metabolic conversion by microorganisms. *Science* 173:49–51
59. Matsumura, F., Boush, G. M. 1971. Metabolism of insecticides by microorganisms. In *Soil Biochemistry,* ed. A. D. McLaren, J. Skujins, 2:320–36. New York: Marcel Dekker. 527 pp.
60. Miller, P. M., Lukens, R. J. 1966. Deactivation of sodium N-methyldithiocarbamate in soil by nematocides containing halogenated hydrocarbons. *Phytopathology* 56:967–70
61. Mills, J. T., Schreiber, K. 1971. Use of latex-coated pellets for control of common root rot of wheat. *Can. J. Plant. Sci.* 51: 347–52
62. Mitchell, J. E., Hagedorn, D. J. 1971. Residual dexon and the persistent effect of soil treatments for control of pea root rot caused by *Aphanomyces euteiches. Phytopathology* 61: 978–83
63. Mughogho, L. K. 1968. The fungus flora of fumigated soils. *Trans. Brit. Mycol. Soc.* 51: 441–59
64. Moje, W., Munnecke, D. E., Richardson, L. T. 1964. Carbonyl sulphide, a volatile fungitoxicant from nabam in soil. *Nature* 202:831–32
65. Monro, H. A. U., Olsen, O. A., Buckland, C. T. 1970. Methyl bromide and ethylene oxide fumigation of *Synchytrium endobioticum. Can. J. Plant. Sci.* 50: 649–58
66. Moore, R. B. 1970. Effects of pesticides on growth and survival of *Euglena gracilis Z. Bull. Environ. Contam. Toxicol.* 5:226–30
67. Mortland, M. M. 1970. Clay-organic complexes and interactions. *Advan. Agron.* 22:75–117
68. Munnecke, D. E. 1967. Fungicides in the soil environment. In *Fungicides,* ed. D. Torgeson, 1: 509–559. New York: Academic. 697 pp.
69. Munnecke, D. E., Moore, B. J. 1967. Fungicidal activity in soil in relation to time, concentration and Penicillium population. *Phytopathology* 57:823
70. Munnecke, D. E., Martin, J. P., Moore, B. 1967. Effect of ammonium humate and clay preparations on release of methylisothiocyanate from soil treated with fungicides. *Phytopathology* 57:572–75
71. Munnecke, D. E., Michail, K. Y. 1967. Thiram persistence in soil and control of damping-off caused by *Pythium ultimum. Phytopathology* 57:969–71
72. Munnecke, D. E., Kolbezen, M. J., Stolzy, L. H. 1969. Factors affecting field fumigation of citrus soils for control of *Armillaria mellea. Proc. 1st Int. Citrus Symp.,* ed. H. D. Chapman, 3: 1273–77. Univ. Calif., Riverside. 1839 pp.
73. Munnecke, D. E., Moore, B. J. 1969. Effect of storage at −18 C of soil infested with *Pythium* or *Fusarium* on damping-off of seedlings. *Phytopathology* 59: 1517–20
74. Munnecke, D. E., Wilbur, W. D., Kolbezen, M. J. 1970. Dosage

response of *Armillaria mellea* to methyl bromide. *Phytopathology* 60:992–93
75. Munnecke, D. E., Moore, B. J., Abu-El-Haj, F. 1971. Soil moisture effects on control of *Pythium ultimum* or *Rhizoctonia solani* with methyl bromide. *Phytopathology* 61:194–97
76. Munnecke, D. E., Kolbezen, M. J. (unpublished)
76a. Munnecke, D. E., Wilbur, W. D., Kolbezen, M. J. (unpublished)
76b. Ohr, H. O. (unpublished)
77. Papavizas, G. C., Lewis, J. A. 1971. Effect of amendments and fungicides on Aphanomyces root rot of peas. *Phytopathology* 61:215–20
78. Pilinskaya, M. A. 1970. Chromosome aberrations in persons exposed to ziram during its production. *Genetika* 6(7):157–63 (Russ). *Chem. Abstr.* 74: 62787y, 1971
79. Prasad, I., Pramer, D. 1968. Genetic effects of ferbam on *Aspergillus niger* and *Allium cepa*. *Phytopathology* 58:1188–89
80. Raski, D. J., Hewitt, W. B., Schmitt, R. V. 1971. Controlling fanleaf virus-dagger nematode disease complex in vineyards by soil fumigation. *Calif. Agric.* 25(4):11–14
81. Richardson, L. T. 1954. The persistence of thiram in soil and its relationship to the microbiological balance and damping-off control. *Can. J. Bot.* 32:335–46
82. Richardson, L. T. 1966. Reversal of fungitoxicity of thiram by seed and root exudates. *Can. J. Bot.* 44:111–12
83. Ross, J. P., Harper, J. A. 1970. Effect of *Endogone* mycorrhiza on soybean yields. *Phytopathology* 60:1552–56
84. Ross, J. P. 1971. Effect of phosphate fertilization on yield of mycorrhizal and nonmycorrhizal soybeans. *Phytopathology* 61:1400–03
85. Rovira, A. D., Bowen, G. D. 1966. The effects of micro-organisms upon plant growth. II. Detoxication of heat-sterilized soils by fungi and bacteria. *Plant Soil* 25:129–42
86. Schreiber, L. R., Hock, W. K., Roberts, R. R. 1971. Influence of planting media and soil sterilization on the uptake of benomyl by American elm seedlings. *Phytopathology* 61:1512–15
87. Smalley, E. B. 1971. Prevention of Dutch elm disease in large nursery elms by soil treatment with benomyl. *Phytopathology* 61: 1351–54
88. Smiley, R. W., Cook, R. J., Papendick, R. I. 1970. Anhydrous ammonia as a soil fungicide against *Fusarium* and fungicidal activity in the ammonia retention zone. *Phytopathology* 60: 1227–32
89. Smith, A. M., Stynes, B. A., Moore, K. J. 1970. Benomyl stimulates growth of a Basidiomycete on turf. *Plant Dis. Reptr.* 54:774–75
90. Snel, M., Fletcher, J. T. 1971. Benomyl and thiabendazole for the control of mushroom diseases. *Plant Dis. Reptr.* 55: 120–21
91. Spanis, W. C., Munnecke, D. E., Solberg, R. A. 1962. Biological breakdown of two organic mercurial fungicides. *Phytopathology* 52:455–62
92. Spencer, W. F. 1970. Distribution of pesticides between soil, water and air. In *Pesticides in the Soil: Ecology, Degradation and Movement,* 120–28. Int. Symp. on Pesticides in Soil. Mich. State Univ. 144 pp.
93. Suzuki, T., Furukawa, K., Tonomura, K. 1968. Studies on the removal of inorganic mercurial compounds in waste by the cell reused method with a mercury resistant bacterium. (Hakko Kogaku Zasshi) *J. Fermentation Technol.* 46:1048–55
94. Thorn, G. D., Richardson, L. T. 1970. Effects of various fungicides in soil on water and amino acid transport in tomato plants. *Can. J. Bot.* 48:2033–36
95. Thorn, G. D., Richardson, L. T. 1971. Some physiological effects of cycloheximide in soil on the tomato plant. *Can. J. Bot.* 49: 1419–24
96. Tonomura, K., Nakagami, T., Futai, F., Maeda, K. 1968. Studies on the action of mercury-resistant microorganisms on mercurials. I. The isolation of mercury-resistant bacterium and the binding of mercurials to the

cells. *J. Fermentation Tech.* 46: 505–12

97. Tonomura, K., Maeda, K., Futai, F., Nakagami, T., Yamada, M. 1968. Stimulative vaporization of phenylmercuric acetate by mercury-resistant bacteria. *Nature* 217:644
98. Tweedy, B. G., De Hertogh, A. A. 1968. The literature of agricultural pesticides. *Advan. Chem.* 78:636–51
99. Waack, R., Alex, N. H., Frisch, H. L., Stannett, V., Szwarc, M. 1955. Permeability of polymer films to gases and vapors. *Ind. Eng. Chem.* 47:2524–27
100. Webley, D. M., Jones, D. 1971. Biological transformation of microbial residues in soil. In *Soil Biochemistry,* ed. A. D. McLaren, J. Skujins, 2:446–81. New York: Marcel Dekker. 527 pp.

100a. Weindling, R. 1932. *Trichoderma lignorum* as a parasite of other soil fungi. *Phytopathology* 22: 837–45

100b. Weindling, R. 1934. Studies on the lethal principle effective in the parasitic action of *Trichoderma lignorum* on *Rhizoctonia solani* and other soil fungi. *Phytopathology* 24:1153–79

101. Wilhelm, S. 1966. Chemical treatments and inoculum potential of soil. *Ann. Rev. Phytopathol.* 4:53–78
102. Wilhelm. S. 1963. Control of Pythium damping-off of melons by simultaneous seeding and soil fumigation with chloropicrin. *Phytopathology* 53:1144
103. Woodcock, D. 1971. Metabolism of fungicides and nematocides in soils. In *Soil Biochemistry,* ed. A. D. McLaren, J. Skujins, 2:337–60. New York: Marcel Dekker. 527 pp.
104. Zentmyer, G. A. 1972. Control of Phytophthora root rot of avocado with *p*-dimethyl aminobenzenediazo sodium sulfonate (dexon). *Phytopathology* 62

FUNGICIDE USAGE ON DECIDUOUS FRUIT TREES[1]

3557

F. H. LEWIS AND K. D. HICKEY

The Pennsylvania State University Fruit Research Laboratory, Biglerville, Pennsylvania, and Department of Plant Pathology & Physiology, Virginia Polytechnic Institute & State University, Winchester Fruit Research Laboratory, Winchester, Virginia

In the United States, research work on fruit fungicides is divided between the chemical industry and various tax-supported institutions, primarily the United States Department of Agriculture and the agricultural experiment stations in states where the fruit industry is of major economic significance. The search for new compounds with fungicidal properties is carried out by chemical companies with the manpower, facilities, experience, and financial resources necessary to patent, label, and market any new product. After a promising new compound is identified, varying numbers of field trials may be carried out by chemical company personnel. In all cases, numerous field trials are carried out by experiment station personnel who are interested in studying the potential value of the fungicide on some crop or disease in which they are interested. The agricultural experiment station worker wishes to acquire and develop information concerning the performance of agricultural chemicals and to make recommendations for effective and practical population management of various pest species. With tests being conducted in several countries or states, it usually can be determined within a few years whether or not a new fungicide can be used in ways that will result in the control of one or more major diseases. Frequently, however, there is a long gap between that point and the use of the product in an effective and practical population management program. With this in mind, we propose to trace the development of fungicide usage on apples and discuss some of the methods now being tested to reduce fungicide usage on that crop. In some instances, it is known that the work is directly applicable to other fruit crops, including Citrus. In other instances, the necessary work has not been done.

[1] The mention of any product does not constitute a recommendation by The Pennsylvania State University, the Virginia Polytechnic Institute & State University, or their personnel.

Improved Disease Control

In 1910 sprayers were 50-gallon barrels equipped with a hand-operated pump. Some were mounted on the bed of a horse-drawn wagon with a wooden tower and a long-handled spray gun to aid in reaching the top of the trees. Spray materials were Bordeaux mixture, lead arsenate, nicotine sulfate, lime sulfur, self-boiled lime sulfur, and crude formulations of elemental sulfur. Trees were sprayed two or three times each year and the fruit sorted at harvest. Critical spray timing did not exist as indicated by the common statement that it didn't matter when the cover spray was applied since the farmer would use only one. The presence of disease and insect damage on some of the fruits was accepted as normal both in domestic and foreign trade.

The situation today is radically different. Neither the processor nor the fresh fruit buyer will accept fruit with obvious disease or insect injury. The tolerance for worms or insect parts in some fruit products is zero: one fruit fly larvae can cause the rejection of an entire truckload of cherries. There have been significant advances in our ability to control all of the fungus diseases, insects, and mites that are problems on leaves and fruit. Fruit growers can now control such important diseases as apple scab (caused by *Venturia inaequalis* [Cke.] Wint.), brown rot of stone fruits (*Monilinia fructicola* [Wint.] Honey), cherry leaf spot (*Coccomyces hiemalis* Higgins), and black knot of plums (*Dibotryon morbosum* [Schw.] Theiss, and Syd.).

Why, then, must we concern ourselves with these problems if they can be controlled? We are concerned because the pathogens are still present and the high costs of their control are causing economic losses to the modern fruit grower as serious as the disease epidemics of 50 years ago.

Changes in pest control programs, fungicides, and equipment.—Modern fruit pest control programs began during the period 1906 to 1920 after lime sulfur was introduced as an apple foliage and fruit fungicide (11) and after spray equipment was improved to the point where it was physically possible to apply several sprays each year. From that time until the early 1940s, much of the research attempted to evaluate the injury caused by the copper fungicides and by lime sulfur, establish the range of usefulness of lime sulfur, improve the elemental sulfur fungicides and sulfur dusts as substitutes for lime sulfur, develop methods of safening mixtures of lime sulfur and lead arsenate, and perfect the high pressure sprayer.

It was obvious from the beginning that lime sulfur was safer than Bordeaux mixture on peaches and that it would provide acceptable control of the major fungus diseases of apples, plums, and cherries. Still, it caused severe injury on all deciduous fruits when used in repeated applications. As increased amounts of lead arsenate became necessary for insect control, the number of sprays was increased and the trees were drenched with sprays applied at pressures up to 800 pounds per square inch (psi). Evidence accumulated showing that lime sulfur decreased the productivity of apple trees and there was no way to use it during the growing season without an undesirable degree of injury.

The use of elemental sulfur fungicides escalated in the 1930s as evidence accumulated regarding the injury caused by lime sulfur. The change was hastened by the introduction of more finely divided dry wettable sulfurs and sulfur pastes. By 1945, most growers of deciduous fruits had good high pressure spray equipment and had some experience with the elemental sulfurs.

The complete abandonment of lime sulfur and the change to elemental sulfur and ferbam came when it still was not possible to time the sprays adequately on large acreages of fruit. The high pressure equipment, while good in quality, required relatively large amounts of manpower to spray 10-12 acres of apple trees per day. The airblast sprayers still were not adequate for the task of spraying large trees; both air volume and velocity were low by present standards.

The control of diseases improved more or less steadily with more effective fungicides, better equipment, and better grower attitudes and educational levels. Great emphasis was placed on thorough spraying and on teaching growers that the spray had to be applied before infection occurred. The work on apple scab during this period was more extensive than that on any other fruit crop. Mills (43, 44) and Hamilton (19–21) made invaluable contributions to our knowledge of how to use fungicides on fruit.

Mills had the problem of developing a system under which protectant fungicides like elemental sulfur and ferbam could be recommended to thousands of growers for scab control on the highly susceptible apple cultivar 'McIntosh.' It was known that protectants, principally sulfur at the time, were moderately effective if all new tree growth was covered shortly before a rain period of one or two days' duration. However, weather forecasts were not completely accurate for an area of several thousand square miles; so it was not feasible to apply all sprays shortly before rain periods. When rain began without adequate fungicidal coverage of the trees, spraying or dusting in the rain sometimes was effective and sometimes failed with no adequate explanation of the failures.

Mills developed a table (published as a series of curves in 1944) giving the approximate number of hours of wetting necessary for primary apple scab infection in an orchard containing an abundance of inoculum. These data were slightly modified as the years passed and additional observations were made. They were tested in experimental plots by Lewis (34) who found that numerous sulfur dusts prevented infection when they were applied shortly before the minimal time for infection as indicated by Mills' table. When the same dusts were applied after the indicated period, scab lesions appeared.

The table, as given to Lewis by Mills in 1942, follows. The curves, as published by Mills in 1944, were slightly more conservative than his table, and perhaps were deliberately so arranged that precise figures were not always obtainable. This was, however, a major step forward in our understand-

TABLE 1. Data of W. D. Mills on the approximate hours of wetting necessary for primary scab infection in an apple orchard containing an abundance of inoculum. Data for odd temperatures omitted.

Temperature °F	Amount of infection		
	Light	Medium	Severe
42	30	40	60
44	22	30	45
46	19	25	38
48	16	22	33
50	15	20	30
52	13	18	27
54	12.5	17	25
56	12	16	24
58	11	14.5	22
60	10	13.5	20.5
62	9.5	13	19.5
64	9	12	18.5
66–75	9	12	18
76	9.5	12.5	19
78	13	17.5	26

ing of the apple scab disease. It helped to eliminate the psychology of crisis governing fungicide recommendations and usage on apples.

Mills' work provided the basis for an effective spray warning service, widely copied in the United States and Europe, whereby growers were informed by telephone and radio of the progress of each infection period and of the time when protectant fungicides could no longer be expected to give control. The end of the period of effectiveness could be predicted on the basis of the time when rainfall began, the prevailing temperature during the rain, and the number of hours required for infection at that temperature.

The use of a spray warning service such as that devised by Mills and later by Roosje (48) clearly implied that spraying or dusting in the rain was satisfactory if the fungicide was applied before infection occurred. This practice was common in northeastern United States by 1937 and is still followed except during periods of heavy rainfall. The elemental sulfur fungicides and all of the common dry wettable organic fungicides have given satisfactory results when applied on wet foliage. Lime sulfur is not effective if diluted quickly on the tree, and there is some question about the use of triarimol and benomyl under such conditions. The rationale for spraying in the rain is that timing and coverage are more important than the amount of fungicide deposited on the leaves and fruit.

The work of Hamilton, concurrent for many years with that of Mills, showed that it was possible to obtain scab control with certain new fungicides when applied after infection had occurred.

Hamilton developed a method of testing fungicides on potted trees in moist chambers and the greenhouse (19). He found that one of the valuable characteristics of some new fungicides was the ability to control scab when applied after periods of wetness sufficiently long to allow abundant infections on unsprayed trees. For example, while 10 hours' wetness at 60°F allowed scab infection, and sulfur, ferbam, and glyodin were not effective after that time, new products such as captan, maneb, dichlone, dodine, and phenyl mercury were effective if applied several hours later.

This work was of significant help in getting sprays applied on time and was one of the factors contributing toward a general improvement in the level of control. Unfortunately, the significance of the work has been partially obscured by increasing problems regarding terminology. In the work on scab eradication during and after infection periods, the period over which fungicide application could be delayed was measured from the beginning of the wet (rain) period because it was not possible to determine readily the exact time when infection occurred. Yet, the entire period has been referred to in nearly all cases (for example, 21) as the period of "eradication", "kickback", "after-infection control" and "curative action". In a recent paper (54), Szkolnik referred to after-infection control and carefully explained that the data included the entire period from the beginning of rainfall, and so included protection up to the time of infection plus an additional time after infection (eradication). Technically, confusion could be created by any slowdown in the infection process caused by the fungicide. The terminology problem becomes acute when one considers that there now is a series of fungicides from which one can choose materials to provide control when applied at any time from a few days before an infection period until several days after the period. Also, some of these fungicides may kill all exposed conidia

Fungicides	Number of hours after beginning rainfall that application can be delayed
captan 50% WP, 2 lb	18–24 hours
dichlone 50% WP, 1/4 lb	30–36 hours
dichlone 50% WP, 1/2 lb	36–48 hours
dodine 65% WP, 3/8 lb	18–24 hours
ferbam 76% WP, 2 lb	None
glyodin 30% solution, 2 pints	None
maneb 80% WP, 1 1/2 lb	18–24 hours
Phybam-S, 4 lb	24–36 hours
Polyram, 2 lb	18–24 hours
sulfurs, 5 lb (actual)	None
thiram 65% WP, 2 lb	15–20 hours

and mycelium on a scab lesion (for example, dodine) or may kill the fungus and the leaf tissue under and around a lesion (for example, benomyl). It is obvious that the term "eradication" as commonly used can have any one of several meanings. For this paper, we are interested in the performance of the fungicides as measured by the number of hours after the beginning of rainfall when a spray application will give scab control (protection plus eradication).

The number of hours after the beginning of rainfall when a spray application will give scab control has been listed in numerous publications and updated as more information was accumulated. With the permission of Dr. Michael Szkolnik, the 1972 listing for New York is reproduced here (personal communication).

The data are given for an infection period at 50–60°F. At 50°F the period is the higher figure given. At 60°F it is the lower time period. Mercury is no longer recommended, but it has given control for up to 72 hours.

In another discussion of this topic, Szkolnik warned that highly concentrated sprays or dusts applied to dry foliage had not been reliable in after-infection control but were satisfactory if applied to wet foliage.

Roosje (48) found that captan, ferbam, thiram, and ziram, either had no curative action or that their curative action was insufficient for practical purposes. He reported that dodine was intermediate between mercury and the above fungicides.

Brown et al (5) applied triarimol to 'McIntosh' apple seedlings at various intervals before and after inoculation with conidia of the apple scab fungus. Disease symptoms were observed on untreated foliage after 96 hours; normally mycelium and conidia would be visible at that time. Triarimol at 10 ppm gave perfect control when applied before or up to 72 hours after inoculation. Leaves sprayed after 96 hours developed small, necrotic lesions, but no fungus was observed on these lesions until 22 days after inoculation. When 25 ppm triarimol was applied 96 hours after inoculation, no fungal growth was observed at any time. In a field trial with triarimol at 35 ppm, 98–100% scab control was obtained although the pre-pink and full-bloom sprays were applied 8–10 days after good scab infection periods. In a later paper (6), Brown & Hall reported that the curative action of triarimol resulted from its ability to arrest fungal growth within leaf tissue. Gilpatrick & Szkolnik (17) found that the first application of triarimol almost completely prevented scab on cluster leaves even though the fungicide was applied 4–8 days after the beginning of the infection period. These results are of special interest because the fungicide is nonphytotoxic to apples, is highly effective on the three major early-season diseases of apples (scab, powdery mildew, rust), and has the ability to stop the development of the scab fungus in apple leaves at any time before lesions appear.

The work of Mills, Hamilton, Roosje and others on apple scab control during and after infection periods was important in improving the control of this disease, and, to a lesser extent, the fungus diseases of all deciduous fruit crops. The organic mercury fungicides were developed and introduced after the work of Howard, Locke & Keil (25). Dichlone, captan, and dodine became available. Airblast sprayers were improved, with many units capable of delivering 90,000 cubic feet of air per minute (cfm). Some units were equipped with centrifugal pumps which operated at pressures of 180–200 pounds per square inch (psi); others had high pressure pumps operated at 300–500 psi. Old types of metal nozzles and nozzle discs were replaced with ceramic or special metal discs that abraded slowly and did not require frequent recalibration to avoid over-spraying. Under the influence of European and Canadian imports, orchard sprayers designed to deliver low gallonages per acre became available.

As a result of these changes, and an increased level of grower education, control of the major fungus diseases of fruits in some fruit growing areas reached a point in the early 1960s where annual disease losses were negligible. The same situation developed in the control of such major insect pests as the codling moth, the redbanded leafroller (*Argyrotaenia velutiana* Walker), the Oriental fruit moth (*Grapholita molesta* Busck), and the plum curculio (*Conotrachelus nenuphar* Herbst). It became evident that we had reached a plateau in pest control, and were preparing pest control recommendations for crisis conditions when in fact the need was for maintenance programs for many orchards with no special pest problems. The European red mite (*Panonychus ulmi* Koch) did remain a special problem. Its prevalence, its ability to develop resistance to new acaricides, and the presence of both insect and mite predators presented an open invitation to try integrated control programs on this pest.

Within the past 10 years, there has been a marked reduction in the use of pesticides on fruit in the Cumberland-Shenandoah Valley area of Pennsylvania, Maryland, West Virginia, and Virginia. Reductions have proceeded through several stages: (*a*) adoption of standards of 400 gallons of dilute spray to be applied on pruned apple trees 18–22 feet in height and of 350 gallons per acre for stone fruits planted in rows 25 feet apart, (*b*) adoption of concentrate or low volume spraying with the amount of pesticides per acre reduced about 20% for apples and 25% for stone fruits, (*c*) the use of large airblast sprayers to apply sprays from alternate row middles at fixed intervals of 7 days with the amount of pesticide per acre per year about 55% of the dilute standard, and (*d*) movement into large scale use of integrated mite control programs with the concentration of all pesticides held at rates low enough to avoid severe injury to the predators. It is estimated that 80% of the apple acreage in southern Pennsylvania is now sprayed with reduced amounts of pesticides. Some phases of this progression in the Cumberland-Shenandoah Valley are similar to those on apples in other parts of the United States, on deciduous fruits other than apple, and on the Citrus fruits.

Some Fruit Fungicides

The performance characteristics of four popular fruit fungicides are listed here as examples of the fact that all fungicides have their good and bad points which must be considered in their use for disease control. Greater detail on these and other fruit fungicides is available in the references cited and in *Fungicide and Nematicide Tests* published by the American Phytopathological Society.

Captan [N-(trichloromethyl)thio-4-cyclohexene-1,2-dicarboximide] is the most widely recommended fruit fungicide.

On apples, captan controls scab, black rot (caused by *Physalospora obtusa* [Schw.] Cke.), white rot (*Botryosphaeria ribis* Gross. and Dug.), bitter rot (*Glomerella cingulata* [Ston.] Spauld. and Schrenk), Brooks spot (*Mycosphaerella pomi* [Pass.] Lind.), and blossom-end rot (*Botrytis cinerea* Pers, ex-Fries) (9, 20). It does not provide adequate protection against cedar apple rust and quince rust (*G. Clavipes* Ck. and Pk.) (34) nor does it control sooty blotch (*Gloeodes pomigena* [Schw.] Colby) and fly speck (*Mycrothyriella rubi* Petr.) on apples harvested more than 40 days after the last spray is applied (23, 36). Powdery mildew (*Podosphaera leucotricha* [Ell. and Ev] Salm.) and several species of mites are often more severe on trees sprayed with captan.

On pears, captan is widely used for control of pear scab, leaf blight and fruit spot (*Fabraea maculata* Atk.), bitter rot and miscellaneous summer diseases.

On stone fruits captan controls brown rot, Botrytis rot, and cherry leaf spot. It is one of the most effective fungicides for the control of brown rot during bloom and at harvest. Scab (*Cladosporium carpophilum* Thuem.) is controlled by captan when spray intervals are not more than 14–17 days.

Captan-treated fruit generally has the excellent finish necessary for top quality grades. It is especially suited to Golden Delicious apples which often russet easily. Severe fruit russeting of Ben Davis and Gano apples and of D'Anjou pears has resulted from captan sprays. It also has caused a necrotic leaf spotting, yellowing, and dropping of leaves when used at full recommended rates early in the season on Delicious and to a lesser extent on Stayman, Winesap, Baldwin, and King cultivars of apple. This injury is more severe under poor drying conditions or when captan is used in combination with sulfur. A similar type of leaf injury is often produced on Emperor Francis, Schmidt, and Giant cultivars of sweet cherry. D'Anjou pear foliage has been stunted and cupped. Necrotic spots on fruit and shot-hole injury to leaves of plums may occur when captan is used in several successive sprays. Shot-hole injury may occur on some cultivars of peach when captan is used during the bloom to shuck-split stages.

Captan should not be mixed with oil, lime, or other alkalies. It should not be used within 5–7 days after an oil application or applied over a Bordeaux mixture residue. Combinations with sulfur, dodine, or malathion may result in increased injury. The effectiveness of malathion has been reduced in such mixtures. In combination with lead arsenate at equal amounts, captan acts as a safener in preventing arsenical injury.

Dodine [n-(dodecyl guanidine acetate)] has provided excellent control of scab on apples (51), pears, and pecans, and of cherry leaf spot, but it has not been satisfactory for control of other diseases on tree fruits. It is a mild eradicant against apple scab and may be used successfully up to 24 hours after the beginning of an infection period. Secondary spread of apple scab may be prevented if dodine is applied to established infections. It does not eradicate the fungus in the lesions but suppresses spore germination and production (1, 22, 26). Recently, failure to control apple scab in several western New York orchards was attributed partially to the development of tolerance by the fungus to dodine (53). This apparent resistance has not been reported in other areas and is attributed to extensive usage of dodine at reduced rates during the past 10 years.

Dodine may russet yellow cultivars of apple, particularly Golden Delicious. Fruit injury to McIntosh and Cortland may occur with combinations of dodine and Kelthane used late in the season (52). Combinations of dodine and captan may cause necrotic leaf spotting on Delicious apples. Combinations of dodine and lead arsenate have been safe on apples when the insecticide was used only at 2 lb per 100 gallons in 3 or 4 sprays. However, when dodine-lead arsenate combinations are used in a full season program on apples or cherries, a safener to prevent arsenical injury should be added.

Dodine is compatible with most pesticides, but not with lime or other alkaline products. It may not mix readily with oil or with insecticides formulated as emulsifiable concentrates or flowable emulsions.

Zineb (zinc ethylene bis dithiocarbamate) is used on apples and the stone fruits. On apples, it is used in early season sprays for rust control but has not been effective enough against apple scab to be used alone. Zineb has given good control of apple blotch (*Phyllosticta solitaria* E. and E.) when used in the sprays soon after the petal fall stage (51). It has a longer residual life than captan or thiram and, for this reason, is more effective against sooty blotch and fly speck (36). Because of its weakness against apple scab, it is often combined with other fungicides for late season use. It is very effective against black knot of plums (37) but cannot be depended upon to give adequate control of brown rot.

Zineb is compatible with all pesticides commonly used on apples and plums. Its use results in a minimum of fruit russet or unsightly residue.

Triarimol [a-2,4-dichlorophenyl-a-phenyl-5-pyrimidinemethanol)] is a new systemic fungicide of promise for use on tree fruits. It has been tested extensively but has not been approved for commercial use in the United States. On apples, it is effective in the control of four major early and mid-season diseases (scab, rust, powdery mildew, black rot (5, 24, 31). It is a very effective eradicant of apple scab and powdery mildew fungi. Apple scab control has been obtained when triarimol was used 8–10 days after infection (5). The residual life of triarimol is relatively short (13, 14) and this may be a major cause for its failure to control the apple fruit rots, sooty blotch, and fly speck (10). It does not appear to have any effect on mites. On the stone fruits, it has given excellent control of brown rot on fruit and blossoms but poor control of peach scab and Rhizopus rot (*Rhizopus stolonifer* [Ehrenb. ex Fr.] Lind.) (33, 55). It has also been effective against the cherry leaf spot fungus and cherry powdery mildew (*Podosphaera oxyacanthae* [DC.] De By.).

Triarimol has not been phytotoxic to tree fruits. It is compatible with many of the pesticides used on tree fruits, but the complete list has not been evaluated. In one test on cherries, it did not act as a safener for lead arsenate.

The four fungicides discussed here (captan, dodine, zineb, and triarimol) illustrate the progress that has been made and indicate that no fungicide is satisfactory for the entire disease control program on any deciduous fruit. When these and other fungicides are combined into a program that takes advantage of the best features of two or more materials, the result can be a disease-free crop.

Economics and Fruit Pesticide Usage

In most discussions on disease control by fungicides, there is no attempt to evaluate costs as a factor in their use. However, the assumption in making disease control recommendations is that the treatment will increase production sufficiently to provide a profit over the cost of the treatment. Also, cost calculations are a convenient way to compare some disease control practices.

Commercial production of fruit crops is a business that requires large capital investments, expert management, and continued input of new knowledge. It is not unusual for a few growers to own several hundreds or thousands of acres of trees, and these growers may produce more than 50% of the crop in any district. Yet, in one area perhaps typical for the deciduous fruits, 80% of the growers own less than 60 acres of apples although many of them grow other fruits. Many of these growers do not own sufficient acreage to minimize pesticide application costs.

The investment in land, buildings, trees, and equipment often exceeds $1000 per acre or $25,000–$50,000 per full time employee. The first returns on the investment may not be available until the trees are 3–7 years of age. After production begins, gross income per acre tends to be high in relation to such crops as corn, wheat, and cotton, but production expenses and labor requirements also are high. Government controls and price supports do not exist in the United States fruit industry.

Zuroske in Washington (56) and Kelly in Pennsylvania (27) illustrated the impact that pest control costs can have on the orchard operations. Zuroske found that the spray materials cost per acre of apples in central Washington was $84.39 in 1945 and $65.90 in 1965. The average yield of fruit for 26 study years was 473 boxes per acre. Thus, the spray material costs were 17.8–13.9 cents per box. Labor costs for spraying increased to a maximum of $54.37 per acre in 1945 and decreased to $6.49 by 1965. These costs can be compared with a price received for apples of 21 cents per box above costs in 7 of the years 1953–1965.

Kelly's records for the years 1959–62 showed spray material costs were $61.53 per acre or 19.7 cents per bushel of apples. This was 20% of the total cost of production and harvesting. Data of Lewis (38) indicate that pesticide application costs are usually $20–$30 per acre for the season. Thus, the average total cost of pest control among a representative group of Pennsylvania apple growers in 1959–62 was in the range of 25–30 cents per bushel. Processor prices for many of these apples frequently were $1.00 per bushel plus or minus about 15 cents. Thus, the average cost of pest control exceeded the profit, if any, from the crop. In the heavy crop, low price years of 1970 and 1971, high costs have continued while hundreds of thousands of bushels of apples could not be sold at a price above picking and hauling costs.

Brann & Steiner (4) quoted data of Stanton & Dominick comparing the cost of spraying apple trees in New York in 1950 and 1962. The costs totaled $100 per acre in 1950 and $102 in 1962. This was 41% of the cost of growing the 1962 crop. Spray material costs were 31% of the total production cost and 75% of the cost of spraying. Brann & Steiner said "In the final analysis (the grower) would do well to check on his present operation to see if there is some way he can safely cut down on the total amount of pesticides he is using."

The 1971 revised edition of *"Agricultural Chemicals in Tree Fruit Production"*, prepared by a committee of 12 research and extension specialists and published by Pennsylvania State University, contains two apple spray schedules. The "standard" spray program is based on the use of 400 gallons of dilute spray per acre for mature trees 18–22 feet in height planted in rows 40 feet apart. The cost of pesticides is $130 per acre plus $30–$40 to apply the 11 dilute sprays. This program is the current version of the old, dilute, "fool-proof" listing of treatments needed by a few growers under severe disease conditions. The second program is a "Reduced Pesticide Spray Program for Apples" designed to take maximum advantage of what we know about close management of pest control in a typical commercial orchard in this area without special pest problems. The cost of this program is about $58 per acre for pesticides or 45% of the cost of the standard program.

Orchard labor and machinery costs have increased dramatically within the past 10 years. In 1961, labor was about $1.50 per hour and a large airblast sprayer could be purchased for $7000. Today, labor is $2.50 or more per hour and a large airblast sprayer is about $11,300.00.

The unit prices of several of the common fruit pesticides have decreased from 10–30% during the past 10 years. Some new products are priced as high as $8.50 per pound, but such products cannot be used in quantity under present economic conditions unless the dosage required is very low or the fruit commands a premium price.

These cost data illustrate the economics confronting the fruit grower in pest control. In a sense, he is in double jeopardy. He must protect his trees and produce a pest-free, attractive product in order to stay in business. To accomplish this, a typical pest control program suggests that he use far more pesticides than he can afford. His survival as a grower depends upon his ability to exceed average yields while cutting his costs.

In view of this situation, we have tried various methods of using pesticides in an effort to reduce costs, labor, and total pesticide usage.

Pesticide Application Methods

It is possible with a high pressure pump and spray gun to direct the discharge to any point until the tree is wet. Coverage then is assumed to be complete. This method saves pesticides when spraying trees that occupy only a small percentage of the total row space. However, when the trees are spaced so that they occupy most of the row, and an airblast sprayer operates continuously from one row end to another, the disadvantages of dilute sprays become obvious. About half of the total time spent in the orchard is used in mixing pesticides and loading the sprayer. A minimum of two men is required to operate the tractor, sprayer, and filler unit. There is an investment of several thousand dollars tied up in the filler unit and in a water system capable of supplying about 2000 gallons of water per hour. There is a large and variable loss of pesticides in run-off and drip from the trees. The spray residue on the tree occurs as smears and relatively large drops.

Work with concentrate sprays was under way in several areas by 1950. It was known that their use could result in reduced labor and costs. During the 1950–71 period, this method of spraying was studied extensively by both pathologists and entomologists and is now used widely in all areas of the United States and in several other countries.

Terminology.—Several terms are used in connection with sprays mixed at concentrations higher than the dilute rate: semi-concentrate, concentrate, high-concentrate, ultra-high concentrate, low volume, ultralow volume, and mist sprays. Originally, the plan was to refer to the concentration of the chemical in the spray tank as dilute or 1X. When it was mixed at twice the dilute rate, it was referred to as 2X concentrate. This system worked reasonably well until people began to calibrate the sprayer for 3X, 6X, etc., mix the pesticide at any rate that seemed proper, and apply different amounts of water per acre.

At a meeting of fruit workers in Winchester, Virginia, in 1971, it was suggested that "concentrate" and "low-volume" be used as synonymous terms, and that "ultra-low volume" be used for volumes of about 2 quarts or less per acre. It was recognized that the varied and confusing terminology, including the use of 3X, 6X, etc. at odd amounts per acre, should be abandoned in favor of a system under which an amount of pesticide is to be applied in a given amount of water per acre or in any one of a range of amounts of water per acre (e.g. 20–110 gpa).

Time and cost studies.—In 1963, Lewis (38) published the results of a series of time and cost studies on pesticide application. The data were updated in 1971 for presentation here. Methods used were similar to those used by Larson, Fairbanks & Fenton (32) in calculating the cost of farm machinery.

The annual cost of owning and operating an $11,300 airblast sprayer used 200 hours each year for 10 years is $2858, or about 25% of the original cost. This includes 200 hours labor for one man. The annual cost of owning and operating a $7500 tractor used 400 hours each year for 10 years is $1675, or about 23% of the original cost. Thus, these two pieces of equipment cost $18.48 per hour if used 200 hours for spraying each year and $13.72 per hour if used 400 hours.

The data in Table 2 show the relative importance of the various factors involved in pesticide application. The total machine and labor costs at various pesticide concentrations are given in Table 3.

These data show that the use of concentrate sprays results in an increase in the proportion of the total work day spent in actually spraying trees, and that the time and cost per acre are decreased for this reason. The greatest increases in efficiency occur when moving from dilute sprays to 3X (109 gpa) and 6X (54 gpa) concentrates. Beyond that point, savings in labor and machine costs are small and must be justified on a basis other than the cost of operating one tractor and one sprayer. For example, acreage to be covered, necessity of a refill unit and its operator, and water system required are valid considerations in deciding upon a certain concentrate level.

TABLE 2. Time required to spray one acre of apple trees planted 20 by 40 feet with the sprayer traveling at 2 mph and spraying in two directions. Lewis data of 1971.

Concentration of chemical in tank	Discharge rate. G.P.M. 2 sides	Minutes to discharge 500 gallons	Refill time[a]	Turning time on row ends	Total time to load and spray 500 gallons	Time per acre in minutes
1X (400 gal/acre)	64.68	7.73	7	1	15.73	12.58
1X (327 gpa)	52.80	9.47	7	1	17.47	10.12
2X (163.5 gpa)	26.40	18.94	7	2	27.94	9.14
3X (109 gpa)	17.60	28.41	7	3	38.41	8.37
4X (81.75 gpa)	13.20	37.88	7	4	48.88	7.99
5X (65.4 gpa)	10.56	47.35	7	5	59.35	7.76
6X (54 gpa)	8.80	56.82	7	6	69.82	7.54
7X (46.7 gpa)	7.54	66.31	7	7	80.31	7.50
8X (40.9 gpa)	6.60	75.76	7	8	90.76	7.42
9X (36.3 gpa)	5.86	85.32	7	9	101.32	7.36
10X (32.7 gpa)	5.28	94.70	7	10	111.70	7.31

[a] Refill time is given as the minimum with use of a refill unit and its operator. Where one man does the entire job, refill time commonly has been 30 minutes.

TABLE 3. Machine plus labor costs per acre in spraying apple trees planted 20 by 40 feet using an $11,300 sprayer and a $7500 tractor. Lewis data of 1971.

Concentration of chemical in tank	Gallons spray per acre	Sprayer operating at 2 miles per hour		Sprayer operating at 2 1/2 miles per hour	
		200 hrs/year	400 hrs/year	200 hrs/year	400 hrs/year
1X	400	$3.87	$2.88	$3.49	$2.59
1X	327	3.12	2.31	3.14	2.33
2X	163.5	2.82	2.09	2.43	1.81
3X	109	2.58	1.91	2.13	1.58
4X	81.75	2.46	1.83	2.03	1.51
5X	65.4	2.39	1.77	1.97	1.46
6X	54	2.32	1.72	1.90	1.41
7X	46.7	2.31	1.72	1.87	1.39
8X	40.9	2.29	1.70	1.85	1.38
9X	36.3	2.27	1.68	1.84	1.37
10X	32.7	2.25	1.67	1.83	1.36

In developing an economic model of this type, it is necessary to assume that there is a reasonable amount of work for the equipment and that the operator could be gainfully employed at other work. Spraying 200 hours each year is perhaps a fair average. Using the same equipment (an $11,300 sprayer and a $7500 tractor) at 2 mph for 200 hours, the cost of $2.32 per acre at 6X could be reduced to $1.41 if the trees could be sprayed at 2.5 mph and the sprayer used 400 hours each year (225 to 250 acres sprayed 10 times). Contrary to widely held opinion, the planting of size-controlled trees in 20 foot rows and spraying them with one man operating a $5000 sprayer and a $5000 tractor, would not necessarily result in reduced cost per acre because of the greater row distance to be traveled. Economy would require either the complete spraying of two rows of trees with each pass through the orchard or a reduction in the amount of pesticide needed per acre. It does appear that size-controlled trees would allow a lower investment in spray equipment and lower upkeep costs.

Pest Control with concentrate sprays.—The fact that disease, insect, and mite control can be obtained with concentrate sprays has been confirmed by their use on hundreds of thousands of acres of fruit and other crops. There are a few specific instances where there is some question about the efficiency of low gallonages per acre; for example, the use of 5 gallons of spray per acre for the control of powdery mildew. However, within the range of about 18–400 gallons of spray per acre for apples, the important question is whether or not the same amount of pesticide per acre is required regardless of the amount of water used. Adequate machinery to handle any given concentration is essential, and it is available.

Brann (2, 3), after a series of studies in New York, stated that airblast sprayers used at the concentrate level can lay down pesticide deposits equivalent to dilute sprayers while using 75–80% as much active ingredients in doing so. Potts (46) found that a concentrate sprayer deposited 27% more insecticide on Norway maple twigs than a hydraulic sprayer using the same amount of the insecticide. Martin, as quoted by Fulton (16), applied a constant rate of 25 gallons of spray per acre and obtained a 40% increase in deposit with 100–150 micron droplets as compared to droplets 200–400 microns in diameter (roughly the range of dilute spray droplets) even though the original fungicide concentration was constant. Dibble (12) stated that the amount of insecticide used in concentrate sprays could be reduced by 25–30% because of lack of run-off. Burrell in New York (8), Klingbeil & Mitchell in Michigan (28), and Brooks, Thompson & Jutras in Florida (7), all indicated that the amount of pesticide used in concentrate sprays could be reduced by 20–25% as compared to standard dilute sprays. Krestensen & Graham (29, 30) used 10–25% less pesticide in concentrate sprays on apples than in dilute sprays and found no significant difference in disease or insect control over a 6 year period.

A related question concerns possible variations in the amount of pesticide needed with concentrate sprays varying from about 20–100 gallons per acre. Moore, as quoted by Fulton (16), applied a constant amount of lime sulfur in 3–24 gallons of spray per acre and found no significant difference in apple scab control. Hickey (unpublished) applied Dikar 6.6 lb per acre to apple trees with one sprayer at 20.5, 41, and 82 gallons per acre and with a second sprayer at 11, 16, and 20.5 gallons per acre; there was no significant difference in control of powdery mildew. When triarimol was applied at a constant rate per acre in 20.5, 41, and 82 gallons per acre, mildew control did not differ significantly but there was a trend toward more disease with lower gallonage. In another test, Hickey applied Dikar at 7 lb per acre in 5, 10, 20, and 100 gallons per acre with no significant difference in mildew control.

Lewis (42 and unpublished) assumed that the water used per acre was important only as a part of the transport mechanism between the sprayer and the target; that its volume was unimportant if adequate spray coverage and deposit were obtained. Apple trees were pruned to a height and width of 9 feet and thinned to approximate commercial orchard conditions. Multiple tree plots were arranged in randomized blocks using only the first and third rows in a 4 row block of trees. An airblast sprayer was used, with a pump operating at 400 psi and sufficient air volume to drive the spray into and well above the tree tops without creating a serious drift problem. Triarimol was applied in 1970 at 514 grams of the 10% formulation per acre in 19, 29, 59, and 113 gallons of water per acre. There were no significant differences in control of apple scab, powdery mildew, or cedar apple rust. The experiment was repeated in 1971 using benomyl at 232 grams per acre in 18, 27, and 54 gallons of water. Again, there were no significant differences in the control of apple scab, powdery mildew, or cedar apple rust.

Asquith (41), used the same amount of Alar (a formulation of succinic acid-2,2 dimethylhydrazide) in 50 and 100 gallons of water per acre with the sprays applied on May 18. Residue samples taken November 1 showed over 50% greater residue with 50 gallons per acre as compared to 100. Measurements on terminal growth indicated a greater response from Alar applied in 50 gallons of water.

Pesticide application from one side of the tree.—In 1963, Lewis & Hickey (39) used a Model 36 Speed Sprayer (45,000 cfm air delivery) and a Model 703 Speed Sprayer (90,000 cfm air delivery) to study the coverage (the area of the tree over which the pesticide is deposited) and deposit (the quantity of pesticide deposited on the tree) obtained when pruned apple trees were sprayed only from one side with a standard spray gallonage and amount of pesticide per acre. Separate row sections were sprayed from middles 3, 5, and 7, thus duplicating conditions in an orchard where spraying was done in calm air from alterate middles with both sides of the sprayer in operation. The fluorescent dye Phosphor 2282 was used at ¾ pound per 100 gallons of spray to allow study of spray coverage and deposit under ultra-violet light. For each tree examined, data were taken on one 5-leaf sample from each cubic yard of space in an area extending from the sprayer to the opposite side of the tree and from a level of 4 feet above the ground to the top of the tree. Each sample was rated for coverage and deposit using a scale of 0 to 10 with 10 indicating "perfect". Each number in Figures 1–8 represents the average rating of the upper surface of 5 leaves. The vertical columns of figures are the ratings obtained at heights of 4, 7, 10, 13, etc. feet above the ground. The tree height in Figures 1–4 was 22 feet; in Figures 5–8 it was 28–31 feet.

The medium sized sprayer (45,000 cfm air delivery) placed a few spray drops on every leaf examined on a tree 22 feet in height, 33 feet in diameter and bearing a heavy crop of fruit. Coverage was relatively poor in the top center of the tree, in unusually dense areas of the tree, and at a height of 4–7 feet directly across the tree at a distance of 27–30 feet from the sprayer. The deposit was light over most of the tree. On trees 28–31 feet in height, coverage and deposit were acceptable only on the side of the tree nearest the sprayer. Deposits were very light above 19–22 feet.

The large sprayer (90,000 cfm air delivery) gave better coverage and deposit. Coverage was improved from an average rating of 6.28–7.63 in the experiment on trees 22 feet in height, and from 5.71–6.76 on trees 28–31 feet in height. Deposit increased from 3.59–5.92 on 22 foot trees and from 4.24–5.07 on 28–31 foot trees. Where the comparison was made only on the basis of data from the lower 19 feet of the trees, where the smaller sprayer

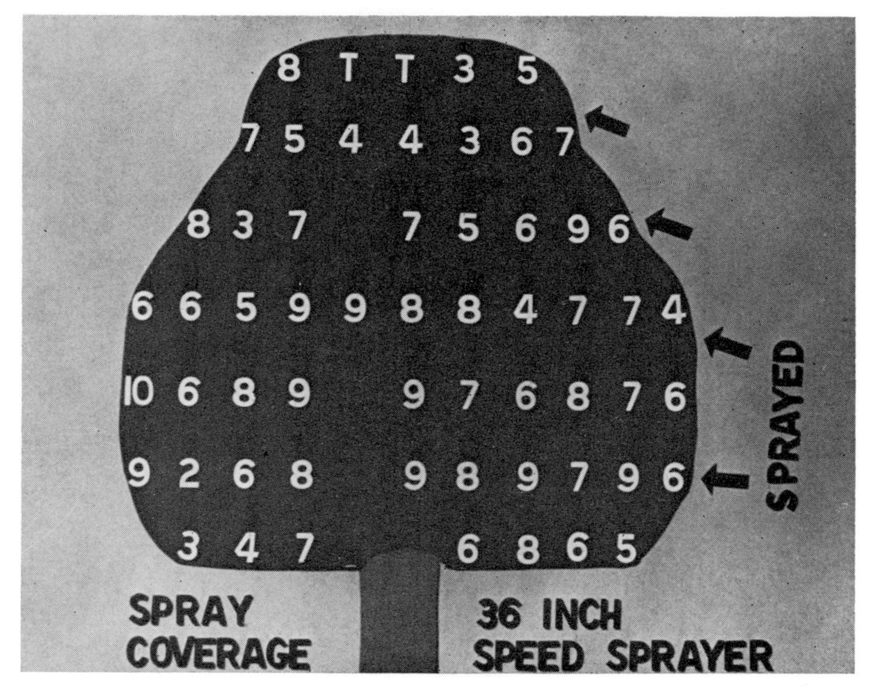

FIG. 1. Average ratings of spray coverage on 22-foot apple trees with air delivery at about 45,000 cubic feet per minute (cfm).

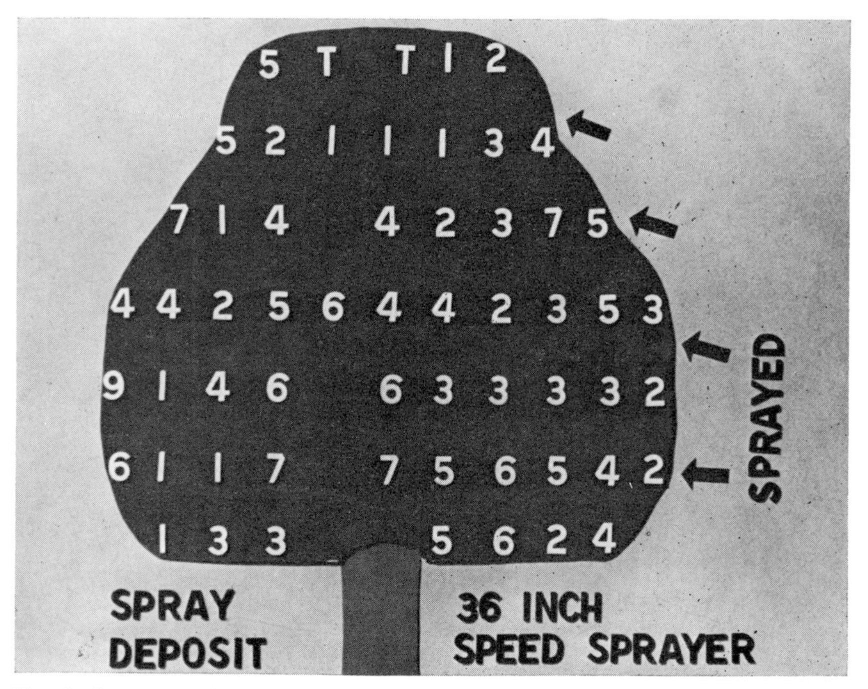

FIG. 2. Average ratings of spray deposit on 22-foot apple trees with air delivery at about 45,000 cfm.

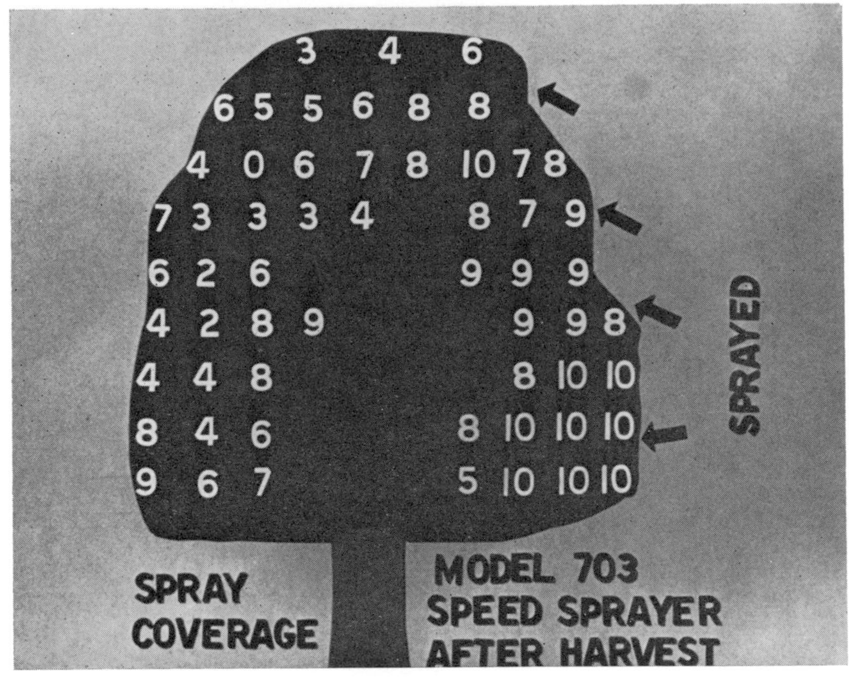

FIG. 7. Average ratings of spray coverage on 28–31-foot apple trees with air delivery at about 90,000 cfm.

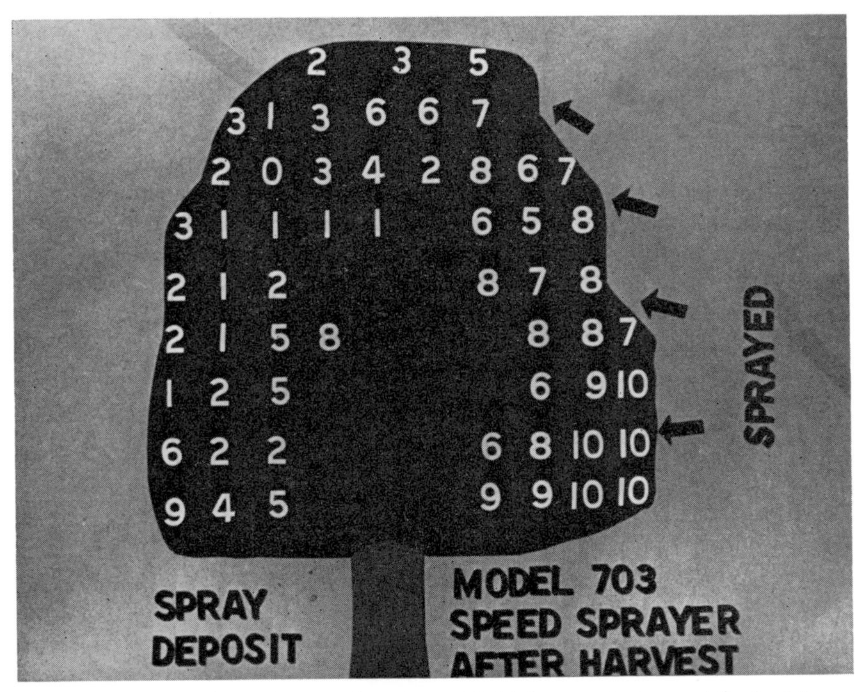

FIG. 8. Average ratings of spray deposit on 28–31-foot apple trees with air delivery at about 90,000 cfm.

would normally be considered reasonably adequate, the larger sprayer improved coverage from an average rating of 6.57 to 7.78 on trees of standard size and from 6.56–7.10 on the lower part of very large trees. Deposit data from the same 19 foot section of the trees were 3.77 versus 6.04 and 5.02 versus 5.51.

Coverage and deposit tended to decrease sharply with tree height. For example, at 10 feet, the two sprayers gave coverage rated at 7.60 and 8.11 with deposit of 4.00 and 6.56. However, at 19–22 feet, coverage was rated at 4.33 and 4.63 with deposit rated at 2.08 and 3.25.

In this work, using the same amount of water and Phosphor 2282 per acre, the use of an airblast sprayer that delivered about 90,000 cubic feet of air per minute at a velocity of about 125 miles per hour resulted in a significant increase in spray coverage and deposit as compared to a sprayer that delivered about 45,000 cfm air at about 125 mph. This is believed to be a significant factor in the amount of pesticide required per acre in a pest control program.

The work discussed up to this point indicates that it is possible to develop reasonable guidelines within which fruit growers can make some rather drastic changes in their methods of using pesticides. Data on the length of life of pesticide residues on the tree suggest the same thing.

Lewis & Hickey (23, 36) studied the control of sooty blotch and fly speck in an abandoned apple orchard where these diseases were a severe problem. They found that the failure of some fungicides to control these diseases was due to the relatively short life of the residue on the tree. Sooty blotch did not appear within 31–39 days after the last captan spray but was present after 45–58 days. It did not appear within 50–60 days after the last spray of zineb, but was present after 65–74 days. Fly speck followed the same pattern, although it was slightly later in its appearance. These data did not appear to be correlated with laboratory tests on spore germination nor with chemical analyses of captan residues.

Thiram was a short-life fungicide for the control of sooty blotch and fly speck. Field experience with Polyram and the data of Drake (13, 14) on triarimol show that their residues are short-lived also. Zineb, folpet, maneb, benomyl, and lead arsenate have relatively long periods of effectiveness against sooty blotch and fly speck, and it is assumed this is due to their long residual life.

Frear et al (15) published disappearance curves showing a rapid loss of chemical residues within a few days after application. From these curves, the estimated percentage loss of some common pesticides within 7 days after spraying was: parathion 94, diazinon 71, dimethoate 45, dodine 60, Guthion 71, and ethion 36. Carbaryl at 27% and DDT at 19% were lost rather slowly.

Groves et al (18), in tests with Karathane for control of powdery mildew on apple, found that 4 ounces applied at 7-day intervals gave as good control as 16 ounces at 14-day intervals. There was little or no residue after 4 days with any concentration used in the tests.

Palmiter (45) compared 6- and 10-day spray intervals in an apple orchard with a high population of the scab fungus. Concentrations were varied so that about the same amount of fungicide per acre was used during the season. With captan 50% WP, the amount of scab with the 6-day interval was less than half as much on the leaves and less than one-third as much on the fruit as with the 10-day interval. Equally striking results were obtained on the foliage with dodine but there were no differences on the fruit. With benomyl, there was 15% leaf scab with the 6-day interval and 24% with the 10-day interval. There were no differences on the fruit.

Methods of Using Airblast Sprayers

The situation on deciduous fruits in the Pennsylvania-Virginia area is one in which improved pesticides, spray equipment, and grower educational levels permit a high level of pest control. Pest control costs are high, and some method is needed to reduce those costs by procedures that can be followed by a large number of growers. Equipment is available to permit one man to spray an acre of mature apple trees in about 8 minutes at a cost of about $2.00 per acre. Many modern pesticides tend to be short-lived on the tree. Within this framework, is there a basis for possible improvement in our procedures? Several approaches have been tried.

A complete spray application every 7 days.—Lewis (40) used large airblast sprayers with air volume of over 90,000 cubic feet per minute to apply sprays in a pest control maintenance program in an orchard of mature apple trees pruned to a height of 18–22 feet and thinned according to normal commercial practice. The orchard was typical of many in the Cumberland-Shenandoah Valley area, having a low population of all pests except for a moderate level of the powdery mildew fungus and a high population of the European red mite. The sprayers were nozzled to deliver 112 gallons of spray per acre while traveling at 2 miles per hour. The pesticides were mixed as for dilute sprays, and were applied as a complete spray every 7 days. Thus, the total use of pesticides was two-thirds the standard amount for concentrate sprays on apples but the application costs were double the usual amount.

The control of all pests was excellent for the first 2 years. Mite counts were less than 1 per leaf although poor control was obtained in other parts of the same orchard with higher concentrations of the same chemicals applied from only one side of the row.

In the third year of the test, pest control continued excellent except for mites. In all blocks of trees on this farm, Morestan was giving successively shorter periods of control, indicative of the appearance of a population tolerant to the miticide. In this part of the orchard, control was poor on one tree by midsummer and a dozen trees were bronzed by September.

Low gallonages of very dilute pesticide mixtures applied at 7-day intervals with large airblast sprayers gave nearly perfect pest control with a reduction of 33% in total pesticide usage, less spray injury, and slightly lower total cost than the standard Pennsylvania 3X concentrate program. The slight cost advantage has now disappeared because of increased prices for labor and equipment. While the procedure has many of the advantages sought, it is not considered generally applicable at this time because of the labor requirement and the danger of hastening the development of miticide resistance in the European red mite.

Spraying two rows of trees with each pass through the orchard.—In each of 2 years, sprayers with air output of over 90,000 cubic feet per minute were used in an attempt to spray completely 2 rows of 18–22 foot apple trees with each pass through the orchard. Tests with fluorescent chemicals showed that coverage was good but the deposit was light on the side of the tree opposite the sprayer. The control of scab and powdery mildew was excellent, but control of the European red mite was adequate only on the side of the tree nearest the sprayer. This method of spraying was abandoned as too unreliable for general use.

Spraying from alternate row middles at fixed intervals.—In the experimental work with alternate middle spraying, it was assumed that most orchards were relatively pest free and that there might be a significant advantage in frequently renewing the coverage and deposit of pesticides with a short residual life. It was also assumed that growers would accept a set of specific guidelines if it could be shown that it was to their advantage to do so. The following conditions were established: (*a*) trees were to be pruned and held at a height of 18–22 feet, (*b*) the sprayer was to have a minimum air output of 90,000 cubic feet per minute, (*c*) the interval between sprays was to be fixed and adhered to except in heavy rainfall or winds above 10 miles per hour, and (*d*) pesticides were to be used at two-thirds the amount usually recommended for concentrate sprays (about 55% of the standard dilute rate calculated at 400 gallons spray per acre).

In trials over a 3-year period, the sprayer was operated from row middles 1, 3, 5, etc. on one date (June 7) with both sides open. This was followed 7 days later by spray applied from middles 2, 4, 6, etc. This alternation continued throughout the trials.

Control of all pests was very good where adequate miticide was included. Powdery mildew was present but the level of control was equal to that in nearby experiments where dilute sprays were used. Where Morestan was applied from alternate row middles on June 7, 14, July 5, 12, August 4, 11, 1965, mite counts showed a maximum population of 1.4 mites per leaf. This was in marked contrast to another treatment plot where the mites reached 15.8 per leaf by July 19. The counts showed higher mite populations on the side of the tree sprayed during the week that the counts were made, but the totals were always low.

The cost of pesticides plus the application of 28 gallons of 4X concentrate per acre applied from alternate row middles every 7 days was 30.7%

below standard concentrate sprays. The total cost of 56 gallons of 2X concentrate applied from alternate middles every 7 days was 29.1% below standard. Leaf injury from Morestan on Golden Delicious was much less than with standard concentrate sprays applied at 14-day intervals.

This method of spraying is now in large scale use in the Cumberland-Shenandoah Valley. It allows the grower to plan his work on a relatively fixed schedule. It allows quick coverage of the orchard, since it is normal to cover about 18 acres of mature apple trees with each tank of spray mixture. It has allowed a substantial decrease in pesticide usage in a practical, controllable procedure any grower can follow. Those with relatively small sprayers can spray one side of each row during one week and from the opposite side during the following week. Alternate middle spraying has been widely used in integrated pest control programs where small amounts of pesticide are needed to reestablish the desired balance between pests and predators.

Discussion

In tree fruit production, disease control requires the use of fungicides applied in a pest control program followed year after year. Disease problems can be so severe that a delay of one or two days in application of a critical spray treatment results in loss of the crop and possibly the trees. Disease resistance is a significant factor in disease severity on several important cultivars of fruit, but resistance to disease is not a primary factor in the choice of cultivars for commercial production. Sanitary measures are useful, but rarely can be applied in a way that will result in satisfactory disease control. Fungicides carry the major part of the burden; and it seems likely that this will continue to be true for the forseeable future.

The development of new fungicides has resulted in great improvement in our ability to control fruit diseases without serious injury to the trees. Ferbam was the first effective rust fungicide for fruit; and its use allowed us for the first time to observe an orchard of healthy apple trees free from injury. Mercury was useful as a highly effective scab fungicide, and was also used occasionally for brown rot blossom blight control on stone fruits. Captan, widely used on most fruits, gave control to many diseases, acted as an arsenical injury preventive, and frequently improved fruit finish on apples and peaches. The development of folpet made available a highly effective fungicide for apple fruit rots. Benomyl is expected to establish new levels of brown rot control in stone fruits, and may allow expansion of the nectarine industry in eastern United States. Triarimol, with its ability to control scab, powdery mildew, the rust diseases, and black rot without affecting the population of the European red mite, will give us a highly effective fungicide for use in integrated pest management programs and may be of great value in semi-arid areas where the occasional spray required for apple scab control can be delayed until after infection has occurred. Fungicide mixtures, such as carbamate plus Karathane or Morocide, are used in many apple orchards because they provide acceptable levels of disease control and delay the build-up of populations of the European red mite.

The development of these and other fungicides, and their use in crop production, is one of the great success stories of Plant Pathology. Many workers around the world have made significant contributions to this success.

In the United States today, we are able to control such diseases as apple scab and brown rot of stone fruits with routine applications of fungicides. Airblast sprayers can cover an acre of mature trees within less than 10 minutes. Labor is a critical problem because of costs. Growers have reduced pesticide usage, in some cases by as much as 50%. Those interested in environmental problems are concerned with the introduction of any pesticides into the environment. There is a need to reexamine present practices, to try other approaches, and to determine whether or not our use of pesticides is at the minimum required for efficient fruit production.

There are still numerous instances where pathogen populations are high and disease control on very susceptible varieties requires the use of standard amounts of fungicides applied before infection occurs. Probably there will continue to be such instances. However, there is a possibility that our recollection of past disease epidemics, our methods of testing fungicides, and human problems related to dealing with many people are all leading us to discount the opportunity to use reduced amounts of pesticides on many thousands of acres of fruit.

The experiment station worker does not have the time to make individual disease case studies except in special instances. The result is a conservative recommendation that is expected to cover all conditions within a wide area. The grower uses the recommendation only as a general guide because he cannot afford and often does not need the amount of fungicide suggested in a typical spray program.

Field trials are conducted under conditions where a high incidence of disease is expected or can be induced; this usually means the worst disease conditions in the area. Uniformity is sought in the experimental plants and in the plot design, the method and rate of application, the insecticides and miticides added, and in the weather conditions under which the treatments are applied. The interval between sprays is varied according to the research worker's judgment of the procedure necessary for control. The treatments usually are applied as dilute sprays with each tree sprayed until it is wet.

This method of testing fungicides has been very useful. Certainly there is a need to have disease pressure on the treatments if they are to be compared in an experiment of one year's duration. The data from such tests have been reasonably adequate to judge the probable performance of a fungicide under commercial conditions. The testing method suggests three points that deserve further study: (*a*) pest populations differ between the test plots and many commercial orchards, (*b*) there is no systematic control of spray timing, and (c) dilute sprays are used in the test plots while low volume, airblasts sprays are standard in the fruit industry.

If a practical, systematic plan could be developed under which university personnel could offer clear guidelines for the use of reduced amounts of pesticides without sacrificing control, there would be no reason to hesitate. The grower either follows the guidelines or assumes complete responsibility for the results. There is adequate evidence to suggest guidelines for further trials on apples and stone fruits in orchards where major disease epidemics are not in progress.

Guideline 1. The tree should be pruned and thinned according to standard commercial practice except that apple trees should be topped at a maximum of 18–19 feet with new growth in the summer reaching a height of 21–22 feet.

Guideline 2. All pesticides should be applied as low volume sprays with the gallonage per acre within the range of 20–100; in most cases about 50 gallons per acre combines the advantages of low cost with minimum mixing and spray drift problems. The use of low volume sprays has been accompanied in nearly all cases with a reduction of 18–20% in fungicide usage on apples and 25% on the stone fruits as compared to dilute sprays at 400 gallons per acre on apples and 350 gallons on the stone fruits.

Guideline 3. There must be some specifications regarding the sprayer to be used in applying low volume sprays on trees of various sizes. All airblast sprayers tend to apply too much pesticide on the portion of the tree near the sprayer and too little in the tops of the trees. There is no satisfactory method known to assess the comparative importance of air volume and velocity in tree fruit spraying. The most useful general guide has been to require that the sprayer have the ability to blow the spray through and several feet above the tops of the trees. We have presented data showing that a sprayer that delivered 90,000 cubic feet of air per minute at a velocity of 125 miles per hour gave significantly better pesticide coverage and deposit as compared to a sprayer of similar type that delivered 45,000 cubic feet of air at 125 miles per hour. In work with alternate middle spraying, it has seemed essential to have large sprayers with an air delivery of 90,000 cfm, an air velocity of about 125 miles per hour, and a pump pressure minimum of 180–200 pounds per square inch.

Guideline 4. The rate of travel of airblast sprayers should be fixed at a maximum of 2 miles per hour for work on large apple and sweet cherry trees, at 2.5 miles per hour for trees of medium size, 3 miles per hour for work on peaches, plums, and most tart cherries.

Guideline 5. The interval between sprays should be short and fixed at a given number of days. It seems probable that this has been one of the major areas where our use of pesticides has been inefficient. It is certain that the residues of several of the most commonly used fruit fungicides disappear rapidly from the tree. Very little is known about the comparative effectiveness of fruit fungicides at varying spray intervals. Indications are that spray programs based on intervals of 14–20 days between sprays must suggest relatively large dosages in an effort to compensate for the relatively short residual life of the pesticides. The cost of the pesticides as compared to the cost of their application and to their efficiency in control should be the guide.

Guideline 6. The practice of spraying from alternate row middles at intervals of 7 days has been highly successful in pest control maintenance programs on apples, peaches, and tart cherries. Above 90% spray coverage has been obtained with a light to moderate spray deposit, and renewed from the alternate side of the row 7 days later. Sometimes the equipment gave light coverage on 3–4 rows of trees on each side of the sprayer. Amounts of pesticides as low as 45–55% of dilute spray standards have been used successfully for several years on thousands of acres of apples where the need was to maintain an already established high level of pest control. The practice should be useful in many more orchards, since quick coverage can be obtained and the concentration of pesticide can be varied according to the need.

There is a need for an airblast sprayer that will provide more uniform spray coverage and deposit while minimizing drift. Present equipment is inefficient because it provides relatively poor coverage and deposit in the top one-third of mature apple trees even though 55–80% of the spray is directed toward that area. Also, Steiner (50) found that 30–45% of the pesticide discharged by an airblast sprayer was not deposited on the apple tree. Further consideration of these problems, plus the trend in the apple industry toward smaller trees, might result in an effort to develop a sprayer that distributes concentrates by nozzles and small fans, or some other delivery system, mounted on a hooded boom that extends to the full height and perhaps halfway across the top of the tree.

LITERATURE CITED

1. Albert, J. J., Lewis, F. H. 1962. Effect of repeated applications of dodine and captan on apple scab foliage lesions. *Plant Dis. Reptr.* 46(3):163–67
2. Brann, J. L., Jr. 1963. Factors affecting the use of airblast sprayers. Paper No. 63–106, 1963 *Ann. Meet. Am. Soc. Agr. Eng.* 12 pp.
3. Brann, J. L., Jr. 1965. Factors affecting the thoroughness of spray applications. *N. Y. State Hort. Soc. Proc.* 110:186–95
4. Brann, J. L., Jr., Steiner, P. 1969. Economic implications of concentrate spraying. *N. Y. State Hort. Soc. Proc.* 114:259–62
5. Brown, I. F., Jr., Hall, H. R., Miller, J. R. 1970. EL-273 a curative fungicide for the control of *Venturia inaequalis. Phytopathology* 60:1013–14
6. Brown, I. F., Jr., Hall, H. R. 1970. A histological study of the curative action of EL-273 against *Venturia inaequalis. Phytopathology* 60:1286 Abstr.
7. Brooks, R. F., Thompson, W. L., Jutras, P. J. 1964. Evaluating spray equipment for Florida Citrus. *The Citrus Industry* 45(1)7–10
8. Burrell, A. B. 1951. Converting large air-blast sprayers for application of semi-concentrates. *N.Y. State Hort. Soc. News Let.* 7(2):2–3
9. Clayton, C. N. 1963. Apple scab, apple bitter rot, white rot, black rot and brown rot. *Fungicide Nematicide Tests* 18:34–35
10. Clayton, C. N. 1971. Bitter rot, white rot, sooty blotch and necrotic leaf blotch. *Fungicide Nematicide Tests* 26:5–6
11. Cordley, A. B. 1908. Lime sulfur spray as a preventive of apple scab. *Rural New Yorker,* March 1, 202. Also *Oregon Agriculturist,* March 1, 1908, 178
12. Dibble, J. E. 1970. Let's concentrate on concentrate spraying. *Am. Fruit Grower* 90(2):13, 14, 60
13. Drake, C. R. 1970. White rot, black rot, scab, sooty blotch, fly speck. *Fungicide Nematicide Tests* 25:9–10
14. Drake, C. R. 1971. Scab, rust, sooty blotch, fly speck, white rot, black rot and leaf blotch on Golden Delicious. *Fungicide Nematicide Tests* 26:7–11
15. Frear, D. E. H. 1963. Pesticide residue investigations on raw agricultural commodities. *Penn. Agr. Exp. Sta. Bull.* 703:1–77
16. Fulton, R. H. 1965. Low volume spraying. *Ann. Rev. Phytopathol.* 3:175–96
17. Gilpatrick, J. D., Szkolnik, M. 1970. Tree fruit fungicide research in western New York in 1969. *N.Y. State Hort. Soc. Proc.* 115:232–45
18. Groves, A. B., Wampler, E. L., Lyon, C. B. 1958. The development of an efficient schedule for the use of Karathane in the control of apple powdery mildew. *Plant Dis. Reptr.* 42(2):252–61
19. Hamilton, J. M. 1931. Studies of the fungicidal action of certain dusts and sprays in the control of apple scab. *Phytopathology* 21: 445–523
20. Hamilton, J. M., Szkolnik, M., Nevill, J. R. 1963. Greenhouse evaluation of fruit fungicides. *Plant Dis. Reptr.* 24(3):224–28
21. Hamilton, J. M., Szkolnik, M. 1964. A decade with dodine. *Cyanograms* 11(1):18–24
22. Heuberger, J. W., Jones, R. K. 1962. Apple Scab II. Effect of serial applications of fungicides on leaf lesions on previously unsprayed trees. *Plant Dis. Reptr.* 46(3):159–62
23. Hickey, K. D. 1960. *The sooty blotch and fly speck diseases of apple with emphasis on variation within* Gloeodes pomigena *(Schw.) Colby,* pp. 1–127. Thesis, Pensylvania State Univ.
24. Hickey, K. D. 1970. Apple powdery mildew and cedar apple rust. *Fungicide Nematicide Tests* 25:17
25. Howard, F. L., Locke, S. B., Keil, H. L. 1944. Synthetic organic fungicides for apples. *Am. Soc. Hort. Sci. Proc.* 45:131–35
26. Jones, R. K., Heuberger, J. W., Bates, J. D. 1963. Apple Scab III. Effect of several applications of fungicides on leaf lesions on previously unsprayed trees—inhibition of conidia, germination, removal (suppression) of the or-

ganism, and subsequent development of late terminal infection. *Plant Dis. Reptr.* 47(5): 420–24
27. Kelly, B. W. 1963. Factors related to the cost of producing apples in Pennsylvania 1959–1962. *Penn. State Univ. Farm Manage.* 11: 1–16
28. Klingbeil, G. C., Mitchell, A. E. 1951. Evaluation of concentrate spray coverage on large apple trees by means of leaf prints. *Mich. Agr. Exp. Sta. Quart. Bull.* 33(4):302–09
29. Krestensen, E. R., Graham, C. 1967. Low volume spraying. *Penn. Fruit News* 46(4):20–24
30. Krestensen, E. R., Graham, C. 1967. Further tests for the control of apple pests with concentrate sprays from 3× to 66×. *Trans. Peninsula Hort. Soc.* 56(5)
31. Lade, D. H. 1970. Apple scab and frog-eye leaf spot. *Fungicide Nematicide Tests* 25:25
32. Larson, G. H., Fairbanks, G. E., Fenton, F. C. 1960. What it costs to use farm machinery. *Kansas Agr. Exp. Sta. Bull.* 417:1–48
33. Latham, A. J., Carlton, C. C. 1971. Peach brown rot, Rhizopus rot and scab. *Fungicide Nematicide Tests* 26:50
34. Lewis, F. H. 1943. Studies on spray and dust schedules for control of apple scab in western New York. *Cornell Univ. Thesis* 1943: 335–38 Abstr.
35. Lewis, F. H. 1954. Control of apple rust. *Fungicide Test Results for 1953*. 11.
36. Lewis, F. H., Hickey, K. D. 1958. Effective life of fungicides as a factor in the control of sooty blotch and fly speck on apple. *Phytopathology* 48: 462 Abstr.
37. Lewis, F. H., Hickey, K. D. 1959. Cherry leaf spot, black knot of plum and brown rot blossom blight. *Fungicide Test Results for 1958*. p. 30–31.
38. Lewis, F. H. 1963. Buying and using an orchard sprayer. *Penn. Fruit News* 42:5, 6, 8, 10, 12, 14, 16
39. Lewis, F. H., Hickey, K. D. 1964. Pesticide application from one side on deciduous fruit trees. *Penn. Fruit News* 43:13, 17, 19–24. Also *N.Y. State Hort. Soc. Proc.* 109:209–13
40. Lewis, F. H. 1967. Methods of using large airblast sprayers on apples. *Penn. Fruit News* 46(4):47, 50, 52–53
41. Lewis, F. H., Asquith, D., Krestensen, E. R., Hickey, K. D. 1969. Calibration of airblast sprayers for use on deciduous fruits. *Penna. Agr. Exp. Sta. Progr. Rept.* 294:1–16
42. Lewis, F. H. 1971. Apple scab. *Fungicide Nematicide Tests* 26: 29–30
43. Mills, W. D. 1944. Efficient use of sulfur dusts and sprays during rain to control apple scab. *Cornell Univ. Ext. Bull.* 630:1–4
44. Mills, W. D., LaPlante, A. A. 1951. Diseases and insects in the orchard. *Cornell Univ. Ext. Bull.* 711:21–27
45. Palmiter, D. H. 1969. Apple disease control with old and new fungicides in 1968. *N. Y. State Hort. Soc. Proc.* 114:253–58
46. Potts, S. F. 1958. Concentrated spray equipment, mixtures and application methods. Dorland Books. Caldwell, N.J. 598 pp.
47. Raffensperger, H. B. 1946. 2000 acres sprayed with phenothiazine. *Virginia Fruit,* 1–3
48. Roosje, G. S. 1963. Research on apple and pear scab in The Netherlands from 1938 until 1961. *Neth. J. Plant Pathol.* 69:132–37
49. Steiner, H. M. 1946. Comparisons of phenothiazine and DDT in handling orchard pests among heavy codling moth infestations of Pennsylvania. *Virginia Fruit,* 1:20
50. Steiner, P. W. 1969. *The distribution of spray material between target and non-target areas of a mature apple orchard by airblast equipment.* MS thesis, Cornell Univ.
51. Struble, F. B., Morrison, L. S. 1961. Control of apple blotch with fungicides. *Plant Dis. Reptr.* 45(6):441–43
52. Szkolnik, M., Hamilton, J. M. 1964. Performance of fruit fungicides in greenhouse and orchard spray tests in 1963. *Proc. 109th Ann. Meet. N. Y. State Hort. Soc.* 214–24.
53. Szkolnik, M., Gilpatrick, J. D. 1969. Apparent resistance of *Venturia inaequalis* to dodine in New York apple orchards. *Plant*

Dis. Reptr. 53(11):861–64
54. Szkolnik, M. 1969. Conditioning growers to apple scab control in 1970. Paper presented Symp. Syracuse, N.Y. 1969–70. 1–11
55. Szkolnik, M. 1971. Brown rot and Rhizopus rot on peach and cherry. *Fungicide Nematicide Tests* 26:51
56. Zuroske, C. H. 1968. Apple production costs and returns. *Wash. Agr. Exp. Sta. Bull.* 696:1–6

ECTOMYCORRHIZAE AS BIOLOGICAL DETERRENTS TO PATHOGENIC ROOT INFECTIONS 3558

DONALD H. MARX

U. S. Department of Agriculture, Southeastern Forest Experiment Station
Athens, Georgia

INTRODUCTION

The infection of feeder roots of most flowering plants by symbiotic fungi and the transformation of these roots into unique morphological structures called mycorrhizae (fungus-roots) undoubtedly constitute one of nature's most widespread, persistent, and interesting examples of parasitism. Most plants of economic importance to man are actually dual organisms—part plant and part symbiotic root-inhabiting fungi.

Two major classes of mycorrhizae are recognized—ectomycorrhizae and endomycorrhizae.[1] A third class, ectendomycorrhizae, an intermediate type, is present on roots of certain tree species under specific ecological situations. The fungal symbionts of ectomycorrhizae penetrate intercellularly and partially replace the middle lamellae between cortical cells of the feeder roots. This hyphal arrangement around the cortical cells is called the Hartig net. Ectomycorrhizal fungi also form a dense, usually continuous, hyphal network—called the fungal mantle—over the feeder root surface. Thickness of the fungal mantle varies from 1–2 hyphal diameters to as many as 30–40, depending on fungal associate, host, and environmental conditions. Most tree species in the Pinaceae, Salicaceae, Betulaceae, Fagaceae, and certain other families, normally form ectomycorrhizae. The majority of ectomycorrhizal fungi are Basidiomycetes, primarily in the Amanitaceae, Boletaceae, Cortinariceae, Russulaceae, Tricholomataceae, Rhizopogonaceae, and Sclerodermataceae. Several orders of ascomycetes, primarily Eurotiales, Tuberales, Pezizales, and Helotiales, contain species suspected of being ectomycorrhizal with trees.

The fungal symbiont in endomycorrhizae intracellularly penetrates cortical cells of the feeder root. Endomycorrhizal fungi may form large vesicules and arbuscules in cortical tissues, called vesicular-arbuscular (VA) mycorrhizae. The endomycorrhizal fungi do not form a dense fungal mantle, but develop on the root surface a loose, intermittent arrangement of mycelium,

[1] The new terminology, suggested by Peyronel et al (62), was accepted by the Mycorrhizae Work Group, Internat. Union Forest Res. Organizations, University of Florida, Gainesville, March 1971.

with numerous large-diameter spores. Endomycorrhizae are formed by most agronomic, horticultural, and ornamental crops, as well as certain forest tree species that do not form ectomycorrhizae. The fungal symbionts of endomycorrhizae are phycomycetes; many belong in the genus *Endogone*.

Ectendomycorrhizae resemble ectomycorrhizae in forming a Hartig net and a fungal mantle, but they also resemble endomycorrhizae in that cortical tissues are intracellularly penetrated by these fungi. The fungal symbionts of ectendomycorrhizae are currently unclassified and very little is known about them. Some may be normally ectomycorrhizal symbionts (92).

The universal presence of mycorrhizae on plants has not been seriously questioned for nearly a century. The influence of this symbiotic association on the plant host, however, has been the subject of much debate since Frank's (16) original description and coinage of the term mycorrhizae in 1885. Subsequent research has supported the view that most forms of mycorrhizae are measurably beneficial to plants and often, especially in the case of *Pinus,* indispensable to their growth. Most research on mycorrhizae has aimed at demonstrating that the plant hosts gain by improved mineral nutrition, i.e., increased absorbing surface of roots and more rapid absorption and accumulation of essential nutrients. The fungal symbionts gain by receiving a supply of essential carbohydrates and other organic metabolites from the plant hosts (21, 23, 25, 26, 52, 53).

Garrett (17) indicated that mycorrhizae are functional longer and are less subject to certain types of pathogenic infections than are nonmycorrhizal feeder roots. Zak (98) suggested in 1964 that ectomycorrhizal fungi, in addition to their physiological influence on plants, may also protect the delicate, unsuberized feeder roots from attack by pathogens.

Mycorrhizal associations and diseases of feeder roots of plants are similar in that both types of parasitism intimately involve the succulent fine feeder roots of their hosts. Ectomycorrhizal fungi are stimulated by host roots, symbiotically infect, and eventually transform feeder roots into dual organs in which the cortical cells are enclosed in the Hartig net and isolated by the fungal mantle from direct contact with soil. During synthesis of ectomycorrhizae, the host responds physiologically to the symbiotic infection, and the fungal symbionts undergo certain transformations. Similarly, feeder-root pathogens—species of *Phytophthora, Pythium, Rhizoctonia,* and *Fusarium*—are stimulated by feeder roots, pathogenically infect, ramify into meristematic and immature cortical tissue, and eventually cause limited or extensive necrosis.

If a pathogen infects and destroys a feeder root prior to symbiotic infection of this root by an ectomycorrhizal fungus, obviously mycorrhizal development on this root cannot take place. If the sequence of parasitic attack is reversed (i.e., if a symbiont infects the feeder root and synthesizes an ectomycorrhiza prior to infection of this root by the pathogen) are the succulent root tissues of this transformed root still vulnerable to attack by the pathogen? A pathogen infecting a nonmycorrhizal feeder root is initially confronted externally only with succulent, thin-walled, epidermal cells, with or without root hairs, and internally with cortical cells which, in most instances, have not undergone thickening of secondary cell walls. On the other hand, a pathogen attacking an ectomycorrhiza is initially confronted externally with the tightly interwoven network of hyphae that makes up the fungal mantle, and then internally with cortex cells with cell walls surrounded by the Hartig net of hyphal tissues. It would appear, therefore, that a pathogen of ectomycorrhizae must penetrate the fungus mantle and Hartig net to establish a pathogenic relationship with root tissue.

If feeder roots in the ectomycorrhizal condition are so physically or chemically altered that they resist infection by pathogens, they are also beneficial to plant health as biological deterrents to feeder-root infection by such pathogens as *Phytophthora, Pythium,* or *Fusarium* (7, 27, 28, 90). Since Zak's review, there has been an impressive amount of research on this subject in various parts of the world.

FIELD OBSERVATIONS RELATING ECTOMYCORRHIZAE TO DECREASED DISEASE OF FEEDER ROOTS

Several workers observed that tree seedlings with ectomycorrhizae were more resistant to feeder-root infections by fungi than were seedlings with few or no ectomycorrhizae. After examination of root diseases of forest-tree nursery stock, Davis et al (12) suggested that ectomycorrhizae are beneficial to tree seedlings by preventing infection of feeder roots by pseudomycorrhizal fungi, and that ectomycorrhizal roots appeared to be less susceptible to root-rot fungi than were nonmycorrhizal roots. Levisohn (38) reported that ectomycorrhizal roots of various *Pinus* spp. and Sitka spruce (*Picea sitchensis*) seedlings were resistant to infection by a *Rhizoctonia* sp. that readily infected nonmycorrhizal feeder roots. She concluded that these ectomycorrhizae provided biological control against the root pathogen, and that soil conditions inhibitory to mycorrhizal development stimulated root infections by the pathogen. After application of various nematocides and fungicides to soil around pecan (*Carya illinoensis*) trees exhibiting symptoms of feeder-root necrosis, Powell et al (64) observed a tremendous prompt increase in ectomycorrhizal development by *Scleroderma bovista.* Most chemicals did not significantly reduce populations of *Pythium* spp. or nematodes which were the cause of feeder root necrosis, but foliar and root symptoms gradually disappeared. They concluded that populations of competing soil microorganisms were reduced by the chemicals, which in turn caused a stimulation in ectomycorrhizal development by *S. bovista,* that the greater ectomycorrhizal development increased nutrient absorption by the trees, and, more importantly, the ectomycorrhizae functioned as deterrents to infection of feeder roots by pathogens still present in significant numbers in the soil. Both functions of ectomycorrhizae, according to these authors, accounted for the disappearance of symptoms of feeder-root disease on the trees. Corte (10) observed that ectomycorrhizae formed by *Suillus granulatus* appeared to protect seedlings of *Pinus excelsa* from root rot caused by *Rhizoctonia* sp., because root rot was less on seedlings with mycorrhizae. After assessment of ectomycorrhizae on loblolly pines (*Pinus taeda*) growing vigorously on pimple mounds and poorly on flats in lowland areas, Napier (55) reported that trees in decline on the flat sites had significantly fewer ectomycorrhizae than healthy trees on the mounds. She concluded that the low numbers of ectomycorrhizae on trees in decline contributed little defense against attack by *Phytophthora cinnamomi* and *Pythium* spp. These pathogens had been implicated previously with the decline of loblolly pines in these lowland areas (40).

Since it is nearly impossible to separate cause and effect of ectomycorrhizae from other factors that interact simultaneously we cannot be sure that ectomycorrhizae on seedlings or trees has not simply brought about a favorable physiological state which may have masked symptoms of feeder-root disease. Comparisons of ectomycorrhizal to nonmycorrhizal plants are at best questionable in the above field observations since poor soil aeration, low organic matter, and other conditions that contribute to development of feeder-root disease are also directly inhibitory to ectomycorrhizal development, and vice versa.

RESEARCH RELATING ECTOMYCORRHIZAE TO CONTROL OF DISEASES OF FEEDER ROOTS

A few investigators have recently shown that ectomycorrhizae decrease the incidence of diseases of feeder roots.

Wingfield (95) observed that ectomycorrhizae formed by *Pisolithus tinctorius* on axenic seedlings of loblolly pine enhanced their survival when growing with the root pathogen, *Rhizoctonia solani.* Pine seedlings without ectomycorrhizae and inoculated with the pathogen exhibited significantly lower survival and vigor. Richard et al (66) found that axenic seedlings of *Picea mariana* inoculated with the ectomycorrhizal fungus, *Suillus granulatus,* grew well, whereas those inoculated with the root pathogen, *Mycelium radicus atrovirens,* were chlorotic and severely stunted. Seedlings in the latter group were initially infected at the root collar and the pathogen eventually was detected in the lateral roots, short roots, and root hairs. The fungus was detected in the feeder-root cortex, often penetrating to the endodermis of the roots. When *S. granulatus* was inoculated simultaneously with the root pathogen, however, the chlorosis and stunting of seedlings caused by the pathogen were eliminated and they grew as well as those with only the ectomycorrhizal fungus.

Since the above research was carried out under germ-free conditions, application of the results to natural systems is limited. A few tests, however, have been made under nonsterile conditions. In a greenhouse study (71), seedlings of the Ocala race of sand pine (*Pinus clausa*) were protected against *Phytophthora cinnamomi* by the presence of ectomycorrhizae formed by *Pisolithus tinctorius.* Nonmycorrhizal pine seedlings were heavily infected by *P. cinnamomi,* and exhibited massive feeder-root necrosis. Only 40% of these seedlings survived after 2 months. Nonmycorrhizal roots on pine seedlings with ectomycorrhizae formed by *P. tinctorius* were also infected by the pathogen. However, 25% of the feeder roots were ectomycorrhizal, thus reducing the amount of susceptible root tissue exposed to the pathogen; this reduction in susceptible tissue contributed to nearly 70% survival of test seedlings. Cortical tissues in the ectomycorrhizal roots were free from *P. cinnamomi,* verifying their resistance to attack by this pathogen. The Choctawhatchee race of sand pine, which was killed by *P. cinnamomi* more rapidly than was Ocala, responded to fertility more vigorously than it, as expressed by rapid production and elongation of lateral roots which therefore outgrew the ectomycorrhizal fungus and its protection against *P. cinnamomi.*

In a similar study Marx (49) found that shortleaf pine (*Pinus echinata*) seedlings with ectomycorrhizae were not affected by *P. cinnamomi.* Nonmycorrhizal shortleaf pine seedlings exposed to the pathogen were significantly lighter in foliar-stem and root dry weights, and had significantly fewer new lateral roots than nonmycorrhizal seedlings grown without the pathogen. The inoculum densities of *P. cinnamomi* in soil with the nonmycorrhizal seedlings did not significantly change during the experiment. Shortleaf seedlings with ectomycorrhizae formed by either *Pisolithus tinctorius* or *Cenococcum graniforme* did not exhibit reduction in foliar-stem or root weights or development of new lateral roots in the presence of *P. cinnamomi,* as did the nonmycorrhizal seedlings. A significant reduction was found in inoculum densities of *P. cinnamomi* in soil with all ectomycorrhizal seedlings at the end of the study. Apparently the high degree of ectomycorrhizal development (70–89%) on those seedlings reduced the amount of susceptible tissue available for attack by *P. cinnamomi,* which in turn caused a decrease in inoculum density of the pathogen and a decrease in development of feeder-root disease. Chlamydospores of the pathogen were apparently stimulated by root exudates from nonmycorrhizal roots, germinated, and either zoospores or vegetative mycelia were able to establish infection. Subsequent to infection of these roots more propagules of the pathogen were formed, thereby preventing depletion of inoculum; however, chlamydospores in the sphere of influence of ectomycorrhizae apparently did not react in the same manner, and inoculum density decreased. Neither of the two ectomycorrhizal fungi used in this experiment produced antibiotics effective against *P. cinnamomi.*

The preceding reports show that plants with ectomycorrhizae did not exhibit reduced top growth, chlorosis, restricted root development, and eventual death, and are therefore more resistant to feeder-root diseases than are nonmycorrhizal plants. Obviously, with an increasing degree of ectomycorrhizal development there is a proportionate reduction in amount of feeder roots susceptible to pathogen attack.

MECHANISMS OF RESISTANCE OF ECTOMYCORRHIZAE TO PATHOGENIC INFECTIONS

Zak (98) postulated several mechanisms by which ectomycorrhizae may afford disease protection to feeder roots of plants. He suggested that ectomycorrhizal fungi may (*a*) utilize surplus carbohydrates in the root thereby reducing the amount of nutrients stimulatory to pathogens, (*b*) provide a physical barrier, i.e., the fungal mantle, to penetration by the pathogen, (*c*) secrete antibiotics inhibitory to pathogens, and (*d*) support, along with the root, a protective microbial rhizosphere population. In addition, Marx (42) suggested that (*e*) inhibitors produced by symbiotically infected host cortical cells may also function as inhibitors to infection and spread of pathogens in ectomycorrhizal roots.

Antibiotic Production by Fungal Symbionts

Wright (96, 97) and others demonstrated that various saprophytic fungi can produce antibiotics in such restricted sites as pieces of straw and seed coats in soil. It is generally accepted (17, 33) that the resulting antibiotic concentrations are sufficient to influence significantly the pattern of saprophytic microbial colonization of these sites. However, the significance of antibiotic production by saprophytes in reducing the inoculum potential of root pathogens and subsequent root disease development is poorly understood. Most attempts at controlling root pathogens in soil by inoculating it with antibiotic-producing saprophytes have failed. The most acceptable explanation for these failures is that antibiotic production is thought to be limited to the immediate substrate or ecological niche of the saprophyte (17). This restricted site of antibiotic production apparently is not of major significance in reducing pathogen inoculum potential in other than the immediate location.

Theoretically, this need not be the fate of antibiotics produced by ectomycorrhizal fungi; the ecological niche of these specialized root parasites is the host root. These fungi, while in mycorrhizal association with roots, are assured of essential metabolites (e.g., carbohydrates, vitamins), for which they need exert only minimal competitive efforts. Any antibiotic thus produced in this niche should be ideally located to produce inhibitory effects on pathogens attempting infection of these ectomycorrhizal roots, and perhaps even adjacent nonmycorrhizal roots.

Antibiotic production in pure culture.—Several workers have investigated antibiotic production by higher Basidiomycetes in pure culture. Trappe (87) associated many of these fungi with ectomycorrhizal relationships, although most authors investigating antibiotic production made no inference to the possible symbiotic nature of the fungi. The following is a list of ectomycorrhizal fungi reported to produce antibiotics either in pure culture or in basidiocarps; it contains only those which have been either experimentally proved to be symbionts, or were associated with ectomycorrhizae according to the compilation by Trappe (87):

Amanita caesaria AB[2] (36, 76, 77), *A. citrina* AB, AF (79, 80), *A. mappa* AB (36), *A. muscaria* AB, AF, AV (31, 36, 76, 77, 89), *A. panther-*

[2] AB = antibacterial activity; AF = antifungal activity; AV = antiviral activity.

ina AB, AF (31, 36, 94), *A. phalloides* AB, AV (36, 89), *A. rubescens* AB, AF (36, 76, 77), *A. solitaria* AB (82), *A. strobiliformis* AB (94), *A. vaginata* AB, AF (31, 82), *A. virosa* AB (94); *Boletinus pictus* AV (89); *Boletus bicolor* AB, AF (36, 76, 77), *B. bovinus* (*Suillus*) AF (31, 32, 80), *B. communis* AB (36), *B. calopus* AB (94), *B. edulis* AB, AF (31, 94), *B. elegans* AF (31), *B. granulatus* AF (31), *B. laki* AB (36), *B. luteus* (*Suillus*) AB, AF (36, 42, 65, 76, 77, 80), *B. rubellus* AB, AF (76, 77), *B. santanus* AB (94), *B. scaber* AF (31), *B. subtomentosus* AF (31), *B. variegatus* (*Suillus*) AB, AF (31, 36, 75, 80, 91); *Cantharellus cibarius* AB (94), *C. tubaeformis* AB (94); *Cenococcum graniforme* AB, AF (37, 45); *Clitocybe aurantiaca* AF (31), *C. candicans* AB (82, 94), *C. diatreta* AB, AF (3), *C. laccata* (*Laccaria*) AB, AF (36, 42, 82), *C. nebuleris* AF (31), *C. odora* AB, AF (3, 94), *C. rivulosa* AB, AF (3, 31, 93); *Clitopilus prunulus* AB, AF (6, 79, 94); *Collybia abutyracea* AF (65), *C. asema* AF (31); *Corticium bicolor* AB (36); *Cortinarius anomalus* AB, AF (31, 94), *C. armeniacus* AB (94), *C. armillatus* AB (94), *C. bolaris* AB (94), *C. caesiocanescens* AB (94), *C. callisteus* AB (82), *C. calochrous* AB (94), *C. cinnabarinus* AB (50), *C. collinithus* AB (94), *C. orichalceus* AB (94), *C. rotundisporus* AB (4), *C. violaceus* AB (82); *Hebeloma crustuliniforme* AB (94), *H. mesophaeum* AB (94), *H. sacchariolens* AB (94), *H. strophosum* AB (94), *H. imbricatum* AB (94), *H. repandum* AB (93, 94); *Hygrophorus chrysodon* AB (94), *H. eburneus* AB (82, 94), *H. nemoreus* AB (94), *H. penarius* AB (94), *H. virgineus* AB (94); *Lactarius aspideus* AB (94), *L. chrysorheus* AB (36, 94), *L. controversus* AB (94), *L. deliciosus* AB, AF (36, 42, 65, 94), *L. helvus* AF (79, 80), *L. necator* AB (94), *L. pallidus* AB (94), *L. quietus* AB (94), *L. vellereus* AB (94), *L.* spp. AB, AF (54, 61); *Lepista nuda* AB, AF (3, 31, 94), *L. personata* AB (94); *Leucopaxillus cerealis* var. *piceina* AB, AF (36, 42, 43); *Marasmius scorodonius* AB (51); *Paxillus involutus* AB (70); *Pisolithus tinctorius* AB (36); *Rhizopogon roseolus* AB, AF (31, 79–81), *R. vinicolor* AF (100); *Rhodophyllis clypeatus* AB (94); *Russula atropurpurea* AB (82), *R. emetica* AB (36), *R. fragilis* AB, AF (79, 80), *R. sanguinea* AB (82); *Scleroderma aurantium* AF (80), *S. bovista* AF (44); *Suillus subolivaceous* AB (36); *Thelephora terrestris* AF (31, 48); *Tricholoma albobrunneum* AB, AF (79, 80, 94), *T. equestre* AB (79), *T. flavobrunneum* AV (89), *T. imbricatum* AB (79, 94), *T. irinum* AB (94), *T. personatum* AB (36), *T. pessundatum* AB, AF (31, 79), *T. psammopodum* AB (94), *T. saponaceum* AB, AF (79–81, 94), *T.* sp. AF (65), *T. ustale* AB (94), *T. vaccinum* AB (79); Unidentified AF (31, 48, 65, 98).

It is obvious from this data that production of antibiotics inhibitory to bacteria, in most instances *Staphylococcus aureus* and *Escherichia coli,* is the most common. However, in many of the studies examination for antifungal activity was not attempted. Since most feeder-root diseases of ectomycorrhizal plants are caused by fungi, it is difficult to implicate antibacterial antibiotics in limiting feeder-root diseases. Assuming that these antibacterial antibiotics are produced by the symbionts in ectomycorrhizal associations, they could have a selective influence on bacterial populations in the rhizosphere of ectomycorrhizae that, in turn, could have a direct or indirect influence on fungal pathogens in the soil.

Wilkins & Harris (94) made extracts from basidiocarps of more than 700 species of higher Basidiomycetes and found that over 24% exhibited antibacterial activity. Many of these fungi have been associated with ectomycorrhizae of trees. Of the seven *Lactarius* spp. reported active by these workers, all are ectomycorrhizal associates. Also approximately 80% of the *Tricholoma* spp., 60% of the *Cortinarius,* and 55% of the *Hygrophorus* spp. that produced antibiotics are also ectomycorrhizal associates. However, none of 43 *Russula* spp. tested by these workers was found to be antibiotically active, and nearly 80% are probably ectomycorrhizal fungi. The literature shows that the genus *Russula,* which contains many symbiont species (87), apparently includes only a few species that produce antibiotics. We can therefore anticipate that certain fungal genera will contain species capable of both forming ectomycorrhizae and producing antibiotics, whereas other symbiotic genera may have species that produce few or no antibiotics at all.

Several workers have looked for antifungal activity in ectomycorrhizal fungi and many have used fungi pathogenic on feeder roots as bioassay organisms. Šašek & Musilek (79–81) found that certain ectomycorrhizal fungi inhibited *Rhizoctonia solani, Pythium debaryanum,* and *Fusarium oxysporum.* Certain strains of symbiotic fungi (mainly of various species of *Suillus*) produced antifungal compounds, while other strains did not. Šašek (78) grew pine seedlings on polyurethane disks floating in liquid medium with several ectomycorrhizal fungi and pathogens, and found that antifungal compounds produced by the fungal symbionts decreased damping-off caused by the pathogens. Ectomycorrhizae were not formed under these culture conditions. *Tricholoma saponaceum* decreased damping-off caused by *R. solani, P. debaryanum,* and *F. oxysporum; Scleroderma aurantium* decreased damping-off caused by the latter two pathogens. *Suillus bovinus* was only effective in limiting damping-off caused by *R. solani,* while *Amanita citrina, Lactarius helvus,* and *Russula fragilis* were only effective against *P. debaryanum.* Although *Fomes annosus* is not a feeder-root pathogen under field conditions (30), several workers have found this fungus to be inhibited by antibiotics produced by several ectomycorrhizal fungi. Šašek & Musilek (80) found that *F. annosus* was only weakly inhibited by a few ectomycorrhizal fungi in pure culture. Hyppel (31) tested 85 isolates of some 42 different species of ectomycorrhizal fungi from Sweden; over 40% of them inhibited *F. annosus* in dual cultures. He also detected variation in antibiotic production by different strains of the same species of ectomycorrhizal fungus. Hyppel (32) demonstrated also that *Boletus bovinus,* a fungal symbiont of Norway spruce (*Picea abies*), protected spruce seedlings from attack by *F. annosus* in greenhouse studies. Although *B. bovinus* did not enter into ectomycorrhizal association with the seedlings due to the short duration of the study, a water-soluble antifungal metabolite produced by the fungal symbiont had an inhibiting effect on *F. annosus,* and thereby considerably reduced seedling mortality caused by the pathogen. Marx (42) found that *Leucopaxillus cerealis* var. *piceina* weakly inhibited *F. annosus,* but strongly inhibited *Cylindrocladium scoparium,* 9 species of *Phytophthora, Polyporus tomentosus* var. *circinatus, Poria weirii,* 24 species of *Pythium,* 5 species of *Rhizoctonia* or *Thanatephorus,* and *Sclerotium bataticola.* The only root pathogens tested that were not inhibited by this fungal symbiont were *Armillaria mellea, Fusarium oxysporum* f. *pini, Rhizoctonia crocorum,* and *Thanatephorus cucumeris.* The antibiotic produced by *L. cerealis* var. *piceina* was identified as diatretyne nitrile, a polyacetylene (43). It inhibited germination of zoospores of *Phytophthora cinnamomi* at 50–70 parts per billion and killed zoospores at 2 ppm. The antibiotic inhibited bacteria from forest soil at 0.5 ppm. *L. cerealis* var. *piceina* also produced diatretyne amide and diatretyne 3, which are reduction products of the nitrile and are antibacterial only. Anchel and associates (3) originally identified the diatretyne antibiotics from culture filtrates of *Clitocybe diatreta, C. odora,* and *Lepista nuda,* known ectomycorrhizal associates of trees. At least 10 other species of basidiomycetes produce polyacetylene antibiotics, but have not been associated with ectomycorrhizae. In addition to reporting antibiotic production by *L. cerealis* var. *piceina,* Marx (42) found other fungal symbionts of pines that could inhibit root pathogens. Interesting differences in biological activity of antibiotics produced by the different fungal symbionts were noted. *Laccaria laccata* inhibited 16 of 21 species of *Pythium* tested, but only 1 of 9 species of *Phytophthora. Lactarius deliciosus,* however, did not inhibit any species of *Pythium,* but did inhibit 6 of the 8 species of *Phytophthora.* Extremes in activity were also found. *Pisolithus tinctorius* did not inhibit any of the 48 root pathogens tested, and *L. cerealis* var. *piceina* inhibited 92%. Intermediate in the biological spectrum of antibiotic activity was *Suillus luteus,* which inhibited over 70% of the pathogens, including all the *Phytophthora* spp. and most of the *Pythium* spp. Marx & Bryan (44) found that *Scleroderma bovista,* a fungal symbiont of pine and pecan, inhibited 5 *Phytophthora* spp. and 4 *Pythium* spp. associated with feeder-root necrosis of pecan. The assumed presence of this antibiotic in ectomycorrhizal roots of pecans in the field was implicated in control of feeder-root disease caused by *Pythium* spp. Park (61) reported that an unidentified *Lactarius* sp., symbiotic on basswood (*Tilia americana*) seedlings, inhibited several pathogens including the feeder-root pathogens *Pythium irregulare, Fusarium solani, F. oxysporum, Cylindrocladium scoparium, Rhizoctonia solani, R. praticola,* and *Sclerotium rolfsii,* and that the culture filtrate of the *Lactarius* sp. was active against damping-off of seed of *Pinus resinosa* caused by *Pythium irregulare* and *Rhizoctonia praticola.* Seed presoaked for 12 hr in the antifungal filtrate had 93% germination and emergence in contrast to nontreated seed with only 7% germination and emergence in the presence of these pathogens. Marx et al (48) found that certain isolates of *Thelephora terrestris,* a widespread fungal symbiont on many tree species, inhibited *Pythium aphanidermatum, P. irregulare* and *P. spinosum,* but not *P. vexans or* four species of *Phytophthora*. Certain unidentified fungal symbionts of shortleaf pine also inhibited the same species of *Pythium*. Pratt (65) examined nearly 50 basidiomyceteous fungi collected from Eucalypt forests in Australia for antagonism to *P. cinnamomi*. Several of these fungi, such as *Boletus* (*Suillus*) *luteus* and *Lactarius deliciosus,* are known ectomycorrhizal symbionts and were found to inhibit vegetative mycelial growth of *P. cinnamomi* in agar-plate studies. Recently Zak (100) reported that *Rhizopogon vinicotor,* a fungal symbiont of Douglas-fir (*Pseudotsuga menziesii* var. *menziesii* and var. *glauca*), was inhibitory to growth of *Phytophthora cinnamomi, Pythium debaryanum, P. sylvaticum, Fomes annosus,* and *Poria weirii.*

Although numerous ectomycorrhizal fungi can produce antibiotics in pure culture or in their basidiocarps, it remains to be shown that they are produced by the symbionts while in ectomycorrhizal association with their hosts.

Antibiotic production in ectomycorrhizal association.—Krywolap et al (37) extracted an antibiotic from ectomycorrhizae formed by *Cenococcum graniforme* on white (*Pinus strobus*) and red (*P. resinosa*) pines and Norway spruce. This antibiotic, active against bacteria but not fungi, was similar in chromatographic and ultraviolet fluorescence analyses to the antibiotic produced in pure cultures of *C. graniforme.* Foliar extracts of the pines with mycorrhizae formed by *C. graniforme* under forest soil conditions contained the antibiotic, indicating that it was readily translocated. However, the antibiotic was detected only in roots and not in foliage of trembling aspen (*Populus tremuloides*), which suggested a lack of translocation. Foliage of nursery-grown pine seedlings without ectomycorrhizae formed by *C. graniforme* also exhibited similar antibiotic activity. They concluded that the seedlings absorbed the antibiotic from sclerotia or hyphae of *C. graniforme* present in great abundance in the nursery soil, and suggested that this antibacterial compound may confer to trees some degree of protection against bacterial pathogens. Grand & Ward (22) also reported this antibiotic in foliage of the same tree species, but could not find a correlation between the number of ectomycorrhizae formed by *C. graniforme* in two different soil types and the amount of antibiotic activity. They attributed lack of correlation to physiological differences between species of hosts which could have inactivated the antibiotic.

Marx & Davey (45) extracted diatretyne nitrile and diatretyne 3 from ectomycorrhizae formed by *L. cerealis* var. *piceina* and from the rhizosphere substrate of the ectomycorrhizae on axenic shortleaf pine seedlings. Neither short roots nor the substrate adjacent to short roots on nonmycorrhizal pine seedlings contained the diatretynes. No attempt was made to detect the diatretyne antibiotics in foliage of seedlings with ectomycorrhizae formed by this symbiont. In earlier work (43), however, diatretyne nitrile was not detected in foliage of young pine seedlings with roots exposed to the antibiotic for 40 days. In experiments designed to determine the susceptibility or resistance of ectomycorrhizae to infection by a pathogen, Marx & Davey (45) demonstrated that the diatretynes present in ectomycorrhizae formed by *L. cerealis* var. *piceina* were functional in the resistance of feeder roots to infection by *Phytophthora cinnamomi*. Not only were the ectomycorrhizae resistant, but nonmycorrhizal short roots adjacent to the ectomycorrhizae which contained the diatretyne antibiotics were only 25% susceptible to infection by zoospores of *P. cinnamomi*. Short roots on control seedlings and on seedlings with ectomycorrhizae formed by either *Laccaria laccata* or *Pisolithus tinctorius* were 100% susceptible to infection. It was not determined whether the diatretynes were translocated to short roots from adjacent ectomycorrhizae or simply absorbed from the rhizosphere. Instead of infecting 100% of short roots, as on nonmycorrhizal seedlings, *P. cinnamomi* infected only 77% of short roots on shortleaf, and 85% of short roots on loblolly pine seedlings adjacent to ectomycorrhizae formed by *Suillus luteus*. They attributed some antibiotic protection here also, since *S. luteus* was found to be antagonistic to *P. cinnamomi* and other related fungi in earlier pure culture tests (42).

Mechanical Barrier Created by Fungal Mantle

A mechanical barrier creates a physical rather than a chemical hindrance to prevent the entrance or spread of a pathogen. The significance of mechanical barriers to host defense against pathogenic attack has received attention in past plant-disease investigations. Several external and internal morphological barriers of a mechanical nature, which may influence either pathogen entrance or its spread into host tissues, are found in plants. Tough outer walls of epidermal cells, suberized root periderm and endodermis, and thick cuticles of leaves have been reported to impede direct penetration of pathogens, and may function as mechanical barriers. Other mechanical barriers often are formed after establishment of infection. In some plants, suberized wound tissue or cicatrical layers develop, limiting the localized lesion of infection. Abscission layers, tyloses, gum deposition, callosites, cellulosic coverings of infective hyphae, and other barriers have been reported to impede physically the spread of certain pathogens. It has been concluded (2) that these various mechanical barriers, especially those of external origin, are relatively ineffectual in the protection of plants in general.

The fungal mantle of ectomycorrhizae, however, creates a unique and totally different kind of potential obstruction to pathogens attempting root penetration. In mature ectomycorrhizae, the fungal mantles are composed of tightly interwoven hyphae, often in well-defined layers, which usually completely cover the root meristem and cortical tissues. This hyphal network, which precludes exposure of root tissue to direct contact with the rhizosphere, usually is complete, i.e., relatively free from voids.

Marx & Davey (45, 46) and Marx (47) have concluded that the fungus mantles of ectomycorrhizae are formidable physical barriers to penetration by *Phytophthora cinnamomi,* based on histological observation of numerous pine ectomycorrhizae formed by several fungal symbionts that had been inoc-

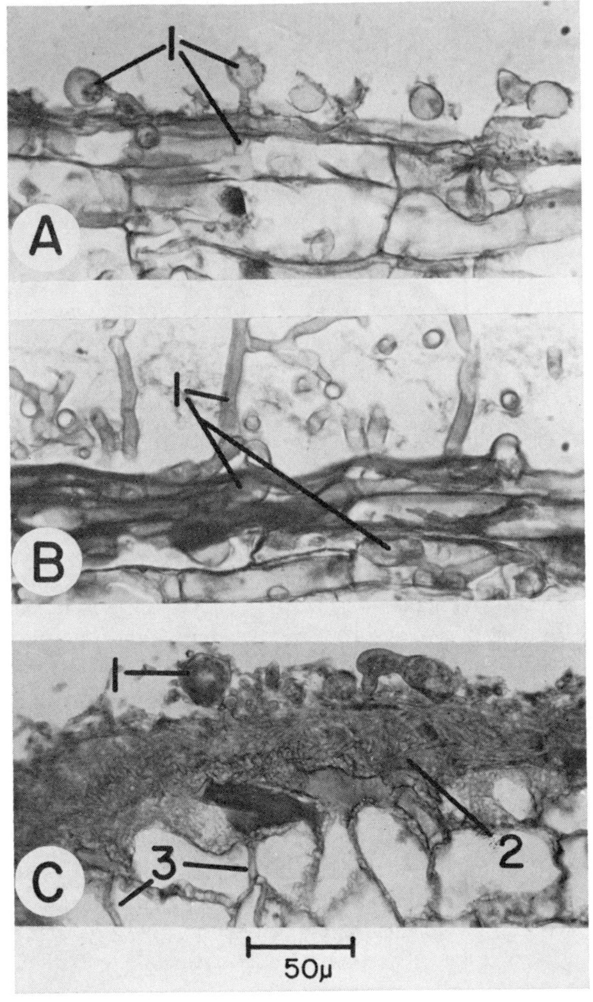

FIGURE 1. Susceptibility and resistance of nonmycorrhizal and ectomycorrhizal feeder roots of pine to *Phytophthora cinnamomi.* Direct penetration of zoospores (A) and vegetative hyphae (B) of *P. cinnamomi* (1) into nonmycorrhizal feeder roots with intracellular cortex infection. Note vesicles and hyphae of *P. cinnamomi* (1) on fungal mantle barrier (2) of ectomycorrhizae in (C) and absence of infection in cortex surrounded by the Hartig net (3). (Marx 47, Marx & Bryan 44).

ulated with either zoospores or vegetative mycelium. Mature ectomycorrhizae with complete fungus mantle, were not infected by *P. cinnamomi*, whereas 100% of all nonmycorrhizal roots used for comparison were infected (Fig. 1). Fungus mantles of ectomycorrhizae formed by nonantibiotic-producing fungal symbionts, which passively covered adjacent nonmycorrhizal root initials on pine seedlings, protected these initials from penetration by *P. cinnamomi*. Histological examination of the root initials revealed the complete absence of fungal-symbiont infection and the apparently passive but protective nature of the mantle. Short root initials not covered by fungal mantle were highly susceptible to infection. There was further evidence that the fungal mantle covering root meristems is a barrier against pathogen penetration. Meristems of ectomycorrhizae of shortleaf and loblolly pines were readily infected by *P. cinnamomi* when the mantle covering was either incompletely formed over the root tip or artificially removed. Infection, however, did not take place in the meristem tissues when the root tips were covered by a complete fungal mantle. The Hartig net surrounding the cortical cells may function as an additional physical barrier since spread of *P. cinnamomi,* originating from either infections of nonprotected meristem tissue without fungus mantle coverings or from infections through artificially excised root tips, was blocked in this region. Figure 2 shows this region located several cortex cells behind the root tips in the area of cell maturation. It was not possible to separate the possible indirect chemical effect of the Hartig net, i.e., inducing the production of chemical inhibitors in cortical cells, from the suggested mechanical effect of this hyphal network.

Chemical Inhibitors Produced by Host

Most plant cells are capable of elaborating inhibitory substances during their metabolic response to pathogen attack. Phenols, quinones, various phytoalexins, and numerous other compounds have been found in tissues of a variety of plants during pathogenesis. Many of these were inhibitory to the pathogen and are considered by many authors to be important in disease resistance (11, 86).

Plant cells exposed to symbiotic parasitism have also been reported to respond by the production of substances inhibitory to the fungal symbiont. Bernard (5), from studies on the symbiosis between *Rhizoctonia repens* and various orchid tubers, concluded that an antifungal compound was formed in response to infection by *R. repens*. Subsequent research (18–20, 58) revealed that tubers of several species of orchid produce orchinol, coumarin, hircinol, and an unidentified phenolic compound in response to infection by several other species of *Rhizoctonia* and other endomycorrhizal and pathogenic fungi. Orchinol could not be found in noninfected tubers. The production of these inhibitory compounds is considered to be the defense mechanism of the orchid which maintains the fungal symbiont in a balanced state; without this defense mechanism, these symbionts would be pathogenic on the orchids. Another consequence of synthesis of these inhibitory compounds, which extends throughout the tuber, is that their presence protects the tissue against infection by pathogenic organisms.

The mycorrhizae of orchids are endomycorrhizal, i.e., the hyphae of the symbiont penetrate the cells and are in intimate association with the cyto-

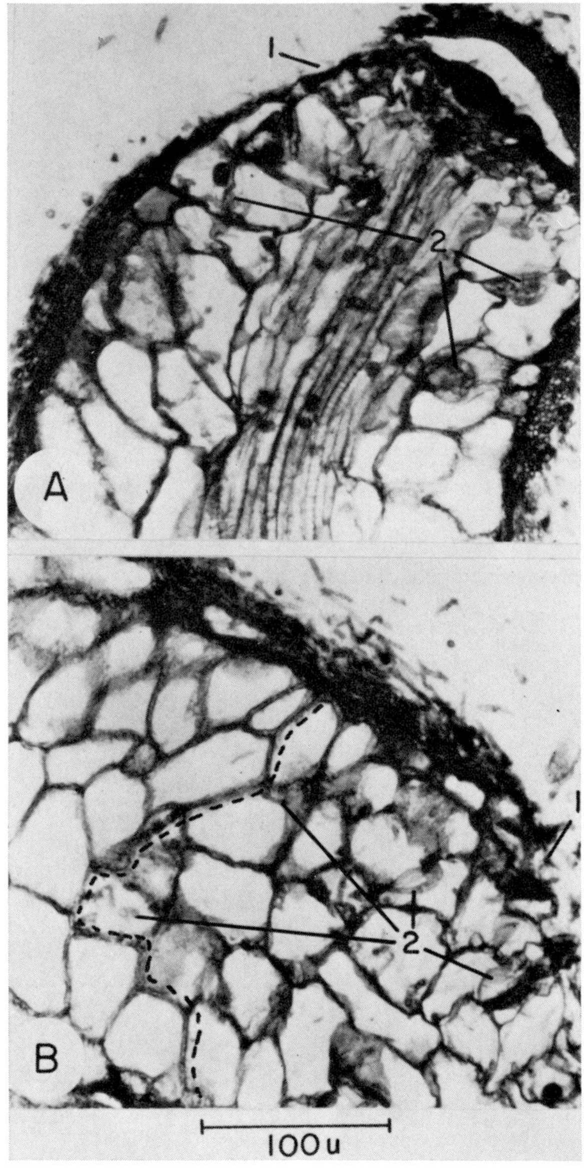

FIGURE 2. *Phytophthora cinnamomi* infection of incomplete ectomycorrhiza of pine. (A) and (B) are sections of the same mycorrhiza taken 60μ apart. Note void in fungal mantle (1), intracellular hyphae of *P. cinnamomi* (2), and boundary of Hartig net (dashed line) which blocked further spread of *P. cinnamomi* (Marx & Davey 45).

plasm. Endomycorrhizal fungi obviously have the enzymatic capacity to degrade cellulosic cell walls. Few ectomycorrhizal fungi have this capacity, and most are limited to the utilization of a few relatively simple carbohydrates (24). The inability of ectomycorrhizal fungi to digest cellulosic cell walls may explain their inability to penetrate cell walls of their hosts. Foster & Marks (14), however, presented electron micrographic evidence that suggests that mycorrhizal fungi exert not only enzymatic action in forming the Hartig net but perhaps mechanical pressure. Why the chemical and mechanical penetration of cortical cell walls by ectomycorrhizal fungi does not occur is conjecture. Host-originated inhibitors may be induced in response to incipient Hartig net development and restrict the ectomycorrhizal symbiont to the balanced state in much the same manner as host-originated inhibitors apparently do in endomycorrhizal orchid tubers. Any cellulases produced by the ectomycorrhizal fungus may also be inhibited by a tannin fraction containing polyphenols present in epidermal cell remanents secreted by the host (15). This tannin fraction must be specific for cellulases and not inhibit pectin-degrading enzymes, since the latter enzymes are functional in forming the Hartig net hyphal development during synthesis of ectomycorrhizae. Since cellulase is an inducible enzyme (41), it is possible that its synthesis is suppressed as long as available carbohydrates are present. Hyphae of ectomycorrhizal fungi in the Hartig net contain much glycogen (14), suggesting that available carbohydrates are not limiting in this region.

Foster & Marks (14) could not detect in electronmicrographs of ectomycorrhizae, apposition of materials by the host cytoplasm that restricted hyphae of the fungal symbiont, as found in pathogenic associations. They found disorganized cytoplasmic organelles, abundant tannin-like materials in cortical cells of mycorrhizae, and a very large vacuole in each cortical cell often lined with polyphenol materials aggregated into large dense masses. Frequently the nuclei of these cells were very large, round, without their usual invaginations, and the chromatin bodies were unusually heavily stained. The cortex cells were also usually devoid of starch and contained amyloplasts. Mitochondria in these cells were elongated and covered with abundant cristae. These cortical cells in the Hartig net association of ectomycorrhizae were quite unlike cells of nonmycorrhizal roots.

Observations (14, 15) suggest that there is a biochemical and cytological reaction of the cortex cells to infection by ectomycorrhizal fungi. Recently, Krupa & Fries (35) found that the fungal symbiont, *Boletus variegatus,* produced in pure culture the fungistatic volatile compounds, isobutanol and isobutyric acid. Other volatile compounds were also identified. Volatile organic compounds were also extracted from the roots of intact seedlings of Scots pine (*Pinus sylvestris*) grown in pure culture with and without *B. variegatus.* However, infection of the roots by the fungal symbiont resulted in the production and accumulation of volatile terpenes and sesquiterpenes in concentration up to 8 times greater than that found in nonmycorrhizal roots. Many of these terpenes and sesquiterpenes are fungistatic, and were considered to be produced as a nonspecific response of the host cells to symbiotic infection. These authors concluded that the nonspecific response of the host to infection by ectomycorrhizal fungi results in the increased production and accumulation of native volatile substances and nonvolatile substances (29). These substances, when present in sufficient concentrations, may restrict the growth of ectomycorrhizal fungi within the host tissue until the symbiotic state finally results. Furthermore, volatile and nonvolatile substances could inhibit pathogens in the root, as volatile substances inhibit pathogens in the rhizosphere. Recently, Krupa (personal communication) found that several monoterpenes extracted from ectomycorrhizae inhibited vegetative growth of *P. cinnamomi* and *F. annosus* by 50% when vapors from 10 microliters of the substances were used. *P. cinnamomi* was sensitive to α-pinene and terpinolene, and *F. annosus* was sensitive to other host products. Krupa also extracted ectomycorrhizae formed in pure culture with southern pines, and found that *Pisolithus tinctorius* stimulated over a 40-fold increase, and *Cenococcum graniforme* a 30-fold increase, in certain fungistatic volatile monoterpenes. Each ectomycorrhizal fungus induced a shift in monoterpene synthesis in comparison to noninfected roots, suggesting that each fungal symbiont elicits a specific response by host roots. Catalfomo & Trappe (8) also found that certain ectomycorrhizal fungi from the genera *Amanita* and *Rhizopogon* and the family Boletaceae produced terpenes which, in their opinion, were involved in the protective role of mycorrhizae.

Host roots apparently respond to ectomycorrhizal infection by the production of inhibitors that also contribute to resistance of ectomycorrhizae to pathogenic infection.

Differences in Chemical Exudation of Ectomycorrhizae

Root exudates of herbaceous plants contain carbohydrates, amino acids, vitamins, organic acids, nucleotides, flavonones, enzymes, and compounds such as hydrocyanic acid, glycosides, and saponins. Root exudates from certain plants stimulate mycelial growth, microsclerotia germination, tactic zoospore activity, and pathogenicity; those from other plants may inhibit the same processes (72).

Slankis et al (83) reported that roots of axenic white pine seedlings with needles exposed to $C^{14}O_2$ liberated a complex mixture of more than 35 radioactive sugars, amides, and organic acids. Agnihotri & Vaartaja (1) isolated and identified 3 sugars and 13 amino acids from exudates of young radicles of red pine seedlings; glucose, fructose, arabinose, asparagine, glutamine, and several amino acids were the most prevalent chemicals. Smith (84) found similar compounds in exudates of young roots of pines and black locust (*Robinia pseudoacacia*), and demonstrated qualitative and quantitative differences in exudates from different species.

The roots of tree seedlings in these investigations were not in an ectomycorrhizal condition. The only differences in exudation patterns between ectomycorrhizal and nonmycorrhizal roots were shown by Krupa and associates on volatile organic constituents of ectomycorrhizae. In my opinion, it is logical to expect differences in exudation patterns between ectomycorrhizal and nonmycorrhizal roots, since ectomycorrhizal fungi derive most if not all their required carbohydrates, amino acids, and vitamins from their intimate association between the cortex cells and the external root surface. Few root exudates could pass through the Hartig net and fungal mantle of ectomycorrhizae without some absorption and utilization by the fungal symbiont. This suggests that exudates of ectomycorrhizal roots are (*a*) those not utilized by the fungal symbiont; (*b*) metabolic by-products of the fungal symbiont; or (*c*) those released as a result of the metabolic interaction of the symbiotic partners. It is reasonable that these changes in root exudations should have some effect on root pathogens. If differences in microbial rhizosphere populations are at least partially due to differences in root exudates, then available circumstantial evidence indicates differences in exudates between ectomycorrhizal and nonmycorrhizal roots, because these different root types harbor different rhizosphere populations.

Lewis & Harley (39) reported that ectomycorrhizae of beech (*Fagus sylvatica*) contained endogenous quantities of glucose, fructose, sucrose, and trehalose together with the acyclic polyol, mannitol, and two cyclic polyols, myo-inositol and an unidentified inositol. Trehalose and mannitol were not extracted from nonmycorrhizal roots of beech, showing their presence to be dependent on ectomycorrhizal infection.

The chemotaxis of motile zoospores of phycomycetous pathogens to plant roots is one of the major means by which these fungi find and attack roots (9). Numerous studies reveal that roots, both excised and on intact plants, and chemicals known to be components of root exudates of these plants, are strongly attractive to zoospores of numerous species of phycomycetes. It is assumed, however, that previous work on chemotaxis has been on nonmycorrhizal roots because they were grown in such a manner (hydroponics, aseptic culture, or closed root containers) as to limit infection by ectomycorrhizal fungi.

The only research on chemotaxis of zoospores to ectomycorrhizal and nonmycorrhizal roots is that of Marx & Davey (45, 46), who used intact ectomycorrhizae formed by several different fungal symbionts and nonmycorrhizal roots on shortleaf and loblolly pine seedlings. Zoospores of *Phytophthora cinnamomi* were not strongly attracted to either nonmycorrhizal or ectomycorrhizal roots. After their encystment on root surfaces, zoospores were observed to germinate faster and more vigorously at the growing tips and the region of cell elongation on nonmycorrhizal roots than on other parts of the root. Encysted zoospores on ectomycorrhizae, however, germinated slowly and with an obvious lack of vigor of germtube elongation comparable to responses of zoospores on heavily suberized root parts. This indicated indirectly that the ectomycorrhizae were not as chemically stimulating to the zoospore germination and germtube growth as were nonmycorrhizal nonsuberized roots. Chemotaxis of zoospores of *P. cinnamomi* did occur on cut feeder root tips of these pine species. The apices of nonmycorrhizal short and lateral roots, and ectomycorrhizal roots, were excised at either 0.1 mm or 1 mm from the growing tip; strong chemotaxis was observed on nonmycorrhizal roots and on ectomycorrhizae with 1 mm of their tips removed. Zoospores germinated faster and with more vigor on cut surfaces of nonmycorrhizal roots than on ectomycorrhizal roots.

Protective Microbial Rhizosphere Populations

Garrett (17) described the rhizosphere as the outermost defense of the plant against attack by root pathogens. This zone normally supports a much greater population of microorganisms than is found in nonrhizosphere soil.

The rhizosphere of ectomycorrhizae actually is the "ectomycorrhizosphere" because the mantle of the fungal symbiont is the external microbial component of the rhizosphere of root cells (15). Tribunskaya (88) found approximately 10 times as many fungi in rhizospheres of ectomycorrhizal pine seedlings as in those of nonmycorrhizal seedlings, and concluded that the fungal symbionts were responsible for the different microflora of the rhizosphere. Katznelson et al (34) showed that ectomycorrhizal roots of yellow birch (*Betula allegheniensis*) increased the numbers of certain physiological groups of soil bacteria and actinomycetes. Bacteria that grew in simple chemical media, and total fungal numbers, appeared to be reduced around ectomycorrhizae. The types of fungi present in the various rhizospheres were different; fungal genera that contain feeder-root pathogens (*Pythium, Fusarium,* and *Cylindrocarpon*) predominated in nonmycorrhizal rhizospheres, while the ectomycorrhizal roots supported *Mycelium radicus, Penicillium* spp., and other rapidly growing fungi. *Pythium* and *Fusarium* spp. were completely absent from the rhizospheres of ectomycorrhizae. These results suggest that the ectomycorrhizae had an inhibiting effect on pathogens in the root zone. Foster & Marks (15) examined ectomycorrhizae of *Pinus radiata* by electron microscopy, and found a distinct spatial distribution of bacteria in the rhizosphere. The largest bacterial populations were found in the outermost fungus mantle layers and in areas of the soil colonized by fungus hyphae that could have originated from the mantle. The bacterial population in the outer layers of the mantle was about 16 times that found in the outermost region of the rhizosphere. Neal et al (56) investigated the rhizosphere microbial population of three morphologically distinct ectomycorrhizae—a white, a grey, and a yellow form—as well as the microbial population of suberized roots and nonrhizosphere soil from a Douglas fir. Each microhabitat contained a distinct microflora. Differences between ectomycorrhizal rhizospheres were attributed to different fungal symbionts. They suggested that influence of specific ectomycorrhizal fungi on the rhizosphere flora may affect the extent of infection of root pathogens, and that some ectomycorrhizae may support a more effective rhizosphere barrier than others. Oswald & Ferchau (60) isolated and identified 51 species of bacteria from coniferous roots. Seven species were found only on nonmycorrhizal roots, 22 only on mycorrhizal roots, and 22 were common to both root types. Other workers (13, 57) have also found that microbial populations of ectomycorrhizal rhizospheres are not only different both qualitatively and quantitatively from those of nonmycorrhizal rhizospheres, but that each morphological type of ectomycorrhiza (presumably formed by different fungal symbionts) harbors in its rhizosphere a different microbial population.

Only Ohara & Hamada (59) have implied that the antagonistic nature of certain fungal symbionts in ectomycorrhizal association is related to population of microorganisms in the rhizosphere. They found that bacteria, especially aerobic and heterotrophic types, as well as actinomycetes, were strongly inhibited around actively growing mycelium of *Tricholoma matsutake* in forest soil or in ectomycorrhizae on *Pinus densiflora.* These microorganisms were found in great abundance in adjacent soil containing neither mycelium nor ectomycorrhizae formed by this fungus; it was inferred that antibiotics were the cause for the inhibition.

Since there are differences in microbial rhizosphere populations between ectomycorrhizal and nonmycorrhizal roots, it may be logically surmised that there are differences in the competitive microbial potential near these roots. It is not known whether these differences influence root pathogen populations and subsequent development of feeder root disease.

INTERACTION OF PLANT PARASITIC NEMATODES AND FUNGAL PATHOGENS ON ECTOMYCORRHIZAE

Many plant-parasitic nematodes are found in forest soils and tree nurseries. Nematodes parasitizing roots normally limit themselves to the feeder roots, the same ones infected by ectomycorrhizal fungi. What influences do the feeding habits of these nematodes have on the resistance of ectomycorrhizae to infection by fungal pathogens? Can nematodes create entry points through the fungal mantle or otherwise reduce the resistance of ectomycorrhizae for fungal pathogens?

Certain nematodes feed directly on hyphae of ectomycorrhizal fungi. Riffle (68) found an *Aphelenchoides* that fed and reproduced on mycelium of *Suillus granulatus* and caused significant reduction in linear growth of the fungus in pure culture. He later (67) observed that *Aphelenchoides cibolensis* fed and reproduced on 53 of 58 ectomycorrhizal fungi in pure culture. The viability of 25 of these symbionts was not affected by the nematodes, but 16 other species failed to revive when transferred to fresh culture medium. Sutherland & Fortin (85) found that *Aphelenchus avenae* fed and reproduced on 7 species of fungal symbionts in pure culture. One symbiont, *Rhizopogon roseolus,* apparently produced a toxin lethal to the nematode. *Aphelenchus avenae* also prevented the formation of ectomycorrhizae by *S. granulatus* on axenic red pine. Because this nematode did not enter the feeder roots, the authors concluded that it prevented ectomycorrhizal development by directly suppressing the fungus prior to symbiotic infection of the roots. Mycophagous nematodes may cause direct inhibition to ectomycorrhizal development, but none has been reported to feed on preformed ectomycorrhizae.

Several plant-parasitic nematodes will feed on ectomycorrhizae. Ruehle (73) found that two endoparasitic nematodes, lance (*Hoplolaimus coronatus*) and pine-cystoid (*Meloidodera floridensis*), penetrated and migrated through ectomycorrhizae of loblolly and slash (*Pinus elliottii*) pines and caused extensive damage to the cortex and vascular tissues. Riffle & Lucht (69) reported root-knot nematodes parasitizing ectomycorrhizae of ponderosa pine (*P. ponderosa*), and Zak (99) found an unidentified *Meloidodera* sp. in 2 of 6 morphologically distinct ectomycorrhizae of Douglas fir. Recently, Ruehle & Marx (74) reported that the lance nematode readily penetrated the fungal mantle of ectomycorrhizae formed by *Pisolithus tinctorius* and *Thelephora terrestris* on shortleaf and loblolly pine seedlings. This nematode confined its feeding to cortical tissues of lateral roots rather than short roots on nonmycorrhizal seedlings, which suggests that short roots after transformation into ectomycorrhizae were more favorable feeding sites for the nematode than nonmycorrhizal short roots.

Although none of the purported mechanisms for resistance of ectomycorrhizae to fungal attack applies to attack by nematodes, what about subsequent infection of these nematode-damaged ectomycorrhizae by fungal pathogens? Barham,[3] using a root-cell technique (45–47), inoculated intact ectomycorrhizae of shortleaf pine formed by *Pisolithus tinctorius* and *Thelephora terrestris* with either spiral (*Helicotylenchus dihystera*) or stunt (*Tylenchorhynchus claytoni*) nematodes and zoospores of *Phytophthora cinnamomi*. Both nematodes penetrated and migrated through the fungal mantle and Hartig net of the ectomycorrhizae. The disruption of structural integrity of the fungus mantle of certain of these ectomycorrhizae by spiral nematodes created infection courts for *P. cinnamomi* and intracellular hyphae and vesicles of the pathogen were found in cortex cells surrounded by the Hartig net (Fig. 3). *P. cinnamomi* did not infect ectomycorrhizae parasitized by stunt nematodes or ectomycorrhizae not inoculated with nematodes. If we assume that host inhibitors of *P. cinnamomi* were present, then perhaps the nematodes destroyed their effectiveness or the leaked inhibitors were diluted or degraded.

Many nematodes may cause root disease or act as predisposing agents (63). Since they can also render normally resistant ectomycorrhizae susceptible to infection by *P. cinnamomi,* they are more important in feeder-root diseases of trees than we had thought.

CONCLUSIONS

Trees and seedlings with significant quantities of ectomycorrhizae growing in soils containing feeder-root pathogens would have less susceptible (nonmycorrhizal) root tissue exposed to attack by the pathogens than those with

[3] Barham, R. O. Master's thesis in preparation, Department of Plant Pathology and Plant Genetics, University of Georgia, Athens.

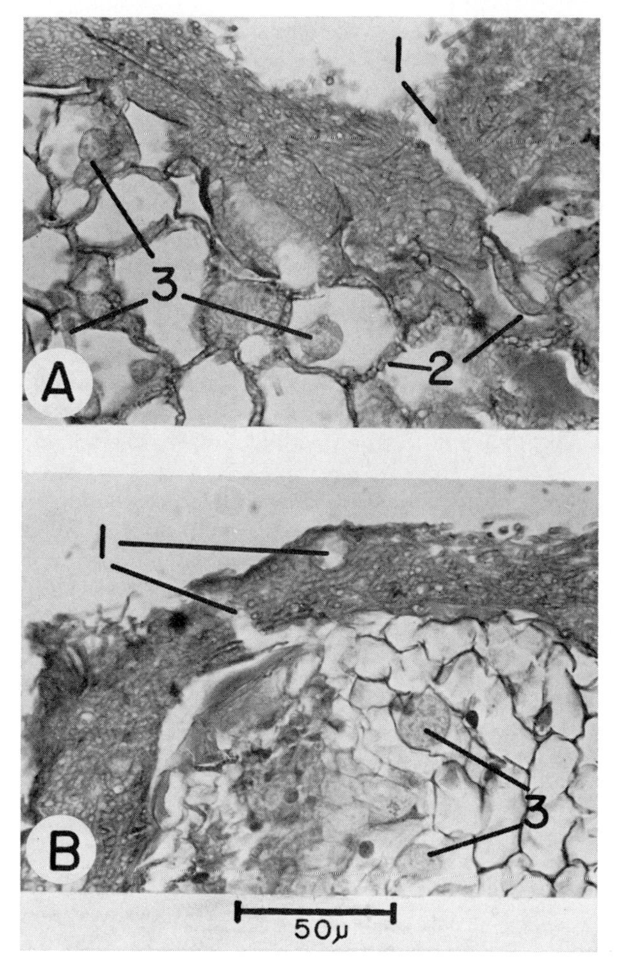

FIGURE 3. Influence of spiral nematodes on susceptibility of ectomycorrhizae to attack by *Phytophthora cinnamomi*. In both photos note the holes in fungal mantle (1) caused by nematodes. In A, note intracellular hyphae of *P. cinnamomi* (3) in cortex cells surrounded by Hartig net of fungal symbiont (2) and in B, note *P. cinnamomi* (3) in meristematic tissue without the Hartig net. (R. O. Barham M.S. thesis in preparation, Dept. Plant Pathol. Plant Genet., Univ. Georgia, Athens.)

few or no ectomycorrhizae. They would also gain from the well-documented physiological benefit of mycorrhizae. In fertile nursery soils containing high populations of pathogens, ectomycorrhizae could be more important in control of feeder-root disease than in nutrition. Ectomycorrhizal development is usually inhibited with high soil fertility (25) and thus, more of the susceptible root tissue (nonmycorrhizal roots) should be present for attack by fungal pathogens.

The protective role of ectomycorrhizae in feeder-root disease appears to be nonsystemic since the presence of a few mycorrhizae on a root system does not control pathogenic root infections on nonmycorrhizal sections of the same root system. A fungal symbiont producing in its ectomycorrhizae a potent antibiotic effective against a feeder-root pathogen might, however, afford protection to adjacent nonmycorrhizal roots by translocation or diffusion of the antibiotic into the nonmycorrhizal roots (45).

The fungal mantle appears to be a mechanical barrier to penetration by pathogens, but even without it the root cortex cells surrounded by the Hartig net also are resistant. This suggests a chemical function, probably of host origin, in this resistance phenomenon. The antibiotic mechanism of resistance is apparently functional since normally susceptible roots adjacent to resistant ectomycorrhizae producing a strong antibiotic may be resistant to attack by pathogenic fungi.

Mycorrhizae support a microbial rhizosphere population different both qualitatively and quantitatively from populations of other ectomycorrhizae, nonmycorrhizal roots, and nonrhizosphere soil. They probably have an effect on microbial competition and, thereby, on root pathogens. Pathogens apparently are not tropically attracted to ectomycorrhizae as they are to nonmycorrhizal roots.

This complex of defense mechanisms acting in concert assures opportunity for biological control of feeder-root pathogens by ectomycorrhizae.

LITERATURE CITED

1. Agnihotri, V. P., Vaartaja, O. 1967. Root exudates from red pine seedlings and their effects on *Pythium ultimum*. *Can. J. Bot.* 45:1031–40
2. Akai, S. 1960. Histology of defense in plants. In *Plant Pathology, an Advanced Treatise,* eds, J. G. Horsfall, A. E. Dimond, 1:391–434. New York: Academic. 674 pp.
3. Anchel, M., Silverman, W. B., Valanju, N., Rogerson, C. T. 1962. Patterns of polyacetylene production. I. The diatretynes. *Mycologia* 54:249–57.
4. Atkinson, N. 1946. Toadstools and mushrooms as a source of antibacterial substances active against *Mycobacterium phlei* and *Bact. typhosum*. *Nature* 157:441
5. Bernard, N. 1911. Sur la fonction fungicide des bulbes d'Ophrydées. *Ann. Sci. Nat. Bot. Biol. Veg.* 14:221–34
6. Bohus, G., Gluz, E. T., Scheiber, E. 1961. The antibiotic action of higher fungi on resistant bacteria and fungi. *Acta Biol. Hung.* 12:1–12
7. Campbell, W. A., Hendrix, F. F., Jr. 1967. *Pythium* and *Phytophthora* species in forest soils in the southeastern United States. *Plant Dis. Reptr.* 51:929–32
8. Catalfomo, P., Trappe, J. M. 1970. Ectomycorrhizal fungi: a phytochemical survey. *Northwest Sci.* 44:19–24
9. Chang-Ho, Y., Hickman, C. J. 1970. Some factors involved in the accumulation of phycomycete zoospores on plant roots. In *Root Diseases and Soil-Borne Pathogens.* eds. T. A. Toussoun, R. V. Bega, P. E. Nelson, 103–09. Berkeley: Univ. Calif. Press. 252 pp.
10. Corte, A. 1969. Research on the influence of the mycorrhizal infection on the growth, vigor,

and state of health of three *Pinus* species. *Archiv. Bot. Biogeogr. Ital.* 45:1–28

11. Cruickshank, I. A. M. 1963. Phytoalexins. *Ann. Rev. Phytopathol.* 1:351–74
12. Davis, W. C., Wright, E., Hartley, C. 1942. Diseases of forest-tree nursery stock. *Fed. Secur. Agency, Civ. Cons. Corp. Forest. Publ.* 9:1–79
13. Fontana, A., Luppi, A. N. 1966. Saprophytic fungi isolated from ectotrophic mycorrhizae. *Allionia* 12:39–46
14. Foster, R. C., Marks, G. C. 1966. The fine structure of the mycorrhizae of *Pinus radiata* D. Don. *Austral. J. Biol. Sci.* 19:1027–38
15. Foster, R. C., Marks, G. C. 1967. Observations on the mycorrhizae of forest trees. II. The rhizosphere of *Pinus radiata* D. Don. *Austral. J. Biol. Sci.* 20: 915–26
16. Frank, A. B. 1885. Ueber die auf Wurzelsymbiose beruhende Ernahrung gewisser Baume durch unterirdische Pilze. *Ber. Deutsch. Bot. Ges.* 3:128–45
17. Garrett, S. D. 1960. *Biology of Root-Infecting Fungi.* Cambridge: Univ. Press. 393 pp.
18. Gäumann, E. 1960. New data on the chemical defence reactions of orchids. *C. R. Acad. Sci.* (Paris) 250:1944–47
19. Gäumann, E., Kern, H. 1959. On chemical defensive reactions in orchids. *Phytopathol. Z.* 36:1–36
20. Gäumann, E., Nuesch, J., Rimpau, R. H. 1960. Further studies on the chemical defense reaction in orchids. *Phytopathol. Z.* 38: 274–308
21. Gerdemann, J. W. 1968. Vesicular-arbuscular mycorrhiza and plant growth. *Ann. Rev. Phytopathol.* 6:397–418
22. Grand, L. F., Ward, W. W. 1969. An antibiotic detected in conifer foliage and its relation to *Cenococcum graniforme* mycorrhizae. *Forest Sci.* 15:286–88
23. Hacskaylo, E. (ed.) 1971. Mycorrhizae. *U.S. Dept. Agr. Misc. Publ.* 1189, 255 pp.
24. Hacskaylo, E. 1971. The role of mycorrhizal associations in the evolution of the higher Basidiomycetes. In *Evolution in the Higher Basidiomycetes,* ed. R. H. Petersen, 217–37. Knoxville: Univ. Tenn. Press. 562 pp.
25. Harley, J. L. 1969. *The Biology of Mycorrhiza.* London: Leonard Hill. 334 pp.
26. Hatch, A. B. 1937. The physical basis of mycotrophy in *Pinus. Black Rock Forest Bull.* 6, 168 pp.
27. Hendrix, F. F., Jr., Campbell, W. A. 1970. Distribution of *Phytophthora* and *Pythium* species in soils in the continental United States. *Can. J. Bot.* 48: 377–84
28. Hendrix, F. F., Jr., Campbell, W. A., Chien, C. Y. 1971. Some phycomycetes indigenous to soils of old growth forests. *Mycologia* 63:283–89
29. Hillis, W. E., Ishikura, N. 1969. The extractives of mycorrhizas and roots of *Pinus radiata* and *Pseudotsuga menziesii. Austral. J. Biol. Sci.* 22:1425–36
30. Hodges, C. S., Rishbeth, J., Yde-Andersen, A. 1970. *Fomes annosus. Proc. III Int. Congr., Internat. Union For. Res. Org., Sect. 24.* Aarhus, Denmark, 1968. 208 pp.
31. Hyppel, A. 1968. Antagonistic effects of some soil fungi on *Fomes annosus* in laboratory experiments. *Stud. Forest. Suec.* 64:1–18
32. Hyppel, A. 1968. Effect of *Fomes annosus* on seedlings of *Picea abies* in the presence of *Boletus bovinus. Stud. Forest. Suec.* 66: 3–16
33. Jackson, R. M. 1965. Antibiosis and fungistasis of soil microorganisms. In *Ecology of Soil-Borne Plant Pathogens,* eds. K. F. Baker, W. C. Snyder, 363–73. Berkeley: Univ. Calif. Press. 571 pp.
34. Katznelson, H., Rouatt, J. W., Peterson, E. A. 1962. The rhizosphere effect of mycorrhizal and nonmycorrhizal roots of yellow birch seedlings. *Can. J. Bot.* 40: 257–76
35. Krupa, S., Fries, N. 1971. Studies on ectomycorrhizae of pine. I. Production of volatile organic compounds. *Can. J. Bot.* 49: 1425–31

36. Krywolap, G. N. 1971. Production of antibiotics by certain mycorrhizal fungi. In *Mycorrhizae,* ed. E. Hacskaylo, 219–221. U.S. Dep. Agr. Misc. Publ. 1189. 255 pp.
37. Krywolap, G. N., Grand, L. F., Casida, L. E., Jr. 1964. The natural occurrence of an antibiotic in the mycorrhizal fungus *Cenococcum graniforme. Can. J. Microbiol.* 10:323–28
38. Levisohn, I. 1954. Aberrant root infections of pine and spruce seedlings. *New Phytol.* 53:284–90
39. Lewis, D. H., Harley, J. L. 1965. Carbohydrate physiology of mycorrhizal roots of beech. I. Identity of endogenous sugars and utilization of exogenous sugars. *New Phytol.* 64:224–37
40. Lorio, P. L. 1966. *Phytophthora cinnamomi* and *Pythium* species associated with loblolly pine decline in Louisiana. *Plant Dis. Reptr.* 50:596–97
41. Mandels, M., Reese, E. T. 1965. Inhibition of cellulases. *Ann. Rev. Phytopathol.* 3:85–102
42. Marx, D. H. 1969. The influence of ectotrophic mycorrhizal fungi on the resistance of pine roots to pathogenic infections. I. Antagonism of mycorrhizal fungi to root pathogenic fungi and soil bacteria. *Phytopathology* 59:153–63
43. Marx, D. H. 1969. The influence of ectotrophic mycorrhizal fungi on the resistance of pine roots to pathogenic infections. II. Production, identification, and biological activity of antibiotics produced by *Leucopaxillus cerealis* var. *piceina. Phytopathology* 59:411–17
44. Marx, D. H., Bryan, W. C. 1969. *Scleroderma bovista,* an ectotrophic mycorrhizal fungus of pecan. *Phytopathology* 59:1128–32
45. Marx, D. H., Davey, C. B. 1969. The influence of ectotrophic mycorrhizal fungi on the resistance of pine roots to pathogenic infections. III. Resistance of aseptically formed mycorrhizae to infection by *Phytophthora cinnamomi. Phytopathology* 59:549–58
46. Marx, D. H., Davey, C. B. 1969. The influence of ectotrophic mycorrhizal fungi on the resistance of pine roots to pathogenic infections. IV. Resistance of naturally occurring mycorrhizae to infections by *Phytophthora cinnamomi. Phytopathology* 59:559–65
47. Marx, D. H. 1970. The influence of ectotrophic mycorrhizal fungi on the resistance of pine roots to pathogenic infections. V. Resistance of mycorrhizae to infection by vegetative mycelium of *Phytophthora cinnamomi. Phytopathology* 60:1472–73
48. Marx, D. H., Bryan, W. C., Grand, L. F. 1970. Colonization, isolation, and cultural descriptions of *Thelephora terrestris* and other ectomycorrhizal fungi of shortleaf pine seedlings grown in fumigated soil. *Can. J. Bot.* 48:207–11
49. Marx, D. H. 1972. Mycorrhizae and feeder root diseases. In *Physiology and Ecology of Ectotrophic Mycorrhizae.* New York: Academic. In press
50. Mathieson, J. 1947. Antibiotics from Victorian Basidiomycetes. *Austral. J. Exp. Biol. Med. Sci.* 24:57–62
51. Melin, E., Wiken, T., Oblom, K. 1947. Antibiotic agents in the substrates from cultures of the genus *Marasmius. Nature* 159:840–41
52. Melin, E. 1953. Physiology of mycorrhizal relations in plants. *Ann. Rev. Plant Physiol.* 4:325–46
53. Melin, E. 1959. Mycorrhiza. In *Handbuch der Pflanzenphysiologie,* ed. W. Ruhland 11:605–38. Berlin: Springer-Verlag
54. Morimoto, M., Iwai, M., Fukumoto, J. 1954. Antibiotic substances from mycorrhizal fungi. I. Isolation of antibiotic-producing strains. *Kagaku to Kogyo* 28:111–16
55. Napier, C. J. 1969. *The occurrence and seasonal variation of the fine roots and mycorrhizae of loblolly pine (Pinus taeda L.) in the West Bay area of Allen Parish, Louisiana.* Master's thesis. Northwestern State Coll. Louisiana, Natchitoches. 43 pp.
56. Neal, J. L., Jr., Bollen, W. B.,

Zak, B. 1964. Rhizosphere microflora associated with mycorrhizae of Douglas fir. *Can. J. Microbiol.* 10:259–65
57. Neal, J. L., Jr., Lu, K. C., Bollen, W. B., Trappe, J. M. 1968. A comparison of rhizosphere microfloras associated with mycorrhizae of red alder and Douglas fir. In *Biology of Alder,* eds. J. M. Trappe, J. F. Franklin, R. F. Tarrant, G. M. Hansen, 57–72. U.S. Dep. Agr. Forest Serv., Northwest Forest & Range Exp. Sta. 292 pp.
58. Nüesch, J. 1963. Defense reactions in orchid bulbs. In *Symbiotic Associations,* eds. P. S. Nutman, B. Mosse, 335–43. London: Cambridge Univ. Press. 356 pp.
59. Ohara, H., Hamada, M. 1967. Disappearance of bacteria from the zone of active mycorrhizas in *Tricholoma matsutake* (S. Ito et Imai) Singer. *Nature* 213:528–29
60. Oswald, E. T., Ferchau, H. A. 1968. Bacterial associations of coniferous mycorrhizae. *Plant Soil* 28:187–91
61. Park, J. Y. 1970. Antifungal effect of an ectotrophic mycorrhizal fungus, *Lactarius* sp., associated with basswood seedlings. *Can J. Microbiol.* 16:798–800
62. Peyronel, B., Fassi, B., Fontana, A., Trappe, J. M. 1969. Terminology of mycorrhizae. *Mycologia* 61:410–11
63. Powell, N. T., Meléndez, P. L., Batten, C. K. 1971. Disease complexes in tobacco involving *Meloidogyne incognita* and certain soil-borne fungi. *Phytopathology* 61:1332–37
64. Powell, W. M., Hendrix, F. F., Marx, D. H. 1968. Chemical control of feeder root necrosis of pecans caused by *Pythium* species and nematodes. *Plant Dis. Reptr.* 52:577–78
65. Pratt, B. H. 1971. Isolation of Basidiomycetes from Australian eucalypt forest and assessment of their antagonism to *Phytophthora cinnamomi. Trans. Brit. Mycol. Soc.* 56:243–50
66. Richard, C., Fortin, J. A., Fortin, A. 1972. Protective effect of an ectomycorrhizal fungus against the root pathogen *Mycelium radicus atrovirens. Can. J. Forest Res.* 1:246–51
67. Riffle, J. W. 1971. Effect of nematodes on root-inhabiting fungi. In *Mycorrhizae,* ed. E. Hacskaylo, 97–113. U.S. Dept. Agr. Misc. Publ. 1189. 255 pp.
68. Riffle, J. W. 1967. Effect of an *Aphelenchoides* species on the growth of a mycorrhizal and a pseudomycorrhizal fungus. *Phytopathology* 57:541–44
69. Riffle, J. W., Lucht, D. D. 1966. Root-knot nematode on ponderosa pine in New Mexico. *Plant Dis. Reptr.* 50:126
70. Robbins, W. J., Hervey, A., Davidson, R. W., Ma, R., Robbins, W. C. 1945. A survey of some wood-destroying and other fungi for anti-bacterial activity. *Torrey Bot. Club Bull.* 72:165–90
71. Ross, E. W., Marx, D. H. 1972. Susceptibility of sand pine to *Phytophthora cinnamomi. Phytopathology.* In press
72. Rovira, A. D. 1965. Plant root exudates and their influence upon soil microorganisms. In *Ecology of Soil-Borne Plant Pathogens.* eds. K. F. Baker, W. C. Snyder, 170–86. Berkeley: Univ. Calif. Press. 571 pp.
73. Ruehle, J. L. 1962. Histopathological studies of pine roots infected with lance and pine cystoid nematodes. *Phytopathology* 52:68–71
74. Ruehle, J. L., Marx, D. H. 1971. Parasitism of ectomycorrhizae of pine by lance nematode. *Forest Sci.* 17:31–4
75. Rypáček, V. 1960. Die gegenseitigen Beziehungen zwischen Mykorrhizapilzen und holzzerstorenden Pilzen. In *Mycorrhiza,* Int. Mykorrhiza Symp. Weimar, 233–40. Gustav Fischer Verlag, Jena 1963. 482 pp.
76. Santoro, T., Casida, L. E., Jr. 1959. Antibiotic production by mycorrhizal fungi. *Bact. Proc. 59th Gen. Meet. Soc. Am. Bacteriol.* p. 16. 137 pp.
77. Santoro, T., Casida, L. E., Jr. 1962. Elaboration of antibiotics by *Boletus luteus* and certain other mycorrhizal fungi. *Can. J. Microbiol.* 8:43–8
78. Šašek, V. 1967. The protective effect of mycorrhizal fungi on the host plant. *Proc. XIV Inter-*

nat. Union For. Res. Org. Congr. (Munich) Sect. 24:182–90

79. Šašek, V., Musilek, V. 1967. Cultivation and antibiotic activity of mycorrhizal Basidiomycetes. *Folia Microbiol.* 12:515–23
80. Šašek, V., Musilek, V. 1968. Two antibiotic compounds from mycorrhizal Basidiomycetes. *Folia Microbiol.* 13:43–5
81. Šašek, V., Musilek, V. 1968. Antibiotic activity of mycorrhizal Basidiomycetes and their relation to the host-plant parasites. *Cesk. Mykol.* 22:50–5
82. Sevilla-Santos, P., Encinas, C. J. 1964. The antibacterial activities of aqueous extracts from Philippine Basidiomycetes. *Philip. J. Sci.* 93:479–98
83. Slankis, V., Runeckles, V. C., Krotkov, G. 1964. Metabolites liberated by roots of white pine (*Pinus strobus* L.) seedlings. *Physiol. Plant.* 17:301–13
84. Smith, W. H. 1969. Release of organic materials from the roots of tree seedlings. *Forest Sci.* 15: 138–43
85. Sutherland, J. R., Fortin, J. A. 1968. Effect of the nematode *Aphelenchus avenae* on some ectotrophic mycorrhizal fungi and on a red pine mycorrhizal relationship. *Phytopathology* 58:519–23
86. Tomiyama, K. 1963. Physiology and biochemistry of disease resistance of plants. *Ann. Rev. Phytopathol.* 1:295–324
87. Trappe, J. M. 1962. Fungus associates of ectotrophic mycorrhizae. *Bot. Rev.* 28:538–606
88. Tribunskaya, A. J. 1955. Investigations of the microflora of the rhizosphere of pine seedlings. *Mikrobiologia* 24:188–92
89. Utech, N. M., Johnson, J. 1950. The inactivation of plant viruses by substances obtained from bacteria and fungi. *Phytopathology* 40:247–65
90. Vaartaja, O., Bumbieris, M. 1964. Abundance of *Pythium* species in nursery soils in South Australia. *Austral. J. Biol. Sci.* 17: 436–45
91. Vaartaja, O., Salisbury, P. J. 1965. Mutual effects in vitro of microorganisms isolated from tree seedlings. *Forest Sci.* 11:160–68
92. Wilcox, H. E. 1971. Morphology of ectendomycorrhizae in *Pinus resinosa*. In *Mycorrhizae*, ed. E. Hacskaylo, 54–68. U.S. Dept. Agr. Misc. Publ. 1189, 255 pp.
93. Wilkins, W. H. 1946. Investigations into the production of bacteriostatic substances by fungi. *Ann. Appl. Biol.* 33: 188–90
94. Wilkins, W. H., Harris, G. C. M. 1944. Investigations into the production of bacteriostatic substances by fungi. VI. Examination of the larger Basidiomycetes. *Ann. Appl. Biol.* 31:261–70
95. Wingfield, E. B. 1968. *Mycotrophy in loblolly pine. I. The role of Pisolithus tinctorius and Rhizoctonia solani in survival of seedlings. II. Mycorrhizal formation after fungicide treatment.* PhD diss. Virginia Polytech. Inst., Blacksburg. 78 pp.
96. Wright, J. M. 1956. The production of antibiotics in soil. III. Production of gliotoxin in wheatstraw buried in soil. *Ann. Appl. Biol.* 44:461–66
97. Wright, J. M. 1956. The production of antibiotics in soil. IV. Production of antibiotics in coats of seed sown in soil. *Ann. Appl. Biol.* 44:561–66
98. Zak, B. 1964. Role of mycorrhizae in root disease. *Ann. Rev. Phytopathol.* 2:377–92
99. Zak, B. 1967. A nematode (*Meloidodera* sp.) on Douglas-fir mycorrhiza. *Plant Dis. Reptr.* 51: 264
100. Zak, B. 1971. Characterization and classification of mycorrhizae of Douglas fir. II. *Pseudotsuga menziesii* + *Rhizopogon vinicolor*. *Can. J. Bot.* 49: 1079–84

PLANT PATHOGENS AS BIOCONTROLS OF AQUATIC WEEDS[1]

3559

F. W. Zettler and T. E. Freeman
Department of Plant Pathology, University of Florida, Gainesville

The papyrus reed when seen for the first time, or carved in stone upon some Egyptian monument, is a beautiful plant with delicate arching fronds making a hieratic pattern against the sky. But when it is multiplied to madness, hundreds of square miles of it spreading away like a green sea on every side, the effect is claustrophobic and sinister . . . (43).

Introduction

Ours differs from many previous articles in the *Annual Review of Phytopathology* for it heralds a topic new to our field rather than re-examines an old one. We concur with Wilson (86) who chose the term "commencement" to summarize his treatise on a similar topic. Plant pathologists are traditionally hired to confront and subdue microbes that would impair the productivity of our crop plants. However, we have largely overlooked those microbes that would be our allies in controlling noxious weed species. In our preoccupation with crop species, virtually all of which are terrestrial, we have also overlooked the many plants that exist in aqueous habitats. Our neglect is a matter of record. Not a single disease is listed in the *Index of Plant Diseases* (32) for our three most notorious waterweeds, water hyacinth (*Eichhornia crassipes*), Eurasian watermilfoil (*Myriophyllum spicatum*), and hydrilla (*Hydrilla verticillata*), currently pests of considerable economic significance. Thus, we have chosen to write a perspective rather than a review. It is our intention to: (*a*) explain the problems created by infestations of waterweeds, (*b*) enumerate the causes of these problems, and (*c*) consider the relationship and potential of plant pathology to the control of waterweeds.

The Problems Created by Waterweeds

As we use ever-increasing quantities of water, perhaps our most important natural resource, we find ourselves on a collision course with rapidly spreading infestations of waterweeds. Aquatic plants in reasonable numbers are not objectionable and are even valuable. Excessive populations, however, create havoc in our waterways. They clog the grids and sluices of hydroelec-

[1] This review was supported in part by Office of Water Resources Research Contract No. D1-14-31-0001-3268, Army Corps of Engineers Contract No. DACW 73-71-C-0002, and Florida Department of Natural Resources Contract No. 3.

tric and irrigation installations, and can render navigation all but impossible on badly infested bodies of water.

Recreational activity on weed-infested lakes is sorely curtailed. Weeds make swimming, boating, and fishing not only unpleasant but hazardous. Fishing is affected because gamefish are at a competitive disadvantage to "trash" species in waters heavily infested with weeds. In many instances, fish populations become exterminated altogether when dissolved oxygen levels are depleted through respiration and decomposition of senescing vegetation. Large accumulations of aquatic vegetation are aesthetically unpleasant, and decomposing masses along shorelines can create odoriferous nuisances. The end result, obviously, is a dramatic depression in values of waterfront properties. In Florida alone, this loss has been estimated to exceed 50 million dollars annually (33).

Excessive waterweed populations can also cause reservoirs and irrigation canals to lose water at disproportionate rates. Rather than conserving moisture by covering the water's surface, weeds such as water hyacinths can, through evapotranspiration, cause reservoirs to lose water at rates many times faster than on open water (31, 66). Timmons (80) reported a loss of 2.425 $\times 10^9$ cubic meters of water annually due to evapotranspiration of aquatic and ditchbank weeds in irrigation systems in 17 western states in the U.S. This loss was conservatively valued at $39,300,000.

Water weeds also compete with cultivated species in areas of the world where lowland rice and other types of subaquatic crops are grown. Paddies allowed to fallow may become so overrun with noxious aquatic vegetation that they must be abandoned (31, 66).

Perhaps most insidiously of all, waterweeds are havens for such dangerous vectors and alternate hosts of human pathogens as mosquitoes and snails. It is unsafe to live near waterbodies choked with weeds in areas of the world where malaria, encephalomyelitis, filariasis, and schistosomiasis occur (20, 31). Plants such as water lettuce (*Pistia stratioites*) further contribute to human misery by providing a clandestine source of air to the *Mansonia* mosquito, a vector of eastern encephalitis virus and rural filariasis. The larvae and pupae of all other mosquitoes must surface to obtain air and thus are subject to suffocation on water coated with oil films. The *Mansonia* mosquito is able to acquire oxygen without ever surfacing, by puncturing the roots of water lettuce. Controlling this mosquito is contingent upon weed control (83).

Aquatic plants occur throughout the world, and many weed species have become cosmopolitan. Although these plants are frequently able to extend their range latitudinally much more readily than terrestrial plants, due to the more constant edaphic conditions of the aquatic environment (66), the most frequent confrontations of man with waterweeds have been in the tropics and subtropics. Waterweeds grow most profusely in regions with long hours of sunshine and mild climates. Areas such as Central Florida and the Bayou region of Louisiana, with numerous shallow lakes and streams, are especially prone to waterweed problems. Waterweed infestations are especially serious in localities where the indigenous population is dependent upon its waterways for survival, but whose economy, technology, or political stability is such that they are unable to cope with the problem. For example, water hyacinth first appeared on the Congo River in 1952, and within 3 years had spread some 1000 miles from Leopoldville to Stanleyville. Their presence blocked the river and drastically reduced the fish populations, depriving the riverine inhabitants of their chief means of transportation as well as their primary source of protein. The Belgian government mustered ships and aircraft, and applied herbicides to these weeds to keep them under partial control. By 1957, the river could again be used. These control efforts unfortunately were interrupted during the tumultous years following the Congo's independence, and the hyacinth again reclaimed the river to bring additional suffering to the inhabitants (31, 36, 66).

The Causes of the Waterweed Problem

Man is chiefly to blame. He has acted as the chief disseminating agent of pestiferous aquatic plants. There is a tendency among water plants to reproduce vegetatively. In fact, many lack the capacity to produce seeds, and therefore may be without a means of aerial transportation over long distances (66). Without man's help, most noxious species would thus be restricted to the continents from which they originated, and sometimes even to finite bodies of water. Man has unwittingly introduced many weeds. Alligatorweed (*Alternanthera philoxeroides*) was unknown in the United States until about 1894 when it arrived from South America as a stowaway in ballast of ships (82, 84). Away from its natural enemies and finding its new habitat favorable, this pest soon became established throughout the southeast, particularly in Louisiana and South Carolina.

Many aquatic plants have been introduced to new habitats deliberately. In his quest for beauty, man has imported aquatic plants from around the world, cultivated them, and carelessly allowed them to escape. The most famous of all aquatic pests imported under the guise of an ornamental is the water hyacinth, introduced into the United States in 1884 when specimens of this plant were distributed to those attending the New Orleans Cotton Centennial Exposition. These plants originally came from Venezuela, and were much admired for their lavender blooms and exotic foliage. Soon they were to be found in garden pools and in farm ponds where they multiplied rapidly. Excess plants were simply discarded in nearby waterways. Water hyacinths were reported in Florida by 1890, and shortly after the turn of the century were found as far north as Virginia and as far west as California (52, 66). Interestingly, water hyacinths are said to have been transported to south Florida in the late 1890s by a cattleman who had the notion that the plant would make nutritious yet inexpensive cattlefood. Unfortunately the plants, though edible, proved low in nutrient value and could not be used for fodder. However, 50 years later they were costing the state 10 million dollars a year for eradication programs (85).

Water pollution is one of the chief reasons we are having major difficulties with aquatic vegetation. Bodies of water age, through natural processes, from oligotrophy to eutrophy. As sediments accumulate, lakes become filled and are eventually transformed into bogs (38). Man has accelerated this process considerably by overnutrifying waterways with human, industrial, and agricultural wastes (26). Florida's 30,000-acre Lake Apopka is an extreme example. Until 1940, this lake had clear water and was nationally famous for its game fishing. Then—encouraged by fertilizers leaching from bordering citrus and vegetable farms, wastes from municipalities, and citrus processing plants—populations of water hyacinth, followed by algae, abounded. By 1965, this once pristine lake had been reduced to a hypereutrophic, sediment-filled body of water almost devoid of gamefish (9, 10, 69).

Man has further compounded the waterweed problem by redesigning nature's waterways. By constructing dams, he thwarts the periodic expelling of excess weed populations seaward during times of heavy rainfall, and thereby he provides placid havens for the proliferation of aquatic vegetation. Man-made lakes throughout the world are infested with noxious water plants. For example, the Tennessee Valley Authority's lakes are severely infested with Eurasian watermilfoil (74), Ghana's Lake Volta is covered with water lettuce, and Nicaragua's Lake Apanas has a severe water hyacinth problem (31, 35, 66). Man-made canals for transportation, and ditches for irrigation and drainage interconnect isolated bodies of water and thus aid the spread of aggressive weed species. The continuity of England's inland waterways enabled Canada elodea (*Elodea canadensis*) to become firmly established throughout that country during the 1880s (66). The interconnected waterways of Florida enabled hydrilla to become established throughout the state within ten years after it was introduced near Miami.

The Status and Potential of Plant Pathology in Solving the Waterweed Problem

Attempts to control aquatic weeds include: (*a*) herbicidal applications, (*b*) removal and disposal with mechanical devices, and (*c*) biological control. Although the first two methods have considerable merit, they alone do not satisfactorily solve the overall aquatic weed problem because of expense and need for continuous treatments. Moreover, these two methods tend to be nonselective in their action. In the case of herbicide applications, the added pollution from their use detracts from the acceptability of this means of control.

Biological control methods may offer the greatest prospects for success by imposing a continual controlling force directly upon the pestiferous plants. Since many of our waterweeds are introduced species, it is logical to expect that searches in the native habitats of these plants would reveal numerous candidates that could be considered as biocontrols. Although several agents have already been evaluated as controls, none presently promises to solve the aquatic weed problem.

Certain snail species such as *Marisa cornuarietis* are promising, but could become pests themselves as they may also devour beneficial plants (6, 31, 67). Other snail species potentially useful as biological controls may be carriers of serious human and animal parasites (20). Herbivorous fish, particularly the white amur (*Ctenopharyngodon idella*), also offer some potential for the control of unwanted plants (6, 31). However, serious problems may result from the introduction of objectionable piscine forms such as the tilapia (*Tilapia melanotheron*). Moreover, there are relatively few herbivorous fish species from which to choose. The manatee (*Trichechus manatus*), although much publicized, offers little hope as it is difficult to breed and is close to extinction. One insect species, the alligatorweed flea beetle (*Agasicles hygrophila*) feeds only on alligatorweed and shows considerable promise for the control of this particular aquatic plant (40, 68). However, insects alone are not likely to control aquatic weed pests because there are relatively few phytophagous species capable of living beneath water. Most aquatic insects are either carnivorous or detrivourous; consequently, the number of insect species with potential to control submersed aquatic weeds is relatively limited.

Several authorities on waterweeds have specifically commented upon the lack of attention given to plant pathogens as biocontrols of aquatic weeds (6, 31, 66). The vast numbers of disease organisms [McNew (42) estimated that there are over 100,000 plant diseases] seemingly offer untapped reservoirs of potential controls for these plants. Advantages of using plant pathogens to control waterweeds would be: (*a*) control applications would presumably require minimal technology and, if successfully established, the pathogen in theory would be self-maintaining, (*b*) the overwhelming number of different plant-pathogenic species from which to choose offers an unmatched versatility in selecting a specific biological control, (*c*) virtually none can attack man or his animals, therefore providing an important advantage over the use of various animals such as snails, which may harbor chordate pathogens, (*d*) plant pathogens, although often killing individuals in a given population, would not be expected to cause the extermination of a species. This attribute is important when considering that the total eradication of one aquatic weed species, such as the water hyacinth, is likely to create an ecological void that in turn may allow a population explosion of a different, more serious species such as hydrilla.

The use of plant pathogens is not without its hazards. Any study undertaken to introduce or test plant pathogens in infested areas must be done with extreme care. The spectacular decline of eelgrass (*Zostera marina*) along the northeastern coast of the United States and in European coastal areas in the 1930s (59, 81) graphically illustrates the potential for destruction that diseases present to plant communities in an aquatic environment. If such a disaster can befall a plant as beneficial as this, we must assume the possibility of the occurrence of a similar event on noxious aquatic plants. This latter event would be of great benefit to man and the possibility of its artificial induction should be seriously considered.

Emersed aquatic plants are probably no less susceptible to plant pathogens than terrestrial plants. In fact, some aquatic plants may have pathogens in common with terrestrial relatives. Numerous viruses, for example, are known to infect the amaranthaceous *Gomphrena globosa* (79), and presumably many of them will be capable of also infecting the related alligatorweed. However, most aquatics are taxonomically unique, having few, if any, terrestrial relatives. Despite their ubiquity throughout the earth's waterways, aquatic plants account for no more than 1% of the known species of angiosperms and 2% of the pteridophytes. Of the 33 families listed by Sculthorpe (66) as consisting more or less exclusively of hydrophytes, 30 have fewer than 10 genera, 17 of these are monogeneric, and 3 are monotypic. Only two families have more than 200 species. Thus, host-specific pathogens such as the rusts and smuts, although perhaps more difficult to locate on these plants, may be ideal as biocontrols since they would not be expected to infect non-target plant species.

Plant pathogens are certainly known to occur in aqueous situations. Nematodes are dependent upon water for their locomotion and survival, and numerous species are to be found in fresh, brackish, or salt waters. Hirschmann (29) cites several reports of *Radopholus gracilis* collected from the roots of such aquatic plants as *Potamogeton, Carex,* and *Phragmites.* That nematodes inflict serious damage to submersed aquatic plants was shown by Smart & Esser (73), who reported that *Aphelenchoides fragariae* inflicted serious damage to *Cabomba, Limnophila,* and other aquatic ornamentals.

Bacteria and fungi are often found in water. Species in the genus *Pseudomonas* are commonly encountered as water inhabitants (8). Among the fungi, Myxomycetes, Ascomycetes, Basidiomycetes, and Fungi Imperfecti all have some aquatic species, and the Phycomycetes have numerous aquatic forms. Sparrow (76) lists the following phycomycete orders as being aquatic: Chytridiales, Blastocladiales, Monoblepharidales, Hypochytridiales, Plasmodiophorales, Saprolegniales, Leptomitales, Lagenidiales, and the pythiaceous Peronosporales. Such zoospore-producing organisms certainly are perfectly adapted to infect submersed plants. Ridings & Zettler (60) implicated a species of *Aphanomyces* as the causal agent of a lethal disease of submersed amazon sword plants (*Echinodorus sp.*) at an aquatic nursery in Florida.

Viruses might be expected to be perpetuated indefinitely in many waterweeds inasmuch as the capacity to produce seed is very much reduced, if not lost, in most vascular aquatics (66); virus-free plants would hence not be forthcoming from this source once plants become infected. Virus vectors can be expected to occur in aquatic environments. The aquatic chytrids, notably *Olpidium brassicae,* are established vectors of such viruses as tobacco necrosis and lettuce big vein (25). Similarly, dorylaim nematodes, species of which are vectors of nematode-transmitted polyhedral-particle viruses (NEPO) and nematode-transmitted tubular-particle viruses (NETU) (11, 25a) are common inhibitors of waterways. Arthropod vectors of viruses, though unlikely to be found beneath the water's surface, could feed and transmit viruses to emergent plant parts. Various aphids have been reported to colonize water plants. *Rhopalosiphum nymphaeae,* a vector of several viruses (34), has been collected from a large variety of aquatic plants including *Marsilea, Potomogeton, Sagittaria, Scirpus, Pistia, Eichhornia, Nuphar, Ceratophyllum, Myriophyllum, Utricularia* (51). Other groups of virus vectors have also been collected from emersed parts of water plants. Silveira-Guido (70) has collected two leafhopper species and an undetermined eriophyid mite from water hyacinths in Uruguay. MacClement & Richards (39a) reported recovering viruses from several aquatic plants (*Lemna minor, Potamogeton crispus, P. pectinatus, Ceratophyllum, Nymphaea*) growing wild at the Royal Botanical Gardens of Hamilton, Ontario.

The Pestiferous Aquatic Plants and Their Diseases

Algae, certain pteridophytes, and various monocotyledonous and dicotyledonous angiosperms all have representatives that have become pestiferous as waterweeds.

Algae.—The most significant algal pests are to be found among the Cyanophyta, Chlorophyta, Charophyta, Euglenophyta, and Chrysophyta. Populations of algae can create unsightly and odoriferous scums on water surfaces and interfere with water clarity. Certain pestiferous charophytes such as *Chara* spp. are macroscopic and can impede water flow. Infestations of other algae typically occur as cyclic "blooms" that tend to materialize within relatively brief periods of time as a result of sudden infusions of nutrients. Although algae and higher plants coexist under normal conditions, population explosions of one tend to occur at the expense of the other, due to competition for nutrients and light. The competition was demonstrated by Hasler & Jones (27), who showed that algae did not develop as well in ponds containing large populations of *Elodea canadensis* and *Potamogeton foliosus* as in identical ponds without these vascular hydrophytes. Conversely, algae can suppress the development of vascular plants.

Plant pathogens infect algae as they do higher plants, but only rarely have they been considered in controlling algal blooms. Various workers have shown blue-green algae to be susceptible to lysogenic viruses closely resembling those affecting bacteria (64), and several have been studied in the United States (62, 64), India (72), Israel (49), Scotland (18), and the Soviet Union (24, 45). Safferman & Morris (63), Daft, Begg & Stewart (18), and Cannon, Shane & Bush (12), have suggested that under natural conditions some cyanophytes that seldom form blooms are prevented from doing so by being continually checked by high populations of viruses, or cyanophages. Much less appears to be known about bacterial pathogens of algae, although Stewart & Brown (77) reported that a species of *Cytophaga* lysed certain blue-green and green algae. Nematodes might be considered potential biocontrols of algae, for numerous marine and fresh water species feed on them. *Dorylaimus ettersbergensis,* was observed by Hollis (30) to consume cells of green and blue-green algae. Aquatic fungi also have potential for controlling algae. Sparrow and others (13–15, 21, 76) list various algae that are hosts of aquatic phycomycetes; among them are species in 13 genera of Cyanophyceae, 67 Chlorophyceae, 4 Characeae, 7 Xanthophyceae, 1 Eugenophyceae, and 30 Bacillariophyceae. Among the phycomycete genera with species infecting algae are *Olpidium,* infecting species in over 25 algal genera (76) and *Aphanomyces,* pathogenic to species of *Mougeotia, Nitella, Spirogyra, Vaucheria,* and *Zygnema* (65).

Vascular Aquatic Weeds.—Whereas algae are principally aquatic forms of life, and have been so since Precambrian times, the progenitors of today's vascular aquatics are descendents of terrestrial plants (66). Water plants are by no means a homogeneous assemblage, as the transition from a terrestrial to an aquatic existence was made repeatedly through time by many different plant groups. Some, like the Isoetaceae and Nymphaeaceae, represent lines that made this transition relatively long ago; others apparently have become aquatic much more recently and still closely resemble their relatives on land.

Vascular aquatic plants can be categorized as either emergent or submergent, with the former the most conspicuous but not necessarily the most troublesome. Emergents may be subdivided into free-floating forms that drift about over the water surface, and anchored emergents attached to the substrate by their roots.

Free-floating plants are raft-like with buoyant foliage and submersed pendent roots. They establish themselves uniformly over waterways and can readily adjust to fluctuations in water levels, but are vulnerable to the caprices of winds and currents and hence are generally restricted to sheltered habitats. They multiply with great rapidity and soon cover the surface of the water, rendering the waterbody meadow-like in appearance.

The most significant of all free-floating plants as weeds are the water hyacinth, water lettuce, and salvinia (*Salvinia auriculata*), all of which are now pantropical. The large stoloniferous forms such as the water hyacinth are generally considered to be of greater significance than diminutive forms such as salvinia.

Water hyacinth is infamous for its prodigious growth rate. In one study, it was calculated that 10 individuals were capable of giving rise to 655,360 plants in a single 8-month growing season (52). This plant is an indigene of Latin America but is now to be found throughout the tropics and subtropics.

Apparently the first disease recorded on water hyacinth was a rust, *Uredo eichhorniae,* reported from the Dominican Republic by Ciferri & Fragoso (17) in 1927. The following year Ciferri (16) reported the occurrence of the smut *Doassansia eichhorniae* from the same area. Neither of these diseases has been studied as biological-control agents.

In 1932, a species of *Fusarium* was reported on water hyacinths from India by Agharkar & Banerjee (1). The fungus induced reddish brown spots on the petioles followed by chlorosis and withering of affected leaves. Interestingly, even at this early date these authors considered utilization of the disease for biological control, as evidenced by their final conclusion: "The infection takes place readily but owing to the high resisting power of the plant, the disease makes very slow progress. From this it may be inferred that this fungus cannot be regarded as a possible remedy against the spread of water hyacinths." The causal agent of this disease was later identified as *Fusarium equiseti* by Banerjee (2). Snyder & Hansen (75) have reduced this species to synonymy with *F. roseum*. This latter species has been found by Rintz & Freeman (61) affecting water hyacinth in Florida.

Recently, a concerted research program on biological control of aquatic weeds was begun at the Commonwealth Institute of Biological Control, Indian Station, in Bangalore. They have investigated the diseases of various aquatic plants in addition to water hyacinth. In addition to the search for new diseases, they have considered the biological-control potential of some previously reported diseases. According to Nag Raj (46), *Cercospora piaropi*, first reported by Thirumalachar & Govindu (78), causes negligible damage and appears of little value in reducing the vigor of hyacinth populations. However, *Cephalosporium eichhorniae* described by Padwick (50) may be of some value. Freeman, Rintz & Zettler (unpublished) have noted a similar leaf-spot disease damaging water hyacinth in Trinidad, Puerto Rico, El Salvador, Louisiana, and Florida, but its relation to that described by Padwick remains to be determined.

Nag Raj & Ponnappa (47) reported in 1967 the occurrence of the *Rhizoctonia* stage of *Corticium solani* on water hyacinth. This presumably is the same *Rhizoctonia solani*-induced blight that had previously been reported by Padwick (50). A closely related fungus, *Hypochnus sasakii*, from rice was found to affect water hyacinth in Taiwan (41) as early as 1933. More recently, Freeman & Zettler (22) reported the isolation of a strain of *R. solani* from anchoring hyacinths (*Eichhornia azurea*) in Panama that can severely affect and kill water hyacinths. The organism produces abundant sclerotia which will survive submersed without loss of virulence for at least 9 months (22). Nag Raj (46) considered *R. solani* to have little use for biological control due to its broad host range, although introducing this pathogen into an aquatic environment would not necessarily increase the already present inoculum in soils around crop plants. The real damage may well be its effect on beneficial aquatic plants. In 1928 Bourn & Jenkins (7) attributed the destruction of large areas (total of about 300 square miles) of aquatic food plants for ducks in Virginia and North Carolina to a physiological strain of *R. solani*. Species of plants affected were *Potamogeton pectinatus, P. perfoliatus, Ruppia maritima, Vallisneria spiralis,* and *Najas flexilis.*

Additional pathogens recorded on water hyacinth by the Indian group include *Myrothecium roridum* var. *eichhorniae* (55), *Marasmiellus inoderma*

(46), *Alternaria eichhorniae* (48), *Helminthosporium bicolor* (58), and *Curvularia clavata* (58). Of these, *M. roridum* var. *eichhorniae* and *A. eichhorniae* appear the most promising for use in biological control. However, Ponnappa (55) believes that the wide host range of *M. roridum* precludes its use. Nag Raj & Ponnappa (48) consider that the narrow host range of *A. eichhorniae,* coupled with the ability of the pathogen to produce a toxin, warrants biological control trials with it.

Although widespread, water lettuce does not rank with water hyacinth as an impediment of waterbodies. It is a relatively fragile plant, prone to damage by natural forces, and hence is most commonly found on relatively placid waterbodies (31, 84). In large exposed waterways such as Guatemala's Lake Izabal, this weed is destroyed by wave action despite continued infusions of fresh plants from nearby tributaries (28). The main hazard from water lettuce is that it harbors the *Mansonia* mosquito which, as noted earlier, is a vector of human diseases.

Water lettuce is affected by *Cercospora* sp. (47), *Sclerotium rolfsii* (47), and *Phyllosticta stratiotes* (56) in India. However, the usefulness of these pathogens for biological control has not been explored. Recently a virus reputedly transmitted by *Rhopalosiphum nymphaeae* has been reported from Africa (53). However, dasheen mosaic virus, an aphid-transmitted virus of several aroids (87), including the aquatic ornamental *Cryptocoryne cordata,* did not infect water lettuce seedlings in Florida (Hartman & Zettler, unpublished).

Salvinia is a diminutive free-floating pteridophyte with pubescent oval leaves about 1 cm long. This species is a native of the neotropics but has become of considerable significance in several areas of the paleotropics, particularly in Ceylon and in Africa's Lake Kariba (31, 66).

Salvinia is affected by *Myrothecium roridum* in India (58). Presumably the use of this pathogen for salvinia control would be objectionable on the same grounds as water hyacinth, i.e., broad host range of the fungus. Also, a cyclic die-back of salvinia associated with species of *Alternaria* and *Spicariopsis* was reported in Africa's Lake Kariba by Loveless (39).

Anchored emergents are normally firmly rooted to the substrate and are thus more limited in habitat than their free-floating counterparts. These plants tend to be restricted to relatively shallow bodies of water, ditches, or along shorelines. However, when they grow profusely, their roots can become tightly interwoven into mats that can float as self-supporting islands, or sudds.

The anchored emergents are an arbitrary assemblage composed of several different taxa, among which are species of Amaranthaceae, Cyperaceae, Gramineae, Polygonaceae, and Typhaceae. These plants are distributed throughout the world and are conspicuous features of the aquatic environment. Sawgrass (*Cladium jamaicensis*), for example, is the dominant plant of the Florida Everglades (84). Although frequently beneficial, they are considered to be pests when they impair navigation, hinder hydroelectric projects, or interfere with fishing or agriculture.

Alligatorweed merits special attention as an anchored emergent. Native to South America, it can now be found in tropical and warm-temperate locales throughout the Western Hemisphere, and in certain areas in the Eastern Hemisphere. It has remarkable versatility, being able to grow equally well in a mat over open water, buoyed by hollow stems, or as a terrestrial plant rooted in soil in a relatively dry field. In Louisiana, this plant, although a weed to most people, is favored by cattlemen as convenient fodder in pastures (40).

Alligatorweed is subject to several diseases, none of which appear to have been investigated as control for this plant. It has been reported to be affected in Louisiana by *R. solani* (19, 71), *Heterodera marioni* (54), and *Anguillulina dihystera* (54). In addition, alligatorweed plants affected by a stunting disease, believed to be virus induced, have been found in the Ortega River of Florida (Hill & Zettler, unpublished). *Alternanthera sessilis,* a near relative of alligatorweed, is affected by *Corticium solani* (47), *Colletotrichum capsici* (56), *Glomerella cingulata* (47, 57), *Phoma* sp. (56), and *Albugo bliti* (57) in India. Goodey, Franklin & Hooper (23) list *Pratylenchus coffeae, Meloidogyne incognita,* and *M. javanica* as infecting several additional species of *Alternanthera.* In addition, Arthur (1a) reported a rust *Uredo nitidula,* infecting alligatorweed plants in Guatemala.

Diseases have also been reported on other anchored emergents. Two of three *Panicum* species of most concern in the United States [maidencane (*P. hemitomum*) and paragrass (*P. purpurescens*)] have had 9 and 7 pathogens reported to attack them, respectively. Eighteen diseases have been reported for the common reed (*Phragmites communis*), three for southern wild rice (*Zizoniopsis miliaceae*), and more than twenty for the grass-like cattails (*Typha* spp.) (32). Other species have not been as thoroughly investigated; no diseases are reported for such conspicuous and important species as torpedograss (*Panicum repens*), water paspalum (*Paspalum fluitans*), and sawgrass (32).

Despite the presence of several pathogens affecting anchored emergents, their use for biological control presents some unique problems. Indeed, in this case biological control may not be feasible because this group of plants is not totally noxious. Certainly we could ill afford to risk the destruction of important waterfowl food plants such as maidencane, southern wild rice, and giant reed. Of no less importance is the use of paragrass as forage in warmer climates and *Typha* spp. as valued ornamentals in aquatic gardens. Thus, it appears that biological control of such anchored emergents as the aquatic grasses may require a degree of specificity in phytopathogens difficult to attain.

Submersed weeds are probably the most serious of all types of aquatic vegetation, and the most difficult to control because, being submersed, they cannot be readily sprayed with herbicides nor can they be easily removed with machines (31). Furthermore, they are immune to predation by many organisms unable to exist under water. The most noxious species have weak fibrous stems incapable of self support, and, except for their flowers, are unable to survive for even brief periods out of water. Although roots are formed, they are of minimal significance as anchoring devices. These plants grow indeterminately and as the stems elongate, they branch in every direction to create an impenetrable labyrinth of green strands capable of converting an unobstructed body of water into a virtual sargasso sea.

Numerous highly specialized species of submersed aquatic plants are regionally notorious for their ability to invade new sites rapidly. Most often cited water pests are as follows: *Ceratophyllum* (Ceratophyllaceae), *Myriophyllum* (Haloragaceae), *Utricularia* (Lentibulariaceae), *Najas* (Najadaceae), *Cabomba* (Nymphaeaceae), *Potamogeton* (Potamogetonaceae), and *Anacharis, Egeria, Elodea,* and *Hydrilla* (Hydrocharitaceae) (31, 66, 84). Because of their beauty, various members of this group have been transported throughout the world as ornamentals, and in many instances have become established in new locales by aquarium plant dealers who introduced them into public waters to be harvested as needed.

Hydrilla, an old-world native introduced into south Florida in 1958–1960, currently ranks second only to water hyacinth as an aquatic pest in that state (5, 44). The rapid spread of hydrilla in Florida is reminiscent of the spread of its new-world relative, Canada elodea, in Europe.

Eurasian watermilfoil is another equally widespread weed in fresh and brackish waters. In the United States, this old-world native has become a nuisance of particular prominence within the last 10-20 years, infesting thousands of acres in the Chesapeake Bay, Tennessee Valley, and Currituck Sound.

In comparison to the diseases reported on free-floating and emersed plants, there is a paucity of reports concerning diseases of submerged plants. This is probably due to lack of investigation rather than absence of diseases affecting these plants. Two disorders, "Northeast Disease" and "Lake Venice Disease" (3, 4), were considered to be causes of a sudden decline in distribution and abundance of Eurasian watermilfoil populations in the Chesapeake Bay in the mid 1960s. No causal agents were ever established for them. Also, milfoil plants did not become infected when inoculated with alfalfa mosaic virus, tobacco mosaic virus, tobacco ringspot virus, potato virus X, or potato virus Y (3). For several years, personnel of the Institute for Plant Protection in Beograd, Yugoslavia, have investigated diseases affecting milfoil under a project supported by PL 480 funds. This group has isolated a variety of fungi from declining milfoil plants and several of them have been reported to be pathogenic to milfoil seedlings. Pathogens reported are: *Alternaria* sp., *Articulospora tetracladia, Botyris* sp., *Dactylella microaquatica, Flagellaspora stricta, Fusarium acuminatum, F. oxysporium, F. poae, F. roseum, F. sporotrichoides, F. tricinctum, Mycelia sterilia, Sclerotium hydrophyllum, and Stemphylium* sp. (37). The nematode, *Ditylenchus dipsachi tobaensis,* has been found on *Myriophyllum verticillatum* (23). Whether any of these can be used on a practical scale for milfoil control remains to be determined.

We have been unable to find reports of diseases affecting species of *Hydrilla, Egeria, Anacharis,* or *Elodea,* although we believe that diseases do affect these hydrocharitaceous plants.

CONCLUSIONS

The plant pathologist may be guilty of tunnel vision by directing most of his research efforts towards terrestrial plants and ignoring the aquatics. Our almost nonexistent research efforts with aquatic plants, particularly the submersed forms, certainly do not reflect their ubiquity and their importance as noxious weeds, food for wildlife, and ornamentals. The lack of information on diseases of aquatic plants is obviously not related to the nonexistence of pathogens. When investigations have been undertaken, pathogens have been found, and in some instances shown to inflict great damage to their hosts. That we have ignored diseases of water plants for so long is surprising. Vascular aquatic plants, having evolved from terrestrial ancestors, adapted themselves in amazing ways to survive in water. It would be intriguing to determine how their pathogens have become adapted for such an existence. Aside from simple curiosity, it may be that our discipline holds the most important key in controlling water weeds. This is reason enough for conducting research in this long-neglected field.

LITERATURE CITED

1. Agharkar, S. P., Banerjee, S. N. 1932. *Fusarium* sp. causing disease of *Eichhornia crassipes.* Solms. *Proc. Indian Sci. Congr.* 19:298

1a. Arthur, J. C. 1920. New species of Uredineae XII. *Torrey Bot. Club. Bull.* 47:465–80

2. Banerjee, S. N. 1942. *Fusarium equiseti* (Cda.) Sacc. (*Fusarium falcatum* App. et Wr.) causing a leaf spot of *Eichhornia crassipes*. *J. Dep. Sci. Calcutta Univ.* 1: 29–37

3. Bayley, S. E. M. 1970. *The ecology and disease of Eurasian water milfoil (Myriophyllum spicatum L.) in the Chesapeake Bay.* Ph.D. thesis, Johns Hopkins Univ., Baltimore, Maryland. 190 pp.

4. Bayley, S. E. M., Rabin, H., Southwick, C. H. 1968. Recent decline in the distribution and abundance of Eurasian milfoil in Chesapeake Bay. *Chesapeake Sci.* 9: 173–81

5. Blackburn, R. D., Weldon, L. W., Yeo, R. R., Taylor, T. M. 1969. Identification and distribution of certain similar appearing aquatic weeds in Florida. *Hyacinth Contr. J.* 8:17–21

6. Blackburn R. D., Sutton, D. L., Taylor, T. 1971. Biological control of aquatic weeds. *J. Irrigation Drainage Div. Proc. Am. Soc. Civil Eng.* 97:421–32

7. Bourn, W. S., Jenkins, B. 1928. *Rhizoctonia* disease on certain aquatic plants. *Bot. Gaz.* 85: 413–26

8. Breed, R. S., Murray, E. G. D., Smith, N. R. 1957. *Bergeys Manual for Determinative Bacteriology.* Baltimore: Williams & Wilkins. 1094 pp.

9. Brezonik, P. L. 1969. Eutrophication in Florida lakes. *Proc. Florida Environ. Eng. Conf. Water Pollution Control.* Gainesville: *Florida Eng. and Ind. Exp. Sta.* 24, *Bull. Ser.* 135:124–29

10. Brezonik, P. L., Morgan, W. H., Shannon, E. E., Putnam, H. D. 1969. Eutrophication factors in north central Florida lakes. *Flor-*

ida Water Resour. Res. Center Publ. 5. Gainesville Florida Eng. and Ind. Exp. Sta. 23, *Bull. Ser.* 134. 101 pp.

11. Cadman, C. H. 1963. Biology of soil-borne viruses. *Ann. Rev. Phytopathol.* 1:143–72
12. Cannon, R. E., Shane, M. S., Bush, V. N. 1971. Lysogeny of a blue-green alga, *Plectonema boryanum. Virology* 45:149–53
13. Canter, H. M. 1950. Fungal parasites of the phytoplankton. I. *Ann. Bot. London* 14:263–89
14. Canter, H. M. 1951. Fungal parasites of the phytoplankton. II. *Ann. Bot. London* 15:129–56
15. Canter, H. M., Lund, J. W. G. 1948. Studies on plankton parasites. I. Fluctuations in the numbers of *Asterionella formosa* Hass. in relation to fungal epidemics. *New Phytol.* 47:238–61
16. Ciferri, R. 1928. Quarta contribuzione allo studio degli Ustilaginales. *Ann. Mycol.* 26:1–68
17. Ciferri, R., Fragoso, R. G. 1927. Hongos parasitos y saprofitos de la Republica Dominicana. *Soc. Espan. Host. Natur.*, Madrid, *Bol.* 27:68–81
18. Daft, M. J., Begg, J., Stewart, W. P. D. 1970. A virus of blue-green algae from fresh-water habitats in Scotland. *New Phytol.* 69: 1029–38
19. Exner, B., Chilton, S. J. P. 1943. Cultural differences among single basidiospore isolates of *Rhizoctonia solani. Phytopathology* 33: 171–74
20. Ferguson, F. F. 1968. Aquatic weeds and man's well-being. *Hyacinth Contr. J.* 7:7–11
21. Fott, B. 1967. *Phycidium scendesmi* spec. nova., a new chytrid destroying mass cultures of algae. *Z. Allg. Microbiol.* 7:97–102
22. Freeman, T. E., Zettler, F. W. 1971. Rhizoctonia blight of water hyacinth. *Phytopathology* 61:892
23. Goodey, J. B., Franklin, M. T., Hooper, D. J. 1965. *T. Goodey's the Nematode Parasites of Plants catalogued under their Hosts.* Commonwealth Agr. Bureaux, Farnham Royal, Bucks, England. 214 pp.
24. Goryushin, V. A., Chaplinskaya, S. M. 1966. Existence of viruses of blue-green algae. *Mikrobiol. Zh. Akad. Nauk Ukr. RSR* 28: 94–7 (English summary)
25. Grogan, R. G., Campbell, R. N. 1966. Fungi as vectors and hosts of viruses. *Ann. Rev. Phytopathol.* 4:29–52

25a. Harrison, B. D. 1964. The transmission of plant viruses in soil. In *Plant Virology,* eds. M. K. Corbett, H. D. Sisler, 118–147. Gainesville: Univ. Florida Press 527 pp

26. Hasler, A. D. 1947. Eutrophication of lakes by domestic drainage. *Ecology* 28:383–95
27. Hasler, A. D., Jones, E. 1949. Demonstration of the antagonistic action of large aquatic plants on algae and rotifers. *Ecology* 30:359–64
28. Hill, H. R., Rintz, R. E. 1972. Observations of declining water lettuce populations in Lake Izabal, Guatemala. *Proc. Southern Weed Sci. Soc.* In press
29. Hirschmann, H. 1955. *Radopholus gracilis* (DeMan, 1880) n. comb. (Synonym-*Tylenchorhynchus gracilis* (DeMan, 1880) Filipjev 1936). *Proc. Helminthological Soc. Wash.* 22:57–63
30. Hollis, J. P. 1957. Cultural studies with *Dorylaimus ettersbergensis. Phytopathology* 47:468–73
31. Holm, L. G. Weldon, L. W., Blackburn, R. D. 1969. Aquatic weeds. *Science* 166:699–709
32. *Index of plant diseases in the United States.* 1960. U.S. Dep. Agr. Handb. 165. 531 pp.
33. Ingalsbe, G. 1969. *Water, Weeds, Trees and Turf* 8(10):8–9
34. Kennedy, J. S., Day, M. F., Eastop, V. F. 1962. *A Conspectus of Aphids as Vectors of Plant Viruses.* London: Commonwealth Inst. Entomol. 114 pp.
35. Lagler, K. F. 1969. *Man-made Lakes: Planning and Development.* Rome: Food & Agr. Organ. UN. 71 pp.
36. Lebrun, J. 1959. La lutte contre le developpement de l'*Eichhornia crassipes. Bull. Agr. Congo Belge* 50:251–52
37. Lekic, M. 1971. *Ann. Rep. Inst. Plant. Prot.*, Beograd. 25 pp.
38. Lindeman, R. L. 1942. The trophic-dynamic aspect of ecology. *Ecology* 23:399–418
39. Loveless, A. R. 1969. The possible

role of pathogenic fungi in local degeneration of *Salvinia auriculata* Aublet on Lake Kariba. *Ann. Appl. Biol.* 63:61–69

39a. MacClement, W. D., Richards, M. G. 1956. Virus in wild plants. *Can. J. Bot.* 34:793–99

40. Maddox, D. M., Andres, L. A., Hennessey, R. D., Blackburn, R. D., Spencer, N. R. 1971. Insects to control alligatorweed. *Bioscience* 21:985–91

41. Matsumoto, T., Yamamoto, W., Hirane, S. 1933. Physiology and parasitism of the fungi generally referred to as *Hypochnus sasakii.* II. Temperature and humidity relations. *J. Soc. Trop. Agr., Taiwan* 5:332–45

42. McNew, G. L. 1966. The nature and cause of disease in plants. *Am. Biol. Teacher* 28:445–61

43. Moorehead, A. 1960. *The White Nile.* London: Hamish Hamilton.

44. Morris, A. 1970. Botanist sees hydrilla as Florida's next big water weed problem. *Press release of January 21, Univ. Florida, Inst. Food & Agr. Sci., Gainesville*

45. Moskovets, S. M. 1969. Utilization of viruses in the fight against pests and causative agents of diseases of agricultural cultures. *Isvestia Akad. Nauk USSR, Seriia Biologi-Cheskaia* 6:875–79 (English summary)

46. Nag Raj, T. R. 1965. Thread blight of water hyacinth. *Curr. Sci.* 34: 618–19

47. Nag Raj, T. R., Ponnappa, K. M. 1967. Some interesting fungi of India. *Commonw. Inst. Biol. Contr. Tech. Bull.* 9:31–43

48. Nag Raj, T. R., Ponnappa, K. M. 1970. Blight of water hyacinth caused by *Alternaria eichhorniae* sp. nov. *Trans. Brit. Mycol. Soc.* 55:123–30

49. Padan, E., Shilo, M., Kislev, N. 1967. Isolation of "Cyanophages" from freshwater ponds and their interaction with *Plectonema boryanum. Virology* 32:234–46

50. Padwick, G. W. 1946. Notes on Indian fungi IV. *Commonw. Mycol. Inst. Mycol. Pap.* 17:1–12

51. Patch, E. M. 1938. Food-plant catalogue of the aphids of the world. *Maine Agr. Exp. Sta. Bull.* 393, 431 pp.

52. Penfound, W. T., Earle, T. T. 1948. The biology of the water hyacinth. *Ecol. Monogr.* 18:447–72

53. Pettet, A., Pettet, S. J. 1970. Biological control of *Pistia stratiotes* in Western State, Nigeria. *Nature* 226:282

54. Plakidas, A. G. 1936. Nematodes on alligatorweed. *Plant Dis. Reptr.* 20:22

55. Ponnappa, K. M. 1970. On the pathogenicity of *Myrothecium roridum-Eichhornia crassipes* isolate. *Hyacinth Contr. J.* 8:18–20

56. Rao, V. P. 1963. *Commonwealth Inst. Biol. Control, Indian Sta., US PL-480 Project Rep.*

57. Rao, V. P. 1964. *Commonwealth Inst. Biol. Control, Indian Sta., US PL-480 Project Rep.*

58. Rao, V. P. 1970. *Commonwealth Inst. Biol. Control, Indian Sta., US PL-480 Project Rep.*

59. Renn, C. E. 1936. The wasting disease of *Zostera marina* I. A phytological investigation of the diseased plant. *Biol. Bull. Marine Lab. Woods Hole* 70:148–58

60. Ridings, W. H., Zettler, F. W. 1972. *Aphanomyces* blight of amazon sword plants. *Phytopathology* 62: In press

61. Rintz, R. E., Freeman, T. E. 1972. *Fusarium roseum* pathogenic to water hyacinth. *Phytopathology* 62: In press

62. Safferman, R. S., Morris, M. E. 1963. Algal virus: isolation. *Science* 140:679–80

63. Safferman, R. S., Morris, M. E. 1964. Growth characteristics of the blue-green algal virus LPP-1. *J. Bacteriol.* 88:771–75

64. Safferman, R. S., Schneider, I. R., Steere, R. L., Morris, M. E., Diener, T. O. 1969. Phycovirus SM-1: a virus infecting unicellular blue green algae. *Virology* 37:386–95

65. Scott, W. W. 1961. A monograph of the genus *Aphanomyces. Virginia Agr. Exp. Sta. Tech. Bull.* 151. 95 pp.

66. Sculthorpe, C. D. 1967. *The Biology of Aquatic Vascular Plants.* London: Arnold. 610 pp.

67. Seaman, D. E., Porterfield, W. A. 1964. Control of aquatic weeds by the snail, *Marisa cornuarietis. Weeds* 12:87–92

68. Selman, B. J., Vogt, G. B. 1971.

Lectotype designations in the South American genus *Agasicles* (Coleoptera: Chrysomelidae), with descriptions of a new species important as a suppressant of alligatorweed. *Ann. Entomol. Soc.* 64:1016–20

69. Sheffield, C. W., Kuhrt, W. H. 1969. Lake Apopka—its decline and proposed restoration. *Proc. Florida Environ. Eng. Conf. Water Pollut. Contr.* Gainesville: *Florida Eng. and Ind. Exp. Sta.* 24, *Bull. Ser.* 135:130–46
70. Silviera-Guido, A. 1965. *Final Rep., Dep. Sanidad Veg., Univ. Repub., Montevideo, Uruguay.* 125 pp.
71. Sims, A. C. 1956. Factors affecting basidiospore development of *Pellicularia filamentosa. Phytopathology* 46:471–72
72. Singh, R. N., Singh, P. K. 1967. Isolation of cyanophages from India. *Nature* 216:1020–21
73. Smart, G. C. Jr., Esser, R. P. 1968. *Aphelenchoides fragariae* in aquatic plants. *Plant Dis. Reptr.* 52:455
74. Smith, G. E. 1971. Resume of studies and control of Eurasian watermilfoil (*Myriophyllum Spicatum* L.) in the Tennessee Valley from 1960 through 1969. *Hyacinth Contr. J.* 9:23–25
75. Snyder, W. C., Hansen, H. N. 1945. The species concept in *Fusarium* with reference to discolor and other sections. *Am. J. Bot.* 32:657–66
76. Sparrow, F. K., Jr. 1960. *Aquatic Phycomycetes.* Ann Arbor: Univ. Michigan Press. 1187 pp.
77. Stewart, J. R., Brown, J. M. 1969. Cytophage that kills or lyses algae. *Science* 164:1523
78. Thirumalachar, M. J., Govindu, H. C. 1954. Notes on some Indian Cercosporae V. *Sydowia* 8:343–48
79. Thornberry, H. H. 1966. *Index of plant virus diseases.* U. S. Dep. Agr. Agr. Handb. 307. 446 pp.
80. Timmons, F. L. 1960. Weed control in western irrigation and drainage systems. *U. S. Dep. Agr., Agr. Res. Service,* ARS 34–14. 22 pp.
81. Tutin, T. G. 1938. The autecology of *Zostera marina* in relation to its wasting disease. *New Phytol.* 37:50–71
82. Weldon, L. W. 1960. A summary review of investigations on alligatorweed and its control. *U. S. Dep. Agr., Agr. Res. Serv., CR* 33–60
83. Weldon, L. W., Blackburn, R. D. 1967. Water lettuce—nature, problem, and control. *Weeds* 15: 5–9
84. Weldon, L. W., Blackburn, R. D., Harrison, D. S. 1969. Common aquatic weeds. *U. S. Dep. Agr. Agr. Handb.* 352. 43 pp.
85. Will, L. E. 1965. *Okeechobee Boats and Skippers.* St. Petersburg, Fla.: Great Outdoors Publishing Co. 72 pp.
86. Wilson, C. L. 1969. Use of plant pathogens in weed control. *Ann. Rev. Phytopathol.* 7:411–34
87. Zettler, F. W., Foxe, M. J., Hartman, R. D., Edwardson, J. R., Christie, R. G. 1970. Filamentous viruses infecting dasheen and other araceous plants. *Phytopathology* 60:983–87

BREEDING FOR DISEASE RESISTANCE IN CUCURBITS

3560

WAYNE R. SITTERLY
Clemson University Truck Experiment Station, Charleston, South Carolina

INTRODUCTION

Members of the Cucurbitaceae are among the important and widespread plants that supply man's food and fiber. Cucurbits are not as important on a world basis as are cereals and legumes, but, from the tropics to the milder portions of the temperate zone, they serve as a source of carbohydrates, as dessert and salad ingredients, and as pickles. Some cucurbits are used as pottery, baskets, insulation, and oil filters. The cultivated cucurbits have needed man for survival since bona fide specimens of the wild counterparts of the cultivated species have apparently never been collected.

Although there are about 90 genera and 750 species in the family Cucurbitaceae, only 6 genera and 12 species are cultivated by man. These genera and species include watermelon (*Citrullus*); cucumber, muskmelon, and gherkin (*Cucumis*); dish-rag and sponge gourds (*Luffa*); white-flowered gourd (*Lagenaria*); squashes, marrow, pumpkin, and figleaf gourd (*Cucurbita*); and chayote (*Sechium*). All the species are frost-sensitive. Within each species there is a wide assortment of sizes, shapes, color variants, flesh textures, flavors, etc (62).

The use of resistant varieties is a simple, effective, and economical means of controlling plant disease (45). A large number of disease resistance factors have been incorporated into cultivated cucurbits, particularly during the past 30 years. Associated with this has been the development of specific desirable horticultural characteristics. These two objectives are inseparable, as it is a bounteous yield of attractive, high quality fruits that is demanded by the consumer. In the present era, cucurbits are being produced commercially in areas where they could not be produced previously because of disease; varieties with resistance to one disease have been replaced by varieties with resistance to many diseases; disease resistance has been a major factor in development of mechanized harvesting; the high cost of disease control has been reduced; disease epidemics have been curtailed; and the use of resistant cultivars has diminished hazards and pollution resulting from use of potentially dangerous fungicides. It is with the theme of a "breakthrough" this discussion is concerned.

Host

Cucurbits have both advantages and disadvantages to the plant breeder. The disadvantages are: plants require much space, which makes large populations expensive; hand pollination is generally necessary for controlled genetic work; usually (except squash) pollination must be done before selection; chromosomes are not easily differentiated from cytoplasm in pollen mother cells; and chromosomes are small and not well separated from each other. The advantages are: plants are easily grown by simple methods; flowers are relatively large and easily pollinated by hand; plants are indeterminate with flowers available over a long time period; the time required to produce a crop is relatively short so that two or more generations can often be produced in one year; fruits are fairly durable; and most fruits yield many seeds.

One must be familiar with his host to understand the "normal" condition and not to be confused by deviations caused by factors of the environment such as temperature or soil fertility. Resistant cucurbit parental material usually occurs in plants obtained from areas where a pathogen is endemic and where resistance has occurred by mutation and natural selection. The major vine crops considered in this paper are thought to have originated in several regions of the world. Watermelon [*Citrullus lanatus* (Thumb.) Mansf.] was shown by DeCandolle (20) to be indigenous to the dry open areas on both sides of the equator in tropical Africa, with perhaps a strong secondary center of diversification in India. Watermelon has been cultivated for centuries by people bordering the Mediterranean Sea. De Candolle (20) also showed cucumber (*Cucumis sativus* L.) to be native to India, where it has been cultivated for over 3000 years. Cucumbers were spread eastward to China and westward where they were enjoyed by the Greeks and Romans. A chromosome count of seven and morphological features such as angular stems separate it from other members of the genus *Cucumis*. The place of origin of cantaloupe (*Cucumis melo* L.) has never been fully resolved (62), but appears to be tropical and subtropical Africa. From there it was spread into Asia, and, when placed in this congenial environment, it exploded into a number of subspecies. The squashes and pumpkins (*Cucurbita* spp.) are indigenous to North America, with the source apparently in central or southern Mexico or the northern part of Central America. As northern migration occurred they developed both xerophytic and humid adaptive types, which the Indians of this area utilized for many of their needs (62).

For the individuals interested in the mechanics of cucurbit plant breeding, Whitaker & Davis (62) and Sitterly (51) offer extensive reviews, which include information on floral morphology and fertilization, specific breeding procedures, methods of harvesting and storing seed, methods of recording data, and procedures for inoculation and screening of plant populations.

Disease resistance must not be looked upon as a goal independent of other objectives. Genotypic balance is a proper objective of all breeding for cucurbit disease resistance. The goal of evolution is toward survival of the plant under prevailing conditions. The goal of plant breeding is the production of a product wanted by man. The genetic balance between the two goals must provide for survival and a desirable product. There is no single character that can be called ideal, but all successful commercial varieties must have good appearance, which results in large part from good genotypic balance. For example, currently all cucumber varieties must have resistance to drouth and temperature extremes. Fruit should have the desired conformation, a small seed cavity surrounded by crisp flesh, slowly developing seed, a waxy surface, and a nonbitter flavor. Pickle cucumbers should have a definite length-diameter ratio according to regional market requirements, and moderate warty protuberances. They should be free of internal cavities due to carpel separation and hollow placentae, and they must have the desired crispness and quality when processed. Slicer cucumbers should be free of warty protuberances and have a dark green skin color. All melon and squash varieties must be productive, transportable, and of high edible quality to be useful in commerce. Melons include a wide assortment of acceptable sizes, shapes, color variants, flesh textures, and flavors.

Diseases

Many diseases attack cucurbits. Some are worldwide in occurrence and cause moderate to severe losses wherever cucurbits are grown. Others are restricted by climatic factors to specific regions. Emphasis in breeding has been placed on diseases that cause significant losses in the region where the research is being done. Little emphasis has been placed on some important diseases simply because a source of resistance in compatible lines has not been discovered. This paper is confined to diseases against which significant progress in development of resistant varieties had been achieved prior to 1970.

Anthracnose.—Anthracnose is caused by *Colletotrichum lagenarium* (Pass.) Ell. & Halst. and is relatively common in humid regions throughout the world. Watermelon, muskmelon, and cucumber are affected, but squash and pumpkin are not seriously affected. This disease has received much attention from plant breeders and pathologists. The anthracnose pathogen produces black stromata bearing conidiophores on the surface of the diseased host. Conidia are produced successively by budding at the tips of the conidiophores and are disseminated by water. The fungus penetrates by an infection peg from an appressorium produced at the tip of a short germ tube. Humid or rainy weather is essential for sporulation, dissemination, and penetration (58). *C. lagenarium* is composed of several physiologic races. Goode (22) demonstrated the presence of three races that were morphologically and culturally indistinguishable. Race 1 was virulent on all cucumber varieties tested; produced slight infection on Charleston Gray, Congo, and Fairfax watermelon varieties; and moderately attacked Butternut squash. Race 2 was virulent on all cucumber and watermelon varieties, and moderately attacked

Butternut squash. Race 3 was identical to race 1 except that Butternut squash was immune. Goode also showed that cucumber Plant Introduction (PI) 163217 and PI 196289 were resistant to race 2. Four more races of the pathogen have been identified more recently. Jenkins, Winstead & McCombs (33) stated that race 4 was virulent on all hosts. Race 5 was weakly virulent on their cucumber differentials and highly virulent on watermelons. Race 6 was weakly virulent on cantaloupe and highly virulent on watermelons. Race 7 was similar to race 3 except only weakly virulent on Pixie cucumber. There is some conflict concerning race 4, as Dutta, Hall & Hayne (21) stated that Model cucumber and Chris Cross watermelon were resistant. Race 1 and race 2 appear to be most common, and there have been attempts to distinguish between them by other means than using differential hosts. Crossan & Lynch (18) demonstrated that qualitative differences do exist in that race 2 contained galactose, glycine, and glutamic acid components that were not present in race 1. Hadwiger & Hall (24) indicated that the physiological difference between races as distinguished by differential host varieties was accompanied by inherent differences in the utilization of growth factors (amino acids, sugars, indolacetic acid) by the respective races. The higher concentration of citrulline in infected tissues accounted for the most consistent and largest amino acid difference. Winstead, McCombs & Lowie (65), however, maintain that differentiation of cucurbit anthracnose fungi, and of races, by the amino acids of the mycelial protein hydrolysates or free amino acids of the culture filtrate, is not feasible. Thus we do not have a clearly established and acceptable physiological basis for differentiation of races.

Since some varieties within a species are resistant and others susceptible to specific pathogenic races, breeders have to work with resistance to these specific races. Barnes & Epps (6) found two types of resistance among PI accessions of cucumber. One type involved an extremely high level of resistance controlled by several genes (PI 197087), while the other involved a moderate degree of resistance controlled by a single dominant gene (PI 175111). The genetic ratio of the mode of inheritance of the polygenic resistance of PI 197087 has not been solved. Busch & Walker (13) confirmed monogenic resistance in PI 175111, but stated the lack of definite segregation ratios suggested the presence of modifiers. They also showed that the parasite penetrated the cells of both resistant and susceptible plants, but hyphal progress was much slower and cells collapsed more slowly in resistant plants. Goode (23) essentially confirmed this by demonstrating that S^{35} accumulated around anthracnose lesions in both resistant and susceptible plants, but not around artificial wounds. Once S^{35} had accumulated in a given area it was not translocatable. In addition, he also reported that resistant plants elaborate toxic substances that inactivate the fungus.

Layton (36) found edible watermelon cultivars from Africa that were highly resistant to anthracnose, and that this resistance was governed by a single dominant gene. The resistant reaction resulted in only slight symptoms and a greatly retarded rate of fungus growth. Winstead, Goode & Barham

(64) evaluated many varieties and introductions of watermelon and found that those resistant to race 1 were also resistant to race 3, but susceptible to race 2. The mode of inheritance of resistance to races 1 and 3 was as described by Layton. Winstead, Goode & Barham also found African Citron W-695 to be resistant to race 2, but the mode of inheritance was not determined.

Sources of cucumber anthracnose resistance are: race 1—PI 163213 (monogenic) and PI 197087 (polygenic); race 2—PI 196289 and PI 197087 (both polygenic) and race 3—PI 197087 and PI 163213. A range of reaction to race 1 and race 2 would be highly resistant—PI 197087 and Poinsett; moderately resistant—PI 175111; slightly resistant—Ashley; and susceptible —Marketer.

Resistance in watermelon to races 1 and 3 can be found in Africa 8, and to race 2 in African Citron W-695. A range of reaction is given by: resistant to races 1 and 3—Charleston Grey and Congo; resistant to race 2—African Citron W-695; susceptible—Klondike.

Although 7 races of anthracnose have been identified, races 1 and 2 are apparently most prevalent and most pathologically destructive. Cucumber plant breeders are not attempting to incorporate resistance to races 3 through 7 into their gene pools. Likewise, watermelon plant breeders aren't attempting to incorporate resistance to races 4 through 7.

The author does not know of any active breeding work in muskmelon for developing anthracnose resistant varieties. If such investigations are occurring, he would be most happy to be enlightened. As mentioned previously, anthracnose is apparently not an important factor in squash production.

Powdery mildew.—Powdery mildew is a widely occurring disease caused by *Erysiphe cichoracearum* Dc. In the United States it is particularly destructive in fields in the Imperial Valley in California, and in the fall crop in the southeast. Powdery mildew is one of the principal diseases of greenhouse culture everywhere in the world. Conidia are continuous, elliptic, hyaline, and borne in chains on short unbranched conidiophores. Conidia are readily detached and borne by air currents. Penetration is confined to the epidermal cells, with most of the fungus remaining on the host surface. Cleistothecia are borne infrequently on the host surface as dark bodies with flexuous, indeterminate appendages. Cleistothecia contain about fifteen asci, each having two hyaline ascospores (58).

Although powdery mildew appears to have many biotypes, the extent of cucurbit specialization hasn't been determined. Some race investigation has been done in cantaloupe by Jagger, Whitaker & Porter (31). These investigators noted that powdery mildew was attacking the previously resistant cantaloupe variety PMR 45 in some areas; in other areas susceptible lines were being attacked, but not PMR 45; and in the nursery some breeding lines were not attacked, but PMR 45 was attacked. On this basis, with PMR 45 as the differential, they established the existence of races 1 and 2. Pryor & Whitaker (46) established differentials for distinguishing the two races; race 1—virulent on Hales Best, Casaba, and Tip Top cantaloupe; race 2—virulent on the preceding varieties plus PMR 45 and PMR 8. The sexual stage of the pathogen does occur occasionally; thus it is possible to obtain hybridization and selection.

Using resistant cantaloupe material from India (PI 79374), Jagger & Scott (30) found that resistance to race 1 was controlled by a single dominant gene, referred to as Pm^1. Using race 2-resistant varieties, also developed from resistant Indian material, Bohn & Whitaker (10) demonstrated that race 2 resistance was controlled by a partly dominant gene, designated Pm^2. They stated that two modifier genes differentiated extreme resistance to race 2, and were epistatic to PM^2 but hypostatic to pm^2. Using Seminole as the resistant stock, Harwood & Markarian (26) demonstrated that resistance in this germ plasm was controlled by two gene pairs whose effects were unequal and partly additive. The major gene showed incomplete dominance while the minor gene showed complete dominance. Both of these genes were different from Pm^1 (race 1) in their resistance reactions but were either identical, allelic, or closely linked to Pm^2.

Smith (52) found that Puerto Rico 37 cucumber was resistant to powdery mildew and, upon crossing it with Abundance, found the F_1 to be susceptible, but found a very small number of resistant individuals in the F_2. This suggested resistance could be polygenic. Barnes (personal communication) states that perhaps three genes are involved, but plants with all three genes may have been discarded due to close linkage with a gene, or genes, that resulted in the plant being unable to utilize manganese, as had previously been demonstrated by Robinson (49). Kooistra (34) states that resistance of PI 1200818 is of the hypersensitive type, controlled by two genes, and is possibly recessive. It would be well to re-emphasize that Kooistra was working with a supersensitive gene derived from some Japanese plant material. Barnes was working with three different genes (entirely different from Kooistra) all of which produce a cumulative effect. This effect is aptly demonstrated in his powdery mildew resistant varieties Ashley (one gene) versus Poinsett (three genes).

When susceptible cantaloupe breeding lines are inoculated with powdery mildew, luxuriant fungus growth occurs on all organs 6–8 days after inoculation (10). There is little apparent early tissue injury, but tissue is dead after about 20 days. The fungus infects resistant plants, but grows sparsely, with sporulation and growth very sensitive to environmental fluctuations. The F_1 reaction is intermediate, but closely approaches that of the resistant parent.

Whitaker (61) showed *Cucurbita lundelliana* Bailey to be resistant to powdery mildew. Rhodes (47, 48), utilizing the interspecific cross technique, developed a common gene pool of divergent germ plasm in a series of crosses between *C. lundelliana, C. pepo, C. moschata, C. mixta,* and *C. maxima.* He demonstrated that resistance in squash is controlled by a single dominant gene that had been transferred from *C. lundelliana* to *C. moschata.* Sitterly (unpublished information) obtained germ plasm from Rhodes and transferred resistance into the bush type *C. pepo.* On susceptible squash plants, powdery mildew grows on both upper and lower leaf surfaces and later spreads to petioles and stems. On resistant plants, fungus growth is more restricted and very small colonies grow only on the upper leaf surfaces. An intermediate reaction consists of luxuriant growth of powdery mildew on mature leaves, and small restricted colonies on younger leaves.

Sources of cantaloupe resistance are PI 79374, PI 124111, and PI 134198. A range of reaction would be: high resistance to races 1 and 2—Planters Jumbo; race 1 resistant—PMR 45; race 2 resistant—Seminole; moderately resistant—Georgia 47; susceptible—Hales Best and Honey Dew. Sources of cucumber resistance are PI 197087 and PI 1200815. A range of resistance would include: one resistant gene—Palmetto and Ashley; two resistant genes—Cherokee; three resistant genes—PI 197087 and Poinsett; susceptible—Model. A source for squash powdery mildew resistance is *Cucurbita lundelliana.* No resistant varieties have been released, although an effort to incorporate this resistance into acceptable varieties of summer squash is now being made.

Downy mildew.—Although the downy mildew pathogen [*Pseudoperonospora cubensis* (Berk. & Curt.) Ros.] cannot overwinter in areas with temperatures below 0°C, it can thrive in both warm and cool temperatures. Thus humidity is the most important factor in establishment, and makes this disease particularly important in the humid environment encountered in the eastern United States and other parts of the world. Mycelium lives intercellularly in the host with haustoria being intracellular. Sporangiophores arise in groups through the stomata, are branched, and produce sporangia on the tips. Sporangia are grayish-purple in mass, ovoid, and have a papilla at the distal end. Sporangia germinate by producing biciliate zoospores, which in turn produce germ tubes that penetrate through the stomata (58).

Cochran (16) crossed the tolerant Indian cucumber cultivar Bangalore with commercial items and stated that resistance was apparently determined by several factors. J. M. Jenkins (32) crossed the highly resistant cultivars Chinese Long and Puerto Rico 37 with commercial varieties and found the F_1 to be intermediate in resistance. Successive selection demonstrated this resistance was also controlled by multiple genes. After stabilization of resistance this material was crossed by Barnes (3) with Cubit and Marketer to produce Palmetto and Ashley. Barnes & Epps (7) reported a second source of a high degree of resistance in PI 197087; it appeared to be polygenic, with inheritance controlled by one or two major genes and one or more minor genes. These investigators also reported a new symptom associated with the physiological response of PI 197087 to downy mildew. The mildew lesions were initially light brown and then turned dark brown without passing through a yellow stage. Resistance was an expression of hypersensitivity.

Genetic resistance to downy mildew has not been fully investigated in cantaloupe. Ivanoff (28), using four tolerant West Indian varieties, found resistance to be partially dominant. By combining these four types and selecting within the progeny he was able to increase resistance. Ivanoff also observed an apparent close relationship of downy mildew resistance with aphid resistance.

Sources of downy mildew resistance in cucumber are Chinese Long and PI 197087. A range of reaction would include: highly resistant—Poinsett; resistant—Ashley; intermediate—Santee; susceptible—Marketer.

Sources of resistance in cantaloupe are Seminole and PI 124112. A range of resistance would include: immune—Seminole; resistant—Edisto 47; moderately resistant—Georgia 47; tolerant—Smiths Perfect; susceptible—Hales Best.

Scab.—The scab organism, *Cladosporium cucumerinum* Ell. & Arth., affects many cucurbits, particularly in world areas with cool growing seasons accompanied by intense fogs and dews (mountain valleys of North Carolina, northern United States and southern Canada, northern Europe, etc.). Penetration of the pathogen into the host occurs through the stomata. Mycelium is hyaline when young and turns olivaceous with age. Oblong, colored, continuous conidia are borne successively on short branched conidiophores. Intermediate cells between the conidiophore and conidia may also be detached and germinate (58). Spread of the pathogen and development of the disease are favored by relatively low night temperatures. The optimum temperature for disease development (below 20°C) is lower than that for either the host or the pathogen (57). At temperatures above 21°C, lesions are rapidly cicatrized by host reaction, and further disease development is stopped.

Bailey & Burgess (2) showed that some selections of the cucumber cultivar Longfellow were resistant to scab. These were self pollinated and the segregating resistant and susceptible plants were crossed. Analysis of the F_2 showed resistance to be controlled by a single dominant gene. Walker (57) confirmed this and showed that plants grown in a moist chamber at 17°C could be accurately rated for resistance. Local lesions developed on resistant lines without rapid stem invasion, while stems and leaves of susceptible items became watersoaked and died rapidly. In very young seedlings resistance is not completely dominant, and thus it is possible at this stage to separate homozygous resistant plants from heterozygous plants. Pierson & Walker (42) demonstrated that at 17°C both resistant and susceptible hosts were invaded, but in the resistant hosts progress was stopped by a series of responses associated with cell wall thickening and cell necrosis. The fungus rarely became intercellular, sunken lesions did not develop, and no gummy exudate was formed. At temperatures above 21°C there was no damage to susceptible lines because of rapid cicatrization. The investigator thus must be careful of misinterpreting effects of temperature fluctuations.

Investigation concerning the biochemistry of cucumber scab resistance has centered around the role of enzyme systems. Both Hussain & Rich (27) and Strider & Winstead (55) demonstrated that host tissues are disintegrated by cellulase produced by the fungus. Kuc (35) also says that proteolytic enzymes are produced and enhance separation and disorganization of cells. The cell wall constituents are modified and made available as a carbon source to the fungus. Translocation pathways are thus blocked by early necrosis of infected tissue, which prevents movement or availability of ammonium ions in leaf tissue for manufacture of such amino acids as asparagine and glutamine (12). Mahadevan, Kuc & Williams (37) maintain that in resistant varieties, the pathogen triggers synthesis of an inhibitor that inactivates the pectinolytic enzymes secreted by the pathogen and thus inhibits growth of the pathogen. This process is fully complete 24 hours after inoculation.

Sources of cucumber slicer and pickle resistance are Main 2 and Highmoor. A range of resistance would include: resistant—Highmoor and Wisconsin SMR 15; susceptible—Chicago Pickling and Marketer.

Although scab is an important disease of squash, particularly summer squash in cool areas, no genetic investigations of scab resistance have been completed on this crop. Strider & Konsler (54) screened available *Cucurbita* accessions for scab resistance and found that none were immune. Eight of the *C. pepo* group were rated as tolerant and crosses were made with Early Prolific Straightneck. The results from these crossings have not been announced. The *C. pepo* accessions tolerant to scab were PI 164957, PI 167136, PI 171622, PI 174183, PI 174184, PI 177376, and PI 227237.

Angular leaf spot.—Angular leaf spot caused by *Pseudomonas lachrymans* E. F. Sm. & Bryan occurs mainly on field grown cucumbers in humid regions with fairly cool temperatures. The other cucurbits are apparently not seriously affected. The angular leaf spot pathogen is a rod-shaped bacterium which has 1–5 polar flagella and forms capsules. The organism overwinters on infected plant residue and is spread by surface water and rain. Penetration of the host is through stomata, and initial development is intercellular. In infected fruit the bacteria are found in the xylem elements of the mesocarp and in the tissue surrounding the developing seeds. Optimum disease development occurs at 24–28°C (58).

Chand & Walker (15), using resistant PI 169400 from Turkey, found when crosses were made with Wisconsin SMR 15 and Wisconsin SMR 18, that the segregation in the F_2 and backcross generations didn't fit any standard ratio. This indicated resistance is possibly polygenic. Resistance gradually increased with continuous selection of resistant individuals. Barnes (4) demonstrated that PI 197087 from southern India had even more resistance than PI 169400 and this resistance was also polygenic. Chand & Walker (14) found that mature leaves of susceptible plants were more resistant than young leaves. Young and old leaves reached maximum bacterial concentration 3–4 days after inoculation, but young leaves had 20–40 times as many bacteria. When inoculated leaves were compared, the multiplication was greatest in susceptible plants, least in resistant plants, and intermediate in a hybrid between resistant and susceptible plants. Van Gundy & Walker (56) showed that as the nitrogen concentration in a plant increases, the plant becomes more susceptible. Plants deficient in potassium are very susceptible because of a correspondingly high nitrogen content. Williams & Keen (63) showed that cell permeability is altered after successful infection. Inorganic ions, sugars, and amino acids were lost rapidly 48 hours after inoculation when lesions became vein-limited.

PI 197087 is a source of angular leaf spot resistance in cucumber. A range of resistance would include: highly resistant—PI 197087 and PI 169400; resistant—Southern Cross; tolerant—Pixie; susceptible—Model and SMR 18.

Fusarium wilts.—Fusarium wilt of watermelon is caused by *Fusarium oxysporium* f. *niveum* (E. D. Sm.) Snyder & Hansen. It occurs world-wide, and is particularly destructive in the southern United States. Fusarium wilt of cantaloupe is caused by *F. oxysporium* f. *melonis* (Leach & Currence) Snyder & Hansen. It is important mainly in the northern United States and southern Canada. Although squash and cucumbers are affected by Fusarium root rot, they apparently are not affected by Fusarium wilt. *Fusarium* produces a septate mycelium in which chlamydospores are formed and which produces conidia. Infection occurs through the root-tip region, wounds, and through the radicle of germinating seeds. The fungus invades the root cortex and then becomes established and moves within the xylem elements. Other tissues are not invaded until the plant dies or approaches death. Seedling injury is severe when soil temperatures are 20–30°C. In older cantaloupe and watermelon plants the optimum soil temperature for the organism is 27°C (58). Higher soil temperatures favor watermelon plant health since the optimum soil temperature for growth of watermelon seedlings is 32°C.

Fusarium does not develop above 33°C. There is a positive correlation between degree of soil infestation and rate of wilting in both cantaloupe and watermelon (58).

Although there are many strains of the watermelon fungus, very few studies of physiologic specialization have been made. Crall (17) showed that two races exist in Florida. Race 1 was most common and produced 100% wilt in the susceptible Florida Giant variety, and scattered wilt in Charleston Gray and Summit. Race 2 produced 100% wilt in susceptible varieties and no wilt in Charleston Gray or Summit—even in seedlings.

Little is known about the inheritance of resistance because of the numerous strains of the fungus and the lack of a precise method for clearly distinguishing between susceptible and resistant individuals in a mixed population. Orton (40) crossed the resistant Stock Citron with the susceptible variety Eden and demonstrated resistance could be transferred by hybridization and selection. Porter & Melhus (43) demonstrated that wilt resistant individuals occur in commercial varieties, thus avoiding the necessity of using the inedible citron in a breeding program. They also showed resistance is polygenic but not completely fixed. Braun (11) attempted to discover the physiological basis of resistance to wilt by demonstrating that acetic acid retards fungus growth, and that a similar chemical was found in the stems of resistant Citron watermelon but not in the roots of either the resistant or the stems and roots of a susceptible variety.

Henderson et al (26a) studied the inheritance of resistance to Fusarium wilt in the Summit, Charleston Gray, and New Hampshire Midget varieties of watermelon. They postulated three gene models—a single locus multiple-allelic model and two two-locus models to explain the relationships among the three varieties. They also proposed that a variety such as Summit whose resistance is inherited in a completely dominant manner "could be of value in obtaining F_1 hybrids highly resistant to Fusarium wilt, as well as aid in development of highly resistant open-pollinated varieties."

Mortensen (38) devised a method for distinguishing between resistant and susceptible cantaloupe genotypes. He showed that the best separation could be made at a constant soil temperature of 30°C. Resistance was dominant to susceptibility, with resistance being controlled by one principal dominant gene (*R*) plus two complementary dominant genes (*A*) and (*B*). The hypothetically resistant genotypes would include: *RAB, RAbb, RaaB, Raabb,* and *rrAB;* those that would be susceptible include: *rrAbb, rraaB,* and *rraabb*.

Sources of watermelon wilt resistance are Stock Citron and Iowa Belle. A range of resistance would include: highly resistant—Summit and Calhoun Gray; moderately resistant—Charleston Gray, Hawksbury, and Garrisonian; susceptible—Florida Giant.

Persian, casaba, and honeydew cantaloupes are naturally quite resistant to Fusarium wilt, but require a long growing season. *Fusarium* resistant cultivars are required in areas with a short growing season. A source of cantaloupe wilt resistance is Golden Gopher. A range of resistance would include: highly resistant—Spartan Rock; resistant—Golden Gopher; susceptible—Hales Best and Gulfstream.

Cucumber mosiac.—Cucumber mosaic virus (CMV) has many variants, which differ in host range and in the type of symptoms on specific hosts, but are similar in physical properties. CMV is world-wide in occurrence. It is not seed transmitted, but survives from season to season in such perennial host plants as *Commelina nudiflora* L. Annuals, such as *Melothria pendula* L., may serve as summer reservoirs of the virus. Transmission from these weeds to the cultivated host or from plant to plant in the field is by the green peach and melon aphids or by mechanical means such as farm machinery.

Porter (44) found the cucumber varieties Chinese Long and Tokyo Long Green to be highly resistant to CMV. Upon crossing these varieties with susceptible material, Shifriss, Myers & Chupp (50) demonstrated that, although resistance was dominant, it involved at least three major genes and possibly some modifiers. They maintained that resistance was associated with the presence (susceptible) or absence (resistant) of chlorosis in the cotyledonary stage, and the degree of tolerance could be determined by the distance from the cotyledons to the leaf in which symptoms first appear. Ratios changed constantly as the plants developed beyond the first true-leaf stage. At this stage, several gene modifiers cause a low frequency of symptomless plants in a mature plant population. Thus it is possible to select plants with varying degrees of tolerance determined by the number of resistant genes involved. Wasuwat & Walker (60), crossing resistant Wisconsin SMR 14 × National Pickling, demonstrated that, in their specific source of resistance, the resistance was controlled by a single dominant gene. The same investigators demonstrated that in the greenhouse the distinction between resistant and susceptible plants was best determined 20 days after inoculation. All resistant plants became infected and showed symptoms, but symptoms were mild and tended to disappear. In susceptible plants the leaves remained smaller, foliage was intensely mottled, internodes became shorter and fruit became mottled. In the field, plant stunting and mottling were the criteria of susceptibility.

Tokyo Long Green and Chinese Long are sources of resistance to CMV. A range of resistance would include: highly resistant—Tablegreen; resistant —Wisconsin SMR 18; susceptible—Model.

Gummy stem blight.—This disease is caused by *Mycosphaerella melonis* (Pass.) Chiu and Walker. It attacks watermelon, cucumber, cantaloupe, and certain squash. This disease has become increasingly severe as chemical and genetic procedures have gradually resulted in the effective control of other cucurbit diseases. The pathogen is most common in the southern United States and in the sub-tropical and tropical areas of the world; it is also found occasionally in northern growing areas.

Black pycnidia in the lesions normally produce both large and small one- or two-celled conidia. In the sexual stage, hyaline, two-celled ascospores are produced within asci located in perithecia. Leaf penetration is either direct or through intercellular spaces around the bases of trichomes. Stem penetration is apparently through wounds. Fruit penetration is either through wounds or possibly through flowers at time of pollination. Cotyledons and young leaves of watermelon and cantaloupe are very susceptible, but those of cucumber and certain squash are resistant and become susceptible only as they mature (58). The presence of races does not appear a problem as yet. Norton & Prasad (39) determined the resistance of cantaloupe PI 140471 to be dominant and controlled by a single gene which they designated *RmRm*. Further investigation showed moderate resistance to be controlled by an independent single gene which they designated *MmMm*. This, of course, ultimately should result in the highly resistant *RmRmMmMm* genotype.

The highest degree of resistance was found in PI 140471, a wild melon native to Texas. A range of resistance would include: highly resistant—PI

140471; moderately resistant—Auburn cantaloupe breeding lines C1 and C8; susceptible—Smith's Perfect, Hales Best Jumbo.

Sowell & Pointer (53) screened many watermelon PI's and found that the inedible PI 189225 of *Citrullus vulgaris* was consistently resistant to gummy stem blight. Although they do not mention the mode of inheritance, I suspect this germ plasm has been utilized in various breeding programs. Breeders may be having difficulty returning to a desirable horticultural fruit type of the edible watermelon.

Alternaria leaf spot.—This disease is caused by *Alternaria cucumerina* (Ell. & Ev.) J. A. Elliot. Although it is widespread in occurrence, little effort has been made to developing resistant varieties. Perhaps the situation is akin to that of gummy stem blight in that the *Alternaria* organism infects later-maturing plant tissues that ordinarily would have been infected earlier by other organisms. Varieties resistant to these other organisms have now been developed. Thus much more leaf tissue is now available to *A. cucumerina* and Alternaria leaf spot may become even more important than at present.

Jackson (29) has demonstrated Alternaria leaf spot on cucurbits differs in size, shape, color of lesions, and in rate of disease development. Conidia produced by mycelium in dead host material may be the primary inoculum in regions where the pathogen overwinters, seed-borne conidia providing the primary inoculum in new areas (8, 29). Mycelium and conidiophores become dark colored with age. Conidia are beaked, muriform, dark colored, and are borne singly or in chains of two (58). Germ tubes penetrate the host directly; lesions result from intercellular growth of hyphae and destruction of leaf tissue. Serious outbreaks occur when temperatures range from 20–32°C. Free moisture apparently is not critical (26, 29).

Little information is available on either the genetics or the nature of resistance. Hughes (personal communication) relates that Dr. C. Hall of the University of Kansas, using the cultivar Hearts of Gold as his resistant parent, obtained a 3:1 genetic ratio with *Alternaria* resistance dominant. Blinn (8) released Pollock in 1905 as a resistant Rocky Ford type of cantaloupe. Hartman & Gaylord (25) released the Purdue 44 cantaloupe as resistant to *Alternaria.*

The most reliable source of *Alternaria* resistance appears to be PI 140471, which is also highly resistant to downy mildew and gummy stem blight but susceptible to powdery mildew. Other satisfactory PI numbers are 145594, 164551, 124109, and 116915 (Hughes, personal communication).

A range of resistance could include PI 140471 as highly resistant; Edisto and Edisto 47 as moderately resistant; and Jumbo Hales Best as susceptible.

Other cucurbit diseases.—Several other cucurbit diseases, each important in its own area, have either had only a few individual efforts to develop resistance, or have had much pathological investigation, but few concrete plant breeding results. The latter has been particularly true of cucurbit viruses. In some cases the time-consuming process of selection and breeding for disease resistance has not kept pace with the development of new diseases; in others, sources of resistance just aren't presently known.

"Breakthrough"

Breeding cucurbits for disease resistance has resulted either directly or indirectly in several spectacular accomplishments in a relatively short period of "plant time." A few examples merit further discussion.

Individual disease resistance.—The most obvious benefit of resistance to individual diseases is commercial production of cucurbit crops in areas where economic production was not feasible because of the presence of a specific disease. Prime examples in the southeastern United States are the growing of watermelons after incorporation of resistance to Fusarium wilt and fall production of cucumbers after incorporation of downy mildew resistance. Another beneficial result is the decrease in cost of crop production as a result of requiring few or no fungicide applications for controlling a disease such as cucumber downy mildew. This also decreases the danger of air, soil, and water pollution with pesticidal chemicals and the hazard of injury to man associated with the application of pesticides and the contamination of foods with toxic residues.

Multiple disease resistance.—As individual diseases are conquered by breeding it is practical to combine the genes for resistance to several specific diseases into multiple-disease-resistant varieties. It has not been easy for plant breeders to combine these characteristics into high quality cultivars. Walker & Pierson (59) achieved one of the earliest successes by combining scab and cucumber mosaic resistance in the Wisconsin SMR pickle series. Barnes (4, 5) has produced several inbred and hybrid varieties of pickle and slicer cucumbers with resistance to seven diseases. Multiple-disease resistant watermelon and cantaloupe varieties also have been developed.

When widely divergent multigenic characteristics are involved, the probability of obtaining a single plant with all the desired horticultural and resistance genes is exceedingly slim. Barnes (5) used the "recombination cross" technique in which plants having desirable complementary characters were crossed in the F_3 and F_5 generations. This circumvented the loss of resistance encountered with backcrossing and thus made the entire gene complex available. Using this technique plants are inoculated with two or three different pathogens at the same time. Individual breeding lines are also inoculated separately to determine the influence of any interaction between the pathogens used simultaneously or successively. Inoculation with virus is made last on the assumption that it may tend to protect the plant from other pathogens.

Another excellent method of combining sources of disease resistance in cucurbits was described by Rhodes (47). It involves interspecific crosses to develop a wide base of genetic diversity. This method consists of developing an interbreeding population, or gene pool, with *C. pepo, C. mixta, C. maxima,* and *C. moschata.* Plants from this pool are used in a series of bridging crosses to transfer specific genes between incompatible species. Fertile interspecific hybrids combining the germ plasm from two to four species can be maintained. Rhodes utilized this method to transfer the bush habit of growth from *C. pepo* to *C. moschata,* and powdery mildew resistance from *C. lundelliana* to the previously mentioned gene pool. It was from this pool that Sitterly (51), as mentioned earlier, obtained powdery mildew resistance to combine with downy mildew and squash mosaic virus resistance in his *C. pepo* breeding lines.

Examples of multiple-disease resistant cucumber varieties are: pickle variety—Chipper (downy mildew, powdery mildew, angular leaf spot, anthracnose, and cucumber mosaic); slicer variety—Poinsett (downy mildew, powdery mildew, angular leaf spot, and anthracnose). Multiple-disease resistant cantaloupe varieties are: Seminole (powdery mildew race 2, downy mildew), and Edisto 47 (powdery mildew and downy mildew). Multiple-disease resistant watermelon varieties are: Charleston Gray and Garrisonian (anthracnose races 1 and 3, Fusarium wilt).

Hybrids and hybrid seed production.—After the development of multiple-disease resistant lines the next logical step was synthesis of hybrids. This procedure has been most effectively developed in cucumber where heterosis appears marked in terms of yield and earliness. Peterson & Wiegle (41), and Barnes (5), using the same source of Korean male sterility, have developed stable male sterile lines with the desired multiple-disease resistance. This has resulted in the economical production of hybrid seed of slicer and pickle cucumbers.

In squash, hybrid vigor is not marked and, likewise, vigor is not too rapidly lost by inbreeding. Curtis (19) demonstrated that second generation seed did not differ significantly from first generation seed, and that, if parental inbreds were similar, the resulting crop wasn't too variable for market if the resulting fruit were sold at an immature stage. This procedure will allow very early commercial production of hybrid quality seed.

Male sterility in the sterile muskmelon described by Bohn & Whitaker (9) is controlled by a single recessive gene and has to be carried in a heterozygous condition. Since this is not practical, and since hand pollination of cantaloupe is very inefficient (10%–50% successful), commercial production of hybrid seed is very expensive.

Due to the need for an abundance of hand labor, most hybrid watermelons are of Japanese origin. Additionally, heterosis isn't too marked. Polyploidy has been used to develop seedless fruit; triploid fruit of a cross between diploid and tetraploid parents do not develop viable seed. The cost of producing hybrid watermelon seed is high.

As stated by Andrus (1), most melon breeders have noted the following in relation to disease resistance of hybrids: (*a*) susceptibility tends to be dominant in the F_1 hybrid in respect to most of the important common diseases; (*b*) disease resistance is rarely dominant, so that heterozygotes tend to be intermediate in resistance; (*c*) the increased load of fruit presumably resulting from heterosis in some F_1 hybrids tends to amplify the damage caused by disease; (*d*) hybrids often appear to be more susceptible than their genetic composition would lead one to expect; (*e*) productive, fast-maturing, and extra early varieties and hybrids tend to develop more severe symptoms of foliage diseases, whereas late-maturing unproductive varieties develop less severe symptoms than their genetic composition would lead one to expect.

Machine harvest.—This area has been most exploited in cucumbers with the introduction of the stable gynoecious character. This characteristic has resulted in the development of pickle and slicer varieties with small vines. These varieties give high yields in once-over harvest systems but show a wide range of resistance to disease.

In the United States, both Michigan State University and North Carolina State University are developing cucumber harvesting machinery. The once-over type of machinery has been fairly well established, particularly in the pickle industry. The multiple-pick type of machinery for both cucumber slicers and pickles is being developed and appears entirely feasible from an economical view. The cost reduction for harvesting is staggering. Machinery for the actual harvesting of cantaloupe, watermelon, and squash is not as well developed.

Goals to be Achieved

Cucurbit plant breeders are continually developing improved varieties that are better adapted to a wider range of growing conditions, have multiple-disease resistance, and meet the fluctuating demands of economics and consumers. With all cucurbits, work must be directed to development of smaller, compact vines for mechanical harvest.

Specific goals in cucumber plant breeding are: better resistance to powdery mildew in the greenhouse and in the tropics; resistance to gummy stem blight, fruit rot, Fusarium wilt, and bacterial wilt; and investigation of the relationship between watermelon mosaic virus and cucumber.

Specific goals for watermelon plant breeding are: resistance to anthracnose (race 2), downy mildew, and gummy stem blight; development of triploid and diploid seedless hybrids; clarification of a physiological basis of resistance to Fusarium wilt; and establishment of a precise method for measuring resistance to Fusarium wilt in both greenhouse and field.

Specific goals for cantaloupe plant breeding are: a higher level of resistance to gummy stem blight, Alternaria leaf spot, and viruses; investigation of the relationship between cantaloupe and powdery mildew; and incorporation of the gynoecious character into cantaloupe for hybrid seed production.

Specific goals for squash and pumpkin plant breeding are: resistance to watermelon, squash, and cucumber mosaic viruses; resistance to downy mildew and powdery mildew; and the incorporation of the gynoecious character into squash for hybrid seed production.

Acknowledgment

The author gratefully acknowledges Dr. W. M. Epps, Department of Plant Pathology and Physiology, Clemson University, for his invaluable editorial comments. Also, acknowledgment is accorded Dr. W. C. Barnes, Dr. C. F. Andrus, and Dr. M. B. Hughes for professional comment.

LITERATURE CITED

1. Andrus, C. F. 1970. *Breeders Notebook*. Unpublished
2. Bailey, R. M., Burgess, I. M. 1935. Breeding cucumbers resistant to scab. *Proc. Am. Soc. Hort. Sci.* 36:645–46
3. Barnes, W. C. 1948. The performance of *Palmetto*, a new downy mildew resistant cucumber variety. *Proc. Am. Soc. Hort. Sci.* 51:437–41
4. Barnes, W. C. 1961. Multiple disease resistant cucumbers. *Proc. Am. Soc. Hort. Sci.* 77:417–23
5. Barnes, W. C. 1966. Development of multiple disease resistant hybrid cucumbers. *Proc. Am. Soc. Hort. Sci.* 89:390–93
6. Barnes, W. C., Epps, W. M. 1952. Two types of anthracnose resistance in cucumber. *Plant Dis. Reptr.* 36:479–80
7. Barnes, W. C., Epps, W. M. 1954. An unreported type of resistance to cucumber downy mildew. *Plant Dis. Reptr.* 38:620
8. Blinn, P. K. 1905. A rust-resisting cantaloupe. *Colorado Agr. Exp. Sta. Bull.* 104
9. Bohn, G. W., Whitaker, T. W. 1949. A gene for male sterility in muskmelon (*Cucumis melo*). *Proc. Am. Soc. Hort. Sci.* 53: 309–14
10. Bohn, G. W., Whitaker, T. W. 1964. Genetics of resistance to powdery mildew race 2 in muskmelon. *Phytopathology* 54:587–92
11. Braun, A. E. 1942. Resistance of watermelon to the wilt disease. *Am. J. Bot.* 27:683–84
12. Burton, C. L., deZeeuw, D. J. 1961. Free amino acid constitutions of healthy and scab infected cucumber foliage. *Phytopathology* 51:776–77
13. Busch, L. V., Walker, J. C. 1958. Studies of cucumber anthracnose. *Phytopathology* 48:302–04
14. Chand, J. M., Walker, J. C. 1964. Relation of age of leaf and varietal resistance to bacterial multiplication in cucumber inoculated with *Pseudomonas lachrymans*. *Phytopathology* 54:49–51
15. Chand, J. M., Walker, J. C. 1964. Inheritance of resistance to angular leafspot of cucumber. *Phytopathology* 54:51–54
16. Cochran, F. D. 1937. Breeding cucumbers for resistance to powdery mildew. *Proc. Am. Soc. Hort. Sci.* 35:541–43
17. Crall, J. M. 1963. Physiologic specialization in *Fusarium oxysporum* v. *niveum*. *Phytopathology* 53:87
18. Crossan, D. F., Lynch, O. L. 1958. A qualitative comparison of the amino acid and sugar content of acid hydrolysates from the mycelium of several anthracnose fungi. *Phytopathology* 48:55–57
19. Curtis, L. C. 1941. Comparative earliness of first and second generation squash (*Cucurbita pepo*) and the possibilities of using second generation seed for commercial planting. *Proc. Am. Soc. Hort. Sci.* 38:596–98
20. DeCandolle, A. 1882. *Origine des planter cultives*. Germes Bailliere, Paris. 337 pp.
21. Dutta, S. K., Hall, C. V., Hayne, E. C. 1960. Observations on physiological races of *Colletotrichum lagenarium*. *Bot. Gaz.* 121:163–66
22. Goode, M. G. 1958. Physiological specialization in *Colletotrichum lagenarium*. *Phytopathology* 48: 79–82
23. Goode, M. G. 1967. Radioautographic evidence for induced resistance to anthracnose in cucumbers. *Phytopathology* 57: 1028–31
24. Hadwiger, L. A., Hall, C. V. 1963. A biochemical study of the host-parasite relationship between *Colletotrichum lagenarium* and cucurbit hosts. *Proc. Am. Soc. Hort. Sci.* 82:378–87
25. Hartman, J. D., Gaylord, F. C. 1944. The Purdue 44 muskmelon. *Indiana Agr. Exp. Sta. Cir.* 295:1–8
26. Harwood, R. R., Markarian, D. 1966. Genetics of resistance to powdery mildew in the Michigan cantaloupe breeding program. *Proc. XVII Int. Hort. Congr.* 1: 454

26a. Henderson, W. R., Jenkins, S. F., Jr., Rawlings, J. O. 1970. The inheritance of *Fusarium* wilt resistance in watermelon, *Citrullus lanatus* (Thunb.) Mansf. *J. Am. Soc. Hort. Sci.* 95:276–82

27. Hussain, A., Rich, S. 1958. Extracellular pectic and cellulolytic enzymes of *Cladosporium cucumerinum*. *Phytopathology* 48: 316–20
28. Ivanoff, S. C. 1944. Resistance of cantaloupes to downy mildew and the melon aphid. *J. Hered.* 35:35–39
29. Jackson, C. R. 1959. Symptoms and host-parasite relations of the *Alternaria* leaf spot disease of cucurbits. *Phytopathology* 49: 731–34
30. Jagger, I. C., Scott, G. W. 1937. Development of *Powdery Mildew Resistant Cantaloupe No. 45. Cir. U. S. Dept. Agr.* 441, 5 pp.
31. Jagger, I. C., Whitaker, T. W., Porter, D. R. 1938. A new biological form of powdery mildew on muskmelons in the Imperial Valley of California. *Plant Dis. Reptr.* 22:275–76
32. Jenkins, J. M., Jr. 1946. Studies on the inheritance of downy mildew resistance and of other characters in cucumber. *J. Hered.* 37:267–71
33. Jenkins, S. F., Winstead, N. N., McCombs, C. L. 1964. Pathogenic comparison of three new and four previously described races of *Glomerella angulata* var. *orbiculare*. *Plant Dis. Reptr.* 48: 619–23
34. Kooistra, E. 1968. Powdery mildew resistance in cucumber. *Euphytica* 17:236–44
35. Kuc, J. 1962. Production of extracellular enzymes by *Cladosporium cucumerinum*. *Phytopathology* 52:961
36. Layton, D. V. 1937. The parasitism of *Colletotrichum lagenarium* (Pass.) Ell. & Holst. *Iowa Agr. Exp. Sta. Res. Bull.* 233:39–67
37. Mahadevan, A., Kuc, J., Williams, E. B. 1965. Biochemistry of resistance in cucumber against *Cladosporium cucumerinum*. I. Presence of a pectinase inhibitor in resistant plants. *Phytopathology* 55:1000–04
38. Mortensen, J. A. 1959. The inheritance of *Fusarium* resistance in muskmelon. *Diss. Abstr.* 19:2209
39. Norton, J. D., Prasad, K. 1966. Inheritance of resistance to *Mycosphaerella melonis* (Pass.) Chiu & Walker (gummy stem blight) in *Cucumis melo* L. muskmelon). *Proc. XVII Int. Hort. Congr.* 1:444
40. Orton, W. A. 1911. The development of disease resistant varieties of plants. *IV Conf. Int. Genet.* Paris, C. R. et Rapp. 247–65
41. Peterson, C. E., Wiegle, J. L. 1958. A new method for producing hybrid cucumber seed. *Mich. Agr. Exp. Sta. Quart. Bull.* 40:960–65
42. Pierson, C. F., Walker, J. C. 1954. Relation of *Cladosporium cucumerinum* to susceptible and resistant cucumber tissue. *Phytopathology* 44:459–65
43. Porter, D. R., Melhus, I. E. 1932. The pathogenicity of *Fusarium niveum*. (E. F. Sm) and the development of wilt-resistant strains of *Citrullus vulgaris* (Schrad.). *Res. Bull. Iowa Agr. Expt. Sta.* No. 149, 183 pp.
44. Porter, R. H. 1928. Further evidence of resistance to cucumber mosaic in the Chinese cucumber. *Phytopathology* 18:143
45. Principles of Plant and Animal Pest Control. Vol. 1: Plant Disease Development and Control. 1968. *Nat. Acad. Sci.* USA. Publ. 1596
46. Pryor, D. E., Whitaker, T. W. 1942. The reaction of cantaloupe strains to powdery mildew. *Phytopathology* 32:995–1004
47. Rhodes, A. M. 1959. Species hybridization and interspecific gene transfer in the genus *Cucurbita*. *Proc. Am. Soc. Hort. Sci.* 74: 546–52
48. Rhodes, A. M. 1964. Inheritance of powdery mildew resistance in the genus *Cucurbita*. *Plant. Dis. Reptr.* 48:54–56
49. Robinson, R. 1960. Genetic studies of disease resistance and a mineral element deficiency in cucumbers. *Proc. Am. Soc. Hort. Sci.* 57:21
50. Shifriss, O. C. Myers, C. H., Chupp, C. 1942. Resistance to the mosaic virus in the cucumber. *Phytopathology* 32:773–84
51. Sitterly, W. R. 1971. *Cucurbits. Plant Disease Control by Hereditary Means.* Pennsylvania State Univ. Press. In press
52. Smith, P. G. 1948. Powdery mildew resistance in cucumber. *Phytopathology* 39:1027–28
53. Sowell, G., Jr., Pointer, G. R. 1962. Gummy stem blight resistance of

introduced watermelons. *Plant Dis. Reptr.* 46:883–86
54. Strider, D. L. Konsler, T. R. 1965. An evaluation of the *Cucurbita* for scab resistance. *Plant Dis. Reptr.* 49:388–92
55. Strider, D. L., Winstead, N. N. 1961. Production of cell-wall dissolving enzymes by *Cladosporium cucumerinum* in cucumber tissue and in artificial media. *Phytopathology* 51:765–68
56. Van Gundy, S. D., Walker, J. C. 1957. Relation of temperature and host nutrition to angular leaf spot of cucumber. *Phytopathology* 77:615–19
57. Walker, J. C. 1950. Environment and host resistance in relation to cucumber scab. *Phytopathology* 40:1094–1102
58. Walker, J. C. 1952. *Diseases of Vegetable Crops.* McGraw-Hill Book Company, Inc.
59. Walker, J. C., Pierson, C. F. 1955. Two new cucumber varieties resistant to scab and mosaic. *Phytopathology* 45:451–53
60. Wasuwat, S. C., Walker, J. C. 1961. Inheritance of resistance in cucumber to cucumber mosaic virus. *Phytopathology* 51:423–24
61. Whitaker, R. W. 1959. An interspecific cross in *Cucurbita* (*C. lundelliana Bailey* × *C. moschata* Duch). *Madrono* 15:4–13
62. Whitaker, T. W., Davis, G. N. 1962. *Cucurbits.* Interscience, New York, 250 pp.
63. Williams, P. H., Keen, N. T. 1967. Relation of cell permeability alterations to water congestion in cucumber angular leaf spot. *Phytopathology* 57:1378–86
64. Winstead, N. N., Goode, M. G., Barham, W. C. 1959. Resistance in watermelon to *Colletotrichum lagenarium,* races 1, 2, and 3. *Plant Dis. Reptr.* 43:570–77
65. Winstead, N. N., McCombs, C. L., Lowie, L. B. 1963. A qualitative comparison of the amino acid content of acid hydrolysates from the mycelium of cucurbit anthracnose fungi. *Phytopathology* 53:1365–67

ECONOMIC IMPACT OF COFFEE RUST IN LATIN AMERICA

3561

EUGENIO SCHIEBER

Formerly: Dirección de Investigación Agrícola, Ministry of Agriculture, Guatemala, Guatemala[1]

INTRODUCTION

Rust is the most important coffee disease of the world. Latin American countries have been much concerned about the discovery of coffee rust in South America in January, 1970 near Itabuna, Bahia in Brazil because all coffee (*Coffea arabica* L.) varieties and selections under cultivation in the Western Hemisphere are susceptible to it. Incited by *Hemileia vastatrix* Berk. & Br., it is a typical rust affecting the leaves and causing defoliation that debilitates the tree and reduces yield. A century ago, without modern means of control, coffee plantations in Ceylon were devastated by the disease.

Rust was found in early 1970 in the northern coffee producing states of Bahia, Espirito Santo, and part of Minas Gerais, on several cultivars of *Coffea arabica* (22, 33, 34, 63, 66, 81, 82). Later in 1971 in São Paulo and Paraná; these are the most important coffee growing states in Brazil.

Discovery of the rust in South America immediately caused mobilization of plant scientists in the American Tropics. Research on the disease is presently in a dynamic and fluid state; Brazil is developing much new information which will soon be published (62).

This review is concerned primarily with the disease as it now occurs in South America, with emphasis on the more recent studies regarding dissemination of the pathogen, environmental influence, possibility for control, and effect on the economy of Latin America. It is not intended to be a complete review of coffee rust and does not include much of the earlier information (20, 71) gained through decades of study of the disease and the fungus in Asia, Africa, and Portugal.

CEYLON 100 YEARS AGO

Since the coffee plant (*Coffea arabica*) is originally from Ethiopia it is believed that the rust originated in the mountains of this African country and Uganda (83). In northeast Africa, it had native names. A British explorer discovered the disease in 1861 in the region of Lake Victoria in East Africa

[1] Present address: P.O. Box 226, Antigua, Guatemala, C.A.

affecting wild coffee. The rust was then reported from Ceylon in 1869, from India in 1870, Sumatra in 1876, Java in 1878, and the Philippines in 1889. During 1913 it crossed from Kenya to the Congo, where it was found in 1918, spreading to West Africa, the Ivory Coast (1954), Liberia (1955), Nigeria (1962–63) and Angola (1966) (20, 83).

In four years it devastated the coffee plantations of Ceylon, and in only 28 years had stopped all exports from that country. Ceylon had been producing approximately 42 million kilograms of coffee a year, and the Hemileia rust reduced its production to less than 3 million kilograms. The 68,787 hectares that Ceylon had in 1866–68 were reduced to 14,170 in 1893–95. Between 1869 and 1878, coffee production decreased by more than 75% and by 1890 almost all of the coffee area was abandoned (20). Planters turned to tea and rubber in place of Arabian coffee.

During those years British scientists began to study the disease. Berkeley & Broome (6) named the pathogen *Hemileia vastatrix* in 1869. Marshal Ward, from 1880 to 1882 studied the disease in detail in Ceylon, including the life cycle of the fungus (75, 76). Two rusts attack the coffee plant; *Hemileia vastatrix* and *H. coffeicola.* Of these only *H. vastatrix* is widespread.

It must be emphasized that at the time Ceylon's coffee industry was destroyed, there were no modern means to reduce the effects of the rust, as are now available.

Discovery in Brazil

Many plant pathologists predicted that coffee rust would cross the Atlantic Ocean and invade the Western Hemisphere (8, 25, 39, 55, 60, 64, 65, 77–79). In 1953, Wellman (78) stated that history had demonstrated that *Hemileia* disease did not remain within the borders of countries or continents. Since the disease was naturally propagated by wind and storms, coffee plants in the Western Hemisphere would certainly be affected in the near future.

Rayner in 1960, wrote "comparatively recent spread of leaf-rust in West Africa, with the inevitable build up of inoculum there, presented a threat to the coffee-growing regions of the new world, as uredospores might be carried there from Africa by the north-east trade winds" (55).

The author wrote in 1960 that, although coffee rust had not invaded the American continent, the danger of introduction and dissemination in the plantations of Latin America was imminent, and that, with the recent increase in international air traffic, spores of the disease could very well cross oceanic barriers and bypass established quarantines (64). Riker, in 1964, went so far as to propose certain eradication measures when it would be found (60).

Mayne (39) stated in 1969 "*Hemileia vastatrix* still poses a problem and a threat. It is impossible to believe that the immunity enjoyed by the Western Hemisphere can be permanent."

Gutiérrez & Bianchini (32) stated in 1968 that, except for Puerto Rico, rust disease had not become established or found in the Western Hemisphere. They did not know that the rust was already well established there.

On January 17, 1970, plant pathologist Arnaldo Medeiros was working on a cacao disease in the Municipio de Aurelino Leal in Bahia, Brazil. He accidentally touched some rust-affected coffee leaves growing at the margin of a cacao plantation and discovered the rust (40). Identification of the rust was soon confirmed by Robbs & Bitancourt from Brazil, and D'Oliveira from Portugal (34, 63, 66, 68). Wellman, Desrosiers, & Schieber visited the affected areas with Sebastiao from Brazil, and reported that the disease was firmly established on the American continent (22, 63, 66, 80, 81).

Dissemination of the Rust in Brazil

There has been some controversy as to how uredospores of *H. vastatrix* are spread.

Spread by wind.—Several scientists stated that natural dissemination of the disease was by wind and storms (55, 66, 78). Rayner (55) speculated on the spread of coffee rust from Africa to Ceylon, stating "It would seem highly probable that it was blown there by the south-west monsoon from the horn of Africa." He thought that the comparatively recent spread of leaf-rust in West Africa, with inevitable build up of inoculum there, presented a threat to the coffee-growing regions of the new world, since uredospores of *H. vastatrix* might be carried there from Africa by the north-east trade winds. Schieber wrote in 1970 that the rapid dissemination of the rust in Brazil suggested that wind played an important role, although it was not known with what facility and to what distances wind disseminated the uredospores of *H. vastatrix* (Figs. 1, 5). He also wrote that in "Brazil, . . . wind currents in the affected region, are from north to south, and it is estimated that it is in this direction that *Hemileia* has been disseminated . . ." (66).

Later the same year (1970), investigations made by the Departamento Nacional de Meteorología of Brazil (5), indicated that during January (Figure 1) aircurrents could take uredospores towards the states of São Paulo and Paraná, then still free of rust. It is of interest that the Hemileia rust was found in the state of São Paulo during January 1971, as expected from the direction of the prevailing winds.

Early in 1971, trapping tests were made in Brazil, using a small airplane (35). The Instituto Brasileiro do Café (IBC) in collaboration with the Instituto Biológico of São Paulo and the Escola Superior de Agronomía de Lavras, made flights at 50, 100, 250, 500, 1000, and 1500 meters altitude, utilizing slides with solid vaseline and with silicone spray. These tests revealed uredospores of *H. vastatrix* up to 1000 meters. Spores were trapped at a distance of 150 km from a diseased area in Jabotical, São Paulo. This study indicated a correlation between the number of spores trapped and the proximity to affected sites. A higher percentage of germination of uredospores was trapped in the silicone spray than on vaseline slides (35).

FIGURE 1. Air movement during January in Brazil.

Martínez (personal communication) recently found uredospores at high altitudes over the state of Paraná (then still free from coffee rust) while the *Hemileia* rust was spreading rapidly over the state of São Paulo.

Investigations by the Departamento Nacional de Meteorología of Brazil in 1970 showed the possibility of *Hemileia* spores being carried by air currents across the Atlantic Ocean from Africa to Brazil (7000 km) by means of the trade winds (5). Air circulation over the Atlantic along the African coast could carry spores originating in areas south of the 20°S. (Figure 2). North of 20°S the winds converge toward the African continent. It was estimated that, with wind currents with a velocity of 20 km an hour, spores would take 15 days to reach the Brazilian Atlantic coast. More recently Bowden et al suggested that trade winds could have brought spores of *Hemileia* across the Atlantic from Angola to Bahia, Brazil in 5–7 days (12).

Studies by Chaves and coworkers in Viçosa, Brazil, will soon provide information on the period of viability of unredospores (19).

It is now known that *H. vastatrix* is truly air-borne, and that the rust has spread throughout the large state of São Paulo in a scattered fashion in 10 months in 1971. This suggests that it is air-borne, similar to the cereal rusts. In the same way "The *Puccinia polysora* rust now reported for Hawaii, has circled the entire world, spreading from west to east, during this century" (70).

Spread by rain water.—Investigators in East Africa reported that unredospore may be dispersed by water over short distances (15, 46, 47, 49).

Rust uredospores must reach the stomata on the lower surface of the leaf to infect. Rayner (55, 57, 58) therefore, studied the wetting of the under-surface of coffee leaves. He found that, although some wetting resulted from raindrops landing directly on the under-surface when coffee leaves were disturbed by turbulent winds, "most of the wetting was by raindrops rebounding from the top surfaces of lower leaves; and . . . such splashes could carry with them spores deposited on these surfaces." Bock (9) found that rain-splash dispersed the uredospores in direct proportion to amount and intensity of rain.

Nutman & Roberts (48, 49) made high-speed flash photographic recordings of rain falling on coffee leaves. They found that, when a rain drop struck the surface of a coffee leaf, it fragmented and was dispersed for several inches in all directions. Some of the minute droplets were intercepted by lower surfaces of neighboring leaves. Small water droplets, deposited delicately on the surface of a pustule, immediately released the spores from their mutual adhesion so that they floated to the surface of the droplet.

Spread by infected seedlings.—Butler (17) stated that the rust pathogen might be carried on planting material, and that its introduction into Asia

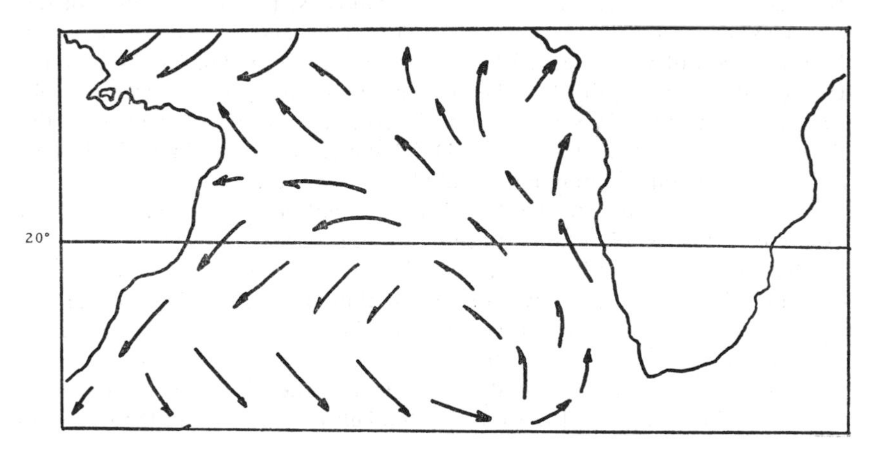

FIGURE 2. Air surface circulation between Africa and Brazil.

probably was a result of movement of seedlings from one island to another across the Indian Ocean. Schieber (66) also suggested that the rust may have come from Africa to Brazil on plant material such as coffee or cacao seedlings, the later one as a carrier. This was reinforced by suggestions of J. M. J. Sebastiao (personal communication) and also on observations made in Brazil, where even very small seedings (100 days old) had rust pustules of *H. vastatrix*. Nutman & Roberts (49) said "There is much more movement of planting material both between countries and within a country than is usually realized."

Spread by insects.—Crowe (21) reported that two species of Hymenoptera were vectors of *H. vastatrix* in East Africa. The Instituto Brasileiro do Café reported in 1970 evidence suggesting that insects such as honey bees might carry uredospores. Amante et al (1) found that uredospores of *H. vastatrix* adhered to the body of *Drosophila* sp. in the coffee region of Franca in São Paulo. This insect, as well as *Anastrepha* sp. and *Ceratitis* sp., is found especially during the time of ripening of the coffee berries. Another possible vector is the Coffee Leaf Miner (*Perileucoptera cofeella*).

Spread by man.—With today's jet-travel, visitors from other countries could carry viable rust uredospores (49, 66). A report in *Nature* (4) thus assumed that the rust had been introduced into Brazil by man.

How Defoliation Occurs

Coffee rust is typically a leaf disease, although pustules are infrequently found on berries. Uredopustules that develop in 3–4 weeks (depending on climatic conditions) on leaf under-surfaces (Figure 3) cause the leaf to fall. The nature of this leaf-shedding has not been completely studied. Nutman & Roberts (49) report that even one pustule can cause premature falling of the leaf. This premature shedding weakens coffee trees and affects, year by year, the production of wood needed to bear the future crop (66, 68, 74). Coffee leaf rust does not usually kill the tree in Latin America, but progressively weakens it, resulting in severe die-back. New leaves are affected after the older ones have fallen (Figure 4). Defoliation takes place in limited areas within a plantation and may not affect all trees at once.

Coffee rust is similar to the American leaf spot disease incited by *Mycena citricolor,* but defoliation takes place in a more spectacular form (Figure 4).

Recent observations by the author (69) at Tres Pontas (Minas Gerais), show that defined rust areas within a plantation are due to microclimate and topography of the land, as well as position of the trees or orientation toward the sun's rays. Rust severity is different on east and west sides of the coffee tree. Coffee trees along the roads or borders of the plantation also showed less defoliation. There is also a relationship between aeration in the tree rows and defoliation.

What must be emphasized is that Hemileia rust is found in well defined small areas within the plantation, due to microecology and topography (69).

Ecological Studies in Brazil

Bock (10) studied the close relationship between climatic conditions of the regions observed in East Africa and intensity of rust development. The disease is more severe in the warm humid regions west of the Rift Valley in Kenya than in the cooler and dryer regions east of the Rift Valley. Dowson observed in 1921 that rust severity decreased at higher altitudes, and Bock (10) also demonstrated that, at 1.830 meters altitude in Kenya, with a rainfall of 830 milimeters, coffee rust was of no economic importance. According to Bock this is due to the effect of lower temperatures affecting disease development. He also stated that when temperature is not the limiting factor, the severity of the disease depends, first on the distribution and intensity of rainfall, followed by the number of spores of the pathogen on the foliage, and finally the degree of defoliation.

Brown & Cocheme (13) reported that disease development is always lo-

FIGURE 3. Uredo-pustules that develop on leaf under-surfaces cause leaf fall.

FIGURE 4. Defoliation caused by rust progresses upward from the first branches.

calized, and its dispersion relatively slow. Because of this, its control is determined by disease development in relation to climatic conditions. This was also observed by the author (67, 69) in Tres Pontas, Minas Gerais, Brazil.

The relationship between climatic conditions and rust development and severity should be considered at the plantation level. This is why the author has compared coffee rust development in the plantation with the American leaf spot disease incited by *Mycena citricolor* (66, 68, 69).

Investigators of the Departamento Nacional de Meteorología in Brazil (5) have compared the climates of Kitale and Kiambu in Kenya with Londrina in

Paraná, Brazil (Figure 6). In the two African regions, disease development is slight because of low temperature (10). The investigators in Brazil wanted to forecast disease development and severity according to climatic conditions, if the rust became established in Londrina, Paraná. A similar study by Páez (52) to forecast coffee rust incidence in Costa Rica (free from rust) indicated that surveys to detect rust in this country should be made during periods of high temperature and rainfall.

Ecological studies in Brazil and in the rest of Latin America would be important in learning where rust would be severe and devastating. Also they would serve as a base for introduction of resistant varieties, while susceptible varieties could be used in regions not favoring the disease. Such studies also would give an idea of how the disease will develop in other countries in the Western Hemisphere.

Brazil has begun to determine areas suited for coffee growing but not favorable for rust development. In determining the incubation period, these studies are based on Rayner's equation: $Y = 90.61 - 0.408\ X_1 - 0.440\ X_2$, that is an estimate of sporulation in 50% of the lesions. Rayner developed this regression equation relating X_1 (mean maximum temperature in °F) and X_2 (mean minimum temperature in °F) to Y (estimated incubation period in days). Matiello determined that all areas in Brazil above 700 meters in elevation are not severely attacked by the rust (38).

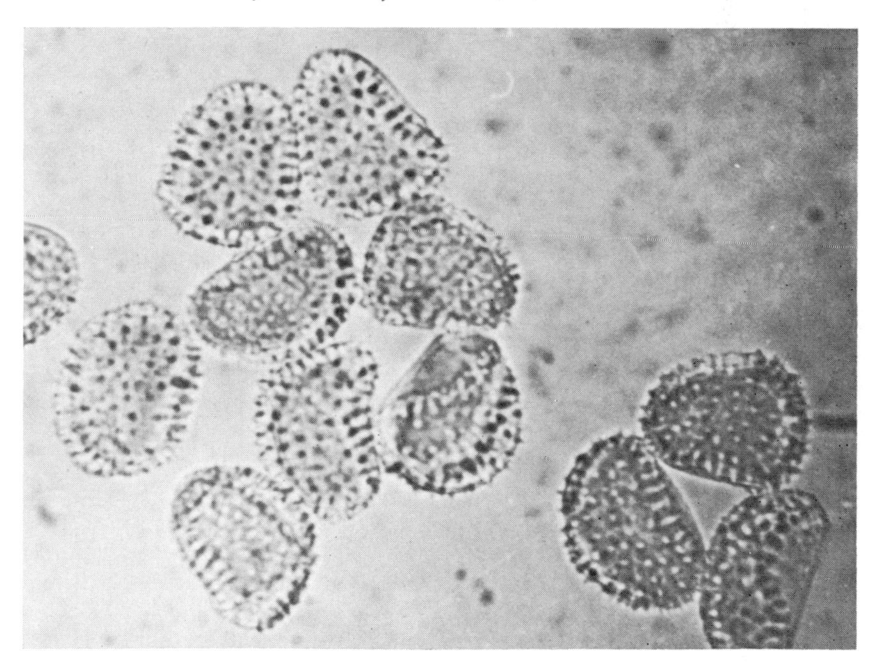

FIGURE 5. The uredospores of coffee rust. (Photograph courtesy of Coffee Rust Research Center, Portugal.)

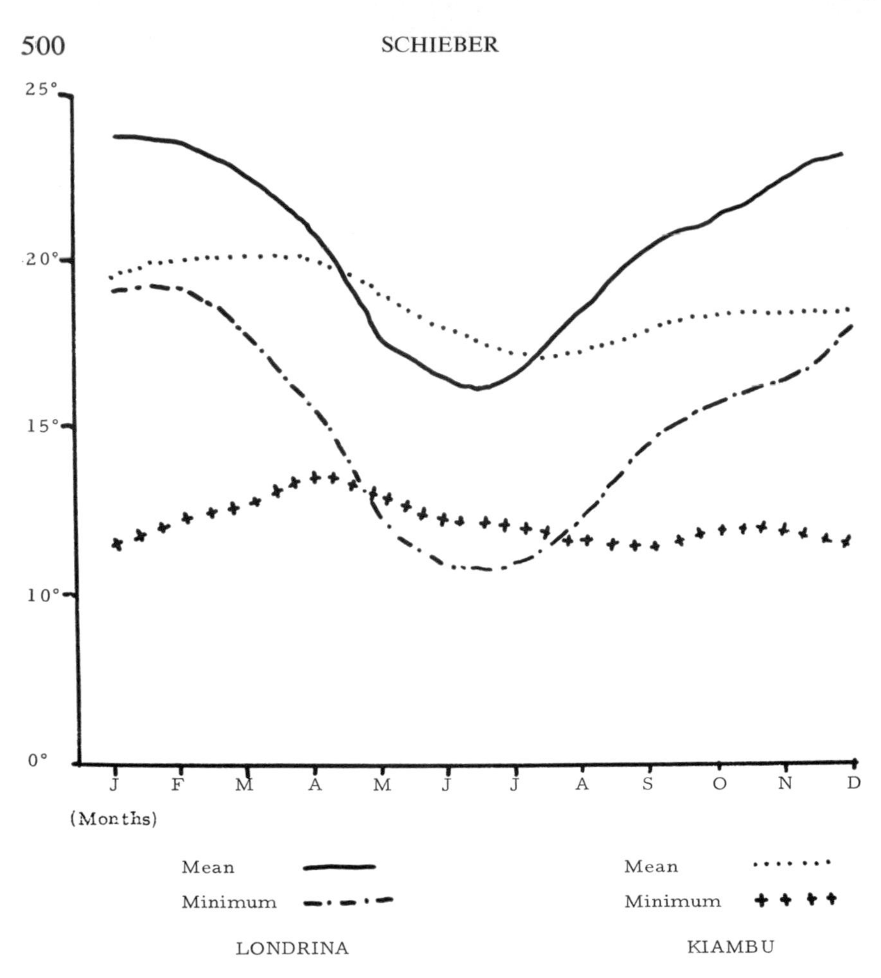

FIGURE 6. Mean and minimum temperatures (C°) in coffee regions in Kenya (Kiambu) and Paraná, Brazil (Londrina). From Departamento Meteorológico Nacional, Brazil; (Reference No. 5).

One study needed is to determine what influence the use of shade in coffee growing has in rust establishment, development, and severity. Coffee-growing countries in Latin America, from Colombia in South America to Mexico, grow coffee under shade conditions. Since Kenya grows coffee without shade, as Brazil does, no studies from East Africa are now available on the effects of shade, a point of great importance to the Western Hemisphere.

Fungicides Against Rust

Since Kenya grows coffee varieties developed from the susceptible *Coffea arabica* L., the use of fungicides against *H. vastatrix* has been studied intensively there. Although rust was reported from Kenya in 1912, it was not until 1921 that use of fungicides against the rust was studied there, although Bordeaux mixture was used as early as 1917 against rust in India (3). Most of the Kenya studies were made from 1958 to 1965, investigating the nature of the fungicide effect, timing, and methods of application (11, 29, 30, 59, 73). It was soon found that copper fungicides were consistently superior to others in rust control. Rayner (54, 59) demonstrated the protective action of copper fungicides by inhibition of uredospore germination.

East African scientists soon found that there was a direct relation between disease increase and rain intensity, this being of importance in planning control measures. The key for rust control is to apply fungicide at the beginning of the rainy season.

Bock (11) concluded from his experiments made west of the Rift Valley in Kenya (where rust is severe) that: (*a*) maximum control of coffee leaf rust was obtained from spraying just before and during the rainy season, (*b*) timing of treatments was critical, and (*c*) efficiency of copper fungicides is reduced with increased interval between sprays during the first rains.

Rayner (59) stressed the importance of redistribution of the copper fungicide by rain-splash to upper leaves. He found that the efficiency of sprayers was related to the amount of fungicide residue left on the leaves. Several investigators in East Africa then reported that 7 kg/acre of a 50% copper fungicide in any volume of water gave best control. Wallis & Firman (73) found that, to obtain good rust control, a copper spray must be applied with a minimum of 60 mg of copper per square meter of foliar surface. Firman & Wallis (30) found low volume sprays effective, using 50 mg of a spray with 2–4 g of a 50% copper fungicide per tree in Kenya, applied with a knapsack mistblower.

The use of coppers in Kenya has demonstrated the "tonic effect" of this fungicide, also resulting in increased yield of the order of 215–325 kg/ha. Wallis & Rayner (72) found that copper sprayings also increased foliar density when used to control rust. Burdekin (14, 16), by testing several fungicides, found that cupric chemicals gave higher yields than zinc fungicides which decreased number of pustules and leaf fall. Park & Burdekin (53) tested the efficiency of coppers in Tanzania with similar results. Griffiths (31) reported that Difolatan and Benlate used against coffee berry disease in Kenya supplemented the coppers against leaf rust.

Newhall & Orillo (44) reported excellent yield increase in the Philippines with 5–10 annual applications of copper fungicides (Bordeaux and Cuprous oxide), but intermediate to poor control with maneb, captan, ferbam, and other organics.

Recent tests carried out by the IBC in Santa Teresa, Espirito Santo, showed that zineb, cuprous oxide, and Bordeaux gave the lowest percentage of rusted leaves (37). Studies are under way comparing coppers and dithane, and testing new systemic fungicides under laboratory, greenhouse, and field conditions in Brazil (36, 37).

Common knapsack sprayers, motorized knapsack sprayers, and spraying equipment attached to tractors, are being used for rust control in East Africa. Although aerial application does not give adequate rust control, it is used against coffee berry disease in East Africa (18, 66).

Testing of equipment, including aerial application, is under study in Brazil. The topography of the land and tree density between and in the rows are being studied in Brazil to obtain effective control. Coffee plantations were established in Brazil with high tree density, not considering the use of fungicides in the future. Pruning would be necessary for the effective use of chemicals against *H. vastatrix* (18, 69).

Regarding costs of applying fungicides, Nutman & Roberts (49) reported that it cost US $810,000 in Kenya to protect 18,750 hectares of coffee with fungicides. They calculated a net return of US $2,910,000 for a yield increase of 240 kg/ha of coffee produced.

Narasimhaswamy (43) calculated cost of spraying in India, and reported that use of fungicides amounted to 10% of total costs, resulting in a yield increase of 98%.

Chaves et al (19) wrote that in Brazil, "Adoption of spraying schedules against coffee rust is largely dependent on farm conditions, growers 'know how', and of course, whether or not sprays are economically feasible."

Mesquita (41) concluded from a study on other coffee diseases in Zona de Mata in Brazil (Minas Gerais), that farmers cannot afford a coffee spray program if yield levels are lower than 620–900 kg/ha.

Camargo (18, 51) stated that to be able to use fungicides against rust in Brazil, production per hectare had to be raised to cover the cost of application of the fungicide. Coffee plantations with an average of 500 kg/ha of processed coffee have a high production cost, but recent work in Brazil has shown that this cost, including the use of fertilizer, easily doubled the yields.

Wallis (74) calculated an annual cost of US $67.00 per hectare for fungicides and cost of application, based on experience in East Africa and 1970 prices in Brazil. He further stated that "If all farmers in the intermediate risk areas carried out a full protective spraying programme, then the total internal expenditure would be about US $74 million; annually, this is equivalent to 9 percent of the total foreign-exchange earning from coffee exported from Brazil in 1968/1969, but of course not all the expenditure would be in foreign exchange."

Change to Resistant Varieties

Search for resistance has been carried out mainly in the internationally known Coffee Rust Research Center at Oeiras, Portugal. Plant pathologist D'Oliveira and collaborators have, since the center was established in 1955, explored sources of resistance in the genus *Coffea* towards many races of *H. vastatrix* (23–26, 45). Some work on resistance has also been carried out in East Africa by Rayner (56) and others.

Resistance to race II has been found, for example on K7, SL-6, and KP 532. Firman & Hanger (28) tested a number of varieties of Arabica coffee under field conditions in what they called rust exposure trials to various races (mainly I and II) of *H. vastatrix* present in Kenya. They studied the behavior of varieties SL-6 and K7, extensively recommended in Kenya where rust is severe. Variety K7, susceptible to race I, is resistant to race II of *H. vastatrix*, but segregated about 25% toward susceptibility to this race. Similar unpublished results have been obtained in Brazil.

Orillo & Valdez (50) reported from the Philippines that some resistant selections of Arabica coffee segregate, producing up to 19% of susceptible seedlings.

It is thought that it will take 10 years to replace a susceptible variety with a resistant one in a specific region. As with rust diseases affecting other cultivated plants, breeding for resistance will be a continuous task, since new races of the pathogen may appear in the future.

Vertical resistance to race II of the pathogen now present in the Western Hemisphere is what Brazil needs in the immediate future; however, horizontal resistance (found already in plants of *C. arabica* in Angola and Brazil) would be desirable if it is possible to utilize it (67). Brazil started searching for resistance in 1952 (Bettencourt & Carvalho 7). Carvalho studied also the resistance towards race II of the pathogen now present in South America. Brazil has started adaptation trials in Caratinga, Minas Gerais; Venda Nova and Marilandia, Espirito Santo; and Jaguaquara in Bahia (IBC, 1971). These regions are all ecologically different.

Several coffee-producing countries in Latin America made immediate introductions of resistant materials from Portugal and Africa. The goal is to find the ideal combination of resistance, good agronomical adaptation, and good quality.

Since coffee is grown in diverse ecological conditions in the American neotropics, adaptation studies will be of prime importance. This means that the change to resistant varieties is going to be a slow process, whose impact will be felt only after several years.

Economic Aspect and Impact

The importance of coffee to Latin America can be seen by the Rockefeller Report of 1970, which stated "It has been calculated that a fall of one cent per pound in the coffee price, signifies a loss of US $55 million in foreign exchange for the 14 countries that produce coffee in the Western Hemisphere" (61). Latin America grows more coffee than Africa and Asia. Coffee production, processing, and marketing provide employment for millions of people. As several countries in the Western Hemisphere depend on coffee sales for their foreign exchange (Table 1), concern about the Hemileia rust now in the American continent is understandable.

As Wallis (74) stated in relation to Brazil "some attempt has to be made

TABLE 1. Foreign exchange provided in Latin America by coffee sales in 1968. (U.S. Dept. Agr. data).

Country	Percentage of all foreign exchange
Colombia	67.7
El Salvador	42.7
Brazil	41.2
Haiti	38.9
Guatemala	33.0
Costa Rica	31.4
Ecuador	17.9
Nicaragua	14.0
Honduras	12.6
Domican Republic	11.9
Mexico	6.3

to anticipate the impact of coffee leaf-rust on coffee production. In any country this would be difficult and for such a huge area as Brazil it is particularly hazardous."

The above statement will apply also to the rest of Latin America when the rust spreads, but some countries will experience an economic effect of different proportion. The disease will sooner or later spread through all coffee-growing areas of Latin America, and the economic impact will vary from one country to another (27), as it does between the northern coffee region in Brazil with that in São Paulo and Paraná in the south.

An economic aspect to be considered is the use of fungicides. In some countries (Costa Rica, El Salvador, and Guatemala), coffee farmers know how to apply chemicals to control some existing important coffee diseases, but certain countries will have to learn how to use fungicides in coffee plantations. Some coffee areas will need to change the type of farm management and the traditional way of planting coffee, especially in relation to tree spacing, to be able to use fungicides. These factors will influence how economically coffee can be grown in certain areas. Wallis (74) considered that the use of fungicides would cost US $67.00 per hectare per year; at the price now received for coffee in Brazil, this would be equivalent to about three 60 kg bags/ha.

It can be predicted that the economic impact on the rest of Latin America will be important since all coffee cultivars grown in Latin America are susceptible to the race II of the rust found in Brazil. It is thought that it will take 10 years to replace a susceptible variety with a resistant one because of the problem of agronomic and qualitative adaptability in the different coffee regions in the American tropics.

Another factor to consider is expansion of new coffee areas to balance the economy based on coffee production. In the case of Brazil, Wallis (74) stated "that practically all land suitable for coffee in Brazil has been used at least once for this crop. Expansion to the south is limited by the risk of frosts and inland to the west rainfall unreliability becomes a critical factor. In the north the lack of soil fertility, high temperatures and now coffee leaf-rust all limit the possibilities for economic coffee production."

Muyshondt (42) made a recent analysis of the possible impact of coffee rust in Central America, Panama, and Mexico. He stated, "It has to be understood, that the importance of an economic loss caused by the rust, directly on the coffee industry activity, would also affect profoundly a depression in the activities of the banking system, industry, commerce and service institutions; consequently indirectly affecting the working class." He analyzed the direct and indirect losses in the coffee industry, using data of the Latin American area from Panama to Mexico still free from rust. Considering total production for the year 1968–69 of 11.0 million 60 kg bags, with an approximate income of US $442 million (calculated at US $40.00/60 kg bag), and hand labor of 155 million man days, he has calculated the effects of 5–30% rust on production (Table 2).

TABLE 2. Possible impact of coffee rust in the area from Panama to Mexico (After Muyshondt, 42).

Percentage of damage caused by *Hemileia vastatrix*	Reduction in production (60 kg bags)	Reduction in income (U.S. Dollars)	Reduction in hand-labor (man/days)
5	550,000	22,100,000	7,750,000
10	1,100,000	44,200,000	15,500,000
15	1,650,000	66,300,000	23,250,000
20	2,200,000	88,400,000	31,000,000
25	2,750,000	110,500,000	38,750,000
30	3,300,000	132,600,000	46,500,000

As Muyshondt stated, "The figures presented in Table II show that even a 5 percent of losses due to the rust would have a true negative impact on the economic and social development of these countries, carrying great disturbances in the internal political order in each of the countries of these areas still free from rust in our Hemisphere."

A more positive effect of rust has been the starting of research to improve coffee technology in Latin America. Shortly after the rust was discovered in Itabuna, the Brazilian Government allocated a sum of over US $8 million for programs to study and combat the disease. Research and extension programs in relation to coffee technology have also been somewhat strengthened in Colombia, Costa Rica, and El Salvador.

ACKNOWLEDGMENTS

The author is indebted to Dr. George A. Zentmyer for advice, suggestions, and revision of the manuscript. He also thanks the following persons who provided him with recent publications and information on the subject: Dr. José María Jorge Sebastiao, of the Instituto Brasileiro do Café; Mr. Robert C. Moncure, of the U.S. Department of Agriculture; Mr. Arnaldo G. Medeiros of Centro de Pesquisas do Cacau, Itabuna, Brazil; Dr. J. A. N. Wallis, of the International Coffee Organization; Mr. William Davis, former Agricultural Attache in Nairobi, Kenya. He is also indebted to Dr. Victoria Rossetti and Dr. Karl M. Silberschmidt of the Instituto Biológico of São Paulo, Brazil for suggestions on the manuscript.

LITERATURE CITED

1. Amante, E., Vulcano, M. A., Abrahao, J. 1971. Observações preliminares sobre a influencia da entomofauna na dispersao dos uredosporos da ferrugem do cafeeiro (*Hemileia vastatrix*). *O Biológico,* 37:102–05
2. Amaral, M., Beduim, C. D. 1970. A ferrugem alaranjada do cafeeiro. *Equipe Técnica de Defesa Sanitaria Vegetal.* Ministerio da Agricultura, Brazil
3. Ananth, K. C. 1969. Timing and frequency of spraying for control of coffee leaf rust in southern India. *Exp. Agr.* 5:117–23
4. Anonymous, 1970. Brazil. Death in the pot. *Nature,* 226:997
5. Anonymous. 1970. Influência dos fatóres meteorológicos na ocorrencia da *Hemileia vastatrix*. *Departamento Nacional de Meteorología,* Brazil
6. Berkeley, M. J. 1869. *Gard. Chron.* 45:1157
7. Bettencourt, A. J., Carvalho, A. 1968. Melhoramento visando a resistência do cafeeiro â ferrugem. *Bragantia* 27:35–68
8. Bitancourt, A. A. 1970. Observações sobre a ferrugem do cafeeiro nos principais paises cafeicolas do mundo. *O Biológico,* 36: 263–70
9. Bock, K. R. 1962. Dispersal of uredospores of *Hemileia vastatrix* under field conditions. *Trans. Brit. Mycol. Soc.* 45:63–74
10. Bock, K. R. 1962. Seasonal periodicity of coffee leaf rust and factors affecting the severity of outbreaks in Kenya Colony. *Trans. Brit. Mycol. Soc.* 45:289–300
11. Bock, K. R. 1962. Control of coffee leaf rust in Kenya Colony. *Trans. Brit. Mycol. Soc.* 45:301–13
12. Bowden, J., Gregory, P. H., Johnson, C. G. 1971. Possible wind transport of coffee leaf rust across the Atlantic Ocean. *Nature* 229: 500–01
13. Brown, L. H., Cocheme, J. 1969. A study of the agroclimatology of the highlands of Eastern Africa. FAO Interag. Proj. Agroclim. *Tech. Rep.,* Rome, 336 pp.
14. Burdekin, D. A. 1960. The effect of Captan and copper sprays on leaf rust and leaf fall of coffee. *Tanganyika Coffee Research Station. Res. Rept.,* 1960, Lyamungu, Tanganyika Coffee Board, pp. 56–69
15. Burdekin, D. A. 1960. Wind and water dispersal of coffee leaf rust in Tanganyika. *Kenya Coffee* 25: 212–213, 219
16. Burdekin, D. A. 1964. The effect of various fungicides on leaf rust, leaf retention and yield of coffee. *E. African Agr. Forest. J.* 30: 101–04
17. Butler, E. J. 1918. *Fungi and disease in plants.* Thatcher, Spink, Co. Calcutta 547 pp.
18. Camargo Vianna, A. C. 1971. Contróle Químico da Ferrugem do Cafeeiro (*Hemileia vastatrix,* Berk. et Br.) Seminario Brasileiro de Radiodifusáo Rural para o Contróle da Ferrugem do Cafeeiro, Campinas, Brazil
19. Chaves, G. M., Filho, J. daC., de Carvalho, M. G., Matsuoko, K., Coelho, D. T., Shimoya, C. 1970. A ferrugem do cafeeiro (*Hemileia vastatrix,* Berk. & Br.) Revisão de literatura com observações e comentários sôbre a enfermidade no Brasil. Seiva *30* (Ed. Espec.): 1–75
20. Cramer, P. J. S. 1957. A review of literature of coffee research in Indonesia. pp. 41–47 *Inter. Am. Inst. Agr. Sci. Misc. Publ.* 15:1–262
21. Crowe, T. J. 1963. Possible insect vectors of the uredospores of *Hemileia vastatrix* in Kenya. *Trans. Brit. Mycol. Soc.* 46:24–26
22. Desrosiers, R. 1970. Coffee rust in Brazil, caused by *Hemileia vastatrix. Report U.S. Agency Int. Dev.* 6 pp.
23. D'Oliveira, B., Rodríguez, C. J. 1959. Progress Report to Ethiopia. *García de Orta* 7:279–92
24. D'Oliveira, B., Rodríguez, C. J. 1960. A survey of the problem of coffee rust. II. Screening for resistance to *Hemileia vastatrix* on *Coffea arabica.* Lisboa. *Junta Exportação Café.* p. 46
25. D'Oliveira, B. 1965. Progress report 1960–1965. *Coffee Rust Res. Center, Oeiras, Portugal.* pp. 7–9
26. D'Oliveira, B. 1970. El trabajo del Centro de Investigaciones de las

Royas del Cafeto de Oeiras, Portugal. Identificación de razas de *Hemileia* y tipos de resisténcia. In *Reunión Técnica sobre las Royas del Cafeto*. Inst. Inter-Am. Cien. Agr. San José, Costa Rica. 6 pp.

27. Fernández, O. 1970. Importancia de la roya del cafeto para la economía Columbiana. Mesa Redonda sobre Roya del Cafeto. *Proc. VII Reunión Latinoamericana de Fitotecnia*. Bogotá, Colombia
28. Firman, I. D., Hanger, B. F. 1963. Resistance to coffee leaf rust in Kenya. *Coffee* (Costa Rica) 5: 49–54
29. Firman, I. D. 1965. A review of leaf rust and coffee berry disease control in Kenya. *Tropical Agr.* 42:111–18
30. Firman, I. D., Wallis, J. A. N. 1965. Low-volume spraying to control coffee leaf rust in Kenya. *Appl. Biol.* 55:123–37
31. Griffiths, E. 1969. Plant Pathology. In *Coffee Res. Found. Kenya. Ann. Rept.* 1968–1969:38–43
32. Gutiérrez, G., Bianchini, C. 1968. Roya del cafeto, una amenaza permanente para el Continente Americano. *Café* (Turrialba) 9: 3–5
33. Instituto Brasileiro do Café. 1970. Ferrugem do Cafeeiro. Características da doenca e providencias para seu controle. Ministério da Indústria e Comércio. Inst. Bras. Café-Grupo Erad. Café-GERCA. 32 pp.
34. Instituto Brasileiro do Café. 1970. A ferrugem do cafeeiro no Brasil. Ministério da Indústria e Comércio. Inst. Bras. Café-Grupo Erad. Café-GERCA. 75 pp.
35. Instituto Brasileiro do Café. 1971. Vento carrega ferrugem. Informativo Inst. Bras. Café-Grupo Erad. Café-GERCA. I(4) 7 p.
36. Instituto Brasileiro do Café. 1971. Brometo inibe esporos da ferrugem. Informativo Inst. Bras. Café-Grupo Erad. Café-GERCA. I(4) 3 p.
37. Instituto Brasileiro do Café. 1971. Novos resultados de contróle químico da ferrugem. Informativo Inst. Bras. Café-Grupo Erad. Café-GERCA. I(6) 8 p.
38. Matiello, B. J. 1970. Estudios preliminares de zonificación del cultivo del cafeto en el Brasil, en función de *Hemileia vastatrix*. Mesa Redonda sobre Roya del Cafeto. *Proc. VII Reunión Latinoamericana de Fitotecnia*. Bogotá, Colombia.
39. Mayne, W. W. 1969. A century of coffee leaf disease, 1869–1969. *Biologist* 16:58–60
40. Medeiros, A. G. 1970. Informe sobre *Hemileia vastatrix* en café, en Bahía, Brasil. *Comi. Exec. Pl. Recup. Econ. Rural Lav.* Cacaueira, Rio de Janeiro
41. Mesquita, A. 1969. A cafeicultura e sua combinacao ótima com outras actividades na Zona da Mata, Minas Gerais, 1968–1969. *Thesis, Escola Superior de Agricultura, Univ. Fed. Vicosa,* Vicosa, Minas Gerais
42. Muyshondt, M. 1971. Posible impacto de la roya del cafeto (*Hemileia vastatrix*) en la economía de los países miembros del Organismo Internacional Regional de Sanidad Agropecuaria (OIRSA). *Phytopathology* 62: In press
43. Narasimhaswamy, R. L. 1961. La herrumbre del café (Hemileia) en la India. *Café* (Turrialba) 3 (9):41–49
44. Newhall, A. G., Orillo, F. T. 1971. Coffee rust control experiment in the Philippines. *Plant Dis. Reptr.* 55:216–19
45. Noronha-Wagner, M, Bettencourt, A. J. 1967. Genetic study of the resistance of *Coffea* spp. to leaf rust. I. Identification and behavior of four factors conditioning disease reaction in *Coffea arabica* to twelve physiologic races of *Hemileia vastatrix. Can. J. Bot.* 45:2021–31
46. Nutman, F. J., Roberts, F. M., Bock, K. R. 1960. Method of uredospore dispersal of the coffee leaf rust fungus, *Hemileia vastatrix. Trans. Brit. Mycol. Soc.,* 43: 509–15
47. Nutman, F. J., Roberts, F. M. 1962. Dispersal of coffee rust *Hemileia vastatrix* B. et Br. *Nature,* 194:1296
48. Nutman, F. J., Roberts, F. M. 1963. Studies on the biology of *Hemileia vastatrix* Berk & Br. *Trans. Brit. Mycol.* 46:27–48
49. Nutman, F. J., Roberts, F. M.

1970. Coffee Leaf Rust. *Pest Articles News Summ.* 16:607–24

50. Orillo, F. T., Valdez, R. B. 1961. The selection of coffee species and varieties resistant to coffee rust in the Philippines. *Philippine Agr.* 45:223–334
51. Ortolani, A. A., Camargo Vianna, A. C., Abreu, R. G., 1971. *Hemileia vastatrix* Berk et Br. *Estudos e observações em regioes da Africa e sugestoes a cafeicultura do Brasil.* Inst. Bras. Café, and Sec. Agr. São Paulo. 228 p.
52. Paéz, G. 1971. Método de muestreo para el reconocimiento de la roya en Costa Rica. *Inst. Inter-Am. Cien Agr. Org. Estad. Am. Turrialba,* Costa Rica. 106 pp.
53. Park, P. O., Burdekin, D. A. 1964. Studies on the ageing of copper fungicides used to control coffee leaf rust. *Ann. Appl. Biol.* 54: 335–47
54. Rayner, R. W. 1957. Leaf Rust. *Coffee Bd. Mo. Bull.* 1935–36: 101–110. Nairobi, Kenya
55. Rayner, R. W. 1960. Rust disease of coffee. II. Spread of the disease. *World Crops,* 12:222–24
56. Rayner, R. W. 1960. Rust disease of coffee. III. Resistance. *World Crops,* 12:261–64
57. Rayner, R. W. 1961. Germination and penetration studies on coffee rust (*Hemileia vastatrix* B. & Br.) *Ann. Appl. Biol.* 49:497–505
58. Rayner, R. W. 1961. Spore liberation and dispersal of coffee rust *Hemileia vastatrix* B. et Br. *Nature* 191:725
59. Rayner, R. W. 1962. The control of coffee rust in Kenya by fungicides. *Ann. Appl. Biol.* 50:245–61
60. Riker, A. J. 1964. *Reunión del Comité Internacional Regional de Sanidad Agropecuaria.* 12a. Reunión. San José, Costa Rica. pp. 15–19
61. Rockefeller, N. *La calidad de la vida en las Américas.* Informe sobre América Latina, presentado por una Misión Presidencial de los Estados Unidos de América. 40 pp.
62. Rossetti, V. 1970. Medidas tomadas en el Estado de São Paulo con relación a la roya del cafeto. Mesa Redonda sobre Roya del Cafeto. *Proc. VII Reunión Latinoamericana Fitotecnia.* Bogotá, Colombia.
63. Sebastiao, J. M. J. 1970. El problema de la roya del cafeto en el Brasil. Enfoque del Instituto Brasileiro del Café. Mesa Redonda sobre Roya del Cafeto. *Proc. VII Reunión Latinoamericana Fitotecnia.* Bogotá, Colombia.
64. Schieber, E. 1960. La herrumbre del café y la cooperación internacional. *El Informador Agrícola,* Min. Agr., Guatemala
65. Schieber, E. 1963. La herrumbre del café y el centro de investigaciones en Portugal. *Revista Cafetalera.* 1963: 21–23
66. Schieber, E. 1970. Viaje al Brasil y el Africa para estudiar y observar el problema de la herrumbre del café. *Rept. Org. Int. Reg. San. Agr.* 109 pp.
67. Schieber, E. 1971. Comparative observations on coffee rust in Brazil and Kenya. *Plant Dis. Reptr.* (55):209–12
68. Schieber, E. 1971. Observaciones sobre la roya del cafeto provocada por *Hemileia vastatrix,* en Brasil y Kenia. *Rept. Org. Int. Reg. San. Agr.* p. 42
69. Schieber, E. 1971. Informe al Secretario General de GERCA-IBC sobre visita al Brasil, donde se observaron algunos problemas fitopatológicos del café (unpublished).
70. Schieber, E. Laemmlen, F., Martínez, A. 1971. *Puccinia polysora* rust on corn, established in Hawaii. *Phytopathology* 62: In press
71. Stevenson, J. A., Beam, R. 1953. An annotated bibliography of coffee rust (*Hemileia* spp.) *Div. Mycol. Dis. Survey, US. Dept. Agr. Spec. Public.* 4:1–80
72. Wallis, J. A. N., Rayner, R. W. 1957. *Ann. Rept. Coffee Res. Sta.* Ruiru, Kenya. p. 33
73. Wallis, J. A. N., Firman, I. D. 1962. Spraying Arabica coffee for the control of leaf rust. *E. African Agr. Forest. J.* 28:89–104
74. Wallis, J. A. N. 1970. Coffee leaf rust in South America. *A Report to the International Coffee Organization.* 49 p.
75. Ward, H. M. 1882. Researches on

the life history of *Hemileia vastatrix,* the fungus of the "coffee-leaf disease." *Linn. Soc. J.* (Bot.) 19:229–335
76. Ward, H. M. 1882. On the morphology of *Hemileia vastatrix* Berk. & Br. (The fungus of the coffee disease of Ceylon.) *Quart. J. Microscop. Sci.* (n.s.) 22:1–11
77. Wellman, F. L. 1952. Peligro de introducción de la *Hemileia* del café a las Américas. *Turrialba* 2(2):47–50
78. Wellman, F. L. 1953. The Americas face up to the threat of coffee rust. *For. Agr.* 17:47–52, 64
79. Wellman, F. L. 1957. *Hemileia vastatrix.* Investigaciones presentes y pasadas en la herrumbre del café y su importancia en la América Tropical. Publicado por *Fed. Cafet. Am.* San Salvador
80. Wellman, F. L. 1970. Rust of Coffee in Brazil. *Plant Dis. Reptr.* 54:355
81. Wellman, F. L., Desrosiers, R., Schieber, E. 1970. The *Hemileia vastatrix* coffee disease established in the American tropics. *Phytopathology* 60:1319
82. Wellman, F. L. 1970. The rust *Hemileia vastatrix* now firmly established on coffee in Brazil. *Plant Dis. Reptr.* 54:539–41
83. Wellman, F. L. 1970. Coffee yellow leaf rust: World History: Minimizing losses in Tropical America. *Proc. Reunión Técnica sobre las Royas del Cafeto.* Inst. Inter-Am. Cien. Agr. Org. Estad. Am., San José, Costa Rica

SOME RELATED ARTICLES IN OTHER *ANNUAL REVIEWS*

From the *Annual Review of Plant Physiology*, Volume 23 (1972)

Plant Cell Protoplasts—Isolation and Development, *E. C. Cocking*, 29–50
Cell Wall Chemistry, *D. H. Northcote*, 113–32
Water Transport Across Membranes, *P. J. C. Kuiper*, 157–72
Auxins and Roots, *Tom K. Scott*, 235–58
Biosynthesis and Mechanism of Action of Ethylene, *Fred B. Abeles*, 259–92
Salt Transport in Relation to Salinity, *D. W. Rains*, 367–88
Mycoplasma, *Richard O. Hampton*, 389–464

From the *Annual Review of Entomology*, Volume 17 (1972)

Mutagenic, Teratogenic, and Carcinogenic Properties of Pesticides, *William F. Durham and Clara H. Williams*, 123–48
Factors Influencing the Effectiveness of Soil Insecticides, *C. R. Harris*, 177–198
Fate of Pesticides in the Environment, *Calvin M. Menzie*, 199–222
Economics of Agricultural Pest Control, *J. C. Headley*, 273–86
Pathogenic Microorganisms in the Regulation of Forest Insect Populations, *Gordon R. Stairs*, 355–72
Transmission of Plant-Pathogenic Viruses by Aphids, *M. A. Watson and R. T. Plumb*, 425–52
A Critique of the Status of Plant Regulatory and Quarantine Activities in the United States, *Halwin L. Jones*, 453–60
Flight Behavior of Aphids, *James B. Kring*, 461–92

From the *Annual Review of Ecology & Systematics*, Volume 2 (1971)

Economic and Ecological Effects of a Stationary Economy, *Robert U. Ayres and Allen V. Kneese*, 1–22
Biological Control of Insects, *Robert van den Bosch*, 45–66
Soils as Components of Ecosystems, *Martin Witkamp*, 85–110

From the *Annual Review of Biochemistry*, Vol. 40 (1971)

Inhibitors of Ribosome Functions, *Sidney Pestka*, 697–710
RNA Polymerase, *Richard R. Burgess*, 711–40
Cyclic AMP, *Jean-Pierre Jost and H. V. Rickenberg*, 741–74
Antibiotics and Nucleic Acids, *Irving H. Goldberg and Paul A. Friedman*, 775–810
Lysogeny: Viral Repression and Site-Specific Recombination, *H. Echols*, 827–5
Regulation of Biosynthetic Pathways in Bacteria and Fungi, *Joseph M. Calvo and Gerald R. Fink*, 943–68

From the *Annual Review of Microbiology,* Volume 25 (1971)

Aggregation and Differentiation in the Cellular Slime Molds, *John Tyler Bonner,* 75–92

DNA Restriction and Modification Mechanisms in Bacteria, *Herbert W. Boyer,* 153–76

The Pathogenicity of Soil Amebas, *Clyde G. Culbertson,* 231–54

Biochemical Ecology of Microorganisms, *Martin Alexander,* 361–92

Microbial Criteria of Environment Qualities, *E. Fjerdingstad,* 563–82

The Bdellovibrios, *Mortimer P. Starr and Ramon J. Seidler,* 649–78

From the *Annual Review of Genetics,* Volume 5 (1971)

Genetics of Disease Resistance in Plants, *A. L. Hooker and K. M. S. Saxena,* 407–24

Pseudomonas Genetics, *B. W. Holloway, V. Krishnapillai, and V. Stanisich,* 425–46

The Origin of Maize, *Walton C. Galinat,* 447–78

AUTHOR INDEX

D

H

I

J

N

O

P

T

SUBJECT INDEX

A

F

H

I

Q

R

T

CUMULATIVE INDEXES

VOLUMES 6 - 10

INDEX OF CONTRIBUTING AUTHORS

INDEX OF CHAPTER TITLES

VOLUMES 6-10